Recent Advancement in Microbial Biotechnology

Recent Advancement in Microbial Biotechnology

Agricultural and Industrial Approach

Edited by

Surajit De Mandal
Postdoctoral Researcher, College of Agriculture, South China Agricultural University, Guangzhou, People's Republic of China

Ajit Kumar Passari
Postdoctoral Scientist, Departmento de Biología Molecular y Biotecnología, Instituto de Investigaciones Biomédicas, Universidad Nacional Autónoma de México (UNAM), Ciudad de México, México

Academic Press is an imprint of Elsevier
125 London Wall, London EC2Y 5AS, United Kingdom
525 B Street, Suite 1650, San Diego, CA 92101, United States
50 Hampshire Street, 5th Floor, Cambridge, MA 02139, United States
The Boulevard, Langford Lane, Kidlington, Oxford OX5 1GB, United Kingdom

Library of Congress Cataloging-in-Publication Data
A catalog record for this book is available from the Library of Congress

British Library Cataloguing-in-Publication Data
A catalogue record for this book is available from the British Library

ISBN 978-0-12-822098-6

For information on all Academic Press publications
visit our website at https://www.elsevier.com/books-and-journals

Publisher: Andre Gerhard Wolff
Acquisitions Editor: Linda Versteeg-Buschman
Editorial Project Manager: Susan Ikeda
Production Project Manager: Niranjan Bhaskaran
Cover Designer: Greg Harris

Typeset by SPi Global, India

Contents

Contributors

Numbers in parenthesis indicate the pages on which the authors' contributions begin.

Saira Abbas (123), Department of Zoology, University of Science and Technology, Bannu, Pakistan

Sharjeel Ahmad (123), National Microbial Culture Collection of Pakistan (NCCP), Bio-resource Conservation Institute (BCI), National Agriculture Research Center (NARC), Islamabad; PirMehr Ali Shah Arid Agriculture University, Rawalpindi, Pakistan

Iftikhar Ahmed (123), National Microbial Culture Collection of Pakistan (NCCP), Bio-resource Conservation Institute (BCI), National Agriculture Research Center (NARC), Islamabad, Pakistan

Azka Asif (309), School of Biological Sciences, University of the Punjab, Lahore, Pakistan

Soma Barman (49), Soil and Agrobio-Engineering Laboratory, Department of Environmental Science, Tezpur University, Tezpur, Assam, India

Unsa Bashir (387), Department of Allied Health Sciences, The Superior College Lahore, Lahore, Pakistan

Chanda Parulekar Berde (27), School of Earth, Ocean and Atmospheric Sciences, Goa University, Taleigão, Goa, India

Vikrant B. Berde (27), Department of Zoology, Arts, Commerce & Science College, Lanja, Maharashtra, India

Satpal Singh Bisht (1,413), Department of Zoology, Kumaun University, Nainital, Uttarakhand, India

Jose Antonio Cervantes-Chávez (171), Queretaro Autonomous University, Basic and Applied Microbiology Unit, Natural Science Faculty, Santiago de Querétaro, Mexico

Debmalya Dasgupta (1,413), Department of Biotechnology, National Institute of Technology, Yupia, Arunachal Pradesh, India

Surajit De Mandal (413), College of Plant Protection, South China Agricultural University, Laboratory of Bio-Pesticide Innovation and Application of Guangdong Province, Guangzhou, PR China

Francisco Javier Delgado-Virgen (171), Plant and Microbial Biotechnology Laboratory, Mexico's National Technologic, Colima Institute of Technology, Colima, México

Raunak Dhanker (339), Department of Biological Sciences, School of Basic and Applied Sciences GD Goenka University, Gurugram, Haryana, India

Kunal Dutta (435), Department of Human Physiology, Vidyasagar University, Midnapore, West Bengal, India

Chandradipa Ghosh (435), Department of Human Physiology, Vidyasagar University, Midnapore, West Bengal, India

Pralay Shankar Gorai (49, 99), Mycology and Plant Pathology Laboratory, Department of Botany, Visva-Bharati, Santiniketan, West Bengal, India

Linee Goswami (99), Mycology and Plant Pathology Laboratory, Department of Botany, Visva-Bharati, Santiniketan, West Bengal, India

Shubham Goyal (339), Amity Institute of Biotechnology, Amity University, Noida, India

Rifat Hayat (123), PirMehr Ali Shah Arid Agriculture University, Rawalpindi, Pakistan

Touseef Hussain (339), Department of Botany, Aligarh Muslim University, Aligarh, Uttar Pradesh, India

Anum Ishaq (387), Department of Allied Health Sciences, The Superior College Lahore, Lahore, Pakistan

Debarati Jana (435), Department of Human Physiology, Vidyasagar University, Midnapore, West Bengal, India

Nazia Kanwal (387), Department of Allied Health Sciences, The Superior College Lahore, Lahore, Pakistan

Monalisha Karmakar (435), Department of Human Physiology, Vidyasagar University, Midnapore, West Bengal, India

Sengodan Karthi (71), Department of Biochemistry, Centre for Biological Sciences, K. S. Rangasamy College of Arts and Science (Autonomous), Tiruchengode, Tamil Nadu, India

Rabia Khalid (123), PirMehr Ali Shah Arid Agriculture University, Rawalpindi, Pakistan

Patcharin Krutmuang (71), Department of Entomology and Plant Pathology, Faculty of Agriculture, Chiang Mai University, Chiang Mai, Thailand

Krishna Kumar (339), Department of Biotechnology, School of Chemical and Life Sciences, New Delhi, India

Kulbhushan Kumar (1), Department of Zoology, Kumaun University, Nainital, Uttarakhand, India

Arabinda Mahanty (413), Crop Protection Division, National Rice Research Institute, Cuttack, Odisha, India

Tushar Kanti Maiti (357), Microbiology Laboratory, Department of Botany, The University of Burdwan, Golapbag, Purba Bardhaman, West Bengal, India

Narayan Chandra Mandal (49,99,357), Mycology and Plant Pathology Laboratory, Department of Botany, Visva-Bharati, Santiniketan, West Bengal, India

Rashi Miglani (1), Department of Zoology, Kumaun University, Nainital, Uttarakhand, India

Rojita Mishra (1,413), Department of Botany, Polasara Science College, Polasara, Ganjam, Odisha, India

Hareem Mohsin (309,387), Department of Allied Health Sciences, The Superior College Lahore, Lahore, Pakistan

Sonia Mumba (71), Department of Agriculture, Livestock Development and Fisheries, Directorate of Fisheries, County Government of Kilifi, Kilifi, Kenya

Rosa Jazmin Osuna-Cisneros (171), Plant and Microbial Biotechnology Laboratory, Mexico's National Technologic, Colima Institute of Technology, Colima, México

Amiya Kumar Panda (435), Department of Chemistry, Vidyasagar University, Midnapore, West Bengal, India

Amrita Kumari Panda (1,413), Department of Biotechnology, Sant Gahira Guru University, Ambikapur, Chhattisgarh, India

Hai The Pham (209), Research group for Physiology and Applications of Microorganisms (PHAM group), GREENLAB, Center for Life Science Research (CELIFE) and Department of Microbiology, Faculty of Biology, VNU University of Science, Vietnam National University, Hanoi, Vietnam

Sarayut Pittarate (71), Department of Entomology and Plant Pathology, Faculty of Agriculture, Chiang Mai University, Chiang Mai, Thailand

Krishnendu Pramanik (357), Mycology and Plant Pathology Laboratory, Department of Botany, Visva-Bharati, Santiniketan; Microbiology Laboratory, Department of Botany, The University of Burdwan, Golapbag, Purba Bardhaman, West Bengal, India

Julius Rajula (71), Department of Entomology and Plant Pathology, Faculty of Agriculture, Chiang Mai University, Chiang Mai, Thailand

Priyanka Raul (435), Department of Human Physiology, Vidyasagar University, Midnapore, West Bengal, India

Prachiti Rawool (27), Department of Microbiology, Gogate Jogalekar College, Ratnagiri, Maharashtra, India

Yasir Rehman (309), Department of Life Sciences, School of Science, University of Management and Technology, Lahore, Pakistan

Syed Abdul Qadir Shah (387), Department of Allied Health Sciences, The Superior College Lahore, Lahore, Pakistan

Malee Thungrabeab (71), Rajamangala University of Technology Lanna, Chiang Mai, Thailand

Laura Leticia Valdez-Velázquez (171), Biological Products Laboratory, Chemical Engineering Laboratory, Chemical Science Faculty, University of Colima, Colima, México

Oscar Fernando Vázquez-Vuelvas (171), Biochemical Engineering and Bioprocessing Laboratory, Chemical Science Faculty, University of Colima, Colima, México

Hina Zain (387), Department of Allied Health Sciences, The Superior College Lahore, Lahore, Pakistan

Chapter 1

Microbial biofertilizers: Recent trends and future outlook

Debmalya Dasgupta[a], Kulbhushan Kumar[b], Rashi Miglani[b], Rojita Mishra[c], Amrita Kumari Panda[d], and Satpal Singh Bisht[b]
[a]*Department of Biotechnology, National Institute of Technology, Yupia, Arunachal Pradesh, India,* [b]*Department of Zoology, Kumaun University, Nainital, Uttarakhand, India,* [c]*Department of Botany, Polasara Science College, Polasara, Ganjam, Odisha, India,* [d]*Department of Biotechnology, Sant Gahira Guru University, Ambikapur, Chhattisgarh, India*

Chapter outline

1 Introduction

The rapid increase in the human population worldwide is coupled with an exponential food demand and food security in developing countries is a matter of concern for farmers and agronomists. There are reports that the food grain demand will reach 28.8 million tons by 2030 and the availability will be limited to 21.6 million tons only leading to a shortage of about 7.2 million tons worldwide. A sharp decline in the availability of biomass in the form of organic fertilizers and demand for more productivity in less time promoted the voluminous production of inorganic fertilizers and their use across the crops. These chemical-based

Recent Advancement in Microbial Biotechnology. https://doi.org/10.1016/B978-0-12-822098-6.00001-X

fertilizers are more expensive and unaffordable to small-scale and marginal farmers, and are not only detrimental to the soil but also cause serious threats to the soil ecology as potential environmental hazards. Extensive and irrational use of synthetic chemicals causes irreversible damage to the soil chemistry and microbial diversity leading to the loss of soil fertility. The continuous use of chemical fertilizers causes air, soil, and underground water pollution by eutrophication of various water bodies (Khosro & Yousef, 2012). It also speeds up soil acidification and severely contaminates groundwater and the atmosphere. So researchers attempt continuously toward the new formulation of nutrient-rich biofertilizers having adequate biosafety (Youssef & Eissa, 2014).

Biofertilizers are a mixture of living or dormant cells containing different microbial strains, viz., nitrogen-fixing, phosphate-solubilizing, etc. They are powerful biological tools for sustainable agriculture and effective alternative to chemical fertilizers with an ability to maintain soil microflora. They are designed to multiply microbes of interest that can supply vital nutrients required by host plants. Biofertilizers could be supplemented with commercial fertilizers to enhance agriculture productivity by maintaining soil health and microbial diversity. Biofertilizers can be applied to the soil directly or via seed and seedling treatment methods (Muthuselvam & Tholkappian, 2008). The microorganisms associated are commonly known as plant growth-promoting microorganisms (PGPM), mainly comprised of *Azospirillum, Azotobacter, Cyanobacteria, Phosphobacteria, and Rhizobia* (Abo-Baker & Mostafa, 2011; Barassi et al., 2007; Hussain, Mujeeb, & Tahir, 2002). Biofertilizer has been identified and recommended as a substitute for chemical fertilizer in sustainable farming practices. High-efficacy biological fertilizers play a pivotal role in productivity and soil sustainability and could act as cost-effective agriculture inputs (Sinha, Valani, & Chauhan, 2014). The benefits of biofertilizers and their contribution toward soil microbial activities, carbon sequestration, and soil fertility make it a safer substitute over chemical fertilizer (Fig. 1).

Biofertilizers augment the soil fertility by various macro- and micronutrients through atmospheric nitrogen fixation, potassium and phosphate solubilization/mineralization, secretion of various plant growth-regulating substances, and decomposition of organic matter. Biofertilizers boost crop productivity by actively regulating nutrient dynamics when used as seed or soil inoculants (Grabber & Galloway, 2008). Numerous parameters need to be considered in biofertilizer preparation such as type of microbe, growth profile and optimum growth conditions of organism used, and formulation of inoculums. The design of inoculants, process of application, and storage of the formulations are all critical for the efficiency of the biological product. Preparation of biofertilizers involves various steps: (i) Screening of active microorganisms, (ii) Isolation of target microbes, (iii) Standardization of propagation methods and carrier material, (iv) Phenotype testing, and (v) large-scale tests.

Simultaneously in the recent past, nanotechnology research significantly contributed to the production of nanobiofertilizer (an amalgamation preparation of nanoparticle and plant growth-promoting bacteria) is more useful to increase

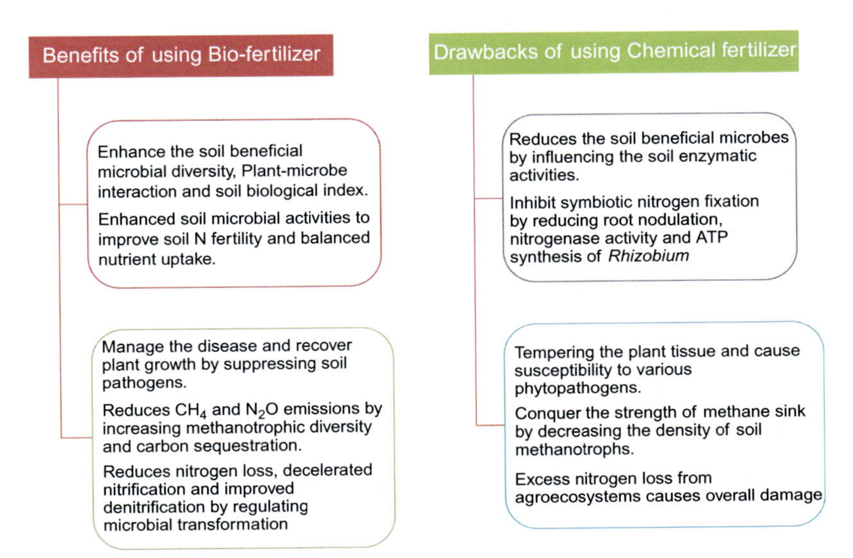

FIG. 1 Advantages of biofertilizers over chemical fertilizers.

soil fertility. The encapsulation of nutrients with nanomaterials, polymers cause sustained and tranquil release of organic nutrients to plants. The nanoparticle coatings of biofertilizer increases the stability of the bioavailable nutrients and accessible to plants at various growth stages (Kumari & Singh, 2019). Various steps involved in the production of nanobiofertilizer are described in Fig. 2. Nanobiofertilizer have a multitude of advantages viz., reduction in soil toxicity, higher efficiency of nutrients utilization, and reduced negative effects associated with the overdose. Hence, nanobiofertilizers is a reasonable method for sustainable agricultural practices in many developing countries (El-Ghamry, Mosa, Alshaal, & El-Ramady, 2018). Seyed Sharifi, Khalilzadeh, Pirzad, and Anwar (2020) studied the effect of nanofertilizer on grain yield of wheat under water deficit conditions and reported that the combined applications of nano Zn-Fe oxide and *Azotobacter* increased wheat yield under severe drought conditions. Rahbar Keykha, Khammari, Dahmardeh, and Forouzandeh (2017) reported the effect of nanobiofertilizer on yield of sesame varieties and found that the applications of nanobiofertilizer increases the seed yield and oil percentage. Jakiene, Spruogis, Romaneckas, Dautart, and Avizienyte (2015) studied the efficiency of nanobiofertilizer on sugar beet and found increase in root biomass (42.6%), leaf area (19.6%), net photosynthetic productivity (15.8%), and sucrose content 1.03%, resulting in 19.2% increase in the sugar yield.

2 Categories of biofertilizers

Biofertilizers are categorized into seven types based on the type of microorganisms used and their functional attributes. The different types of biofertilizers depending upon their activity and function are summarized in the following sections.

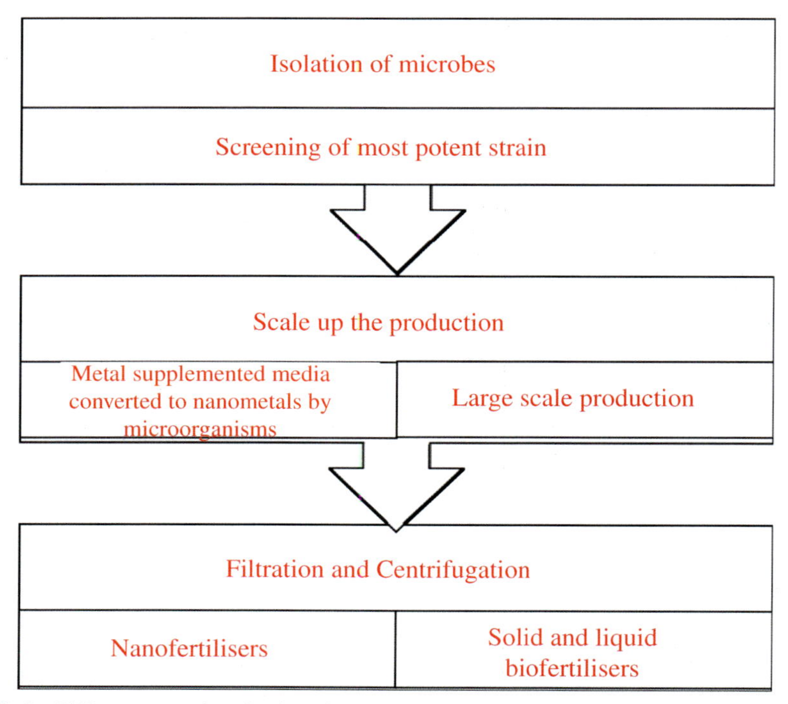

FIG. 2 Different steps of production of nanobiofertilizer.

2.1 Nitrogen-fixing biofertilizers

Nitrogen-fixing biofertilizers (NFBs) include *Rhizobium* spp., *Azospirillum* spp., and blue-green algae, which fix atmospheric N_2 accessible to plants. They are crop-specific biofertilizers and can be applied as inoculant directly (as seed coating) or indirectly (inoculant applied in soil). NFB accounts for the largest share of global biofertilizer market. China embraces more than 511 biofertilizers-based products followed by Canada with 150 NFB-based products. During 2017, China held around 43.2% of the biofertilizers market share for Asia-Pacific region because of many inflexible regulations enforced on chemical fertilizers (Soumare et al., 2020).

2.2 Phosphate-solubilizing biofertilizer

Phosphate solubilizing biofertilizers (PSBs) include *Bacillus* spp., *Pseudomonas* spp., and*Aspergillus* spp., which solubilize the insoluble phosphate into the soil. Soil bacteria and fungi help in converting these insoluble phosphates to their soluble form. These organisms produce organic acids that decrease soil pH and cause the release of phosphate forms leading to P solubilization.

The mechanism of P solubilization consists of four steps: (i) Production of organic acid causes a decrease of soil pH, (ii) cation chelation, (iii) compete with P sorption, and (iv) formation of soluble metal complexes. Organic acids secreted by the microorganisms chelates divalent cations Ca^{2+} that releases phosphates from insoluble complexes. Organic acids form metal-soluble complexes co-complexed with insoluble P, thereby secreting the P moiety.

2.3 Phosphate mobilizing biofertilizers

Phosphate mobilizing biofertilizers (PMBs) include *Mycorrhiza* that scavenges phosphates from the soil and mobilize the insoluble phosphorus by recruiting alkaline phosphatase, which in turn increases P mineralization (Zhang et al., 2016). There are reports that mycorrhiza changes the soil community composition and these mycorrhizal hyphae-linked microbiome increases mineralization of organic P in soil (Zhang et al., 2018). PMBs contribute significantly to the orthophosphate pool of the soil biome by transforming insoluble inorganic or organic phosphates into soluble orthophosphate (Pi). The mechanisms of the mobilization of sparingly soluble phosphates by mycorrhizal hyphae involves the secretion of H^+ following the exploitation of NH_4^+ leading to a decrease in soil pH.

2.4 Plant growth-promoting biofertilizer

Pseudomonas spp. are excellent examples of plant growth-promoting rhizobacteria (PGPR) capable of producing phytohormones, antimetabolites etc. that promote root growth and organic matter decomposition that in turn helps in mineralization. Plant growth-promoting biofertilizers (PGPBs) are crop-specific biofertilizers and the commercialization of PGPR-based inoculants comprises the following steps: (i) Isolation and screening of PGPB from rhizospheric and phyllospheric region of plants, (ii) field screening for a variety of crops, soil types, and topographical locations, (iii) assessment of the potential combinations of microbial strains, (iv) assessment of experimental evidences for the eco-toxicological effects, (v) formulation (solid or liquid form), and (vi) product registration and regulatory consent.

2.5 Potassium-solubilizing biofertilizer

Potassium-solubilizing biofertilizers (KSBs) include *Aspergillus niger* and *Bacillus* spp. Potassium in the soil mainly exists as silicate minerals that are not available to plants. KSBs are broad-spectrum biofertilizers and solubilize rock minerals to increase available phosphate in the soil. The solubilization of minerals mainly occurs due to organic acid production such as gluconic acid, 2-ketogluconic acid, etc. Potassium-solubilizing microorganisms releases K^+ from K-bearing minerals by various mechanisms. The major mechanism of K solubilization is acidolysis i.e., by production the organic and inorganic acids and protons that convert the insoluble K (muscovite, mica, and biotite feldspar) to soluble forms of K (Etesami, Emami, & Alikhani, 2017).

2.6 Potassium-mobilizing biofertilizer

Bacillus spp. mainly responsible for the channeling of potassium more specifically the silicates into the soil.

2.7 Sulfur-oxidizing biofertilizer

These biofertilizers act by oxidizing sulfur to sulfates, which can be trapped by plants; the best example of sulfur-oxidizing microorganisms is *Thiobacillus* spp. Studies have shown that sulfur oxidation takes place by following two major pathways viz., (i) S4 intermediate pathway involving the formation and oxidation of tetrathionate or polythionate and sulfur from thiosulfate and (ii) paracoccus sulfur oxidation that converts thiosulfate directly to sulfate (Friedrich, Rother, Bardischewsky, Quentmeier, & Fischer, 2001). Few commercially available microbial biofertilizers as summarized in Table 1 have been introduced into the market and efforts have been made for farmers awareness on their potential use for sustainable agriculture.

3 Symbiotic nitrogen-fixing bacteria

3.1 Rhizobium

Rhizobium is a nonspore-forming rod-shaped, motile, aerobic, gram-negative soil bacterium able to colonize in the rhizospheric region of leguminous plants and symbiotically fixes atmospheric nitrogen. They were initially classified based on the cross-inoculation group concept (Subba, 1993) and later based on a polyphasic approach i.e., *Rhizobium* with 14 root-nodulating nitrogen-fixing species and 04 nonnodulating nonnitrogen fixing species, *Sinorhizobium* with 09 species, *Mezorhizobium* with 08 species, *Bradyrhizobium* with 04 species, and 01 species of *Azorhizobium*. Recently 02 more genera viz., *Methylobacterium* with 01 species under α-proteobacteria and *Burkholderia* under β-proteobacteria have been included in the rhizobia group. To date, rhizobia consist of 61 species, belonging to 13 genera, namely *Rhizobium*, *Mesorhizobium*, *Bradyrhizobium*, *Azorhizobium*, *Sinorhizobium*, *Allorhizobium*, *Methylobacterium*, *Phyllobacterium*, *Ochrobactrum*, *Herbaspirillum*, *Burkholderia*, *Cuprivaidus*, and *Devosia* (Passricha, Saifi, Sikka, & Tuteja, 2020).

They have a diverse range of morphology, biochemistry, and physiology based on the situations they thrive and found to be the most competent biofertilizer based on the amount of nitrogen they fix. Most plants belonging to Leguminosae family have a strong symbiotic relationship with this bacterium. They

TABLE 1 Few commercial microbial biofertilizers.

Product	Microorganism	Formulation	Crop	Country	Benefits	References
BioGro	*Pseudomonas fluorescens* *Citrobacter freundii* *Klebsiella pneumoniae*	Solid	Rice	Australia	Increased shoot and root weight in greenhouse trials	Kecskés et al. (2016)
	Bacillus amyloliquefaciens *Pseudomonas fluorescens* *Bacillus subtilis* *Candida tropicalis*			Mekong Delta, Vietnam	Improve rice yield	Nguyen et al. (2017)
Soil-Life	Actinomycetes Fungi Lactic acid bacteria Photosynthetic bacteria Yeasts	Liquid	Sugarcane	Australia	Enhance soil physical and biological health, favorable microbial communities, and induce crop yield	Berg et al. (2019)

Continued

TABLE 1 Few commercial microbial biofertilizers—cont'd

Product	Microorganism	Formulation	Crop	Country	Benefits	References
Bioboots	*Delftia acidovorans*, *Bradyrhizobium* sp.	Liquid	Canola crop and soybean	Canada		Gracia-Fraile, Menendez, and And (2015)
EVL coating	*Lactobacillus helveticus*	Solid	–	Canada	Plant growth-promoting bacteria and biostimulant	Gracia-Fraile et al. (2015)
Nitrofix	*Azospirillum* sp.	Solid	Sugarcane	Cuba	Fix nitrogen and produce phytohormones and also promotes the growth of crops	Gracia-Fraile et al. (2015)
Mamezo, R-processing seed, Hyper coating seeds	*Rhizobium* sp.	Mixed with peat		Japan	Process legume seeds and grass	Kretzschmar and Manefield (2015)

perform their role via different mechanisms such as exudation of flavonoids by the host upon infection (Hassan & Mathesius, 2012). Purine derivatives play an important role to initiate nodule formation (Giraud et al., 2007), while some *Rhizobium* strain invades plant roots through "crack entry" without any kind of infection inside root hairs of the host (Madsen et al., 2010). It has been predicted that 40–250 kg N/ha/year is fixed by various legume plants due to colonization of *Rhizobium. Rhizobia* can make a useful association with crop plants such as rice, wheat, maize, and barley millets without any disease symptoms or nodule-like structures and can be a useful component in the formulation of crop-enhancing biofertilizers. It increases the yield of *Cicer arietinum* (Giri & Joshi, 2010). Many new rhizobium strains viz., *Rhizobium leguminosarum* strain named PEPV16 has been identified and can be applied to nonleguminous plants i.e., lettuce and carrots (Flores-Félix et al., 2013) as well as *Agrobacterium radiobacter* strain 204 developed in Russia for cereal crop production (Humphry et al., 2007).

Some *Rhizobium* strains in connection with arbuscular mycorrhizal fungi have been established to be efficient biofertilizers that could improve the vegetative growth of faba bean (Abd-Alla, El-Enany, Nafady, Khalaf, & Morsy, 2014) and *Trigonella foenum graecum* L. (Nagananda, Das, Bhattacharya, & Kalpana, 2010). Reports reveal that pilot-scale biofertilizer production using *Rhizobium* and *Azobacter* is a profit-making venture that can magnetize new entrepreneurs (Sethi & Adhikary, 2012).

3.2 Free-living nitrogen-fixing bacteria

3.2.1 Azotobacter

They are one of the most extensively used PGPMs because of their beneficial activity to a wide variety of crops. These are polymorphic, possess peritrichous flagella, produce polysaccharides, sensitive to acidic pH, high salts, and temperatures above 35°C, grow on a nitrogen-free medium and fix atmospheric nitrogen. Various types of azotobacteria groups have been reported from rhizosphere viz., *Azotobacter armenicus, Azotobacter chroococcum, Azotobacter nigricans, Azotobacter paspali, Azotobacter salinestris*, and *Azotobacter vinelandi. Azotobacter chroococcum* is most frequently found in Indian soils (Wani, Chand, Wani, Ramzan, & Hakeem, 2016). Besides nitrogen fixation, they produce siderophores, antifungal substances, and plant growth regulators (Suneja, Narula, Anand, & Lakshminarayana, 1996).

Production of phytohormones such as auxin, cytokinin by *Azotobacter chroococcum*, have been reported (Verma, Kukreja, Pathak, Suneja, & Narula, 2001). Due to the production of an ample amount of slime, this bacterium is useful in soil aggregation. They have been recognized as effective biofertilizers for various crops such as mulberry (*Morus* spp.) (Sudhakar, Chattopadhyay, Gangwar, & Ghosh, 2000), mangroves (Ravikumar,

Kathiresan, Ignatiammal, Selvam, & Shanthy, 2004), Canola (*Brassica napus* L.) (Yasari, Azadgoleh, Pirdashti, & Mozafari, 2008), cucumber (*Cucumis sativus*) (Alhia, 2010), and corn (Peng et al., 2013). Besides, they are reportedly useful in freshwater aquaculture (Tripathy & Ayyappan, 2005). This bacterium usually favors soil with elevated organic matter.

3.2.2 Azospirillum

Azospirillum is a type of *Rhizobacteria* known for its phytohormone production and nitrogen-fixing ability. It was first isolated and originally named as *Spirillum lipoferum*. They are gram-negative, vibroid to straight rods, sometimes curved, highly pleomorphic with an abundant cytoplasmic accumulation of poly β-hydroxybutyrate, motile, inclined to salts and organic acids as carbon source, and oxidase-positive. It is considered as the safest bacteria that can be commercially exploited as biofertilizer for several crops having economic viability viz., cereals, rice-wheat, etc. (Mehnaz, 2015). *Azospirillum* is comprised of seven species i.e., *Azospirillum amazonense*, *Azospirillum brasilense*, *Azospirillum doebereinnerae*, *Azospirillum halopraeferens*, *Azospirillum irakense*, *Azospirillum largimobile*, and *Azospirillum lipoferum*. These are isolated from the rhizospheric and phyllospheric regions of many crop plants. The organisms reproduce under both aerobic and anaerobic conditions, but it is preferably microaerophilic in the absence or presence of combined nitrogen in the medium. Apart from nitrogen fixation, production of growth-promoting substances, disease resistance, and drought tolerance are the additional benefits due to *Azospirillum* inoculation (Yadav & Mowade, 2005).

Novel strains of *Azospirillum* viz., Az6 has been isolated that reportedly improves rice and maize yields (Malle, Kassogue, Babana, de Oliveira Paiva, & Murriel, 2020). Some reports also suggest that certain diseases such as fusarium wilts in bananas cannot be effectively controlled by the existing strains of *Azospirillum* (Widyantoro, 2020). Besides, there are many other novel strains of *Azospirillum* viz., *Azospirillum amazonanse* (Magalhaes, Baldani, Souto, Kuykendall, & Dobereiner, 1983) and *Azospirillum irakense* (Khammas, Ageron, Grimont, & Kaiser, 1989). Many reports suggested that the majority of this rhizobacteria is effective against rice (*Oryza sativa*) (Asiloglu et al., 2020; Chamam et al., 2013; Elbeltagy et al., 2001; Fukami, Cerezini, & Hungria, 2018; Razie & Anas, 2008; Steenhoudt & Vanderleyden, 2000; Watanabe & Lin, 1984).

4 Phosphorus-solubilizing biofertilizers

4.1 *Bacillus*

They are gram-positive, ubiquitous, and recovered from all niches in the environment. They are a dominant genus and widely used as biofertilizer. Aline, the first commercial bacterial biofertilizer, is reportedly developed from a *Bacillus* strain (Kilian et al., 2000). They are one of the most widely used bacteria in the biopesticide market in North America (Borriss, 2011) and many agricultural, industrial, and medicinal products are also being prepared using this genus (Lyngwi & Joshi, 2014). They enhance the plant-available forms of nutrients in rhizospheres, put a check on disease-causing pathogens, and also initiate pest defense systems (Gracia-Fraile et al., 2015; Kang, Radhakrishnan, & Lee, 2015).

Some endospore-forming *Bacillus* strains viz., *Bacillus amyloliquefaciens* form biofilms and assist plant growth by suppressing rhizosphere-based plant pathogens (Borriss, 2011) and *Bacillus licheniformis* biofertilizer reportedly increases the antioxidant activity of tomato *Solanum lycopersicum* L. var. Sheva (Ochoa-Velasco et al., 2016). Many novel strains viz., *Bacillus methylotrophicus* strain NKG-1 has proven to be a potential biofertilizer (Ge et al., 2016). *Bacillus halotolerans* when applied as a biofertilizer reportedly increased phenolic contents in *Coriandrum sativum* L. (Jimenez-Gomez, García-Estevez, García-Fraile, Escribano-Bailón, & Rivas, 2020). Some other *Bacillus* strains viz., *Bacillus cereus* increases the productivity of *Cicer arietinum* L (Kaur, Singh, & Sharma, 2020). *Bacillus subtilis* biofertilizer not only reduces excess agricultural ammonia emission up to 44% and also helps in the high yield of crops (Sun et al., 2020) but also compensates for nitrogen loss in soil due to the use of excess fertilizer (Sun et al., 2020). *Bacillus amyloliquefaciens* FZB42 has been reported as a promising agent to reduce soil-borne plant pathogens (Borriss, 2020). Examples include control of *Fusarium* wilt in banana by using *Bacillus velezensis* H-6 as a biofertilizer (Huang et al., 2019).

4.2 *Pseudomonas*

They are diverse and ecologically versatile bacteria capable to colonize in varied habitats viz., terrestrial, freshwater, and marine (Yarzábal, Monserrate, Buela, & Chica, 2018). They can withstand different kinds of stress. *P. fluorescens* reportedly used to control bacterial wilt (Liu, Li, Tang, & Zeng, 1999) as well as supports the growth of soyabean plants grown in shade conditions (Miftahurrohmat & Sutarman, 2020). *Pseudomonas fluorescens* increases the plant biomass and uptake of nutrient in black pepper (Paul, Srinivasan, Anandaraj, & Sarma, 2003). Biofertilizers supplemented with *Pseudomonas putida promotes* the growth and increase alkaloid content of*Rauwolfia serpentine* (Rai, Kumar, Bauddh, Singh, & Singh, 2017). Some novel bacterium viz., *Pseudomonas aeruginosa* PA-3 has the potential to convert kitchen waste oil into biofertilizers (Li, Cui, Wang, & Ge, 2018). Some strains of *Pseudomonas* can be used as active biofertilizers in cold regions especially for crops such as wheat *Triticum aestivum*) (Yarzábal et al., 2018). *Pseudomonas putida* strain PT has proven to be a multipurpose biofertilizer that could remove pollutants viz., cadmium from the soils (Khashei, Etemadifar, & Rahmani, 2020).

Recently, *Pseudomonas* sp. strain TK35 has been reported to facilitate the development and growth of tobacco roots (Cao et al., 2020).

Some strains of *Pseudomonas* are used as potent zinc-based fertilizers that can compensate for zinc insufficiencies caused by chemical fertilizers (Shahid, Tariq, & Mehnaz, 2020).

5 Free-living nitrogen-fixing cyanobacteria

Various free-living cyanobacteria or blue-green algae are capable of nitrogen fixation. They are abundantly used by several rice-growing countries such as China, Vietnam, and India (Lumpkin & Plucknett, 1982; Venkataraman, 1972). The reason being that paddies favor similar conditions viz., an amount of sunlight, water, temperature, humidity, and nutrients that are essential for the growth of cyanobacteria (Kumar, Kaushal, Saraf, & Singh, 2018). Prominent ones include *Nostoc*, *Anabaena*, *Aulosira*, *Cylindrospermum*, *Totypothrix*, and *Stigonema*. Cyanobacteria add organic matter as well as extra nitrogen to the soil due to their photosynthetic nature and also fix atmospheric nitrogen. They are exceptionally cost-effective biofertilizers and require an additional supplement of phosphate, potassium, and molybdenum.

Many reports on the use of such biofertilizers based on nitrogen-fixing cyanobacteria have been reported in rice (Kannaiyan, Aruna, Kumari, & Hall, 1997; Mishra & Pabbi, 2004; Pereira et al., 2009; Vaishampayan et al., 2001). They secrete plant growth hormones and aid in nutrient transport from the underground soil to plants. This is one of the inexpensive sources of nitrogen fixator for underprivileged farmers who are unable to capitalize on nitrogen fertilizers (Issa, Abd-Alla, & Ohyama, 2014). They may either be in the unicellular form such as *Aphanothece, Chroococcidiopsis, Dermocapsa, Synechococcus, Gloeocapsa (Gloeothece), Myxosarcina, Pleurocapsa, Xenococcus*, or filamentous form viz., heterocystous cyanobacteria are specialized for their ability to fix nitrogen into ammonia (NH_3) or nitrates (NO^{-3}), which can be absorbed by plants and converted into proteins. (*Anabaena, Anabaenopsis, Aulosira, Calothrix, Camptylonema, Chlorogloea, Chlorogloeopsis, Cylindrospermum, Fischerella, Gloeotrichia, Hapalosiphon, Mastigocladus, Nodularia, Nostoc, Nostochopsis, Rivularia, Scytonema, Scytonematopsis, Stigonema, Tolypothrix, Westiella, Westiellopsis*) or Nonheterocystous cyanobacteria (*Lyngbya, Chthonoplastes, Microcoleus, Oscillatoria, Myxosarcina, Plectonema boryanum, Schizothrix, Pseudanabaena, Trichodesmium*) (Joshi, Shourie, & Singh, 2020). Presently many engineered cyanobacterial biofertilizers are proposed for improving crop productivity especially rice (Shamim, Mahfooz, Hussain, & Farooqui, 2020).

6 Potassium-solubilizing microbes

Potassium solubilization is mainly done by a variety of saprophytic microbes that are from different taxonomic subgroups such as bacteria, actinomycetes, and fungi (Etesami et al., 2017). Microbial solubilization of K is regulated by various factors such as pH, oxygen, the type of bacteria and K-bearing minerals (Sheng & Huang, 2001). Microorganisms such as *Bacillus mucilaginosus, Pseudomonas, Bacillus edaphicus, Acidothiobacillus ferrooxidans, Burkholderia, and Bacillus circulans* releases potassium in usable form in the soil. A large variety of soil microorganisms have been identified, which can be used to solubilize silicate minerals such as *Arthrobacter* sp., *Enterobacter hormaechei, Paenibacillus mucilaginosus, Paenibacillus frequentans, Burkholderia*, and *Paenibacillus glucanolyticus* (Bashir et al., 2017). The most significant mechanisms known today, which are used by potassium solubilizing microbes are decreasing the pH; by increasing chelation of the cations bound to K; and acidosis of the surrounding area of microorganism (Meena, Maurya, & Verma, 2014). Potassium-solubilizing biofertilizers increase germination percentage, seedling vigor, plant growth, and yield (Etesami et al., 2017).

7 Mycorrhiza

Mycorrhiza usually exists in symbiotic association with the roots of higher plants and mobilize trace elements such as P, Fe, Zn, etc. The symbiotic association is maintained as the plants provide sugars to the fungal partner and the fungi provide essential nutrients i.e., phosphorus to the host plants. Mycorrhizal fungi absorb, accumulate, and transport phosphate within their hyphae and release it to plant cells in the root tissue. The mycorrhizal roots are characterized by dense woolly growth of mycorrhizal hyphae on the surface with the lack of root hairs and root cap.

Glomus sp. is one of the important fungal partners of mycorrhiza, plants inoculated with mycorrhiza are reportedly more resilient to root diseases. Mycorrhiza increases the higher branching of plant roots facilitating contact with a wider area of the soil surface, which results in a rise of the absorbing area for water and nutrients. Mycorrhiza increases higher branching of roots facilitating contact with a wider area of the soil surface resulting from an increase of net absorbing surface for nutrients and water.

Mycorrhizae are of two types based on the habitation of the fungus viz., ectomycorrhiza and endomycorrhiza.

7.1 Ectomycorrhiza

They form a mantle on the root surface and lies in the cortex of intercellular spaces. The fungal hyphae depend on food ingredients and sugars present in the intercellular spaces that are released by the root cells. Absorption of water, the release of inorganic nutrients, solubilization of organic matter, direct absorption of minerals from the soil are the primary roles of ectomycorrhiza. Plants with ectomycorrhiza mutualism absorb 2–3 times additional phosphorus, nitrogen, potassium, and calcium. They are found in association with various trees such as *Eucalyptus*, oak (*Quercus*), peach, pine, etc.

7.2 Endomycorrhiza

These fungal hyphae live in the root cortex, especially in the intercellular spaces with some hyphal tips passing inside the cortical cells e.g., crop plants, grasses, orchids, and some woody plants. The fungal hyphae provide nutrition by forming nutrient-rich pelotons at the seedling stage of orchids.

7.2.1 Vesicular arbuscular mycorrhiza

They are obligate intercellular, endosymbionts having prolonged root system upon establishment. These fungi possess special structures known as vesicles and arbuscular. They also gather various micro, macronutrients for the host in addition to the collection of moisture from deeper niches of the soil. Vesicular arbuscular mycorrhiza facilitates phosphorus availability, mobility, and ameliorates the uptake of Co, Zn, P, and water.

They have established to be beneficial in the regeneration of many medicinal and crop plants such as *Foeniculum vulgare* Mill (Darzi, Ghalavand, & Rejali, 2008), some legume species (Djebali, Turki, Zid, & Hajlaoui, 2010), *Arundo donax* L. (Tauler & Baraza, 2015), American ginseng (*Panax quinquefolius* L.) (Liu et al., 2020), cashew seedlings (Trisilawati, 2020), sugarcane (Kumar & Kumar, 2020), and linseed (Salahi, Yadavi, Salehi, & Balouchi, 2020). Besides this, in combination with some inoculated bacteria, they have shown to be effective against rice (*Oryza sativa*) (Pathirana & Yapa, 2020). Many approaches have been designed nowadays to synthesize next-generation biofertilizers by applying cell-free supernatants and exopolysaccharides. These modern-day biofertilizers are stable under storage conditions and promote plant growth by enhancing nodulation and growth of rhizospheric microflora.

8 Role of microbial fertilizers toward sustainable agriculture

Microbial fertilizers are superlative fertilizers and is an exemplified way to perk up ecological agriculture, which encourage its use toward sustainable agriculture. Biofertilizers promote the growth of crop plants by enhancing the nutrient availability directly through nitrogen fixation, phosphorous solubilization, and iron uptake, or indirectly by altering the levels of phytohormones, antibiotic production, etc. There are reports that application of biofertilizers improves growth and abiotic, biotic stress tolerance in crops (Azmat et al., 2020). Few examples of biofertilizer applications for sustainable agriculture are summarized in Table 2. Biofertilizers can reduce the toxic effects of pesticides by producing lytic enzymes. Many bacterial strains such as *Serratia, Paenibacillus, Gordonia, Azospirillum, Klebsiella, and Azotobacter* has the capacity to detoxify the pesticides (Shaheen & Sundari, 2013). *Actinomycetes* also help in biotransformation of pesticides. Plant growth-promoting microbes produce lytic enzymes, oxidases, and glutathione S transferases, which plays important role in pesticide degradation (Ortiz-Hernandez, Sanchez-Salinas,

TABLE 2 **Various applications of microbial biofertilizers in sustainable agriculture.**

Inoculated biofertilizer	Plant	Applications	Mechanism	References
Pseudomonas PS01	*Arabidopsis thaliana*	Improved survival and germination of *Arabidopsis thaliana* plants under salt stress	Biofilm produced by PS01 stimulated plant growth under salt stress by reducing plant Na^+ uptake	Chu, Tran, and Hoang (2019)
Bacillus subtilis Rhizo SF 48	*Solanum lycopersicum*	Sustainable tomato production in arid and semiarid regions under drought stress	ACC deaminase produced by Rhizo SF 48 protect tomato plants against oxidative damage caused due to drought stress	Gowtham et al. (2020)
Pseudomonas lini *Serratia plymuthica*	*Ziziphus jujuba*	Significantly increased plant height, shoot and root dry matter, and relative water content	Antioxidant enzyme activities increased	Zhang et al. (2020)
Pseudomonas sp. strain S3	*Solanum lycopersicum*	Improved root/shoot length, fresh and dry weight, photosynthetic pigment content, increased accumulation of osmolytes, enhanced activities of catalase and peroxidase and phenolic content in salt-stressed plants	ACC deaminase activity counteracts the salt stress-induced growth inhibition by regulating the ethylene synthesis	Pandey and Gupta (2020)
Pseudomonas fluorescens strain FB-49	*Acacia abyssinica*	Improved shoot height, root length, shoot/root dry weight	Biofilm formers and extracellular polymeric substances (EPS) producers play protective roles under stress conditions	Getahun, Muleta, Assefa, and Kiros (2020)
Mesorhizobium panacihumi DCY119^T	*Panax ginseng*	Higher biomass and higher levels of antioxidant genes	Siderophore-producing rhizobacteria confer heavy metal resistance	Huo et al. (2021)

Continued

TABLE 2 Various applications of microbial biofertilizers in sustainable agriculture—cont'd

Inoculated biofertilizer	Plant	Applications	Mechanism	References
Paenibacillus lentimorbus B-30488 (B-30488), *Bacillus amyloliquefaciens* SN13	Rice var. *Sarju-52*	Altered carbohydrate and fatty acid metabolic pathways	Metabolic re-programming in rice var. *Sarju-52* enhances nutrient use efficiency, tolerance, and growth under suboptimum nutrient conditions	Bisht and Chauhan (2020)
Trichoderma longibrachiatum KH	*Solanum lycopersicum*	Increased root volume, enhanced shoot potassium uptake	Organic acids produced by *Trichoderma* isolates accelerate the release of potassium from K-bearing minerals	Khoshmanzar et al. (2020)
Trichoderma longibrachiatum T6	*Triticum aestivum* L.	Enhancing tolerance to NaCl stress	Increased ACC-deaminase activity and indole acetic acid (IAA) production modulate the IAA and ethylene synthesis gene expression in wheat seedling roots under salt stress and minimize ionic toxicity by disturbing the intracellular ionic homeostasis	Zhang, Gan, and Xu (2019)
Bacillus pumilus strain DH-11	*Solanum tuberosum*	Protected plant against various abiotic stress such as salinity, water deficit, and heavy-metal toxicity	Increased expression levels of ROS-scavenging enzymes (APX, SOD, CAT, DHAR, and GR), trigger abiotic stress-related defense pathways	Gururani et al. (2013)

Dantan-Gonzalez, & Castrejon-Godínez, 2013; Ramakrishnan, Megharaj, Venkateswarlu, Sethunathan, & Naidu, 2011). Plants primed with microbial biofertilizers are able to counter phytopathogens through ethylene and jasmonate signaling pathways (Pangesti et al., 2016). The application of *Bacillus endophyticus* and *Pseudomonas aeruginosa* increases the tolerance of tomato plant against *Spodoptera litura* by producing salicylic acid, abscisic acid, etc. (Kousar, Bano, & Khan, 2020).

Plant growth mainly depends upon the rate of photosynthesis and microbial biofertilizers, which promotes the photosynthesis rate in stress conditions (Cohen et al., 2015; Gururani et al., 2013). Photosynthesis rate increases due to the production of photosynthetic pigments and chlorophyll content in the leaves (Heidari & Golpayegani, 2012). For example, *Rhizobia* helps in increasing the rate of photosynthesis in rice and banana plants (Mia & Shamsuddin, 2010). The use of *Azospirillum brasilens, Bacillus lentus,* and *Pseudomonas* sp. confers the plants water stress tolerance capacity. Plants can scavange reactive oxygen species (ROS) by producing more antioxidant molecules. Bioremediation is a best strategy to remove heavy metal toxicity and inorganic pollutants from soil (Lim et al., 2014). Studies by Mehnaz (2013), Shinwari et al. (2015), and Gontia-Mishra, Sapre, Kachare, and Tiwari (2017) revealed that there are many PGPBs that help in bioremediation. One of the finest mechanism toward heavy metal tolerance is production of 1-aminocyclopropane-1-carboxylate deaminase (ACC deaminase) and iron-chelating compounds known as siderophores (Mehnaz, 2013). PGPR is associated with the production of various types of aminoacids. They produce a variety of aminoacids depending upon the type of plant growth-promoting bacteria associated with roots (Kang, Kim, Yun, & Chang, 2010). Microbial biofertilizers have some nontargeted effects such as effect on biogeochemical cycle, soil properties, soil texture, etc. (Pereg & McMillan, 2015). Before releasing into market these can be studied thoroughly regarding its impact to the ecosystem. Nowadays, genetically modified strains are also used for biofertilizer formulation and can impact the metabolic pathways of plant growth-promoting substance production thus enhancing the yield. Genetically modified microorganisms offer plenteous advantages as particular biochemical pathways could be engineered with novel functions. Genetically modified microbial biofertilizers have been introduced with huge attainment with their specific action and persistence rates. The field trials with genetically modified (GM) biofertilizers are limited, due to the nontarget effects. Genetically modified microbes contribute to an amended nutrient availability in plants and aid in plant development.

9 Constraints and future outlook

Biofertilizers have congregated significant attention over the last few decades owing to their utility in a large number of crops, its widespread application has some potential constraints that need urgent attention. These constraints may either be physical, chemical, biological, technological, financial, and marketing. The agroclimatic conditions are the key factor that determines the class of biofertilizer produced. The superiority of biofertilizers is the most imperative factor that is indispensable for them to gain market potential. The superiority of biofertilizers also depends upon the type of microorganism used along with their metabolic activity under and after field applications. The selection of inoculum from reliable sources determines the commercial biofertilizer quality. A soil temperature of about 45–50°C at a depth of 5 cm is another constraint in the application of biofertilizer. Severe drought conditions are the hindrance as the number of microbial colonies drastically reduces under such arid conditions. Several biological constraints also restrict the biofertilizer's practices viz., occurrence of harmful parasites/predators in soil/fields, competition with the native population and predatory effects of certain nematodes on soil microbes. The interaction of biofertilizers with various plant species varies. Hence, the viewpoint of a universal biofertilizer is quite uttered that needs to be answered among the agronomists and scientists worldwide. The inherent competitive ability of a biofertilizer against the indigenous flora is one of the primary factor that determine the attainment of a biofertilizer.

An in-depth understanding of the multifaceted synergistic environment among the microorganism and its associated environments should be studied carefully for designing a potential biofertilizer candidate for sustainable agriculture. Advanced technologies must be recruited for the enormous production of biofertilizers and quality checks at each phase of production must be thoroughly assessed. The appropriate use and practices need to be disseminated among the farmers at the grass-root level so that they would be trained and encouraged to set up their regional biofertilizer-producing units at small scale level, which can harbor specific indigenous bacterial strains and scientific communities must take positive enthusiasm regarding the development of strain-specific and species-specific biofertilizers (Etesami et al., 2017). Future research must be done in this regard for sustainable agriculture.

The demand of biofertilizer production and organic food among the health-sensible society is gaining a momentum. The integrated farming practices such as application of microbial fertilizers, vermicomposting, etc., have an edge over the chemical counterparts due to their cost-effective inputs. The use of microbial fertilizers not only promote healthy grown organic food but also uphold sustainable environment. Various initiatives and adequate steps need to be taken by government to further stimulate the biofertilizer market countrywide. There are several limitations at the production level, marketing and field trials that reduce the application of microbial fertilizers at a larger scale. Production level constraints such as raw material, survivability of strains at diverse agroclimatic regions, biological, technical, and economic constraints act as a limitation for the biofertilizer production. At the marketing level storage, transport facilities and lower demand reduce the adoption of sustainable practice (Debnath, Rawat, Mukherjee, Adhikary, & Kundu, 2019). The toxic element present in the soil, drought, waterlogging conditions, variation in temperature, and poor organic content is accountable for field-level applications (Fierer & Jackson, 2006; Mazid & Khan, 2014). Against all odds, it is widely accepted that biofertilizers are the future fertilizers to restore and sustain agriculture by managing soil fertility and productivity.

References

Abd-Alla, M. H., El-Enany, A. W. E., Nafady, N. A., Khalaf, D. M., & Morsy, F. M. (2014). Synergistic interaction of *Rhizobium leguminosarum* bv. viciae and arbuscular mycorrhizal fungi as a plant growth promoting biofertilizers for faba bean (*Vicia faba* L.) in alkaline soil. *Microbiological Research, 169*(1), 49–58.

Abo-Baker, A. A., & Mostafa, G. G. (2011). Effect of bio-and chemical fertilizers on growth, sepals yield and chemical composition of *Hibiscus sabdariffa* at new reclaimed soil of South Valley area. *Asian Journal of Crop Science, 3*(1), 16–25.

Alhia, B. M. H. (2010). *The effect of Azotobacter chrococcum as nitrogen biofertilizer on the growth and yield of Cucumis sativus*. The Islamic University. http://hdl.handle.net/20.500.12358/21551.

Asiloglu, R., Shiroishi, K., Suzuki, K., Turgay, O. C., Murase, J., & Harada, N. (2020). Protist-enhanced survival of a plant growth promoting rhizobacteria, *Azospirillum* sp. B510, and the growth of rice (*Oryza sativa* L.) plants. *Applied Soil Ecology, 154*, 103599.

Azmat, A., Yasmin, H., Hassan, M. N., Nosheen, A., Naz, R., Sajjad, M., … Akhtar, M. N. (2020). Co-application of bio-fertilizer and salicylic acid improves growth, photosynthetic pigments and stress tolerance in wheat under drought stress. *PeerJ, 8*, e9960.

Barassi, C. A., Sueldo, R. J., Creus, C. M., Carrozzi, L. E., Casanovas, E. M., & Pereyra, M. A. (2007). *Azospirillum* spp., a dynamic soil bacterium favourable to vegetable crop production. *Dynamic Soil, Dynamic Plant, 1*(2), 68–82.

Bashir, Z., Zargar, M. Y., Husain, M., Mohiddin, F. A., Kousar, S., Zahra, S. B., … Rathore, J. P. (2017). Potassium solubilizing microorganisms: Mechanism and diversity. *International Journal of Pure and Applied Bioscience, 5*(5), 653–660. https://doi.org/10.18782/2320-7051.5446.

Berg, S., Dennis, P. G., Paungfoo-Lonhienne, C., Anderson, J., Robinson, N., Brackin, R., & Schmidt, S. (2019). Effects of commercial microbial biostimulants on soil and root microbial communities and sugarcane yield. *Biology and Fertility of Soils*, 1–16.

Bisht, N., & Chauhan, P. S. (2020). Comparing the growth-promoting potential of *Paenibacillus lentimorbus* and *Bacillus amyloliquefaciens* in *Oryza sativa* L. var. Sarju-52 under suboptimal nutrient conditions. *Plant Physiology and Biochemistry, 146*, 187–197.

Borriss, R. (2011). Use of plant-associated Bacillus strains as biofertilizers and biocontrol agents in agriculture. In *Bacteria in agrobiology: Plant growth responses* (pp. 41–76). Berlin, Heidelberg: Springer.

Borriss, R. (2020). Phytostimulation and biocontrol by the plant-associated Bacillus amyloliquefaciens FZB42: An update. In *Phyto-microbiome in stress regulation* (pp. 1–20). Singapore: Springer.

Cao, Y. Y., Ni, H. T., Ting, L. I., Lay, K. D., Liu, D. S., He, X. Y., … Qiu, L. J. (2020). *Pseudomonas* sp. TK35-L enhances tobacco root development and growth by inducing HRGPnt3 expression in plant lateral root formation. *Journal of Integrative Agriculture, 19*(10), 2549–2560.

Chamam, A., Sanguin, H., Bellvert, F., Meiffren, G., Comte, G., Wisniewski-Dyé, F., … Prigent-Combaret, C. (2013). Plant secondary metabolite profiling evidences strain-dependent effect in the *Azospirillum–Oryza sativa* association. *Phytochemistry, 87*, 65–77.

Chu, T. N., Tran, B. T. H., & Hoang, M. T. T. (2019). Plant growth-promoting rhizobacterium *Pseudomonas* PS01 induces salt tolerance in *Arabidopsis thaliana*. *BMC Research Notes*, *12*(1), 11.

Cohen, A. C., Bottini, R., Pontin, M., Berli, F. J., Moreno, D., Boccanlandro, H., ... Piccoli, P. N. (2015). *Azospirillum brasilense* ameliorates the response of *Arabidopsis thaliana* to drought mainly via enhancement of ABA levels. *Physiologia Plantarum*, *153*, 79–90.

Darzi, M. T., Ghalavand, A., & Rejali, F. (2008). Effect of mycorrhiza, vermicompost and phosphate biofertilizer application on flowering, biological yield and root colonization in fennel (*Foeniculum vulgare* Mill.). *Iranian Journal of Crop Sciences*, *10*(137), 88–109.

Debnath, S., Rawat, D., Mukherjee, A. K., Adhikary, S., & Kundu, R. (2019). Applications and constraints of plant beneficial microorganisms in agriculture. In *Biostimulants in plant science* (pp. 1–26). Intech Open. https://doi.org/10.5772/intechopen.89190.

Djebali, N., Turki, S., Zid, M., & Hajlaoui, M. R. (2010). Growth and development responses of some legume species inoculated with a mycorrhiza-based biofertilizer. *Agriculture and Biology Journal of North America*, *1*, 748–754.

El-Ghamry, A., Mosa, A., Alshaal, T., & El-Ramady, H. (2018). Nanofertilizers vs. biofertilizers: New insights. *Environment, Biodiversity and Soil Security*, *2*, 51–72. https://doi.org/10.21608/jenvbs.2018.3880.1029.

Elbeltagy, A., Nishioka, K., Sato, T., Suzuki, H., Ye, B., Hamada, T., ... Minamisawa, K. (2001). Endophytic colonization and in planta nitrogen fixation by a *Herbaspirillum* sp. isolated from wild rice species. *Applied and Environmental Microbiology*, *67*(11), 5285–5293.

Etesami, H., Emami, S., & Alikhani, H. A. (2017). Potassium solubilizing bacteria (KSB): Mechanisms, promotion of plant growth, and future prospects—A review. *Journal of Soil Science and Plant Nutrition*, *17*(4), 897–911. https://doi.org/10.4067/S0718-95162017000400005.

Fierer, N., & Jackson, R. B. (2006). The diversity and biogeography of soil bacterial communities. *Proceedings of the National Academy of Sciences of the United States of America*, *103*, 626–631.

Flores-Félix, J. D., Menéndez, E., Rivera, L. P., Marcos-García, M., Martínez-Hidalgo, P., Mateos, P. F., ... Rivas, R. (2013). Use of *Rhizobium leguminosarum* as a potential biofertilizer for *Lactuca sativa* and *Daucus carota* crops. *Journal of Plant Nutrition and Soil Science*, *176*(6), 876–882.

Friedrich, C. G., Rother, D., Bardischewsky, F., Quentmeier, A., & Fischer, J. (2001). Oxidation of inorganic sulfur compounds by bacteria: Emergence of a common mechanism? *Applied and Environmental Microbiology*, *67*, 2873–2882.

Fukami, J., Cerezini, P., & Hungria, M. (2018). *Azospirillum*: Benefits that go far beyond biological nitrogen fixation. *AMB Express*, *8*, 1–12.

Ge, B., Liu, B., Nwet, T. T., Zhao, W., Shi, L., & Zhang, K. (2016). *Bacillus methylotrophicus* strain NKG-1, isolated from Changbai Mountain, China, has potential applications as a biofertilizer or biocontrol agent. *PLoS One*, *11*(11), e0166079.

Getahun, A., Muleta, D., Assefa, F., & Kiros, S. (2020). Plant growth-promoting Rhizobacteria isolated from degraded habitat enhance drought tolerance of acacia (*Acacia abyssinica Hochst.* ex Benth.) seedlings. *International Journal of Microbiology*, *2020*.

Giraud, E., Moulin, L., Vallenet, D., Barbe, V., Cytryn, E., Avarre, J. C., ... Sadowsky, M. (2007). Legumes symbioses: Absence of Nod genes in photosynthetic bradyrhizobia. *Science*, *316* (5829), 1307–1312.

Giri, N., & Joshi, N. C. (2010). Growth and yield response of chick pea (*Cicer arietinum*) to seed inoculation with *Rhizobium* sp. *Nature and Science*, *8*(9), 232–236.

Gontia-Mishra, I., Sapre, S., Kachare, S., & Tiwari, S. (2017). Molecular diversity of 1-aminocyclopropane-1-carboxylate (ACC) deaminase producing PGPR from wheat (*Triticum aestivum* L.) rhizosphere. *Plant and Soil*, *414*, 213–227.

Gowtham, H. G., Singh, B., Murali, M., Shilpa, N., Prasad, M., Aiyaz, M., … Niranjana, S. R. (2020). Induction of drought tolerance in tomato upon the application of ACC deaminase producing plant growth promoting rhizobacterium *Bacillus subtilis* Rhizo SF 48. *Microbiological Research*, *234*, 126422.

Grabber, N., & Galloway, J. V. (2008). *An earth system of the global nitrogen cycle* (pp. 293–296). Nature Publishers.

Gracia-Fraile, P., Menendez, E., & And, R. R. (2015). Role of biofertilizers in agriculture and forestry. *Bioengineering*, *2*(3), 183–205.

Gururani, M. A., Upadhyaya, C. P., Baskar, V., Venkatesh, J., Nookaraju, A., & Park, S. W. (2013). Plant growth-promoting rhizobacteria enhance abiotic stress tolerance in Solanum tuberosum through inducing changes in the expression of ROS-scavenging enzymes and improved photosynthetic performance. *Journal of Plant Growth Regulation*, *32*(2), 245–258.

Hassan, S., & Mathesius, U. (2012). The role of flavonoids in root–rhizosphere signalling: Opportunities and challenges for improving plant–microbe interactions. *Journal of Experimental Botany*, *63*(9), 3429–3444.

Heidari, M., & Golpayegani, A. (2012). Effects of water stress and inoculation with plant growth promoting rhizobacteria (PGPR) on antioxidant status and photosynthetic pigments in basil (*Ocimum basilicum* L.). *Journal of the Saudi Society of Agricultural Sciences*, *11*, 57–61.

Huang, J., Pang, Y., Zhang, F., Huang, Q., Zhang, M., Tang, S., … Li, P. (2019). Suppression of *Fusarium* wilt of banana by combining acid soil ameliorant with biofertilizer made from Bacillus velezensis H-6. *European Journal of Plant Pathology*, *154*(3), 585–596.

Humphry, D. R., Andrews, M., Santos, S. R., James, E. K., Vinogradova, L. V., Perin, L., … Cummings, S. P. (2007). Phylogenetic assignment and mechanism of action of a crop growth promoting rhizobium radiobacter strain used as a biofertiliser on graminaceous crops in Russia. *Antonie Van Leeuwenhoek*, *91*(2), 105–113.

Huo, Y., Kang, J. P., Ahn, J. C., Kim, Y. J., Piao, C. H., Yang, D. U., & Yang, D. C. (2021). Siderophore-producing rhizobacteria reduce heavy metal-induced oxidative stress in Panax ginseng Meyer. *Journal of Ginseng Research*, *45*(2), 218–227.

Hussain, N., Mujeeb, F., & Tahir, M. (2002). Effectiveness of *Rhizobium* under salinity stress. *Asian Journal of Plant Science*, *4*, 124–129.

Issa, A. A., Abd-Alla, M. H., & Ohyama, T. (2014). Nitrogen fixing cyanobacteria: Future prospect. *Advances in Biology and Ecology of Nitrogen Fixation*, *2*, 24–48.

Jakiene, E., Spruogis, V., Romaneckas, K., Dautart, A., & Avizienyte, D. (2015). The bio-organic nano fertilizer improves sugar beet photosynthesis process and productivity. *Zemdirbyste-Agriculture*, *102*, 141–146.

Jimenez-Gomez, A., García-Estevez, I., García-Fraile, P., Escribano-Bailón, M. T., & Rivas, R. (2020). Increase in phenolic compounds of *Coriandrum sativum* L. after the application of a *Bacillus halotolerans* biofertilizer. *Journal of the Science of Food and Agriculture*, *100* (6), 2742–2749.

Joshi, H., Shourie, A., & Singh, A. (2020). *Cyanobacteria* as a source of biofertilizers for sustainable agriculture. In *Advances in cyanobacterial biology* (pp. 385–396). Academic Press.

Kang, B. G., Kim, W. T., Yun, H. S., & Chang, S. C. (2010). Use of plant growth-promoting rhizobacteria to control stress responses of plant roots. *Plant Biotechnology Reports*, *4*, 179–183.

Kang, S. M., Radhakrishnan, R., & Lee, I. J. (2015). *Bacillus amyloliquefaciens* subsp. plantarum GR53, a potent biocontrol agent resists Rhizoctonia disease on Chinese cabbage through hormonal and antioxidants regulation. *World Journal of Microbiology and Biotechnology, 31*(10), 1517–1527.

Kannaiyan, S., Aruna, S. J., Kumari, S. M. P., & Hall, D. O. (1997). Immobilized cyanobacteria as a biofertilizer for rice crops; Intl. Conference on applied algology, Knysna, South Africa, April 1996. *Journal of Applied Phycology, 9*(2), 167–174.

Kaur, D., Singh, G., & Sharma, P. (2020). Symbiotic parameters, productivity and profitability in Kabuli chickpea *(Cicer arietinum* L.) as influenced by application of phosphorus and biofertilizers. *Journal of Soil Science and Plant Nutrition*, 1–16.

Kecskés, M. L., Choudhury, A. T. M. A., Casteriano, A. V., Deaker, R., Roughley, R. J., Lewin, L., ... Kennedy, I. R. (2016). Effects of bacterial inoculant biofertilizers on growth, yield and nutrition of rice in Australia. *Journal of Plant Nutrition, 39*(3), 377–388.

Khammas, K. M., Ageron, E., Grimont, P. A. D., & Kaiser, P. (1989). *Azospirillum irakense* sp. nov., a nitrogen-fixing bacterium associated with rice roots and. *Research in Microbiology, 140*(8), 679–693.

Khashei, S., Etemadifar, Z., & Rahmani, H. R. (2020). Multifunctional biofertilizer from *Pseudomonas putida* PT: A potential approach for simultaneous improving maize growth and bioremediation of cadmium-polluted soils. *Biological Journal of Microorganism, 8*(32).

Khoshmanzar, E., Aliasgharzad, N., Neyshabouri, M. R., Khoshru, B., Arzanlou, M., & Lajayer, B. A. (2020). Effects of Trichoderma isolates on tomato growth and inducing its tolerance to water-deficit stress. *International journal of Environmental Science and Technology, 17* (2), 869–878.

Khosro, M., & Yousef, S. (2012). Bacterial bio-fertilizers for sustainable crop production: A review. *Journal of Agricultural and Biological Science, 7*(5), 237–308.

Kilian, M., Steiner, U., Krebs, B., Junge, H., Schmiedeknecht, G., & Hain, R. (2000). FZB24® *Bacillus subtilis*–mode of action of a microbial agent enhancing plant vitality. *Pflanzenschutz-Nachrichten Bayer, 1*(00), 1.

Kousar, B., Bano, A., & Khan, N. (2020). PGPR modulation of secondary metabolites in tomato infested with Spodoptera litura. *Agronomy, 10*(6), 778.

Kretzschmar, A. L., & Manefield, M. (2015). The role of lipids in activated sludge floc formation. *AIMS Environmental Science, 2*(2), 122–133. https://doi.org/10.3934/environsci.2015.2.122.

Kumar, M., & Kumar, N. (2020). Effect of integrated nutrient management with mycorrhizal biofertilizer on physical and bio-chemical properties of soil in planted and ratoon crop of sugarcane. *Journal of Soil and Water Conservation, 19*(1), 91–99.

Kumar, A., Kaushal, S., Saraf, S., & Singh, J. S. (2018). Microbial bio-fuels: A solution to carbon emissions and energy crisis. *Frontiers in Bioscience (Landmark Ed), 23*, 1789–1802.

Kumari, R., & Singh, D. P. (2019). Nano-biofertilizer: An emerging eco-friendly approach for sustainable agriculture. *Proceedings of the National Academy of Sciences, India Section B: Biological Sciences*, 1–9.

Li, Y., Cui, T., Wang, Y., & Ge, X. (2018). Isolation and characterization of a novel bacterium *Pseudomonas aeruginosa* for biofertilizer production from kitchen waste oil. *RSC Advances, 8*(73), 41966–41975.

Lim, S. P., Pandikumar, A., Huang, N. M., Lim, H. N., Gu, G., & Ma, T. L. (2014). Promotional effect of silver nanoparticles on the performance of N-doped TiO_2 photoanode-based dye-sensitized solar cells. *RSC Advances, 4*, 48236–48244.

Liu, Q. G., Li, Z., Tang, Z., & Zeng, X. M. (1999). Control of tobacco bacterial wilt with antagonistic bacteria and soil amendments. *Chinese Journal of Biological Control, 15*, 94–95.

Liu, N., Shao, C., Sun, H., Liu, Z., Guan, Y., Wu, L., … Zhang, B. (2020). Arbuscular mycorrhizal fungi biofertilizer improves American ginseng (*Panax quinquefolius* L.) growth under the continuous cropping regime. *Geoderma*, *363*, 114155.

Lumpkin, T. A., & Plucknett, D. L. (1982). *Azolla as a green manure: Use and management in crop production*. Westview Press, Inc.

Lyngwi, N. A., & Joshi, S. R. (2014). Economically important *Bacillus* and related genera: A mini review. In *Vol. 3. Biology of useful plants and microbes* (pp. 33–43). New Delhi, India: Narosa Publishing House.

Madsen, L. H., Tirichine, L., Jurkiewicz, A., Sullivan, J. T., Heckmann, A. B., Bek, A. S., … Stougaard, J. (2010). The molecular network governing nodule organogenesis and infection in the model legume *Lotus japonicus*. *Nature Communications*, *1*(1), 1–12.

Magalhaes, F. M., Baldani, J. I., Souto, S. M., Kuykendall, J. R., & Dobereiner, J. (1983). *New acid-tolerant Azospirillum species*. Anais-Academia Brasileira de Ciencias.

Malle, I., Kassogue, A., Babana, A. H., de Oliveira Paiva, C. A., & Murriel, I. I. (2020). A Malian native Azospirillum sp. Az6-based biofertilizer improves growth and yield of both rice (*Oryza sativa* L.) and maize (*Zea mays* L.). *African Journal of Microbiology Research*, *14*(7), 286–293.

Mazid, M., & Khan, T. A. (2014). Future of bio-fertilizers in Indian agriculture: An overview. *International Journal of Agricultural and Food Research.*, *3*, 10–23.

Meena, V. S., Maurya, B. R., & Verma, J. P. (2014). Does a rhizospheric microorganism enhance K + availability in agricultural soils? *Microbiological Research*, *169*, 337–347.

Mehnaz, S. (2013). Secondary metabolites of *Pseudomonas aurantiaca* and their role in plant growth promotion. In N. K. Arora (Ed.), *Plant microbe symbiosis: Fundamentals and advances* (pp. 373–393). New Delhi: Springer India. https://doi.org/10.1007/978-81-322-1287-4_14.

Mehnaz, S. (2015). *Azospirillum*: A biofertilizer for every crop. In *Plant microbes symbiosis: Applied facets* (pp. 297–314). New Delhi: Springer.

Mia, M. B., & Shamsuddin, Z. (2010). Nitrogen fixation and transportation by rhizobacteria: A scenario of rice and banana. *International Journal of Botany*, *6*, 235–242.

Miftahurrohmat, A., & Sutarman. (2020). Utilization of Trichoderma sp. and Pseudomonas fluorescens as biofertilizer in shade-resistant soybean. *IOP Conference Series: Materials Science and Engineering*, *821*, 012002.

Mishra, U., & Pabbi, S. (2004). Cyanobacteria: A potential biofertilizer for rice. *Resonance*, *9*(6), 6–10.

Muthuselvam, K., & Tholkappian, P. (2008). Vermicompost: A potential carrier material for bacterial bioinoculants. *Plant Archives*, *8*(2), 895–898.

Nagananda, G. S., Das, A., Bhattacharya, S., & Kalpana, T. (2010). In vitro studies on the effects of biofertilizers (*Azotobacter* and *Rhizobium*) on seed germination and development of *Trigonella foenum-graecum* L. using a novel glass marble containing liquid medium. *International Journal of Botany*, *6*(4), 394–403.

Nguyen, T. H., Phan, T. C., Choudhury, A. T. M. A., Rose, M. T., Deaker, R. J., & Kennedy, I. R. (2017). BioGro: A plant growth-promoting biofertilizer validated by 15 years' research from laboratory selection to rice farmer's fields of the Mekong Delta. *Agro-Environmental Sustainability*, 237–254.

Ochoa-Velasco, C. E., Valadez-Blanco, R., Salas-Coronado, R., Sustaita-Rivera, F., Hernández-Carlos, B., García-Ortega, S., & Santos-Sánchez, N. F. (2016). Effect of nitrogen fertilization and *Bacillus licheniformis* biofertilizer addition on the antioxidants compounds and antioxidant activity of greenhouse cultivated tomato fruits (*Solanum lycopersicum* L. var. Sheva). *Scientia Horticulturae*, *201*, 338–345.

Ortiz-Hernandez, M. L., Sanchez-Salinas, E., Dantan-Gonzalez, E., & Castrejon-Godínez, M. L. (2013). Pesticide biodegradation: Mechanisms, genetics and strategies to enhance the process. In R. Chamy, & F. Rosenkranz (Eds.), *Biodegradation—Life of science* (pp. 251–287). IntechOpen. https://doi.org/10.5772/56098. Available from: https://www.intechopen.com/books/biodegradation-life-of-science/pesticide-biodegradation-mechanisms-genetics-and-strategies-to-enhance-the-process.

Pandey, S., & Gupta, S. (2020). Evaluation of Pseudomonas sp. for its multifarious plant growth promoting potential and its ability to alleviate biotic and abiotic stress in tomato (Solanum lycopersicum) plants. *Scientific Reports*, *10*(1), 1–15.

Pangesti, N., Reichelt, M., van de Mortel, J. E., Kapsomenou, E., Gershenzon, J., van Loon, J. J., … Pineda, A. (2016). Jasmonic acid and ethylene signaling pathways regulate glucosinolate levels in plants during rhizobacteria-induced systemic resistance against a leaf-chewing herbivore. *Journal of Chemical Ecology*, *42*, 1212–1225. https://doi.org/10.1007/s10886-016-0787-7.

Passricha, N., Saifi, S. K., Sikka, V. K., & Tuteja, N. (2020). Multilegume biofertilizer: A dream. In *Molecular aspects of plant beneficial microbes in agriculture* (pp. 35–45). Academic Press.

Pathirana, B. K. W., & Yapa, P. N. (2020). Evaluation of different carrier substances for the development of an effective pelleted biofertilizer for rice (*Oryza sativa* L.) using co-inoculated bacteria and arbuscular mycorrhizal fungi. *Asian Journal of Biotechnology and Bioresource Technology*, 1–10.

Paul, D., Srinivasan, V., Anandaraj, M., & Sarma, Y. R. (2003). *Pseudomonas fluorescens* mediated nutrient flux in the black pepper rhizosphere microcosm and enhanced plant growth. In *6th International PGPR workshop, 5–10 October 2003, Calicut, India* (pp. 18–24).

Peng, S. H., Wan-Azha, W. M., Wong, W. Z., Go, W. Z., Chai, E. W., Chin, K. L., & H'ng, P. S. (2013). Effect of using agro-fertilizers and N-fixing Azotobacter enhanced biofertilizers on the growth and yield of corn. *Journal of Applied Sciences*, *13*(3), 508–512.

Pereg, L., & McMillan, M. (2015). Scoping the potential uses of beneficial microorganisms for increasing productivity in cotton cropping systems. *Soil Biology and Biochemistry*, *80*, 349–358.

Pereira, I., Ortega, R., Barrientos, L., Moya, M., Reyes, G., & Kramm, V. (2009). Development of a biofertilizer based on filamentous nitrogen-fixing cyanobacteria for rice crops in Chile. *Journal of Applied Phycology*, *21*(1), 135–144.

Rahbar Keykha, F., Khammari, E., Dahmardeh, M., & Forouzandeh, M. (2017). Effect of nano biofertilizer and chemical fertilizer application on quantitative and qualitative yield of sesame cultivars. *Crop Science Research in Arid Regions*, *1*(2), 177–190.

Rai, A., Kumar, S., Bauddh, K., Singh, N., & Singh, R. P. (2017). Improvement in growth and alkaloid content of Rauwolfia serpentina on application of organic matrix entrapped biofertilizers (*Azotobacter chroococcum, Azospirillum brasilense* and *Pseudomonas putida*). *Journal of Plant Nutrition*, *40*(16), 2237–2247.

Ramakrishnan, B., Megharaj, M., Venkateswarlu, K., Sethunathan, N., & Naidu, R. (2011). In D. M. Whitacre (Ed.), *Vol. 211. Reviews of environmental contamination and toxicology* (pp. 63–120). New York: Springer New York. https://doi.org/10.1007/978-1-4419-8011-3_3.

Ravikumar, S., Kathiresan, K., Ignatiammal, S. T. M., Selvam, M. B., & Shanthy, S. (2004). Nitrogen-fixing azotobacters from mangrove habitat and their utility as marine biofertilizers. *Journal of Experimental Marine Biology and Ecology*, *312*(1), 5–17.

Razie, F., & Anas, I. (2008). Effect of Azotobacter and Azospirillum on growth and yield of rice grown on tidal swamp rice field in South Kalimantan. *Jurnal Ilmu Tanah dan Lingkungan*, *10*(2), 41–45.

Salahi, T., Yadavi, A., Salehi, A., & Balouchi, H. (2020). The effect of mycorrhiza biofertilizer on yield and yield components of linseed (*Linum usitatissimum* L.) and fenugreek (*Trigonella foenum-graecum* L.) in intercropping. *Journal of Agricultural Science and Sustainable Production*, *29*(4), 1–17.

Sethi, S. K., & Adhikary, S. P. (2012). Cost effective pilot scale production of biofertilizer using Rhizobium and Azotobacter. *African Journal of Biotechnology*, *11*(70), 13490–13493.

Seyed Sharifi, R., Khalilzadeh, R., Pirzad, A., & Anwar, S. (2020). Effects of biofertilizers and nano zinc-iron oxide on yield and physicochemical properties of wheat under water deficit conditions. *Communications in Soil Science and Plant Analysis*, 1–14.

Shaheen, S., & Sundari, K. (2013). Exploring the applicability of PGPR to remediate residual organophosphate and carbamate pesticides used in agriculture fields. *International Journal of Agriculture Food Science and Technology*, *4*, 947–954.

Shahid, I., Tariq, K., & Mehnaz, S. (2020). Zinc solubilizing fluorescent pseudomonads as biofertilizer for tomato (Solanum lycopersicum L.) under controlled conditions. *Asian Journal of Plant Science and Research*, *10*(1), 1–7.

Shamim, A., Mahfooz, S., Hussain, A., & Farooqui, A. (2020). Ability of Al-acclimatized immobilized *Nostoc muscorum* to combat abiotic stress and its potential as a biofertilizer. *Journal of Pure and Applied Microbiology*, *14*(2), 1377–1386.

Sheng, X., & Huang, W. (2001). Mechanism of potassium release from feldspar affected by the sprain Nbt of silicate bacterium. *Acta Pedologica Sinica*, *39*, 863–871.

Shinwari, K. I., Shah, A. U., Afridi, M. I., Zeeshan, M., Hussain, H., Hussain, J., & Ahmad, O. (2015). Application of plant growth promoting rhizobacteria in bioremediation of heavy metal polluted soil. *Asian Journal of Multidisciplinary Studies*, *3*, 179–185.

Sinha, R. K., Valani, D., & Chauhan, K. (2014). Embarking on a second green revolution for sustainable agriculture by vermiculture biotechnology using earthworms. *International Journal of Agricultural Health Safety*, *1*, 50–64.

Soumare, A., Diedhiou, A. G., Thuita, M., Hafidi, M., Ouhdouch, Y., Gopalakrishnan, S., & Kouisni, L. (2020). Exploiting biological nitrogen fixation: A route towards a sustainable agriculture. *Plants*, *9*(8), 1011.

Steenhoudt, O., & Vanderleyden, J. (2000). Azospirillum, a free-living nitrogen-fixing bacterium closely associated with grasses: Genetic, biochemical and ecological aspects. *FEMS Microbiology Reviews*, *24*(4), 487–506.

Subba, R. (1993). *Biofertilizers in agriculture and forestry* (3rd ed.). International Science Publisher.

Sudhakar, P., Chattopadhyay, G. N., Gangwar, S. K., & Ghosh, J. K. (2000). Effect of foliar application of Azotobacter, Azospirillum and Beijerinckia on leaf yield and quality of mulberry (*Morus alba*). *The Journal of Agricultural Science*, *134*(2), 227–234.

Sun, B., Gu, L., Bao, L., Zhang, S., Wei, Y., Bai, Z., … Zhuang, X. (2020). Application of biofertilizer containing Bacillus subtilis reduced the nitrogen loss in agricultural soil. *Soil Biology and Biochemistry*, *148*, 107911.

Sun, B., Bai, Z., Bao, L., Xue, L., Zhang, S., Wei, Y., … Zhuang, X. (2020). Bacillus subtilis biofertilizer mitigating agricultural ammonia emission and shifting soil nitrogen cycling microbiomes. *Environment International*, *144*, 105989.

Suneja, S., Narula, N., Anand, R. C., & Lakshminarayana, K. (1996). Relationship of Azotobacter chroococcum siderophores with nitrogen fixation. *Folia Microbiologica*, *41*(2), 154–158.

Tauler, M., & Baraza, E. (2015). Improving the acclimatization and establishment of *Arundo donax* L. plantlets, a promising energy crop, using a mycorrhiza-based biofertilizer. *Industrial Crops and Products*, *66*, 299–304.

Tripathy, P. P., & Ayyappan, S. (2005). Evaluation of Azotobacter and Azospirillum as biofertilizers in aquaculture. *World Journal of Microbiology and Biotechnology*, *21*(8–9), 1339–1343.

Trisilawati, O. (2020). Effect of arbuscular mycorrhizal fungi biofertilizer on the growth of cashew seedling. *Jurnal Penelitian Tanaman Industri*, *17*(4), 150–155.

Vaishampayan, A., Sinha, R. P., Hader, D. P., Dey, T., Gupta, A. K., Bhan, U., & Rao, A. L. (2001). Cyanobacterial biofertilizers in rice agriculture. *The Botanical Review*, *67*(4), 453–516.

Venkataraman, G. S. (1972). *Algal biofertilizers and rice cultivation*. Today & Tommorrow's Printers & Publishers.

Verma, A., Kukreja, K., Pathak, D., Suneja, S., & Narula, N. (2001). In vitro production of plant growth regulators (PGRs) by. *Indian Journal of Microbiology*, *41*, 305–307.

Wani, S. A., Chand, S., Wani, M. A., Ramzan, M., & Hakeem, K. R. (2016). *Azotobacter chroococcum*—A potential biofertilizer in agriculture: An overview. In *Soil science: Agricultural and environmental prospectives* (pp. 333–348). Cham: Springer.

Watanabe, I., & Lin, C. (1984). Response of wetland rice to inoculation with Azospirillum lipoferum and Pseudomonas sp. *Soil Science and Plant Nutrition*, *30*(2), 117–124.

Widyantoro, A. (2020). Biological control of *Fusarium* wilt on banana plants using biofertilizers. *Biodiversitas Journal of Biological Diversity*, *21*(5).

Yadav, A. K., & Mowade, S. M. (2005). *Handbook of microbial technology*. Nagpur, India: Regional Centre for Organic Farming. 236 p.

Yarzábal, L. A., Monserrate, L., Buela, L., & Chica, E. (2018). Antarctic *Pseudomonas* spp. promote wheat germination and growth at low temperatures. *Polar Biology*, *41*(11), 2343–2354.

Yasari, E., Azadgoleh, A. E., Pirdashti, H., & Mozafari, S. (2008). *Azotobacter* and *Azospirillum* inoculants as biofertilizers in canola (Brassica napus L.) cultivation. *Asian Journal of Plant Sciences*, *7*(8), 490–494.

Youssef, M. M. A., & Eissa, M. F. M. (2014). Biofertilizers and their role in management of plant parasitic nematodes: A review. *Biotechnology Pharmaceutical Resources*, *5*(1), 1–6.

Zhang, L., Xu, M. G., Liu, Y., Zhang, F., Hodge, A., & Feng, G. (2016). Carbon and phosphorus exchange may enable cooperation between an arbuscular mycorrhizal fungus and a phosphate-solubilizing bacterium. *The New Phytologist*, *210*, 1022–1032.

Zhang, L., Shi, N., Fan, J., Wang, F., George, T. S., & Feng, G. (2018). Arbuscular mycorrhizal fungi stimulate organic phosphate mobilization associated with changing bacterial community structure under field conditions. *Environmental Microbiology*, *20*(7), 2639–2651.

Zhang, S., Gan, Y., & Xu, B. (2019). Mechanisms of the IAA and ACC-deaminase producing strain of Trichoderma longibrachiatum T6 in enhancing wheat seedling tolerance to NaCl stress. *BMC Plant Biology*, *19*(1), 22.

Zhang, M., Yang, L., Hao, R., Bai, X., Wang, Y., & Yu, X. (2020). Drought-tolerant plant growth-promoting rhizobacteria isolated from jujube (Ziziphus jujuba) and their potential to enhance drought tolerance. *Plant and Soil*, *452*, 423–440. https://doi.org/10.1007/s11104-020-04582-5.

Chapter 2

Phosphate-solubilizing bacteria: Recent trends and applications in agriculture

Chanda Parulekar Berde[a], Prachiti Rawool[b], and Vikrant B. Berde[c]
[a]*School of Earth, Ocean and Atmospheric Sciences, Goa University, Taleigão, Goa, India,*
[b]*Department of Microbiology, Gogate Jogalekar College, Ratnagiri, Maharashtra, India,*
[c]*Department of Zoology, Arts, Commerce & Science College, Lanja, Maharashtra, India*

Chapter outline

1 Introduction

Organized agriculture involving the use of plants and animals was developed several years ago. With the advancement in time, agricultural practices saw improved methods such as irrigation, organic farming, use of chemical fertilizers, resistant crops, etc. However, the exponential increase in population demanded an increase in agricultural produce from the available land. To suffice the food demands from existing agricultural land, fertilizers are used in an unwarranted manner, in turn, leading to declining soil health and fertility (Tilak et al., 2005).

As the world's human population unceasingly increases, furnishing food to this growing population will be one of the utmost challenges. To face this challenge, we need to focus on the soil biotic system and the agro-ecosystem as a whole. We need to better understand the complex mechanisms and interactions

Recent Advancement in Microbial Biotechnology. https://doi.org/10.1016/B978-0-12-822098-6.00004-5

that control agricultural land stability. At present, there is a limitation on the timely distribution of food produced in sufficient amounts, due to the high-input of the green revolution in agriculture. However, a rise in population at high frequency has threatened global food security. To withstand the pressure of the increasing population, food production has to be increased significantly, which will require another green revolution (Leisinger, 1999; Vasil, 1998).

Fertilizers add nutrients to soil increasing soil fertility and plant growth. However, the food value of plants is reduced due to the use of chemical fertilizers. Using exorbitant amounts of chemical fertilizers not only affects the quality of food but also gives rise to many diseases such as stomach cancer, goiter, and several vector-borne diseases. It also leads to groundwater contamination (Savci, 2012). After harvesting the crops, the nutrient reservoirs in the soil reduces. To remedy this deficiency of nutrients, more chemical fertilizers are applied, and this cycle continues, further worsening the soil condition. Reduction in the usage of chemical fertilizers and the improvement of soil health are the problems requiring immediate handling. Biological control is considered as an alternative way of reducing the use of chemicals in agriculture.

Phosphorus (P), an important nutrient in terms of plant requirements and uptake, is found in two forms in the soil: organic and inorganic. Being available abundantly in insoluble forms, it is, however, inaccessible to the plants. Hence, to suffice the nutritional requirements of the crops, P is supplemented through chemical fertilizers containing nitrogen, phosphorus, and potassium (NPK). A substitute for the use of chemical fertilizers is the use of biofertilizers, which include organisms that can convert inorganic phosphorus present in the soil to soluble forms that plants can assimilate. These are the phosphate-solubilizing microorganisms (PSMs).

2 Phosphorus in soil

Phosphate is an important element for plant growth, next to nitrogen and potassium. Plants require P for macromolecular biosynthesis, signal transduction, and respiration, photosynthesis, etc. (Khan, Zaidi, Ahemed, & Wani, 2010). Plants are unable to use phosphate directly as most, i.e., 95%–99% of phosphate in the soil is in an unavailable form, i.e., insoluble or precipitated form (Pandey & Maheshwari, 2007). This unavailable form of phosphate is converted to available form by phosphate-solubilizing bacteria.

According to Goldstein, Rogers, and Mead (1993), chemical phosphatic fertilizer production being a highly energy-intensive process for meeting the needs of the world, energy worth the US $4 billion yearly is required. Fertilizers added to the soil get precipitated as metal complexes (Stevenson, 2005), further making it unavailable for plant uptake. This accumulated phosphate is sufficient to sustain maximum agricultural phosphorus requirements for another 10 decades (Goldstein et al., 1993).

Phosphorus is found in organic forms in soil surface layers, i.e., the topsoil. The most abundant component of this organic form is phytate (Richardson, 1994). While in the bulk soil, phosphorus being less abundant, its uptake by plants is limited. The most soluble minerals, such as potassium, travels through the soil via bulk flow and diffusion, but P moves mainly by diffusion. According to Daniel, Schachtman, and Ayling (1998), because of the low diffusion rate of P, increased P uptake rates of plants result in a P depletion zone around the roots. Plants obtain their P from the soil solution as H_2PO_4 and HPO_4^{-2}. The uptake of HPO_4^{-2} by the plant is slower than the uptake of H_2PO_4 (Fig. 1).

3 Phosphate solubilization by plant growth-promoting microorganisms in plant rhizosphere

According to Kloepper and Schroth (1981), plant growth-promoting rhizobacteria (PGPR) by the production of substances required for plant growth, influence the rhizosphere flora. There are two modes for promoting plant growth, namely, direct and indirect. In the direct mechanism, the PGPR aids in the supply of nutrients by making available nitrogen, phosphorus, potassium, and essential minerals to the plants. While in the indirect mode, PGPR protects from pathogens by preventing their colonization, by acting as biocontrol agents. Microbes present in the soil are mainly encountered in nutrient-rich regions such as the topsoil layer and the rhizosphere, i.e., around the plant root. Some microbial strains can induce plant growth, augment soils, remove pollutants, and protect plants against pathogens (Tripathi, Nagarajan, & Verma, 2002). The PGPR are found in rhizosphere soil, but their count is not high enough to gain dominance over other microflora present in the rhizosphere. Therefore, the plant rhizosphere needs to be supplemented with high concentrations of

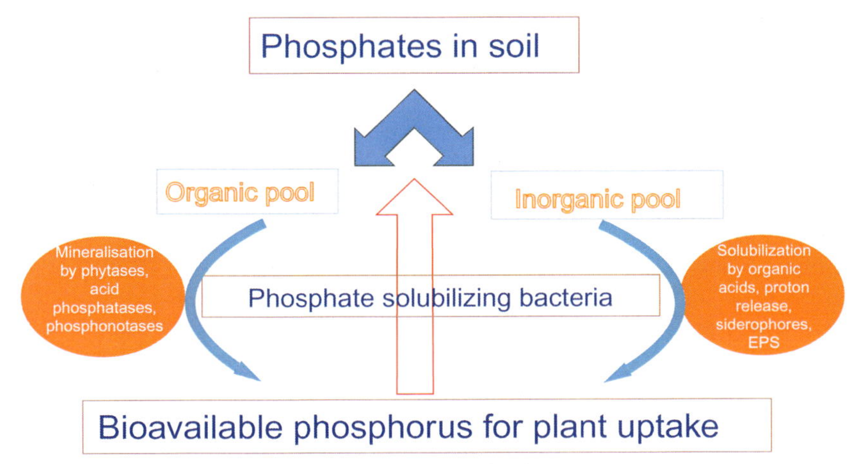

FIG. 1 Mechanisms for solubilization of phosphates in soil.

PGPR. Inoculation of PGPR in the soil is essential because it leads to increased nutrient availability as well as improves the physicochemical and biological properties of soil.

PGPR are categorized as extracellular PGPR (ePGPR) and intracellular PGPR (iPGPR) (Viveros, Jorquera, Crowley, Gajardo, & Mora, 2010). The ePGPR are found on the rhizoplane, in the rhizosphere, and in the intercellular spaces of root cortex cells. The iPGPR resides in nodule-like structures present in the root cells. Numerous bacterial species belonging to the genera *Agrobacterium, Arthrobacter, Azotobacter, Azospirillum, Bacillus, Burkholderia, Caulobacter, Chromobacterium, Erwinia, Flavobacterium, Micrococcus, Pseudomonas,* and *Serratia* are included in ePGPR category (Ahemad & Kibret, 2014).

The microbial flora that colonizes the rhizosphere is also actively engaged in phosphate solubilization and mobilization in soil. Soil microorganisms play a vital role in providing phosphate to plants by solubilizing inorganic P and mineralizing organic P present in the soil (Adhya et al., 2015). Richardson, Hadobas, and Hayes (2001) have documented the beneficial effects of P-solubilizing bacteria on agronomic crops. According to Villegas and Fortin (2002), some bacterial species in soil having good P-solubilizing activities belong to the genus *Mesorhizobium, Rhizobium, Klebsiella, Acinetobacter, Enterobacter, Erwinia, Achrobacter, Micrococcus, Pseudomonas,* and *Bacillus*. Along with the bacterial strains, fungi such as *Penicillium* and *Aspergillus* are also efficient P-solubilizers (Whitelaw, 2000).

The use of PGPR as inoculants helps to increase growth, yield, and uptake of nutrients by crop plants. The field and pot trials with PGPR have been reported to show increased yields as well as increased P uptake from marginal to a significant level (10%–27%) (Altomare, Norvell, Bjorkman, & Harman, 1999). Inoculation of *Pseudomonas striata* and *Bacillus polymyxa* in soil resulted in an increase in nodulation, nitrogenase activity, dry matter yield, P uptake, and grain yield of chickpea plants as compared to uninoculated controls. An increase in grain yield of maize and wheat crops, with increased P uptake, was observed under field conditions as reported by Himani and Reddy (2011).

4 Phosphate-solubilizing bacteria as biofertilizers

The application of phosphatic fertilizers can result in phosphate availability to the crops and subsequent increase in crop yields. However, the cost of chemical fertilizer being high and its aftereffects on the soil health makes chemical fertilizers undesirable and unaffordable. However, with the increasing population and increasing food demand, balancing the demand, cost-effectiveness, and environmental safety is a challenging task. Agronomists are therefore looking for an alternative source for the chemical fertilizers that will give higher yields, be environmentally safe, and cost-effective also. The application of PGPR in agricultural practices is an effective alternative. The application of PSM in-field practices has a number of advantages such as increased nutrient availability, improvement in fertility of the soil, better plant growth and higher crop yield, protection against pathogens, environmental pollution-free, soil health, and conditioning, cost-effective technology (Saber, Nahla, & Chedly, 2005).

The naturally occurring population of P-solubilizing bacteria and fungi is present in almost all rhizospheres. Most of the P-containing inorganic compounds, for example, di- and tri-calcium phosphate, hydroxyapatite, etc., are insoluble and are solubilized by bacterial species present in the soil. Among the reported microbial strains having solubilizing ability include mycorrhizal fungi, *Aspergillus, Penicillium* (Khan & Bhatnagar, 1977), while among bacterial genera include *Agrobacterium, Bacillus, Rhizobium, Burkholderia, Achromobacter, Micrococcus, Pseudomonas, Aereobacter, Flavobacterium,* etc. (Igual, Valverde, Cervantes, & Velázquez, 2001; Kucey, Janzen, & Leggett, 1989; Subbarao, 1988; Whitelaw, 2000). Among the soil phosphate solubilizers, bacteria constitute 10%–50% and P-solubilizing fungi are a meagre 0.1%–0.5% (Chen, Rekha, Arunshen, Lai, & Young, 2006).

Most of the studies on P-solubilizers reported the isolation of phosphate solubilizers from the soil, their preliminary characterization, and evaluating the solubilizing efficiency of microorganisms under laboratory conditions and pot assays. Studies on P-solubilizing activity were carried out mostly using agar media assays showing zones of clearing indicating the solubilization, such as on Pikovaskaya's agar media and colorimetric assays for quantifying the solubilized phosphates. Some researchers, however, reported a lack of consistency and correlation between plate assays and phosphate estimations in liquid cultures (Sharma, Sayyed, Trivedi, & Gobi, 2013). Due to the low number of naturally occurring P-solubilizers in the soils, the P concentrations released in the soil by the activity of these solubilizers are not sufficient to support plant growth. Hence, a vital necessity has arisen for research involving isolation and efficiency checking of PSMs. Best field applicants can be used for P-solubilizing activity in the soil to take advantage of this property for plant yield enhancement (Bhattacharyya & Jha, 2012).

5 Mechanisms of phosphate solubilization

Soil microorganisms are capable of solubilizing available P sources to forms assimilated by the plants (Toro, 2007; Wani, Khan, & Zaidi, 2007). Bacteria capable of P-solubilization belong to the genera *Bacillus, Burkholderia, Rhizobium, Pseudomonas, Achromobacter, Agrobacterium, Micrococcus, Enterobacter, Flavobacterium,* and *Erwinia.* Some bacterial species are also able to carry out both the processes, i.e., mineralize organic P and solubilize inorganic P (Hilda & Fraga, 2000; Khiari & Parent, 2005). The number of P-solubilization mechanisms found in microorganisms is quite large. It has also been reported that many saprophytic bacteria and fungi perform phosphorus-solubilizing activity through chelation-mediated mechanisms (Whitelaw, 2000).

5.1 Inorganic P solubilization

Studies on the solubilization of phosphoric compounds such as dicalcium phosphate, tricalcium phosphate, rock phosphate, etc., containing an insoluble form of P, by bacteria have been described (Goldstein, 1986). PSMs produce substances such as organic acids, which are responsible for the solubilizing ability. Acidification of the medium leads to the release of organic acids by the P-solubilizing bacteria (Goldstein, 1995; Halvorson, Keynan, & Kornberg, 1990; Kim, Jordan, & Donald, 1998; Kim, Jordan, & Krishnan, 1997; Maliha, Samina, Najma, Sadia, & Farooq, 2004). Phosphate-bound cations are chelated by the action of organic acids, especially hydroxylic and carboxylic groups, converting the phosphate to soluble form (Kpomblekou & Tabatabai, 1994; Sagoe, Ando, Kouno, & Nagaoka, 1998). Cations associated with bound phosphate are Al, Fe, Ca, etc., and the release of these cations due to the lower pH of soil observed (Kpomblekou & Tabatabai, 1994; Stevenson, 2005). According to Zaidi, Khan, Ahemad, and Oves (2009), the solubilization takes place by direct oxidation pathway on the outer surface of the cytoplasmic membrane. The quality of the acids produced determines the degree of solubilization; the quantity is less important (Scervino et al., 2010).

Apart from the release of organic acid, which causes the release of bound phosphate, various microbial mechanisms, including proton extrusion, have been reported to be involved in the solubilization processes (Dutton & Evans, 1996; Nahas, 1996; Surange, Wollum, Kumar, & Nautiyal, 1995). Goldstein (1995) and Deubel and Merbach (2000) have reported the production of gluconic and keto-gluconic acids, which are low-molecular-weight acids, that acidify rhizosphere soil leading to the release of P. Lowering of rhizosphere pH is due to the biological production of proton/bicarbonate release that is responsible for the anionic-cationic balance and gaseous (O_2/CO_2) exchanges. It is observed that the P-solubilization by bacteria has a direct correlation with the pH of the medium. According to Hinsinger (2001), root exudates released in the soil also are responsible for altering the P levels.

The phosphate-solubilizing microflora produces organic acids in the process of solubilization of bound inorganic phosphate. The following changes occur as the result of the organic acid production (Halder, Mishra, Bhattacharyya, & Chakrabartty, 1990; Nahas, 1996) acting alone or in the presence of other solubilizing compounds. Reduction of soil pH, i.e., acidic pH, cation chelation, blockage of adsorption sites for binding of soil particles, the formation of metal complexes such as calcium phosphate that are soluble, and thereby releasing the P.

5.2 Organic phosphate mineralization by PSM

Organic phosphorus present in the soil constitutes 40%–90% of the total phosphorus found in the soil. Organic acids and phosphatase production are the basic mechanisms in mineral phosphate solubilization and organic phosphorous mineralization in soil (Hilda & Fraga, 2000). Almost 50% of microflora found in rhizosphere soil harbor the potential to mineralize phosphates. This potential is because of the production of phosphatases (Cosgrove, 1967; Goldstein, 1994; Tarafdar, Rao, & Bala, 1988). The enzyme phosphatase or phosphohydrolases can mineralize insoluble phosphates. The phosphatases hydrolyze phosphodiester bonds present in phosphate complexes. The acid phosphatases have an optimal activity in the pH range between acidic to neutral, while the alkaline phosphatases have an optimal activity in the alkaline pH range. Phosphatases show substrate specificity and may be further classified as specific or nonspecific types. As reported by Rossolini et al. (1998), nonspecific phosphatases are produced in elevated amounts by some strains of bacteria.

Organic phosphate is the substrate for alkaline and acid phosphatases that convert it to inorganic insoluble phosphate (Beech, Paiva, Caus, & Coutinho, 2001). Rhizosphere microflora produces organic anions, siderophores, and acid phosphatases in the soil surrounding plant roots, as reported by Yadaf and Tarafdar (2001), while mineralization of organic phosphorus present in the soil is brought about by alkaline phosphatase enzymes as noted by Tarafdar and Claasen (1988). Kim et al. (1998) reported the process of phosphorus discharge from hydroxyapatite by *Enterobacter agglomerans*. While *Bacillus, Streptomyces, Pseudomonas,* etc., when used together as mixed cultures, are found to give the better activity of organic P-mineralization (Molla, Chowdhury, Islam, & Hoque, 1984). Strains from genera *Arthrobacter, Bacillus, Enterobacter, Flavobacterium, Pseudomonas, Rhizobium, Rhodococcus, Serratia,* etc., are included as the most efficient P-solubilizers among bacteria (Bhattacharyya & Jha, 2012). PGPR uses different approaches to make use of the unavailable form of phosphorous. Mechanism of phosphate solubilization employed by PGPR includes the production of compounds like organic acids, which will aid in the dissolution of inorganic P, production of enzymes like phytases and degradation of complexes holding the P (Sharma et al., 2013).

6 Effect of phosphate solubilizers on plant growth and crop yield

Most of the studies demonstrating the increased uptake of P and grain yield following inoculation of seedlings with PSM were worked on showing the actual effects of bioinoculants on plant growth (Gerretsen, 1948). An increase in the uptake of PO_4 from both soil and fertilizer sources was reported by Sundara Rao, Bajpai, Sharma, and Subbaiah (1963) in the case of seeds inoculated with *Bacillus megaterium*. The benefits of using PSMs in the soil, along with the application of insoluble forms of P sources such as rock phosphate, tricalcium phosphate, and bone meal was initially reported by some workers (Ahmed & Jha, 1977; Loheuarete & Berthelin, 1988). Ahmed and Jha (1977), who reported hydroxylapatite and rock PSMs. They further reported that an increase in P uptake and yield being observed due to the inoculation of soil with *Bacillus megaterium* and *Bacillus circulans*. Increased P uptake resulting in increased plant growth, has been reported in calcareous soil inoculated with PSMs (Khalafallah, Saber, Abel-El, & Maksound, 1982). The inoculation of rice seedlings with mixed cultures of *Azotobacter chroococcum*, *Pseudomonas striata*, and *Aspergillus awamori* resulted in increased N, P uptake, and also subsequent increase in the yield of grain and straw, as observed by Kundu and Gaur (1984).

In a greenhouse experiment, increased P uptake was observed along with increased plant height in finger millet. The soil was inoculated with *Bacillus circulans* and 32P-labeled superphosphate and tricalcium phosphate (Raj, Bagyaraj, & Manjunath, 1981). It is difficult to prove the effects of phosphate solubilization on greenhouse soil or field soil compared to studies carried out in the laboratory conditions. However, plant growth responses to PSM addition have been reported in many studies (Kucey et al., 1989).

Studies on the effect of P-solubilizing *Penicillium* sp., on the P composition of soil as well as the growth and yield in maize, were described by Praveen, Kuligod, Hebsur, Patil, and Kulkarni (2012). According to this report, a 20%–23% increase in yield, over the control, was seen in maize crop after dual inoculation of P-solubilizers. The successful application of PSMs for different crops is summarized in Table 1.

TABLE 1 Phosphate-solubilizing microorganisms used as growth promotors for different agricultural crops/plants.

Phosphate-solubilizing microorganism	Crop/plant	Reference
Azospirillum sp.	Maize, sorghum, wheat	Kapulnik, Gafny, and Okon, 1985, Baldani, Baldani, and Döbereiner (1987), and Sarig, Okon, and Blum (1990)
Aspergillus niger	Wheat	Xiao, Zhang, Fang, and Chi (2013)
Aspergillus awamori S29	Mung bean	Jain, Saxena, and Sharma (2012)
Azotobacter chroococcum	Wheat	Islam et al. (2007)
Azotobacter chroococcum and *Bacillus subtilis*	Wheat	Kumar, Bauddh, Barman, and Singh (2014)
Azotobacter; Bacillus	Wheat	Kloepper, Lifshitz, and Zablotowicz (1989)

TABLE 1 Phosphate-solubilizing microorganisms used as growth promotors for different agricultural crops/plants—cont'd

Phosphate-solubilizing microorganism	Crop/plant	Reference
Azotobacter chroococcum, Saccharomyces cerevisiae, and *Bacillus megaterium*	*Moringa oleifera*	Zayed (2012)
Aspergillus niger, Penicillium aculeatum	Chinese cabbage	Wang et al. (2015)
Bacillus circulans and *Cladosporium herbarum*	Wheat	Islam et al. (2007) and Singh and Kapoor (1999)
Bacillus megaterium and *Azotobacter chroococcum*	Wheat	Rodríguez and Fraga (1999)
Bacillus sp. and *Pseudomonas* sp.	Sesame	Jahan, Mahallati, Amiri, and Ehyayi (2013)
Bacillus megaterium and *Azotobacter chroococcum*	Wheat	Brown (1974)
Bacillus thuringiensis	Rice	David, Raj, Linda, and Rhema (2014)
Bacillus spp.	Peanut, potato, sorghum, wheat	Broadbent, Baker, Franks, and Holland (1977), Burr, Schroth, and Suslow (1978), and Capper and Campbell (1986)
Bradyrhizobium + *Glomus fasciculatum* + *Bacillus subtilis*	Green gram	Zaidi and Khan (2006)
Burkholderia cepacia	Maize	Zhao et al. (2014)
Burkholderia gladioli	Oil palm	Istina, Widiastuti, Joy, and Antralina (2015)
Burkholderia gladioli	Sweet leaf	Mamta et al. (2010)
Mesorhizobium mediterraneum	Chickpea and barley	Peix et al. (2001)
Mycorrhiza + *Pseudomonas putida*	Barley	Mehrvarz, Chaichi, and Alikhani (2008)

Continued

TABLE 1 Phosphate-solubilizing microorganisms used as growth promotors for different agricultural crops/plants—cont'd

Phosphate-solubilizing microorganism	Crop/plant	Reference
Pseudomonas sp.	Soybean crop	Son, Diep, and Giang (2006)
Pseudomonas putida and *Pseudomonas fluorescens*	Canola, lettuce, and tomato	Hall, Pierson, Ghosh, and Glick (1996) and Glick, Changping, Sibdas, and Dumbroff (1997)
Pseudomonas putida and *Pseudomonas fluorescens*	Potato, radishes, rice, sugar beet, tomato, lettuce, apple, citrus, beans, ornamental plants, and wheat	Suslov (1982), Lemanceau (1992), Kloepper (1994), and Kloepper, Lifshitz, and Schroth (1988)
Pseudomonas	*Zea mays* L	Walpola and Yoon (2012) and Bano and Fatima (2009)
Pseudomonas	Soybean	Walpola and Yoon (2012) and Son et al. (2006)
Pseudomonas putida	Moss	Tani, Akita, Murase, and Kimbara (2011)
Pseudomonas chlororaphis and *Pseudomonas putida*	Soybean	Islam et al. (2007) and Singh and Kapoor (1999)
Pseudomonas fluorescent	Peanut	Dey, Pal, Bhatt, and Chauhan (2004)
Pseudomonas striata and *Glomus fasciculatum*	Soybean, wheat	Mahanta et al. (2014)
Paenibacillus favisporus TG1R2	Soybeans	Fernández Bidondo et al. (2011)
Pantoea agglomerans (PSB-1) and *Burkholderia anthina* (PSB-2)	Mung bean	Walpola and Yoon (2013)
Rhizobium tropici CIAT899	Beans	Tajini, Trabelsi, and Drevon (2012)
Serratia sp.	Wheat	Swarnalakshmi et al. (2013)

7 PSB application methods in agriculture

The commonly used method for the application of P-solubilizing microbial inoculants is seed surface application. Traditionally, this method is used before the sowing of seeds and is one of the easiest methods to follow. Application of PSMs in a proper systematic way such that each seed is coated with the microorganism is very important. A sticker solution helps to adhere the bacteria to the seeds and may be added during the application process. Gum arabic is an example of a sticker that is used for adherence to the phosphate-solubilizers on the seed (Khan, Zaidi, & Wani, 2007). PSM may be added to the soil instead of applying to the seeds. This mode of application is used when the seeds are pretreated with pesticides, which may interfere with the PSMs if applied together.

The PSM may be applied alone or maybe co-inoculated together with other bioinoculants. Application of PSM to the soil will increase the PSM count per unit area, thus making soluble phosphates available to the plants. However, there are a number of disadvantages of the application process, such as the availability of enough bioinoculant quantity for total seed surface, contact with chemicals, bacterial movement away from plant roots, or seeds after planting.

8 Recent developments

The application of phosphate solubilizers has yielded good results in a number of crops. Wheat crop showed an increase in grain yield by 17%–18% (Suleman et al., 2018), while a 10%–20% increase in yield was reported for maize, sorghum, and wheat, using a combination of bio-inoculants (Rodríguez & Fraga, 1999). An improved sugarcane yield has been recorded by Sundara, Natarajan, and Hari (2002). The field applications of a number of P-solubilizers such as *Azotobacter* sp., *Bacillus megaterium*, *Bacillus circulans*, *Bacillus subtilis*, and *Pseudomonas striata*, have been carried out with satisfactory results (Bano & Fatima, 2009; Pandey, Trivedi, Kumar, & Palni, 2006; Rodríguez & Fraga, 1999; Satyaprakash, Nikitha, Reddi, Sadhana, & Vani, 2017; Vikram & Hamzehzarghani, 2008). *Penicillium bilaii*, a P-solubilizer, is commercially available as JumpStart (Satyaprakash et al., 2017). N- and P-deficient agrosystems can be treated P-solubilizers in combination with nitrogen-fixing bacteria as an integrated treatment system (Bargaz, Lyamlouli, Chtouki, Zeroual, & Dhiba, 2018). Similar studies by Afzal and Bano (2008) and Yousefi, Khavazi, Moezi, Rejali, and Nadian (2011) showed increased yields in the wheat crop when PSM was used in combination with nitrogen fixers and arbuscular mycorrhizal (AM) fungi. Some studies have also been focused on the application of P-solubilizers for soil reclamation, mostly for the degraded minefields (Chen & Liu, 2019; Liang, Liu, Jia, et al., 2020), which could be used for agricultural practices.

A lot of research on the genetic and molecular aspects of phosphate solubilization has been carried out in recent years. These studies were focused on understanding the mechanisms of the solubilization process. Sashidhar and Podile (2010) have reported studies on proton transport and involvement of the enzyme glucose dehydrogenase in the solubilization process by direct oxidation pathway. Liang et al. (2020) have used the metagenomic approach for studying the effects of P-solubilizer application and the changes occurring in the course of the process during the reclamation of mining soils. During the study, the predominance of genes involved in phosphate solubilization and mineralization, such as gcd gene, was observed. In another study carried out by Zeng, Wu, Wang, and Ding (2017), the molecular mechanisms of P-solubilization, gene expression, and effects of P-solubilization on the growth of *Burkholderia multivorans* WS-FJ9, were investigated and reported. Similar studies were carried out by using molecular tools to understand the ecology, presence, abundance, survival, and interactions of P-solubilizers in different environments that have been carried out (Jorquera, Crowley, Marschner, et al., 2011; Oliveira et al., 2009; Richardson & Simpson, 2011).

Several phosphatase-encoding genes involved in the solubilization of P, both inorganic and organic, were cloned and studied for their efficiency (Rodríguez, Fraga, Gonzalez, & Bashan, 2006). Goldstein and Liu (1987) first reported the cloning of P-solubilization genes in Gram-negative bacteria *Erwinia herbicola*. Fraga, Rodriguez, and Gonzalez (2001) reported the genetic manipulation of a bioinoculant *Burkholderia cepacia* IS-16. The napA phosphatase gene from *Morganella morganii* was inserted in this bioinoculant strain, using the vector pRK293. Thus, the rhizobacteria occurring in nature can be made more efficient by inserting genes for P-solubilization in their genome.

The PSMs can be genetically manipulated and expressed in rhizobacterial strains, thus increasing their efficiency to solubilize P and making them more effective agricultural inoculants. Cloning and expression of genes for phosphate solubilization in rhizobacterial strains, such as using appropriate promoters, has been used as a successful technology. By applying the techniques of gene manipulation, P-solubilizers can further improve their plant growth-promoting ability.

Rossolini et al. (1998) reported the isolation of genes responsible for encoding acid phosphatases from different bacterial species. Thaller and group have worked on the sequence analysis of phosphatase genes, which were cloned and classified into three families, namely, class A, class B, and class C phosphatases (Thaller, Berlutti, Schippa, Lombardi, & Rossolini, 1994; Thaller, Giovanna, Serena, & Rossolini, 1995; Thaller, Schippa, Bonci, Cresti, & Rossolini, 1997). Several genes have been isolated from *Escherichia coli*, responsible for encoding enzymes involved in phosphate solubilization, such as *ush*A, which encodes for 5′-nucleotidase (Burns & Beacham, 1986), *agp,* which encodes enzyme, acid glucose-1-phosphatase (Pradel & Boquet, 1990) and *cpd*B, encoding the 2′-3′ cyclic phosphodiesterase (Beacham & Garrett, 1980).

Several authors have well documented some recent relevant research on P-solubilization (Buch, Archana, & Naresh Kumar, 2008; Ghosh, Barman,

Mukherjee, & Mandal, 2016; Jha, Dafale, & Purohit, 2019; Li, Wu, Ye, & Yang, 2018; Liu et al., 2020; Park, Lee, Jung, et al., 2010; Yang et al., 2016; Zeng, Wu, & Wen, 2016). With the genetic manipulation of rhizosphere bacteria and the development of technologies for the application of these bacteria in the field, sustainable agricultural practices promise a rise in productivity.

9 Conclusions

The use of biological agents is the only alternative to avoid use of chemical fertilizers and their harmful effects. Biofertilizers can completely replace the chemical fertilizers leading to sustainable agricultural practices. Crop production can be boosted using biofertilizers, including the P-solubilizers, thus contributing to sustainable agriculture. The effectiveness of PSM in the plant rhizosphere will depend on its ability to colonize, compete, survive, and proliferate, in the presence of other microflora. The application and successful colonization of PSM in the soil are essential for plant growth and, ultimately, for sustainable agriculture. Application of PSMs to soils deficient in plant-available phosphate can mobilize the bound phosphates for plant uptake. They should be efficient enough and should survive and thrive in the rhizosphere, postapplication. Field trials should be undertaken adequately for maximum exploitation of the effective strains. The genetic manipulation of the PSM with phosphate-solubilizing genes can help in developing the desirable bioinoculants.

References

Adhya, T. K., Kumar, N., Reddy, G., Podile, R. A., Bee, H., & Samantaray, B. (2015). Microbial mobilization of soil phosphorus and sustainable P management in agricultural soils. Special section: Sustainable phosphorus management. *Current Science*, *108*, 1280–1287.

Afzal, A., & Bano, A. (2008). *Rhizobium* and phosphate solubilizing bacteria improve the yield and phosphorus uptake in wheat (*Triticum aestivum* L.). *International Journal of Agriculture and Biology*, *10*, 85–88.

Ahemad, M., & Kibret, M. (2014). Mechanisms and applications of plant growth promoting rhizobacteria: Current perspective. *Journal of King Saud University Science*, *26*, 1–20.

Ahmed, N., & Jha, K. K. (1977). Effect of inoculation with phosphate solubilizing organisms on the yield and P uptake of gram. *Journal of the Indian Society of Soil Science*, *25*, 391–393.

Altomare, C., Norvell, W. A., Bjorkman, T., & Harman, G. E. (1999). Solubilization of phosphorous and micronutrients by plant growth promoting biocontrol fungus, *Trichoderma harzianum*. *Applied and Environmental Microbiology*, *65*, 26–29.

Baldani, V. L. D., Baldani, J. I., & Döbereiner, J. (1987). Inoculation on field-grown wheat (*Triticum aestivum*) with *Azospirillum* spp. in Brazil. *Biology and Fertility of Soils*, *4*, 37–40.

Bano, A., & Fatima, M. (2009). Salt tolerance in *Zea mays* (L) following inoculation with *Rhizobium* and *Pseudomonas*. *Biology and Fertility of Soils*, *45*, 405–413.

Bargaz, A., Lyamlouli, K., Chtouki, M., Zeroual, Y., & Dhiba, D. (2018). Soil microbial resources for improving fertilizers efficiency in an integrated plant nutrient management system. *Frontiers in Microbiology*, *9*, 1606.

Beacham, I. R., & Garrett, S. (1980). Isolation of *Escherichia coli* mutants (cpdB) deficient in periplasmic 2-cyclic phosphodiesterase and genetic mapping of the cpdB locus. *Journal of General Microbiology, 119*, 31–34.

Beech, I. B., Paiva, M., Caus, M., & Coutinho, C. (2001). Enzymatic activity and within biofilms of sulphate-reducing bacteria. In P. G. Gilbert, D. Allison, M. Brading, J. Verran, & J. Walker (Eds.), *Biofilm community interactions: Chance or necessity?* (pp. 231–239). Cardiff: BioLine.

Bhattacharyya, P. N., & Jha, D. K. (2012). Plant growth-promoting rhizobacteria (PGPR): Emergence in agriculture. *World Journal of Microbiology and Biotechnology, 28*, 1327–1350.

Broadbent, P., Baker, K. F., Franks, N., & Holland, J. (1977). Effect of *Bacillus* spp. on increased growth of seedlings in steamed and in non-treated soil. *Phytopathology, 67*, 1027–1034.

Brown, M. E. (1974). Seed and root bacterization. *Annual Review of Phytopathology, 12*, 181–197.

Buch, A., Archana, G., & Naresh Kumar, G. (2008). Metabolic channeling of glucose towards gluconate in phosphate-solubilizing *Pseudomonas aeruginosa* P4 under phosphorus deficiency. *Research in Microbiology, 59*, 635–642.

Burns, D. M., & Beacham, I. R. (1986). Nucleotide sequence and transcriptional analysis of the *Escherichia coli* ushA gene, encoding periplasmic UDP-sugar hydrolase (5′-nucleotidase): Regulation of the ushA gene, and the signal sequence of its encoded protein product. *Nucleic Acids Research, 14*, 4325–4342.

Burr, T. J., Schroth, M. N., & Suslow, T. (1978). Increased potato yields by treatment of seed pieces with specific strains of *Pseudomonas fluorescens* and *Pseudomonas putida. Phytopathology, 68*, 1377–1383.

Capper, A. L., & Campbell, R. (1986). The effect of artificially inoculated antagonistic bacteria on the prevalence of take-all disease of wheat in field experiment. *The Journal of Applied Bacteriology, 60*, 155–160.

Chen, Q., & Liu, S. (2019). Identification and characterization of the phosphate-solubilizing bacterium *Pantoea* sp. S32 in reclamation soil in Shanxi, China. *Frontiers in Microbiology, 10*, 2171.

Chen, Y. P., Rekha, P. D., Arunshen, A. B., Lai, W. A., & Young, C. C. (2006). Phosphate solubilizing bacteria from subtropical soil and their tricalcium phosphate solubilizing abilities. *Applied Soil Ecology, 34*, 33–41.

Cosgrove, D. J. (1967). Metabolism of organic phosphates in soil. In A. D. Mclaren, & G. H. Peterson (Eds.), *Vol. I. Soil biochemistry* (pp. 216–228). New York: Marcel & Dekker.

Daniel, P., Schachtman, R. J. R., & Ayling, S. M. (1998). Phosphorus uptake by plants: From soil to cell. *Plant Physiology, 116*, 447–453.

David, P., Raj, R. S., Linda, R., & Rhema, S. B. (2014). Molecular characterization of phosphate solubilizing bacteria (PSB) and plant growth promoting rhizobacteria (PGPR) from pristine soils. *International Journal of Innovative Science Engineering and Technology, 1*, 317–324.

Deubel, A. G., & Merbach, W. (2000). Transformation of organic rhizodeposits by rhizoplane bacteria and its influence on the availability of tertiary calcium phosphate. *Journal of Plant Nutrition and Soil Science, 163*, 387–392.

Dey, R., Pal, K. K., Bhatt, D. M., & Chauhan, S. M. (2004). Growth promotion and yield enhancement of peanut (*Arachis hypogaea* L.) by application of plant growth-promoting rhizobacteria. *Microbiological Research, 159*, 371–394.

Dutton, V. M., & Evans, C. S. (1996). Oxalate production by fungi: Its role in pathogenicity and ecology in the soil environment. *Canadian Journal of Microbiology, 42*, 881–895.

Fernández Bidondo, L., Silvani, V., Colombo, R., Pérgola, M., Bompadre, J., & Godeas, A. (2011). Pre-symbiotic and symbiotic interactions between *Glomus intraradices* and two *Paenibacillus* species isolated from AM propagules. *In vitro* and *in vivo* assays with soybean (AG043RG) as plant host. *Soil Biology and Biochemistry*, *43*, 1866–1872.

Fraga, R., Rodriguez, H., & Gonzalez, T. (2001). Transfer of the gene encoding the NapA acid phosphatase from *Morganella morganii* to *Burkholderia cepacia* strain. *Acta Biotechnologica*, *21*, 359–369.

Gerretsen, F. C. (1948). The influence of microorganisms on the phosphate intake by the plant. *Plant and Soil*, *1*, 51–81.

Ghosh, R., Barman, S., Mukherjee, R., & Mandal, N. C. (2016). Role of phosphate solubilizing *Burkholderia* spp. for successful colonization and growth promotion of *Lycopodium cernuum* L. (Lycopodiaceae) in lateritic belt of Birbhum district of West Bengal, India. *Microbiological Research*, *183*, 80–91.

Glick, B. R., Changping, L., Sibdas, G., & Dumbroff, E. B. (1997). Early development of canola seedlings in the presence of the plant growth-promoting rhizobacterium *Pseudomonas putida* GR12-2. *Soil Biology and Biochemistry*, *29*, 1233–1239.

Goldstein, A. H. (1986). Bacterial solubilization of mineral phosphates: Historical perspectives and future prospects. *American Journal of Alternative Agriculture*, *1*, 57–65.

Goldstein, A. H. (1994). Involvement of the quinoprotein glucose dehydrgenase in solubilization of exogenous phosphates by gram-negative bacteria. In A. Torriani-Gorini, E. Yagil, & S. Silver (Eds.), *Phosphate in microorganisms: Cellular and molecular biology*. Washington, DC: ASM Press.

Goldstein, A. H. (1995). Recent progress in understanding the molecular genetics and biochemistry of calcium phosphate solubilization by Gram negative bacteria. *Biological Agriculture and Horticulture*, *12*, 185–193.

Goldstein, A. H., & Liu, S. T. (1987). Molecular cloning and regulation of a mineral phosphate solubilizing gene from *Erwinia herbicola*. *Biotech*, *5*, 72–74.

Goldstein, A. H., Rogers, R. D., & Mead, G. (1993). Mining by microbe. *Biotechnology*, *11*, 1250–1254.

Halder, A. K., Mishra, A. K., Bhattacharyya, P., & Chakrabartty, P. K. (1990). Solubilization of rock phosphate by *Rhizobium* and *Bradyrhizobium*. *The Journal of General and Applied Microbiology*, *36*, 81–92.

Hall, J. A., Pierson, D., Ghosh, S., & Glick, B. R. (1996). Root elongation in various agronomic crops by the plant growth promoting rhizobacterium *Pseudomonas putida* GR12-2. *Israel Journal of Plant Science*, *44*, 37–42.

Halvorson, H. O., Keynan, A., & Kornberg, H. L. (1990). Utilization of calcium phosphates for microbial growth at alkaline pH. *Soil Biology and Biochemistry*, *22*, 887–890.

Hilda, R. R., & Fraga, R. (2000). Phosphate solubilizing bacteria and their role in plant growth promotion. *Biotechnology Advances*, *17*, 319–359.

Himani, S., & Reddy, M. S. (2011). Effect of inoculation with phosphate solubilizing fungus on growth and nutrient uptake of wheat and maize plants fertilized with rock phosphate in alkaline soils. *European Journal of Soil Biology*, *47*, 30–34.

Hinsinger, P. (2001). Bioavailability of soil inorganic P in the rhizosphere as affected by root induced chemical changes: A review. *Plant and Soil*, *237*, 173–195.

Igual, J. M., Valverde, A., Cervantes, E., & Velázquez, E. (2001). Phosphate-solubilizing bacteria as inoculants for agriculture: Use of updated molecular techniques in their study. *Agronomie*, *21*, 561–568.

Islam, T. M., Deora, A., Hashidoko, Y., Rahman, A., Ito, T., & Tahara, S. (2007). Isolation and identification of potential phosphate solubilizing bacteria from the rhizoplane of *Oryza sativa* L. cv. BR29 of Bangladesh. *Verlag der Zeitschrift für Naturforschung*, *62c*, 103–110.

Istina, I. N., Widiastuti, H., Joy, B., & Antralina, M. (2015). Phosphate solubilizing microbe from Saprists peat soil and their potency to enhance oil palm growth and P uptake. *Procedia Food Science, 3*, 426–435.

Jahan, M., Mahallati, M. N., Amiri, M. B., & Ehyayi, H. R. (2013). Radiation absorption and use efficiency of sesame as affected by biofertilizers inoculation in a low input cropping system. *Industrial Crops and Products, 43*, 606–611.

Jain, R., Saxena, J., & Sharma, V. (2012). Effect of phosphate-solubilizing fungi *Aspergillus awamori* S29 on mungbean (*Vigna radiata* cv. RMG 492) growth. *Folia Microbiologica, 57*, 533–541.

Jha, V., Dafale, N. A., & Purohit, H. J. (2019). Regulatory rewiring through global gene regulations by PhoB and alarmone (p) ppGpp under various stress conditions. *Microbiological Research, 227*, 126309.

Jorquera, M. A., Crowley, D. E., Marschner, P., et al. (2011). Identification of β-propeller phytase-encoding genes in culturable *Paenibacillus* and *Bacillus* spp. from the rhizosphere of pasture plants on volcanic soils. *FEMS Microbiology Ecology, 75*, 163–172.

Kapulnik, J., Gafny, R., & Okon, Y. (1985). Effect of *Azopirillum* spp. inoculation on root development and NO-3 uptake in wheat (*Titicum aestivum* cv. Miriam) in hydroponic systems. *Canadian Journal of Botany, 63*, 627–631.

Khalafallah, M. A., Saber, M. S. M., Abel-El, & Maksound, H. K. (1982). Influence of phosphate dissolving bacteria on the efficiency of superphosphate in calcareous soil cultivated with Vicia-faba Z. pflangernachr. Bodenkel. *Journal of Plant Nutrition and Soil Science, 145*, 455–459.

Khan, J. A., & Bhatnagar, R. M. (1977). Studies on solubilization of insoluble phosphates by micro-organisms: Part I. Solubilization of Indian phosphate rocks by Aspergillus niger and Penicillium spp. *Fertilizer Technology, 14*, 329–333.

Khan, M. S., Zaidi, A. M., Ahemed, M. O., & Wani, P. A. (2010). Plant growth promotion by phosphate solubilizing fungi—Current perspective. *Archives of Agronomy and Soil Science, 56*, 73–98.

Khan, M. S., Zaidi, A., & Wani, P. A. (2007). Role of phosphate-solubilizing microorganisms in sustainable agriculture—A review. In *Vol. 27. Agronomy for sustainable development* (pp. 29–43). Springer Verlag/EDP Sciences/INRA.

Khiari, L., & Parent, L. E. (2005). Phosphorus transformations in acid light-textured soils treated with dry swine manure. *Canadian Journal of Soil Science, 85*, 75–87.

Kim, K. Y., Jordan, D., & Donald, G. A. M. (1998). Effect of phosphate-solubilizing bacteria and vascular-arbuscular mycorrhizae on tomato growth and soil microbial activity. *Biology and Fertility of Soils, 26*, 79–87.

Kim, K. Y., Jordan, D., & Krishnan, H. B. (1997). *Rahnella aqualitis*, a bacterium isolated from soybean rhizosphere, can solubilize hydroxyapatite. *FEMS Microbiology Letters, 153*, 273–277.

Kloepper, J. W. (1994). Plant growth promoting bacteria (other systems). In J. Okon (Ed.), *Azospirillum/plant association* (pp. 137–154). Boca Raton, FL: CRC Press.

Kloepper, J. W., Lifshitz, K., & Schroth, M. N. (1988). *Pseudomonas* inoculants to benefit plant production. In *ISI atlas of science: Animal and plant sciences* (pp. 60–64). Philadelphia: Institute for Public Information.

Kloepper, J. W., Lifshitz, K., & Zablotowicz, R. M. (1989). Free-living bacterial inocula for enhancing crop productivity. *Trends in Biotechnology, 7*, 39–43.

Kloepper, J. W., & Schroth, M. N. (1981). Relationship of *in vitro* antibiosis of plant growth promoting rhizobacteria to plant growth and the displacement of root microflora. *Phytopathology, 71*, 1020–1024.

Kpomblekou, K., & Tabatabai, M. A. (1994). Effect of organic acids on release of phosphorus from phosphate rocks. *Soil Science*, *158*, 442–453.

Kucey, R. M. N., Janzen, H. H., & Leggett, M. E. (1989). Microbially mediated increased in plant available phosphorus. *Advances in Agronomy*, *42*, 198–228.

Kumar, S., Bauddh, K., Barman, S. C., & Singh, R. P. (2014). Amendments of microbial bio fertilizers and organic substances reduces requirement of urea and DAP with enhanced nutrient availability and productivity of wheat (*Triticum aestivum* L.). *Ecological Engineering*, *71*, 432–437.

Kundu, B. S., & Gaur, A. C. (1984). Rice response to inoculation with N fixing and P solubilizing microorganisms. *Plant and Soil*, *79*, 227–234.

Leisinger, K. M. (1999). Biotechnology and food security. *Current Science (India)*, *76*, 488–500.

Lemanceau, P. (1992). Effects benefiques de rhizobacteries sur les plantes: exemple des *Pseudomonas* spp. fluorescent. *Agronomie*, *12*, 413–437.

Li, G. X., Wu, X. Q., Ye, J. R., & Yang, H. C. (2018). Characteristics of organic acid secretion associated with the interaction between *Burkholderia multivorans* WS-FJ9 and poplar root system. *BioMed Research International*, *2018*, 9619724.

Liang, J., Liu, J., Jia, P., et al. (2020). Novel phosphate-solubilizing bacteria enhance soil phosphorus cycling following ecological restoration of land degraded by mining. *The ISME Journal*, *14*, 1600–1613.

Liu, Y. Q., Wang, Y. H., Kong, W. L., Liu, W. H., Xie, X. L., & Wu, X. Q. (2020). Identification, cloning and expression patterns of the genes related to phosphate solubilization in *Burkholderia multivorans* WS-FJ9 under different soluble phosphate levels. *AMB Express*, *10*, 108.

Loheuarete, F., & Berthelin, J. (1988). Effect of phosphate solubilizing bacteria on maize growth and root exudates over four levels of labile phosphorus. *Plant and Soil*, *105*, 11–17.

Mahanta, D., Rai, R. K., Mishra, S. D., Raja, A., Purakayastha, T. J., & Varghese, E. (2014). Influence of phosphorus and biofertilizers on soybean and wheat root growth and properties. *Field Crops Research*, *166*, 1–9.

Maliha, R., Samina, K., Najma, A., Sadia, A., & Farooq, L. (2004). Organic acids production and phosphate solubilization by phosphate solubilizing microorganisms under *in vitro* conditions. *Pakistan Journal of Biological Sciences*, *7*, 187–196.

Mamta, R. P., Pathania, V., Gulati, A., Singh, B., Bhanwra, R. K., & Tewari, R. (2010). Stimulatory effect of phosphate-solubilizing bacteria on plant growth, stevioside and rebaudioside—A contents of *Stevia rebaudiana* Bertoni. *Applied Soil Ecology*, *46*, 222–229.

Mehrvarz, S., Chaichi, M. R., & Alikhani, H. A. (2008). Effects of phosphate solubilizing microorganisms and phosphorus chemical fertilizer on yield and yield components of barely (*Hordeum vulgare* L.). *American-Eurasian Journal of Agricultural & Environmental Sciences*, *3*, 822–828.

Molla, M. A. Z., Chowdhury, A. A., Islam, A., & Hoque, S. (1984). Microbial mineralization of organic phosphate in soil. *Plant and Soil*, *78*, 393–399.

Nahas, E. (1996). Factors determining rock phosphate solubilization by microorganism isolated from soil. *World Journal of Microbiology and Biotechnology*, *12*, 18–23.

Oliveira, C. A., Sa, N. M. H., Gomes, E. A., Marriel, I. E., Scotti, M. R., Guimaraes, C. T., et al. (2009). Assessment of the mycorrhizal community in the rhizosphere of maize (*Zea mays* L.) genotypes contrasting for phosphorus efficiency in the acid savannas of Brazil using denaturing gradient gel electrophoresis (DGGE). *Applied Soil Ecology*, *41*, 249–258.

Pandey, P., & Maheshwari, D. (2007). Bioformulation of *Burkholderia* sp. MSSP with a multispecies consortium for growth promotion of *Cajanus cajan*. *Canadian Journal of Microbiology*, *53*, 213–222.

Pandey, A., Trivedi, P., Kumar, B., & Palni, L. M. S. (2006). Characterization of a phosphate solubilizing and antagonistic strain of*Pseudomonas putida* (B0) isolated from a sub-alpine location in the Indian central Himalaya. *Current Microbiology*, *53*, 102–107.

Park, K. H., Lee, O. M., Jung, H. I., et al. (2010). Rapid solubilization of insoluble phosphate by a novel environmental stress-tolerant *Burkholderia vietnamiensis* M6 isolated from ginseng rhizospheric soil. *Applied Microbiology and Biotechnology*, *86*, 947–955.

Peix, A., Rivas-Boyero, A. A., Mateos, P. F., Rodriguez-Barrueco, C., Martinez-Molina, E., & Velazquez, E. (2001). Growth promotion of chickpea and barley by a phosphate solubilizing strain of M*esorhizobium mediterraneum* under growth chamber conditions. *Soil Biology and Biochemistry*, *33*, 103–110.

Pradel, E., & Boquet, P. L. (1990). Nucleotide sequence and transcriptional analysis of the *Escherichia coli* agp gene encoding periplasmic acid glucose-1-phosphatase. *Journal of Bacteriology*, *172*, 802–807.

Praveen, M. P., Kuligod, V. B., Hebsur, N. S., Patil, C. R., & Kulkarni, G. N. (2012). Effect of phosphate solubilizing fungi and phosphorus levels on growth, yield and nutrient content in maize (*Zea mays* L.). *Karnataka Journal of Agricultural Sciences*, *25*, 58–62.

Raj, J., Bagyaraj, D. J., & Manjunath, A. (1981). Influence of soil inoculation with vesicular arbuscular mycorrhizal fungus and a phosphate dissolving bacterium on plant growth and ^{32}P uptake. *Soil Biology and Biochemistry*, *13*, 105–108.

Richardson, A. E. (1994). Soil microorganisms and phosphorus availability. In C. E. Pankhurst, B. M. Doube, V. V. S. R. Gupta, & P. R. Grace (Eds.), *Soil biota, management in sustainable farming systems* (pp. 50–62). Melbourne: CSIRO.

Richardson, A. E., Hadobas, P. A., & Hayes, J. E. (2001). Extracellular secretion of *Aspergillus phytase* from *Arabidopsis* roots enables plants to obtain phosphorus from phytate. *The Plant Journal*, *25*, 641–649.

Richardson, A. E., & Simpson, R. J. (2011). Soil microorganisms mediating phosphorus availability update on microbial phosphorus. *Plant Physiology*, *156*, 989–996.

Rodríguez, H., & Fraga, R. (1999). Phosphate solubilizing bacteria and their role in plant growth promotion. *Biotechnology Advances*, *17*, 319–339.

Rodríguez, H., Fraga, R., Gonzalez, T., & Bashan, Y. (2006). Genetics of phosphate solubilization and its potential applications for improving plant growth-promoting bacteria. *Plant and Soil*, *287*, 15–21.

Rossolini, G. M., Shippa, S., Riccio, M. L., Berlutti, F., Macaskie, L. E., & Thaller, M. C. (1998). Bacterial nonspecific acid phosphatases: Physiology, evolution, and use as tools in microbial biotechnology. *Cellular and Molecular Life Sciences*, *54*, 833–850.

Saber, K., Nahla, L. D., & Chedly, A. (2005). Effect of P on nodule formation and N fixation in bean. *Agronomy for Sustainable Development*, *25*, 389–393.

Sagoe, C. I., Ando, T., Kouno, K., & Nagaoka, T. (1998). Relative importance of protons and solution calcium concentration in phosphate rock dissolution by organic acids. *Soil Science & Plant Nutrition*, *44*, 617–625.

Sarig, S., Okon, Y., & Blum, A. (1990). Promotion of leaf area development and field in *Sorghum bicolor* inoculated with *Azospirillum brasilense*. *Symbiosis*, *9*, 235–245.

Sashidhar, B., & Podile, A. R. (2010). Mineral phosphate solubilization by rhizosphere bacteria and scope for manipulation of the direct oxidation pathway involving glucose dehydrogenase. *Journal of Applied Microbiology*, *109*, 1–12.

Satyaprakash, M., Nikitha, T., Reddi, E. U. B., Sadhana, B., & Vani, S. S. (2017). A review on phosphorous and phosphate solubilising bacteria and their role in plant nutrition. *International Journal of Current Microbiology and Applied Sciences*, *6*, 2133–2144.

Savci, S. (2012). An agricultural pollutant: Chemical fertilizer. *International Journal of Environmental Science and Development*, *3*, 77–80.

Scervino, J. M., Mesa, M. P., Monica, I. D., Recchi, M., Moreno, N. S., & Godeas, A. (2010). Soil fungal isolates produce different organic acid patterns involved in phosphate salts solubilization. *Biology and Fertility of Soils*, *46*, 755–763.

Sharma, S. B., Sayyed, R. Z., Trivedi, M. H., & Gobi, T. A. (2013). Phosphate solubilizing microbes: Sustainable approach for managing phosphorus deficiency in agricultural soils. *SpringerPlus*, *2*, 587.

Singh, S., & Kapoor, K. K. (1999). Inoculation with phosphate solubilizing microorganisms and a vesicular arbuscular mycorrhizal fungus improves dry matter yield and nutrient uptake by wheat grown in a sandy soil. *Biology and Fertility of Soils*, *28*, 139–144.

Son, T. T. N., Diep, C. N., & Giang, T. T. M. (2006). Effect of bradyrhizobia and phosphate solubilizing bacteria application on soybean in rotational system in the Mekong delta. *Omonrice*, *14*, 48–57.

Stevenson, E. J. (2005). *Cycles of soil: Carbon, nitrogen, phosphorus, sulfur, micronutrients*. New York: John Wiley and Sons.

Subbarao, N. S. (1988). Phosphate solubilizing microorganism. In *Biofertilizer in agriculture and forestry* (pp. 133–142). Hissar, India: Regional Biofert. Dev. Centre.

Suleman, M., Yasmin, S., Rasul, M., Yahya, M., Atta, B. M., & Mirza, M. S. (2018). Phosphate solubilizing bacteria with glucose dehydrogenase gene for phosphorus uptake and beneficial effects on wheat. *PLoS One*, *13*, e0204408.

Sundara, B., Natarajan, V., & Hari, K. (2002). Influence of phosphorus solubilizing bacteria on the changes in soil available phosphorus and sugarcane and sugar yields. *Field Crops Research*, *7*, 43–49.

Sundara Rao, W. V. B., Bajpai, P. D., Sharma, J. P., & Subbaiah, B. V. (1963). Solubilization of phosphates by phosphate solubilizing organisms using 32 P as tracer and influence of seed bacteriazation on the uptake by the crop. Indian Soc. *Soil Science*, *11*, 209–218.

Surange, S., Wollum, A. G., Kumar, N., & Nautiyal, C. S. (1995). Characterization of *Rhizobium* from root nodules of leguminous trees growing in alkaline soils. *Canadian Journal of Microbiology*, *43*, 891–894.

Suslov, T. V. (1982). Role of root-colonizing bacteria in plant growth. In M. S. Mount, & G. H. Lacy (Eds.), *Phytopathogenic prokariotes* (pp. 187–223). London: Academic Press.

Swarnalakshmi, K., Prasanna, R., Kumar, A., Pattnaik, S., Chakravarty, K., Shivay, Y. S., et al. (2013). Evaluating the influence of novel cyanobacterial biofilm biofertilizers on soil fertility and plant nutrition in wheat. *European Journal of Soil Biology*, *55*, 107–116.

Tajini, F., Trabelsi, M., & Drevon, J. J. (2012). Combined inoculation with *Glomus intraradices* and *Rhizobium tropici* CIAT899 increases phosphorus use efficiency for symbiotic nitrogen fixation in common bean (*Phaseolus vulgaris* L.). *Saudi Journal of Biological Sciences*, *19*, 157–163.

Tani, A., Akita, M., Murase, H., & Kimbara, K. (2011). Culturable bacteria in hydroponic cultures of moss *Racomitrium japonicum* and their potential as biofertilizers for moss production. *Journal of Bioscience and Bioengineering*, *112*, 32–39.

Tarafdar, J. C., & Claasen, N. (1988). Organic phosphorus compounds as a phosphorus source for higher plants through the activity of phosphatases produced by plant roots and microorganisms. *Biology and Fertility of Soils*, *5*, 308–312.

Tarafdar, J. C., Rao, A. V., & Bala, K. (1988). Production of phosphatases by fungi isolated from desert soils. *Folia Microbiologica*, *33*, 453–457.

Thaller, M. C., Berlutti, F., Schippa, S., Lombardi, G., & Rossolini, G. M. (1994). Characterization and sequence of PhoC, the principal phosphate-irrepressible acid phosphatase of *Morganella morganii*. *Microbiology*, *140*, 1341–1350.

Thaller, M. C., Giovanna, L., Serena, S., & Rossolini, G. M. (1995). Cloning and characterization of the NapA acid phosphatase phosphotransferase of *Morganella morganii*: Identification of a new family of bacterial acid phosphatase-encoding genes. *Microbiology*, *141*, 147–154.

Thaller, M. C., Schippa, S., Bonci, A., Cresti, S., & Rossolini, G. M. (1997). Identification of the gene (aphA) encoding the class B acid phosphatase/phosphotransferase of *Escherichia coli* MG 1655 and characterization of its product. *FEMS Microbiology Letters*, *146*, 191–198.

Tilak, K. V. B. R., Ranganayaki, N. L., Pal, K. K., De, R., Saxena, A. K., Nautiyal, C. S., et al. (2005). Diversity of plant growth and soil health supporting bacteria. *Current Science*, *89*, 136.

Toro, M. (2007). Phosphate solubilizing microorganisms in the rhizosphere of native plants from tropical savannas: An adaptive strategy to acid soils? In C. Velazquez, & E. Rodriguez-Barrueco (Eds.), *Developments in plant and soil sciences* (pp. 249–252). Dordrecht: Springer.

Tripathi, A. K., Nagarajan, T., & Verma, S. C. (2002). Inhibition of biosynthesis and activity of nitrogenase in *Azospirillum brasilense* Sp7 under salinity stress. *Current Microbiology*, *44*, 363–367.

Vasil, I. K. (1998). Biotechnology and food security for 21st century: A real world perspective. *Nature Biotechnology*, *16*, 399–400.

Vikram, A., & Hamzehzarghani, H. (2008). Effect of phosphate solubilizing bacteria on nodulation and growth parameters of green gram (*Vigna radiata* L. Wilczek). *Research Journal of Microbiology*, *3*, 62–72.

Villegas, J., & Fortin, J. A. (2002). Phosphorus solubilization and pH changes as a result of the interactions between soil bacteria and arbuscular mycorrhizal fungi on a medium containing NO_3 as nitrogen source. *Canadian Journal of Botany*, *80*, 571–576.

Viveros, O. M., Jorquera, M. A., Crowley, D. E., Gajardo, G., & Mora, M. L. (2010). Mechanisms and practical considerations involved in plant growth promotion by rhizobacteria. *Journal of Soil Science and Plant Nutrition*, *10*, 293–319.

Walpola, B. C., & Yoon, M. (2012). Prospectus of phosphate solubilizing microorganisms and phosphorus availability in agricultural soils: A review. *African Journal of Microbiology Research*, *6*, 6600–6605.

Walpola, B. C., & Yoon, M. (2013). Phosphate solubilizing bacteria: Assessment of their effect on growth promotion and phosphorous uptake of mung bean (*Vigna radiate* [L.] R. Wilczek). *Chilean Journal of Agricultural Research*, *73*, 275.

Wang, H., Liu, S., Zhal, L., Zhang, J., Ren, T., Fan, B., et al. (2015). Preparation and utilization of phosphate biofertilizers using agricultural waste. *Journal of Integrative Agriculture*, *14*, 158–167.

Wani, P. A., Khan, M. S., & Zaidi, A. (2007). Synergistic effects of the inoculation with nitrogen fixing and phosphate-solubilizing rhizobacteria on the performance of field grown chickpea. *Journal of Plant Nutrition and Soil Science*, *170*, 283–287.

Whitelaw, M. A. (2000). Growth promotion of plants inoculated with phosphate solubilizing fungi. *Advances in Agronomy*, *69*, 99–151.

Xiao, C., Zhang, H., Fang, Y., & Chi, R. (2013). Evaluation for rock phosphate solubilization in fermentation and soil–plant system using a stress-tolerant phosphate-solubilizing *Aspergillus niger* WHAK1. *Applied Microbiology and Biotechnology*, *169*, 123–133.

Yadaf, R. S., & Tarafdar, J. C. (2001). Influence of organic and inorganic phosphorus supply on the maximum secretion of acid phosphatase by plants. *Biology and Fertility of Soils*, *34*, 140–143.

Yang, M., Wang, C., Wu, Z., Yu, T., Sun, H., & Liu, J. (2016). Phosphate dissolving capacity, glucose dehydrogenase gene expression and activity of two phosphate solubilizing bacteria. *Wei Sheng Wu Xue Bao*, *56*, 651–663.

Yousefi, A., Khavazi, K., Moezi, A., Rejali, F., & Nadian, H. (2011). Phosphate solubilizing bacteria and arbuscular mycorrhizal fungi impacts on inorganic phosphorus fractions and wheat growth. *World Applied Sciences Journal*, *15*, 1310–1318.

Zaidi, A., & Khan, M. S. (2006). Co-inoculation effects of phosphate solubilizing microorganisms and *Glomus fasciculatumon* on green gram—*Bradyrhizobium* symbiosis. *Turkish Journal of Agriculture and Forestry*, *30*, 223–230.

Zaidi, A., Khan, M. S., Ahemad, M., & Oves, M. (2009). Plant growth promotion by phosphate solubilizing bacteria. *Acta Microbiologica et Immunologica Hungarica*, *56*, 263–284.

Zayed, M. S. (2012). Improvement of growth and nutritional quality of *Moringa oleifera* using different biofertilizers. *Annals of Agricultural Science*, *57*, 53–62.

Zeng, Q., Wu, X., Wang, J., & Ding, X. (2017). Phosphate solubilization and gene expression of phosphate-solubilizing bacterium *Burkholderia multivorans* WS-FJ9 under different levels of soluble phosphate. *Journal of Microbiology and Biotechnology*, *27*, 844–855.

Zeng, Q., Wu, X., & Wen, X. (2016). Effects of soluble phosphate on phosphate-solubilizing characteristics and expression of gcd gene in *Pseudomonas frederiksbergensis* JW-SD2. *Current Microbiology*, *72*, 198–206.

Zhao, K., Penttinen, P., Zhang, X., Ao, X., Liu, M., Yu, X., et al. (2014). Maize rhizosphere in Sichuan, China, hosts plant growth promoting *Burkholderia cepacia* with phosphate solubilizing and antifungal abilities. *Microbiological Research*, *169*, 76–82.

Chapter 3

Trichoderma spp.—Application and future prospects in agricultural industry

Soma Barman[a], Pralay Shankar Gorai[b], and Narayan Chandra Mandal[b]
[a]Soil and Agrobio-Engineering Laboratory, Department of Environmental Science, Tezpur University, Tezpur, Assam, India, [b]Mycology and Plant Pathology Laboratory, Department of Botany, Visva-Bharati, Santiniketan, West Bengal, India

Chapter outline

1 Introduction

The genus *Trichoderma* belongs to Ascomycetes (Phylum: Ascomycota; class: Sordariomycetes; order: Hypocreales; family—Hypocreaceae) is a diverse group of fungi. It was first described in the year 1794, comprising anamorphic fungi, mostly isolated from soil and decomposing several organic matters (Persoon, 1794). They are free-living, avirulent plant symbionts, commonly present in all kinds of soils inhabiting root ecosystems (Harman, Howell, Viterbo, Chet, & Lorito, 2004). Sometimes they are opportunistic and parasites on other groups of fungi (Blaszczyk, Siwulski, Sobieralski, Lisiecka, & Jedryczka, 2014). *Trichoderma* is generally characterized by rapid mycelial growth, profuse production of conidia, as well as sclerotia. Moreover, they produce several colored pigments, ranging from a colorless to greenish yellow, sometimes a reddish touch. The conidia produced by the genus also have diverse pigmentation, ranging from colorless to different kinds of green to grey or sometimes brownish tinges (). *Trichoderma* spp. are among those fungal microorganisms, generally used as biological control agents. These days, they

Recent Advancement in Microbial Biotechnology. https://doi.org/10.1016/B978-0-12-822098-6.00008-2

are promoted as active components of biofertilizers, biopesticides, growth enhancers by stimulating natural resistance. This is owing to their capability of protecting plants by improving their vegetative growth, controlling pathogens under variable agricultural conditions. Moreover, they act as soil amendments for nutrient enrichment, decomposition, and biodegradation (Woo et al., 2014). Apart from biocontrol trait, they stimulated plants by solubilizing essential nutrients, remediate toxic organic pollutants, including heavy metals, thereby impart abiotic stress tolerance. There are many market-available microbe-enriched commercial products, but the instability under field conditions hampers their usefulness. Nowadays, the practice of biological control agents has increased owing to acceptance by the farmers in several agricultural sectors due to growing environmental concerns and the mandate for toxic organic produce.

2 Competency in the rhizosphere and plant root colonization

The abilities of any microorganism to compete with others for essential nutrients and exudates secreted by the plant roots and their capability of colonizing into the root surface of host plants are termed rhizosphere competence (Akter, Ahmed, & Alam, 2019). Plant rhizosphere is a composite ecological niche where massive biological interactions occur and struggle to survive. The root exudate of the plant contains various organic compounds that interact with a number of beneficial microflora of the soil. *Trichoderma* secreted some cysteine-rich hydrophobin-like proteins, e.g., TasHyd1 and Qid74 that helped them to colonize in the plant rhizosphere (Samolski, Rincon, Pinzón, Viterbo, & Monte, 2012; Sood et al., 2020). *Trichoderma* spp. produced some iron (Fe)-chelating substances (also known as siderophores). It increases the iron availability to the plants by binding to the insoluble iron (Fe^{3+}) of soil and plant rhizosphere. Simultaneously, siderophores deplete iron availability in the soil and hinder the growth of pathogenic microorganisms (Srivastava, Gupta, & Sharm, 2018). Occasionally, *Trichoderma* spp. secretes expansin-like proteins with their cellulose-binding modules assists penetration into the root of the plant (Zhang et al., 2013). Swollenin proteins expressively boosted the successful colonization of *Trichoderma* into the roots of plants (Brotman, Briff, Viterbo, & Chet, 2008) or invading the plant root barriers. They are also supported over the soil and inhabited into new niches. This type of signaling transduction with plants rhizosphere leads to enhanced proliferation of roots, thereby supporting growth and development, safeguarding the plants against applied toxic chemicals. After successful entry to the host plant, *Trichoderma* spp. suppress defense systems of plants and reside as a multifunctional endophytic plant symbiont (Morán-Diez et al., 2009). Some species of *Trichoderma* triggered induced systemic resistance (ISR) in different crop plants (Table 1).

TABLE 1 Induced systemic resistance (ISR) in crop plants triggered by different species of *Trichoderma*.

Different species of *Trichoderma*	ISR triggered in plant	Work against pathogen	Results	References
Trichoderma harzianum T-39	Bean (*Phaseolus vulgaris*)	*Colletotrichum lindemuthianum, Botrytis cinerea*	Application of T-39 on roots, no disease symptoms on leaves	Bigirimana, De Meyer, Poppe, Elad, and Höfte (1997)
	Tomato (*Solanum lycopersicum*), tobacco (*Nicotiana tabacum*), lettuce (*Lactuca sativa*), pepper (*Piper nigrum*), bean (*Phaseolus vulgaris*),	*Botrytis cinerea*	Application of T-39 on roots, no infection on leaves	De Meyer, Bigirimana, Elad, and Höfte (1998)
	Grape (*Vitis vinifera*)	*Plasmopara viticola*	Defense related mechanisms were activated	Perazzolli et al. (2012)
	Tomato (*Solanum lycopersicum*)	*Botrytis cinerea*	84% decline in disease severity due to application of 0.4% T39 drench	Meller et al. (2013)
	Cucumber (*Cucumis sativus*), bean, tomato, and strawberry (*Fragaria ananassa*)	*Botrytis cinerea* and *Podosphaera xanthii*	Various helpful rhizospheric microorganisms were activated by directly or indirectly and inhibited different kind of foliar diseases.	Okon Levy et al. (2015)
T. harzianum T-1 and T22; *T. virens* T-3	Cucumber (*Cucumis sativus*)	*Green-mottle, mosaic virus*	When present in roots no infection was observed in leaves	Lo, Liao, and Deng (2000)

Continued

TABLE 1 Induced systemic resistance (ISR) in crop plants triggered by different species of *Trichoderma*—cont'd

Different species of *Trichoderma*	ISR triggered in plant	Work against pathogen	Results	References
T. virens GT3-2	Cucumber (*Cucumis sativus*)	*Colletotrichum orbiculare*, *Pseudomonas syringae* pv. lachrymans	Stimulated genes responsible for superoxide molecules production and lignifications	Koike, Hyakumachi, Kageyama, Tsuyumu, and Doke (2001)
T. harzianum T-22	Tomato (*Solanum lycopersicum*)	*Alternaria solani*	Application of T-22 on roots, no infectious leaves	Seaman (2002)
T. asperellum (T203)	Cucumber (*Cucumis sativus*)	*Pseudomonas syringae* pv. *lachrymans*	Expression of jasmonic acid/ethylene signaling pathway related proteins were modulated	Shoresh, Yedidia, and Chet (2005)
T. harzianum	Pepper (*Piper nigrum*)	*Phytophthora capsici*	Increased the yield of phytoalexins capsidiol	Ahamed and Vermette (2009)
T. virens G-6, G-6-5 and G-11	Cotton (*Gossypium hirsutum*)	*Rhizoctonia solani*	Stimulate terpenoid phytoalexins	Howell, Hanson, Stipanovic, and Puckhaber (2000)
T. asperellum SKT-1	*Arabidopsis thaliana* (L.) Heynh.	*Pseudomonas syringae* pv. *tomato* DC3000	SKT-1 directly as well as cell culture filtrate induced systemic resistance	Yoshioka, Ichikawa, Naznin, Kogure, and Hyakumachi (2012)

	Arabidopsis thaliana	*Cucumber mosaic virus* (CMV)	Fight against CMV by improving defense mechanism	Elsharkawy, Shimizu, Takahashi, Ozaki, and Hyakumachi (2013)
T. harzianum Tr6, and *Pseudomonas* sp. Ps14	Cucumber (*Cucumis sativus*) and *Arabidopsis thaliana*	Fight against *Fusarium oxysporum* f. sp. *radicis cucumerinum* in cucumber and in *Arabidopsis thaliana* against *Botrytis cinerea*	A set of defense-related genes were activated by both the Ps14 and Tr6	Alizadeh et al. (2013)
T. virens and *T. atroviride*	Tomato	*Pseudomonas syringae* pv. *tomato* (Pst DC3000), *Alternaria solani* and *Botrytis cinerea*	Systemic acquired resistance is stimulated by the secreted proteins-Sm1 and Epl1	Salas-Marina et al. (2015)
T. harzianum	Cucumber	*Phytophthora melonis*	Transcription level of the four defensive genes i.e., Phenylalanine ammonialyase, Cucumber pathogen-induced 4, lipoxygenase and galactinol synthase. Were induced	Sabbagh, Roudini, and Panjehkeh (2017)
T. harzianum TH12	*Brassica napus*	*Sclerotinia sclerotiorum*	Induce Salicylic acid (SA)- and jasmonic acid/ethylene(JA/ET)-hormone	Alkooranee et al. (2017) DOI:https://doi.org/10.1371/journal.pone.0168850
T. viride	Cucumber	*Alternaria cucumerina*	Inhibited the pathogen growth of 100% and expression levels of MYC-2 gene (a marker for JA signaling) and PR-2 gene (a marker for SA signaling) were increased	Alkooranee and Kadhum (2019)

3 *Trichoderma* in bioremediation

Bioremediation primarily mediates through specific microorganisms such as fungi, bacteria, algae, and plants that are involved in the biotransformation or biodegradation of toxic contaminants into nonhazardous or less-hazardous by-products. Bioremediation involves the enzymatic breakdown of toxic pollutants by bioconversion into less harmful products. It employs the faster growth of microbes and, therefore, the rapid humiliation of toxic pollutants. *Trichoderma* spp. has multifarious plant growth-promoting effects, which lead to bioremediation by improving soil fertility, suppression of disease helps in biocomposting (Contreras-Cornejo, Macías-Rodríguez, Cortés-Penagos, & López-Bucio, 2009; Lorito, Woo, Harman, & Monte, 2010). Moreover, it secretes several organic acids such as citric acid, fumaric acid, and gluconic acid, which lower the soil pH and thereby promote phosphate solubilization, as well as dissolution of another essential macro- and micronutrients viz., iron, magnesium, and manganese (Hasan, 2016). It can modify the niche for beneficial plant rhizospheric microbes (Ociepa, 2011). The biomass of *Trichoderma* plays a crucial role in the bioremediation and environmental cleanup of polluted soils and therefore applied in phytoremediation. It has also been found that the biosorption ability of the *Trichoderma* helps to remove and concentrate toxic ions including Pb, Cd, Cu, Zn, and Ni (Srivastava et al., 2011; Yazdani, Yap, Abdullah, & Tan, 2009). Srivastava et al. (2011) reported that two different species of *Trichoderma* viz., *Trichoderma viride* and *Trichoderma asperellum* removed arsenic through biovolatilization in vitro. *T. viride* can be successfully applied for bioremediation of Cd and Pb in vitro (Gorai, Barman, Gond, & Mandal, 2020; Sahu, Mandal, Thakur, Manna, & Rao, 2012). *Trichoderma* spp. facilitates the uptake of nitrates and several other toxic ions in the rhizospheric region, thereby assisting the uptake of several metalloids and phytoextraction (Cao et al., 2008). *T. atroviride* was reported to stimulate uptake and simultaneous translocation of Cd, Zn, and Ni in *Brassica juncea*; however *T. harzianum* supported the growth of *Salix fragilis* in heavy metal-polluted soil (Adams, De-Leij, & Lynch, 2007). Several strains of *T. harzianum* can detoxify the hazardous compound potassium cyanide (KCN) and thereby stimulate the root growth and branching of *Pteris vittata*, an arsenic hyperaccumulating fern (Lynch & Moffat, 2005). Apart from that, there are several toxic as well as hazardous organic pollutants viz., cyanides, nitrates, phenols, polycyclic aromatic hydrocarbons (PAHs), artificial dyes such as dichlorodiphenyl trichloroethane (DDT), endosulfan, pentachlorophenol, which were biodegraded by *Trichoderma* spp. (Huang et al., 2018; Sood et al., 2020). Immobilized biomass of *Trichoderma* sp. BSCR02 in Ca-alginate beads helps in the biosorption of Cr (VI) (Smily & Sumithra, 2017). Therefore, bioremoval carried out by *Trichoderma* spp. could serve as a cost-effective agent of treating effluents and the contaminated water bodies with heavy metals.

4 *Trichoderma* in organic agriculture

Several strains of *Trichoderma* used as biocontrol agents in organic agriculture can act by: (a) inhabiting in the soil and/or residing as an endophyte in particular portions of the plant; therefore, dwelling in a particular space in the plant body and escaping the plant from the proliferation of the pathogens; (b) releasing fungal cell-wall degrading enzymes such as chitinase and beta-glucanase against the pathogens; (c) secretes several antibiotics-like compounds that can kill the pathogens; and (d) inducing the defense mechanisms of the host plant and therefore promoting the overall growth and development of the plant (Monte & Llobell, 2003). Harzianic acid produced by *T. harzianum* was effective in controlling *Pythium irregulare*, *Rhizoctonia solani,* and *Pythium irregulare* in vitro (Manganiello et al., 2018). Gupta, Mur, and Brotman (2014) reported suppression of reactive oxygen species (ROS) such as NO, H_2O_2 generation by roots of *Arabidopsis* in *Fusarium oxysporum* by *T. asperelloides.* Mitogen-activated protein (MAP) kinases transport defense-related signals linked to *Trichoderma*-intermediated immunity from receptors to responses in plants such as cotton, rice, and *Arabidopsis thaliana* (Jagodzik, Tajdel-Zielinska, Ciesla, Marczak, & Ludwikow, 2018). The mechanism of action of *Trichoderma* against the pathogens was presented in Fig. 1. They are effective against oomycetes, ascomycetes, and basidiomycetes group of pathogenic fungi (Benítez, Rincón, Limón, & Codon, 2004; Monte, 2001). Moreover, their influence on some soil-dwelling nematodes was reported earlier (Dababat, Sikora, & Hauschild, 2006; Goswami, Pandey, Tewari, & Goswami, 2008; Kyalo, Affokpon, Coosemans, & Coynes, 2007). The activity

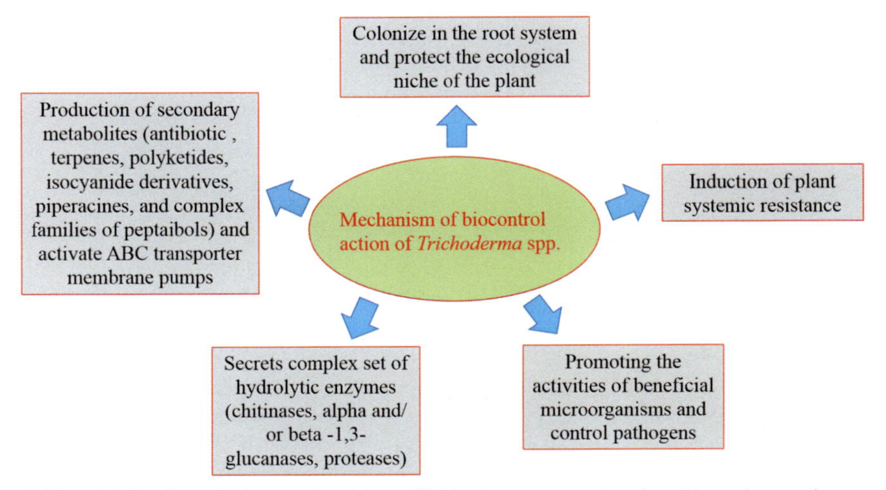

FIG. 1 Mechanisms of biocontrol activity of *Trichoderma* spp. against the pathogenic organisms.

of *Trichoderma* spp. to control root-knot nematodes, i.e., *Meloidogyne incognita* was reported by Márquez-Dávila, Arévalo-López, Gonzáles, Vega, and Meza (2020). The specifics of the biocontrol activity covering the host ranges of *Trichoderma* were enlisted in Table 2. Different strains of *Trichoderma*

TABLE 2 Ability of *Trichoderma* spp. to control different plant diseases by killing the pathogens of crop plants.

Crop plants	Pathogens	References
Cicer arietinum L.	*Rhizoctonia solani*	Khan and Rehman (1997)
	Macrophomina phaseolina, Rhizoctonia solani	Khan and Gupta (1998)
	Rhizoctonia solani, Fusarium oxysporum	Verma, Yadav, Tiwari, and Jaiswal (2014)
	Chaetomium spp., *Fusarium oxysporum, Penicillium* spp., *Sclerotium rolfsii, Macrophomina phaseolina*	Haware et al. (1999), Mukherjee, Haware, and Raghu (1997), Pandey, Pandey, and Pant (2003), and Poddar, Singh, and Dubey (2004)
Lycopersicum esculentum L.	*Fusarium oxysporum* f. sp. *lycopersici*	Khan and Akram (2000) and Alwathnani and Perveen (2012)
	Alternaria alternata, Alternaria solani, Septoria lycopersici	Rathee, Gautam, Sharma, and Verma (2006)
	Pythium aphanidermatum	Kumar and Hooda (2007) and Hazarika, Sarmah, Paramanick, Hazarika, and Phookan (2000)
	Sclerotium rolfsii, Rhizoctonia solani	Dutta and Das (2002) and Jayaraj, Radhakrishnan, and Velazhahan (2006)
Solanum tuberosum L.	*Phytophthora infestans*	Gogoi, Saikia, Helim, and Ullah (2007), Gupta, Singh, and Mohan (2004), and Basu, Konar, Mukhopadhyay, and Chettri (2001)
	Rhizoctonia solani	Gogoi et al. (2007), Singh, Sharma, and Srivastava (2001), and Tsror, Barak, and Sneh (2001)
	Fusarium sp., *Phoma* sp.	Gogoi et al. (2007)

TABLE 2 Ability of *Trichoderma* spp to control different plant diseases by killing the pathogens of crop plants.—cont'd

Crop plants	Pathogens	References
Capsicum annum L	*Fusarium oxysporum, Sclerotium rolfsii, Rhizoctonia solani pseudokoningii* 2013, *Pythium* spp.	Rini and Sulochana (2007), Vasanthakumari and Shivanna (2013), and Kapoor (2008)
	Phytophthora capsici	Tomah, Abd Alamer, Li, and Zhang (2020)
	Alternaria alternata	Kapoor (2008)
Glycine max (L.) Merr.	*Sclerotium rolfsii, Rhizoctonia solani, Macrophomina phaseolina, Sclerotiana sclerotiorum*	Bohra and Mathur (2004)
Brassica oleracea L.	*Pythium aphanidermatum, Rhizoctonia solani*	Ahuja et al. (2012) and Sharma and Sain (2004, 2005)
Oryza sativa L.	*Rhizoctonia solani, Fusarium* spp.	Bhramaramba and Nagamani (2013), Chakravarthy, Nagamani, Ratnakumari, and Bramarambha (2011), Gangwar, Sharma, and Singh (2013), and Subrata and Datta (2013)
	Rhizoctonia solani	Chakravarthy and Nagamani (2007)
Vigna mungo L.	*Macrophomina phaseolina, Alternaria alternate*	Raguchander, Rajappan, and Samiappan (1997) and Mishra, Gupta, Prajapati, and Singh (2011)
	Rhizoctonia solani, Colletotrichum truncatum	Shailbala and Tripathi (2004)
	Macrophomina phaseolina	Sajeena, Salalrajan, Seetharaman, and Mohan Babu (2004)
Solanum melongena L.	*Fusarium solani, Fusarium oxysporum* f. sp.	Jadon (2011) and Balaji and Ahir (2011)
	Pythium aphanidermatum	Ramesh (2004)
Triticum aestivum L.	*Chaetomium globosum*	Selvakumar, Srivastava, Aggarwal, and Singh (2001)
	Alternaria triticina	Kumar and Parveen (2002)

Continued

TABLE 2 Ability of *Trichoderma* spp to control different plant diseases by killing the pathogens of crop plants.—cont'd

Crop plants	Pathogens	References
Sesamum indicum L	*Curvularia lunata, Aspergillus flavus, Pythium notatum, Fusarium moniliforme, Pythium chrysogenum, Fusarium oxysporum, Macrophomina phaseolina, Rhizoctonia nigricans*	Sankar and Jeyarajan (1996), Tamimi and Hadwan (1985), and Jeyalakshmi, Rettinassababady, and Nema (2013)
	Fusarium oxysporum f. sp. sesame	Sangle and Bambawale (2004)
Cucumis sativa L.	*Pythium ultimum* and *Rhizoctonia solani*	Roberts et al. (2005)
Vigna sinensis L.	*Rhizoctonia solani*	Pan and Das (2011)
Cajanus cajan L.	*Fusarium udum*	Hukma and Pandey (2011)
Arachis hypogaea L.	*Thielaviopsis basicola, Rhizoctonia solani, Sclerotium rolfsii* Sacc., *Pythium aphanidermatum, Aspergillus niger, Macrophomia phaseolina*	Kishore, Pande, Rao, and Podile (2001), Rakholiya and Jadeja (2010), Bagwan (2011), and Sreedevi, Devi, and Saigopal (2011)
	Sclerotium rolfsii	Roy and Pan (2005)
Saccharum offcinarum L.	*Ceratocystis paradoxa*	Achuta, Rao, and Rao (2004)
Allium cepa L.	*Sclerotium cepivorum*	Rivera-Méndez, Obregón, Morán-Diez, Hermosa, and Monte (2020)

showed varying activity when applied independently or in the mixture. Different strains act synergistically as a plant protection agent in arrays of crops (Chirino-Valle et al., 2016). The increase of plant growth by *T. harzianum* may also be promoted by the solubilization of organic and inorganic bound phosphates, iron/copper/manganese oxide, metallic Zn (Altomare, Norvell, Björkman, & Harman, 1999; Li et al., 2015).

5 *Trichoderma* formulations

Protein extracts of wild-type strains of *Trichoderma* (TPE) contain a high concentration of chitinase and glucanase. These protein extracts are applied as effective biofungicide for controlling pathogens. Recently they were combined with lower concentrations of chemical fungicides (viz., carbendazim, iprodione) and showed synergistic action. The combined product is more stable for commercial applications. The activity of the newly developed product with fungicidal action were checked in in vitro as well as in vivo conditions. Further research suggested that minimum single enzyme from cellulase (endocellulase), protease, *endo*- and/or *exo*-chitinase, endo- and/or exoglucanase (β-1,3 plus β-1,6) [CWDEs] were recommended for the development of protein formulation. Contrarily, more than two enzymes from the aforementioned classes in a formulation were unable to deliver any antifungal action. This may be due to the unstable formation of the enzyme complex. The in vivo field trials with only TPE (chitinases and glucanases) in large quantities showed no detectable effect in plants compared to the combined application. Fungal cell-wall degrading enzymes (CWDEs) were proved to be safe and do not have any ecotoxicity for animals and humans. Degradation of CWDEs into environmentally friendly residues suggest *Trichoderma* spp. for use as biocontrol agents in India, United States, and European Union. The market-available commercial products of *Trichoderma* spp. (Fraceto et al., 2018) were listed in Table 3. Therefore, they have applied as biocontrol agents in postharvest practices. For large-scale production, the responsive genes coding for protein and enzymes can be introduced into suitable organisms by genetic engineering for the development of cell factories.

Micro- and nanotechnological techniques can be employed to develop new biological control techniques by increasing the viability of biological control agents (Ahluwalia, Kumar, Sisodia, Shakil, & Walia, 2014; Ma et al., 2015; Rathore, Desai, Liew, Chan, & Heng, 2013). Microencapsulation creates a close microenvironment, thereby protecting the applied microorganism. The physical barrier between the microorganism and external environment protects from mechanical stress, UV radiation, high and low temperatures, and other external stimuli (McLoughlin, 1994; Paulo & Santos, 2017). Therefore, the encapsulated microorganism is able to survive for a greater period of time (Cassidy, Lee, & Trevors, 1996). Two different types of microencapsulation techniques were employed viz., spray drying (physical) and coacervation, and ionic gelation (physico-chemical). The coacervation was used for the preservation of microbial cells and enzymes (John, Tyagi, Brar, Surampalli, & Prévost, 2011; Oliveira et al., 2007; Park & Chang, 2000). Conidia of *T. harzianum* was encapsulated in polymeric carbohydrate matrices to increase its shelf life during field applications (Muñoz-Celaya et al., 2012). The enzymatic activity (cellulose and chitinase) of *T. harzianum* was improved after encapsulation. The encapsulated

TABLE 3 ***Trichoderma*** **based commercial products available in the market used to control various diseases.**

Commercial/ trade name	Organism used	Antagonistic against pathogen/diseases	Name of supplier company/ distributor
Trichostar	*Trichoderma harzianum*	*Macrophomina* spp.	Green Tech, Agro Products, Rajaji Road, Coimbatore
Gliostar	*T. virens*	*Rhizoctonia, Fusarium, Sclerotium, Pythium*	GBPUAT, Pantnagar
Plant Shield	*T. harzianum*	*Rhizoctonia, Pythium, Cylindrocladium, Thielaviopsis; Fusarium, Botrytis*	BioWorks, Inc., United States (http://www.bioworksbiocontrol.com)
Promot Plus WP Promot Plus DD	*Trichoderma* spp., *T. koningii, T. harzianum*	Root rot diseases, Wilt	Tan Quy, Vietnam
Anti-Fungus	*Trichoderma* spp.	Root rot	Grondoontsmettingen De Ceuster, Belgium
Superesivit	*Trichoderma* spp.	Root rot and wilt	Bioplant, Denmark
Antagon	*Trichoderma* spp.	Damping-off diseases	De Ceuster Meststoffen N.V. (DCM), Belgium (http://www.agreoBiologicals.com)
Trichogourd	*T. viride*	Damping off	Anu Biotech International Ltd. India
Harzian 20, Harzian 10	*T. harzianum*	Wilt	Natural Plant Protection, Nogueres, France
Binab T	*T. polysporum, T. harzianum*	Wood rots causing internal decay, or originating from pruning wounds; *Chondrostereum, Didymella, Heterobasidion, Verticillium, Botrytis, Fusarium, Pythium, Rhizoctonia, Phytophthora*	BINAB Bio-Innovation AB, Sweden (http://www.algonet.se)

Trichojec Trichodowe Trichopel	*T. harzianum* + *T. viride*	Wilt	Agrimm, Technologies Ltd., New Zealand
Pant biocontrol Agent-I	*T. harzianum*	Root rot, wilt	Dept. of Plant Pathology, G.B. Pant University of Agriculture & Technology, Panatnagar, Uttarakhand
Bip T	*T. viride*	Wilt	Poland
Ecoderma	*T. viride* + *T. harzianum*	Root rot and wilt	Morgo Biocontrol Pvt. Ltd., India
Fulhumaxin 5.15SC	*Trichoderma* spp.	Root rot	An Hung Tuong, Vietnam
Bio Fit	*T. viride*	*Pythium, Fusarium, Rhizoctonia, Sclerotium,* other root rots	Ajay Biotech (India) Ltd. India (http://www.ajaybio.com)
Antagon TV	*T. viride*	*Macrophomina* spp.	Green Tech, Agro Products, Rajaji Road, Coimbatore
TRICO-DHCT	*Trichoderma* spp.	Sheath blight	Can Tho University, Vietnam
Biocon	*T. viride*	Root rot, wilt	Tocklai Experimental Station Tea Research Association, Jorhat (Assam), India
Biospark *Trichoderma*	*T. pseudokoningii, T. parceramosum*	Wilt	Biospark Corporation, Philippines
Tricho-X	*T. viride*	Root rot	Excel Industries Ltd., India
T-22B, T-22G	*T. harzianum*	Root rot	TGT Inc. New York, United States
Soil Gard	*T. virens* GL-21	Root rot	Certis, United States

Trichoderma showed greater activity against *Sclerotinia sclerotiorum* (Maruyama, Bilesky-José, de Lima, & Fraceto, 2020).

6 *Trichoderma* in biofuels

The extreme rise of energy cost in the scenario of forthcoming climate change and the alarming condition of fossil fuels motivated the researchers toward alternative biofuels. Being the potent producers of cellulose, *T. reesei* plays a vital role in bioconversion or breakdown of cellulosic wastes into simple sugars like glucose. The simple sugars are then fermented into bioethanol by yeast (Seidl, Seibel, Kubicek, & Schmoll, 2009). So, the application of *T. reesei* can develop alternative sources for biofuel. Thus *T. reesei* represents an ideal organism for studying any CWDEs. *T. reesei* was used as the most efficient biological tool for the production of biofuels (Zhang et al., 2019). The strain was also used in the regulation and production of lignocellulolytic enzymes like cellulose biofuels (Zhang et al., 2019). Chen et al. (2020) studied some crucial genetic alterations for increased cellulase production in *T. reesei*. Besides *T. reesei*, these genetic changes were very similar to ACE3-723 truncation in some other *Trichoderma* species (Chen et al., 2020). In addition, industrial strain improvement of *T. reesei* through the sexual crossing at *MAT*-loci facilitated advanced research in agriculture and human health. In the Asia-Pacific regions, there are adequate occurrences of sweet potato flour, but its commercial value in fuel production has not been fully discovered. Swain, Mishra, and Thatoi (2013) investigated a coculture technique for fermentation of *Ipomoea batatus* with *Saccharomyces cerevisiae* and *Trichoderma* spp. The ethanol production was 65% higher compared to the single culture of the microorganisms.

7 Conclusion and future prospectives

After successful colonization in the plant rhizosphere, *Trichoderma* secretes a wide range of antimicrobial substances that suppress several plant pathogens. Those antibiotic-like substances protect the plants by stimulating the systemic resistance of plants. In addition, several plant growth-promoting compounds activate signal molecules that regulate and induce the defense of the host. Considering the plant health and environmental concerns, the applicability of *Trichoderma* in agricultural sectors have been increased day by day. The efficacy of biosorption and resistance of toxic heavy metals offers an advantageous tool for monitoring the bioremediation from the environment. Other industrial applications of the genus in organic agriculture, production of biofuels motivated the researchers and scientists to develop its formulations with suitable carrier materials for field applications.

By applying the multifarious properties of *Trichoderma*, a green agricultural economy should develop. Future research will give in-depth knowledge for proper applications using the biocontrol activity of *Trichoderma*. Moreover, the applicability of *Trichoderma* for the replacement of chemical fungicides helps to maintain ecofriendly modern agriculture. *Trichoderma*-enriched solid and/or liquid formulations successfully control the plant diseases and helps in integrated disease/pest management (IDM/IPM) for the development of sustainable agriculture. Subsequently, genomes of *Trichoderma* can serve as a useful source for the production of transgenics to combat any type of abiotic and biotic stresses. The application of *Trichoderma* in the agricultural industry should be encouraged as an effective alternative to any commercially available chemical product that works via multidimensional protection to the agriculture and environment.

Acknowledgment

Authors were thankful to SERB-National Postdoctoral Fellowship (File Number: PDF/2017/002639) for financial support.

References

Achuta, M., Rao, R., & Rao, G. V. N. (2004). Biological control of *Ceratocystis paradoxa*-the incitant of pineapple disease of sugarcane. *Journal of Mycology and Plant Pathology*, *34*, 105–106.

Adams, P., De-Leij, F. A., & Lynch, J. M. (2007). Trichoderma harzianum Rifai 1295-22 mediates growth promotion of crack willow (*Salix fragilis*) saplings in both clean and metal-contaminated soil. *Microbial Ecology*, *54*(2), 306–313.

Ahamed, A., & Vermette, P. (2009). Effect of culture medium composition on *Trichoderma reesei*'s morphology and cellulase production. *Bioresource Technology*, *100*(23), 5979–5987.

Ahluwalia, V., Kumar, J., Sisodia, R., Shakil, N. A., & Walia, S. (2014). Green synthesis of silver nanoparticles by *Trichoderma harzianum* and their bio-efficacy evaluation against *Staphylococcus aureus* and *Klebsiella pneumonia*. *Industrial Crops and Products*, *55*, 202–206.

Ahuja, D. B., Ahuja, U. R., Srinivas, P., Singh, R. V., Malik, M., Sharma, P., et al. (2012). Development of farmer-led integrated management of major pests of cauliflower cultivated in rainy season in India. *Journal of Agricultural Science*, *4*(2), 79.

Akter, F., Ahmed, G. U., & Alam, M. F. (2019). Trichoderma: A complete tool box for climate smart agriculture. *Madridge Journal of Agriculture and Environmental Sciences*, *1*(1), 40–43.

Alizadeh, H., Behboudi, K., Ahmadzadeh, M., Javan-Nikkhah, M., Zamioudis, C., Pieterse, C. M., et al. (2013). Induced systemic resistance in cucumber and Arabidopsis thaliana by the combination of *Trichoderma harzianum* Tr6 and *Pseudomonas* sp. Ps14. *Biological Control*, *65*(1), 14–23.

Alkooranee, J. T., Aledan, T. R., Ali, A. K., Lu, G., Zhang, X., Wu, J., et al. (2017). Detecting the hormonal pathways in oilseed rape behind induced systemic resistance by *Trichoderma harzianum* TH12 to *Sclerotinia sclerotiorum*. *PLoS One*, *12*(1), e0168850.

Alkooranee, J. T., & Kadhum, N. N. (2019). Induce systemic resistance in cucumber by some bioelicitors against alternaria leaf blight disease caused by *Alternaria cucumerina* fungus. *Plant Archives*, *19*(1), 747–755.

Altomare, C., Norvell, W. A., Björkman, T., & Harman, G. E. (1999). Solubilization of phosphates and micronutrients by the plant-growth-promoting and biocontrol fungus *Trichoderma harzianum* Rifai 1295-22. *Applied and Environmental Microbiology*, *65*(7), 2926–2933.

Alwathnani, H. A., & Perveen, K. (2012). Biological control of fusarium wilt of tomato by antagonist fungi and cyanobacteria. *African Journal of Biotechnology*, *11*(5), 1100–1105.

Bagwan, N. B. (2011). Evaluation of biocontrol potential of *Trichoderma* species against *Sclerotium rolfsii*, *Aspergillus niger* and *Aspergillus flavus*. *International Journal of Plant Protection*, *4*, 107–111.

Balaji, L. P., & Ahir, R. R. (2011). Evaluation of plant extracts and biocontrol agents against leaf spot disease of brinjal. *Indian Phytopathology*, *64*(4), 378–380.

Basu, A., Konar, A., Mukhopadhyay, S. K., & Chettri, M. (2001). Biological management of late blight of potato using talc-based formulations of antagonists. *Journal of the Indian Potato Association (India)*, *28*, 80–81.

Benítez, T., Rincón, A. M., Limón, M. C., & Codon, A. C. (2004). Biocontrol mechanisms of *Trichoderma* strains. *International Microbiology*, *7*(4), 249–260.

Bhramaramba, S., & Nagamani, A. (2013). Antagonistic *Trichoderma* isolates to control bakanae pathogen of rice. *Agricultural Science Digest-A Research Journal*, *33*(2), 104–108.

Bigirimana, J., De Meyer, G., Poppe, J., Elad, Y., & Höfte, M. (1997). Induction of systemic resistance on bean (*Phaseolus vulgaris*) by *Trichoderma harziamum*. *Mededelingen van de Faculteit Landbouwkundige en Toegepaste Biologische Wetenschappen*, *62*, 1001–1007. Universiteit Gent.

Blaszczyk, L., Siwulski, M., Sobieralski, K., Lisiecka, J., & Jedryczka, M. (2014). *Trichoderma* spp.–application and prospects for use in organic farming and industry. *Journal of Plant Protection Research*, *54*(4), 309–317.

Bohra, B., & Mathur, K. (2004). Biocontrol agents and neem formulations for suppression of *Fusarium* solani root rot in soybean. *Journal of Mycology and Pant Pathology*, *34*(2), 408–409.

Brotman, Y., Briff, E., Viterbo, A., & Chet, I. (2008). Role of swollenin, an expansin-like protein from *Trichoderma*, in plant root colonization. *Plant Physiology*, *147*(2), 779–789.

Cao, L., Jiang, M., Zeng, Z., Du, A., Tan, H., & Liu, Y. (2008). *Trichoderma atroviride* F6 improves phytoextraction efficiency of mustard (*Brassica juncea* (L.) Coss. var. foliosa Bailey) in Cd, Ni contaminated soils. *Chemosphere*, *71*(9), 1769–1773.

Cassidy, M. B., Lee, H., & Trevors, J. T. (1996). Environmental applications of immobilized microbial cells: A review. *Journal of Industrial Microbiology*, *16*(2), 79–101.

Chakravarthy, S. K., & Nagamani, A. (2007). Efficacy of non-volatile and volatile compounds of *Trichoderma* species on *Rhizoctonia solani*. *Journal of Mycology and Pant Pathology*, *37*, 82–86.

Chakravarthy, K. S., Nagamani, A., Ratnakumari, Y. R., & Bramarambha, S. (2011). Antagonistic ability against *Rhizoctonia solani* and pesticide tolerance of *Trichoderma* strains. *Advances in Environmental Biology*, *5*(9), 2631–2638.

Chen, Y., Wu, C., Fan, X., Zhao, X., Zhao, X., Shen, T., et al. (2020). Engineering of Trichoderma reesei for enhanced degradation of lignocellulosic biomass by truncation of the cellulase activator ACE3. *Biotechnology for Biofuels*, *13*, 1–14.

Chirino-Valle, I., Kandula, D., Littlejohn, C., Hill, R., Walker, M., Shields, M., et al. (2016). Potential of the beneficial fungus *Trichoderma* to enhance ecosystem-service provision in the biofuel grass *Miscanthus* x *giganteus* in agriculture. *Scientific Reports*, *6*(1), 1–8.

Contreras-Cornejo, H. A., Macías-Rodríguez, L., Cortés-Penagos, C., & López-Bucio, J. (2009). *Trichoderma virens*, a plant beneficial fungus, enhances biomass production and promotes lateral root growth through an auxin-dependent mechanism in Arabidopsis. *Plant Physiology*, *149* (3), 1579–1592.

Dababat, A. A., Sikora, R. A., & Hauschild, R. (2006). Use of *Trichoderma harzianum* and *Trichoderma viride* for the biological control of *Meloidogyne incognita* on tomato. *Communications in Agricultural and Applied Biological Sciences*, *71*(3 Pt B), 953–961.

De Meyer, G., Bigirimana, J., Elad, Y., & Höfte, M. (1998). Induced systemic resistance in *Trichoderma harzianum* T39 biocontrol of *Botrytis cinerea. European Journal of Plant Pathology, 104*(3), 279–286.

Dutta, P., & Das, B. C. (2002). Management of collar rot of tomato by *Trichoderma* spp. and chemicals. *Indian Phytopathology, 55*(2), 235–237.

Elsharkawy, M. M., Shimizu, M., Takahashi, H., Ozaki, K., & Hyakumachi, M. (2013). Induction of systemic resistance against cucumber mosaic virus in *Arabidopsis thaliana* by *Trichoderma asperellum* SKT-1. *The Plant Pathology Journal, 29*(2), 193.

Fraceto, L. F., Maruyama, C. R., Guilger, M., Mishra, S., Keswani, C., Singh, H. B., et al. (2018). *Trichoderma harzianum*-based novel formulations: Potential applications for management of next-gen agricultural challenges. *Journal of Chemical Technology & Biotechnology, 93*(8), 2056–2063.

Gangwar, O. P., Sharma, P., & Singh, U. D. (2013). Growth and survival of *Trichoderma harzianum* and *Pseudomonas fluorescens* on different substrates and their temporal and spatial population dynamics in irrigated rice ecosystem. *Indian Phytopathology, 66*(3), 252–257.

Gogoi, R., Saikia, M., Helim, R., & Ullah, Z. (2007). Management of potato diseases using *Trichoderma viride* formulations. *Journal of Mycology and Pant Pathology, 37*(2), 227–230.

Gorai, P. S., Barman, S., Gond, S. K., & Mandal, N. C. (2020). Trichoderma. In N. Amaresan, M. Senthil Kumar, K. Annapurna, K. Kumar, & A. Sankaranarayanan (Eds.), *Beneficial microbes in agro-ecology: Bacteria and fungi* (pp. 571–591). Elsevier, Academic Press.

Goswami, J., Pandey, R. K., Tewari, J. P., & Goswami, B. K. (2008). Management of root knot nematode on tomato through application of fungal antagonists, *Acremonium strictum* and *Trichoderma harzianum. Journal of Environmental Science and Health Part B, 43*(3), 237–240.

Gupta, K. J., Mur, L. A., & Brotman, Y. (2014). *Trichoderma asperelloides* suppresses nitric oxide generation elicited by *Fusarium oxysporum* in Arabidopsis roots. *Molecular Plant-Microbe Interactions, 27*(4), 307–314.

Gupta, H., Singh, B. P., & Mohan, J. (2004). Biocontrol of late blight of potato. *Potato Journal, 31* (1–2), 39–42.

Harman, G. E., Howell, C. R., Viterbo, A., Chet, I., & Lorito, M. (2004). *Trichoderma* species—Opportunistic, avirulent plant symbionts. *Nature Reviews Microbiology, 2*(1), 43–56.

Hasan, S. (2016). Potential of *Trichoderma* sp. in bioremediation: A review. *Journal of Basic and Applied Engineering Research, 3*(9), 776–779.

Haware, M. P., Mukherjee, P. K., Lenne, J. M., Jayanthi, S., Tripathi, H. S., & Rathi, Y. P. S. (1999). Integrated biological-chemical control of *Botrytis* gray mould of chickpea. *Indian Phytopathology, 52*(2), 174–176.

Hazarika, D. K., Sarmah, R., Paramanick, T., Hazarika, K., & Phookan, A. K. (2000). Biological management of tomato damping off caused by *Pythium aphanidermatum. Indian Journal of Plant Pathology, 18*(1/2), 36–39.

Howell, C. R., Hanson, L. E., Stipanovic, R. D., & Puckhaber, L. S. (2000). Induction of terpenoid synthesis in cotton roots and control of *Rhizoctonia solani* by seed treatment with *Trichoderma virens. Phytopathology, 90*(3), 248–252.

Huang, Y., Xiao, L., Li, F., Xiao, M., Lin, D., Long, X., et al. (2018). Microbial degradation of pesticide residues and an emphasis on the degradation of cypermethrin and 3-phenoxy benzoic acid: A review. *Molecules, 23*(9), 2313.

Hukma, R., & Pandey, R. N. (2011). Efficacy of bio-control agents and fungicides in the management of wilt of pigeon pea. *Indian Phytopathology, 64*(3), 269–271.

Jadon, K. S. (2011). Eco-friendly management of brinjal collar rot caused by *Sclerotium rolfsii* Sacc. *Indian Phytopathology, 62*(3), 345.

Jagodzik, P., Tajdel-Zielinska, M., Ciesla, A., Marczak, M., & Ludwikow, A. (2018). Mitogen-activated protein kinase cascades in plant hormone signaling. *Frontiers in Plant Science*, *9*, 1387.

Jayaraj, J., Radhakrishnan, N. V., & Velazhahan, R. (2006). Development of formulations of *Trichoderma harzianum* strain M1 for control of damping-off of tomato caused by *Pythium aphanidermatum*. *Archives of Phytopathology and Plant Protection*, *39*(1), 1–8.

Jeyalakshmi, C., Rettinassababady, C., & Nema, S. (2013). Integrated management of sesame diseases. *Journal of Biopesticides*, *6*(1), 68.

John, R. P., Tyagi, R. D., Brar, S. K., Surampalli, R. Y., & Prévost, D. (2011). Bio-encapsulation of microbial cells for targeted agricultural delivery. *Critical Reviews in Biotechnology*, *31*(3), 211–226.

Kapoor, A. S. (2008). Biocontrol potential of *Trichoderma* spp. against important soilborne diseases of vegetable crops. *Indian Phytopathology*, *61*(4), 492–498.

Khan, M. R., & Akram, M. (2000). Effects of certain antagonistic fungi and rhizobacteria on wilt disease complex of tomato caused by *Meloidogyne incognita* and *Fusarium oxysporum* f. sp. *lycopersici*. *Nematologia Mediterranea*, *28*(2), 139–144.

Khan, M. R., & Gupta, J. (1998). Antagonistic efficacy of *Trichoderma* species against *Macrophomina phaseolina* on eggplant/Antagonistische Wirksamkeit von *Trichoderma*-Arten gegen *Macrophomina phaseolina* an Auberginen. *Zeitschrift für Pflanzenkrankheiten und Pflanzenschutz/Journal of Plant Diseases and Protection*, *105*, 387–393.

Khan, M. R., & Rehman, Z. (1997). Biomanagement of root rot of chickpea caused by *Rhizoctonia*. *Vasundhara*, *3*, 22–26.

Kishore, G. K., Pande, S., Rao, J. N., & Podile, A. R. (2001). Biological control of crown rot of groundnut by *Trichoderma harzianum* and *T. viride*. *International Arachis Newsletter*, *21*, 39–40.

Koike, N., Hyakumachi, M., Kageyama, K., Tsuyumu, S., & Doke, N. (2001). Induction of systemic resistance in cucumber against several diseases by plant growth-promoting fungi: Lignification and superoxide generation. *European Journal of Plant Pathology*, *107*(5), 523–533.

Kumar, R., & Hooda, I. (2007). Evaluation of antagonistic properties of *Trichoderma* species against *Pythium aphanidermatum* causing damping off of tomato. *Journal of Mycology and Pant Pathology*, *37*(2), 240–243.

Kumar, V. R., & Parveen, S. (2002). Integrated disease management of leaf blight of wheat. *Annals of Plant Protection Sciences*, *10*(2), 302–307.

Kyalo, G., Affokpon, A., Coosemans, J., & Coynes, D. L. (2007). Biological control effects of *Pochonia chlamysdosporia* and *Trichoderma* isolates from Benin (West-Africa) on root-knot nematodes. *Communications in Agricultural and Applied Biological Sciences*, *72*(1), 219.

Li, R. X., Cai, F., Pang, G., Shen, Q. R., Li, R., & Chen, W. (2015). Solubilisation of phosphate and micronutrients by *Trichoderma harzianum* and its relationship with the promotion of tomato plant growth. *PLoS One*, *10*(6), e0130081.

Lo, C. T., Liao, T. F., & Deng, T. C. (2000). Induction of systemic resistance of cucumber to cucumber green mosaic virus by the root-colonizing *Trichoderma* spp. *Phytopathology*, *90* (Suppl), S47.

Lorito, M., Woo, S. L., Harman, G. E., & Monte, E. (2010). Translational research on *Trichoderma*: From'omics to the field. *Annual Review of Phytopathology*, *48*, 395–417.

Lynch, J. M., & Moffat, A. J. (2005). Bioremediation–prospects for the future application of innovative applied biological research. *Annals of Applied Biology*, *146*(2), 217–221.

Ma, X., Wang, X., Cheng, J., Nie, X., Yu, X., Zhao, Y., et al. (2015). Microencapsulation of *Bacillus subtilis* B99-2 and its biocontrol efficiency against *Rhizoctonia solani* in tomato. *Biological Control*, *90*, 34–41.

Manganiello, G., Sacco, A., Ercolano, M. R., Vinale, F., Lanzuise, S., Pascale, A., et al. (2018). Modulation of tomato response to *Rhizoctonia solani* by *Trichoderma harzianum* and its secondary metabolite harzianic acid. *Frontiers in Microbiology*, *9*, 1966.

Márquez-Dávila, K., Arévalo-López, L., Gonzáles, R., Vega, L., & Meza, M. (2020). *Trichoderma* and *Clonostachys* as biocontrol agents against *Meloidogyne incognita* in sacha inchi1. *Pesquisa Agropecuária Tropical (Agricultural Research in the Tropics)*, *50*, e60890.

Maruyama, C. R., Bilesky-José, N., de Lima, R., & Fraceto, L. F. (2020). Encapsulation of *Trichoderma harzianum* preserves enzymatic activity and enhances the potential for biological control. *Frontiers in Bioengineering and Biotechnology*, *8*, 225.

McLoughlin, A. J. (1994). Controlled release of immobilized cells as a strategy to regulate ecological competence of inocula. In *Biotechnics/wastewater* (pp. 1–45). Berlin, Heidelberg: Springer.

Meller, H. Y., Haile, M. Z., David, D., Borenstein, M., Shulchani, R., & Elad, Y. (2013). Induced systemic resistance against grey mould in tomato (*Solanum lycopersicum*) by benzothiadiazole and *Trichoderma harzianum* T39. *Phytopathology*, *104*, 150–157.

Mishra, D. S., Gupta, A. K., Prajapati, C. R., & Singh, U. S. (2011). Combination of fungal and bacterial antagonists for management of root and stem rot disease of soybean. *Pakistan Journal Botany*, *43*(5), 2569–2574.

Monte, E. (2001). Understanding *Trichoderma*: Between biotechnology and microbial ecology. *International Microbiology*, *4*(1), 1–4.

Monte, E., & Llobell, A. (2003, October). *Trichoderma* in organic agriculture. In *Proceeding V World Avocado Congress (Actas V Congreso Mundial del Aguacate)* (pp. 725–733).

Morán-Diez, E., Hermosa, R., Ambrosino, P., Cardoza, R. E., Gutiérrez, S., Lorito, M., et al. (2009). The ThPG1 endopolygalacturonase is required for the *Trichoderma harzianum*–plant beneficial interaction. *Molecular Plant-Microbe Interactions*, *22*(8), 1021–1031.

Mukherjee, P. K., Haware, M. P., & Raghu, K. (1997). Induction and evaluation of benomyl-tolerant mutants of *Trichoderma* viridefor biological control of *Botrytis* grey mould of chickpea. *Indian Phytopathology*, *50*(4), 485–489.

Muñoz-Celaya, A. L., Ortiz-García, M., Vernon-Carter, E. J., Jauregui-Rincón, J., Galindo, E., & Serrano-Carreón, L. (2012). Spray-drying microencapsulation of *Trichoderma harzianum* conidias in carbohydrate polymers matrices. *Carbohydrate Polymers*, *88*(4), 1141–1148.

Ociepa, E. (2011). The effect of fertilization on yielding asnd heavy metals uptake by maize and Virginia fanpetals (*Sida hermaphrodita*). *Archives of Environmental Protection*, *37*(2), 123–129.

Okon Levy, N., Meller Harel, Y., Haile, Z. M., Elad, Y., Rav-David, E., Jurkevitch, E., et al. (2015). Induced resistance to foliar diseases by soil solarization and *Trichoderma harzianum*. *Plant Pathology*, *64*(2), 365–374.

Oliveira, A. C., Moretti, T. S., Boschini, C., Baliero, J. C. C., Freitas, O., & Favaro-Trindade, C. S. (2007). Stability of microencapsulated *B. lactis* (BI 01) and *L. acidophilus* (LAC 4) by complex coacervation followed by spray drying. *Journal of Microencapsulation*, *24*(7), 685–693.

Pan, S., & Das, A. (2011). Control of cowpea (*Vigna sinensis*) root and collar rot (*Rhizoctonia solani*) with some organic formulations of *Trichoderma harzianum* under field condition. *Journal of Plant Protection Sciences*, *3*(2), 20–25.

Pandey, G., Pandey, R. K., & Pant, H. (2003). Efficacy of different levels of *Trichoderma viride* against root-knot nematode in chickpea (*Cicer arietinum* L.). *Annals of Plant Protection Sciences*, *11*(1), 101–103.

Park, J. K., & Chang, H. N. (2000). Microencapsulation of microbial cells. *Biotechnology Advances*, *18*(4), 303–319.

Paulo, F., & Santos, L. (2017). Design of experiments for microencapsulation applications: A review. *Materials Science and Engineering: C*, *77*, 1327–1340.

Perazzolli, M., Moretto, M., Fontana, P., Ferrarini, A., Velasco, R., Moser, C., et al. (2012). Downy mildew resistance induced by *Trichoderma harzianum* T39 in susceptible grapevines partially mimics transcriptional changes of resistant genotypes. *BMC Genomics*, *13*(1), 660.

Persoon, C. H. (1794). Disposita methodical fungorum. *Romers Neues Magazine Botany*, *1*, 81–128.

Poddar, R. K., Singh, D. V., & Dubey, S. C. (2004). Integrated application of *Trichoderma harzianum* mutants and carbendazim to manage chickpea wilt (*Fusarium oxysporum* f. sp. *ciceris*). *Indian Journal of Agricultural Sciences*, *74*(6), 346–348.

Raguchander, T., Rajappan, K., & Samiappan, R. (1997). Evaluating methods of application of biocontrol agent in the control of mungbean root rot. *Indian Phytopathology*, *50*(2), 229–234.

Rakholiya, K. B., & Jadeja, K. B. (2010). Effect of seed treatment of biocontrol agents and chemicals for management of stem and pod rot of groundnut. *International Journal of Plant Protection*, *3* (2), 276–278.

Ramesh, R. (2004). Management of Damping-off in Brinjal using biocontrol agents. *Journal of Mycology and Plant Pathology*, *34*, 666–670.

Rathee, V. K., Gautam, G., Sharma, K. C., & Verma, S. (2006). Chemical and biological control of foliar and fruit rot diseases of tomato. *Plant Disease Research-Ludhiana*, *21*(1), 53.

Rathore, S., Desai, P. M., Liew, C. V., Chan, L. W., & Heng, P. W. S. (2013). Microencapsulation of microbial cells. *Journal of Food Engineering*, *116*(2), 369–381.

Rini, C. R., & Sulochana, K. K. (2007). Management of seedling rot of chilli (*Capsicum annuum* L.) using *Trichoderma* spp. and fluorescent pseudomonads (*Pseudomonas fluorescens*). *Journal of Tropical Agriculture*, *44*, 79–82.

Rivera-Méndez, W., Obregón, M., Morán-Diez, M. E., Hermosa, R., & Monte, E. (2020). *Trichoderma asperellum* biocontrol activity and induction of systemic defenses against *Sclerotium cepivorum* in onion plants under tropical climate conditions. *Biological Control*, *141*, 104145.

Roberts, D. P., Lohrke, S. M., Meyer, S. L., Buyer, J. S., Bowers, J. H., Baker, C. J., et al. (2005). Biocontrol agents applied individually and in combination for suppression of soilborne diseases of cucumber. *Crop Protection*, *24*(2), 141–155.

Roy, A., & Pan, S. (2005). Effect of fungistasis on germinability of wild and mutant isolates of *Trichoderma harzianum* and *Gliocladium virens*. *Journal of Mycology and Plant Pathology*, *35*(2), 319–323.

Sabbagh, S. K., Roudini, M., & Panjehkeh, N. (2017). Systemic resistance induced by *Trichoderma harzianum* and *Glomus mossea* on cucumber damping-off disease caused by *Phytophthora melonis*. *Archives of Phytopathology and Plant Protection*, *50*(7–8), 375–388.

Sahu, A., Mandal, A., Thakur, J., Manna, M. C., & Rao, A. S. (2012). Exploring bioaccumulation efficacy of *Trichoderma* viride: An alternative bioremediation of cadmium and lead. *National Academy Science Letters*, *35*(4), 299–302.

Sajeena, A., Salalrajan, F., Seetharaman, K., & Mohan Babu, R. (2004). Evaluation of biocontrol agents against dry root rot of blackgram (*Vigna mungo*). *Journal of Mycology and Plant Pathology*, *34*(2), 341–343.

Salas-Marina, M. A., Isordia-Jasso, M. I., Islas-Osuna, M. A., Delgado-Sánchez, P., Jiménez-Bremont, J. F., Rodríguez-Kessler, M., et al. (2015). The Epl1 and Sm1 proteins from *Trichoderma atroviride* and *Trichoderma virens* differentially modulate systemic disease resistance against different life style pathogens in *Solanum lycopersicum*. *Frontiers in Plant Science*, *6*, 77.

Samolski, I., Rincon, A. M., Pinzón, L. M., Viterbo, A., & Monte, E. (2012). The qid74 gene from *Trichoderma harzianum* has a role in root architecture and plant biofertilization. *Microbiology*, *158*(1), 129–138.

Sangle, U. R., & Bambawale, O. M. (2004). New strains of *Trichoderma* spp. strongly antagonistic against *Fusarium oxysporum f.* sp. *sesami*. *Journal of Mycology and Plant Pathology*, *34*(1).

Sankar, P., & Jeyarajan, R. (1996). Seed treatment formulation of *Trichoderma* and *Gliocladium* for biological control of *Macrophomina phaseolina* in sesamum. *Indian Phytopathology*, *49*(2), 148–151.

Seaman, A. (2002). *Efficacy of OMRI-approved products for tomato foliar disease control*. New York State IPM Program.

Seidl, V., Seibel, C., Kubicek, C. P., & Schmoll, M. (2009). Sexual development in the industrial workhorse *Trichoderma reesei*. *Proceedings of the National Academy of Sciences*, *106*(33), 13909–13914. https://doi.org/10.1073/pnas.0904936106.

Selvakumar, R., Srivastava, K. D., Aggarwal, R., & Singh, D. V. (2001). Biocontrol of spot blotch of wheat using *Chaetomium globosum*. *Annals of Plant Protection Sciences (India)*, *9*(2), 286–291.

Shailbala, & Tripathi, H. S. (2004). Seed treatment with fungicides and biocontrol agent on pathogens in urdbean seeds. *Journal of Mycology and Plant Pathology*, *34*, 851–852.

Sharma, P., & Sain, S. K. (2004). Induction of systemic resistance in tomato and cauliflower by *Trichoderma* spp. against stalk-rot pathogen, *Sclerotinia sclerotiorum* (Lib) de Bary. *Journal of Biological Control*, *18*(1), 21–28.

Sharma, P., & Sain, S. K. (2005). Use of biotic agents and abiotic compounds against damping off of cauliflower caused by Pythium aphanidermatum. *Indian Phytopathology*, *58*(4), 395.

Shoresh, M., Yedidia, I., & Chet, I. (2005). Involvement of jasmonic acid/ethylene signaling pathway in the systemic resistance induced in cucumber by *Trichoderma asperellum* T203. *Phytopathology*, *95*(1), 76–84.

Singh, R. B., Sharma, K. M., & Srivastava, K. K. (2001). Management of black scurf and stem necrosis disease in potato. *Journal of the Indian Potato Association (India)*, *29*, 78–79.

Smily, J. R. M. B., & Sumithra, P. A. (2017). Optimization of chromium biosorption by fungal adsorbent, *Trichoderma* sp. BSCR02 and its desorption studies. *HAYATI Journal of Biosciences*, *24*(2), 65–71. https://doi.org/10.1016/j.hjb.2017.08.005.

Sood, M., Kapoor, D., Kumar, V., Sheteiwy, M. S., Ramakrishnan, M., Landi, M., et al. (2020). *Trichoderma*: The "secrets" of a multitalented biocontrol agent. *Plants*, *9*(6), 762.

Sreedevi, B., Devi, M. C., & Saigopal, D. V. R. (2011). Induction of defense enzymes in *Trichoderma harzianum* treated groundnut plants against *Macrophomina phaseolina*. *Journal of Biological Control*, *25*(1), 33–39.

Srivastava, M. P., Gupta, S., & Sharm, Y. K. (2018). Detection of siderophore production from different cultural variables by CAS-agar plate assay. *Asian Journal of Pharmacy and Pharmacology*, *4*(1), 66–69.

Srivastava, P. K., Vaish, A., Dwivedi, S., Chakrabarty, D., Singh, N., & Tripathi, R. D. (2011). Biological removal of arsenic pollution by soil fungi. *Science of the Total Environment*, *409*(12), 2430–2442.

Subrata, B., & Datta, M. (2013). Evaluation of biological control agents against sheath blight of rice in Tripura. *Indian Phytopathology*, *66*(1), 77–80.

Swain, M. R., Mishra, J., & Thatoi, H. (2013). Bioethanol production from sweet potato (*Ipomoea batatas* L.) flour using co-culture of *Trichoderma* sp. and Saccharomyces cerevisiae in solid-state fermentation. *Brazilian Archives of Biology and Technology*, *56*(2), 171–179.

Tamimi, K. M., & Hadwan, H. A. (1985). Biological effect of *Neurospora sitophila* and *Trichoderma harzianum* on the growth of a range of sesamum wilt causing fungi in vitro. *Indian Phytopathology*, *38*(2), 292–296.

Tomah, A. A., Abd Alamer, I. S., Li, B., & Zhang, J. Z. (2020). A new species of *Trichoderma* and gliotoxin role: A new observation in enhancing biocontrol potential of *T. virens* against *Phytophthora capsici* on chili pepper. *Biological Control*, *145*, 104261.

Tsror, L., Barak, R., & Sneh, B. (2001). Biological control of black scurf on potato under organic management. *Crop Protection*, *20*(2), 145–150.

Vasanthakumari, M. M., & Shivanna, M. B. (2013). Biological control of anthracnose of chilli with rhizosphere and rhizoplane fungal isolates from grasses. *Archives of Phytopathology and Plant Protection*, *46*(14), 1641–1666.

Verma, J. P., Yadav, J., Tiwari, K. N., & Jaiswal, D. K. (2014). Evaluation of plant growth promoting activities of microbial strains and their effect on growth and yield of chickpea (*Cicer arietinum* L.) in India. *Soil Biology and Biochemistry*, *70*, 33–37.

Woo, S. L., Ruocco, M., Vinale, F., Nigro, M., Marra, R., Lombardi, N., et al. (2014). *Trichoderma*-based products and their widespread use in agriculture. *The Open Mycology Journal*, *8*(1), 71–126.

Yazdani, M., Yap, C. K., Abdullah, F., & Tan, S. G. (2009). *Trichoderma atroviride* as a bioremediator of cu pollution: An in vitro study. *Toxicological & Environmental Chemistry*, *91*(7), 1305–1314.

Yoshioka, Y., Ichikawa, H., Naznin, H. A., Kogure, A., & Hyakumachi, M. (2012). Systemic resistance induced in *Arabidopsis thaliana* by *Trichoderma asperellum* SKT-1, a microbial pesticide of seedborne diseases of rice. *Pest Management Science*, *68*(1), 60–66.

Zhang, F., Bunterngsook, B., Li, J., Zhao, X., Champreda, V., Liu, C., et al. (2019). Regulation and production of lignocellulolytic enzymes from *Trichoderma* reesei for biofuels production. *Advances in Bioenergy*, *4*, 79–119.

Zhang, F., Yuan, J., Yang, X., Cui, Y., Chen, L., Ran, W., et al. (2013). Putative *Trichoderma harzianum* mutant promotes cucumber growth by enhanced production of indole acetic acid and plant colonization. *Plant and Soil*, *368*(1–2), 433–444.

Chapter 4

Current status and future prospects of entomopathogenic fungi: A potential source of biopesticides

Julius Rajula[a], Sengodan Karthi[b], Sonia Mumba[c], Sarayut Pittarate[a], Malee Thungrabeab[d], and Patcharin Krutmuang[a]

[a]*Department of Entomology and Plant Pathology, Faculty of Agriculture, Chiang Mai University, Chiang Mai, Thailand,* [b]*Department of Biochemistry, Centre for Biological Sciences, K.S. Rangasamy College of Arts and Science (Autonomous), Tiruchengode, Tamil Nadu, India,* [c]*Department of Agriculture, Livestock Development and Fisheries, Directorate of Fisheries, County Government of Kilifi, Kilifi, Kenya,* [d]*Rajamangala University of Technology Lanna, Chiang Mai, Thailand*

Chapter outline

1 Introduction

Agriculture is the main source of food and plays a crucial role in the economy of the world. However, from the onset, agriculture has been on the receiving end as

Recent Advancement in Microbial Biotechnology. https://doi.org/10.1016/B978-0-12-822098-6.00013-6

far as pests and diseases are concerned. Fungi, weeds, and insects have been identified as a severe problem of agriculture. These have contributed to tremendous yield losses for centuries. Fortunately, the advent of chemical pesticides helped in agricultural productivity, but it had a number of drawbacks (Gupta & Dikshit, 2010). Moreover, the world's population is expanding rapidly, which has greatly increased the demand for food. Therefore, it is important to implement effective pest management measures to protect the crop from pests and diseases. In the past, there has been an overdependence on synthetic chemicals, which has had a negative effect on the soil, nontarget organisms, and humans who interact with them (Aktar, Sengupta, & Chowdhury, 2009; Gupta & Dikshit, 2010). Various factors, such as resistance to insecticide, a resurgence of pests, a negative impact on biodiversity, etc., have forced scientists to conduct various studies to develop better and sophisticated pest control strategies. Additionally, the environment has been adversely affected by synthetic chemicals used to control diseases and pests in agricultural farms and forests. These toxic pollutants eventually contaminate the aquatic environment resulting in a devastating effect on the ecosystem. The potential of biopesticides for sustainable agriculture has been known for several decades. Besides, organic farming is becoming very popular due to the growing demand for safe and healthy food and environmental protection; hence, the use of biopesticides is inevitable (Kumar, 2012). Today, the world is focusing more on sustainability, which leads to the creation of a pollution-free and safe environment. It should certainly be appreciated that the concept of microbial control was conceived many centuries ago. Studies on the diseases of silkworm *Bombyx mori* are some of the works that have largely contributed to this ideology. Notably, entomopathogenic fungi play a vital role in several biopesticide markets. For example, it is documented that almost half of the biopesticides registered in Brazil are mycoinsecticides and mycoacaricides, which comprise hypocrealean fungi (Moura et al., 2019). Additionally, methods of using entomopathogenic fungi in combination with synthetic chemicals are still being explored. Recently, a study was conducted to understand the compatibility of *Beauveria bassiana* with synthetic fungicides, and the results indicate that some of these synthetic chemicals may work seamlessly with entomopathogenic fungi, a characteristic that is critical in integrated pest management (IPM) (Krutmuang, Pittarate, Rahman, Rajula, & Thungrabeab, 2019). It has been established through various studies that it is difficult to get rid of some notorious insects using a single strategy, which makes it necessary to use a combined strategy. The use of entomopathogenic fungi in IPM is also important because it is less expensive and has no or minimal impact on the ecosystem (Skinner, Parker, & Kim, 2014).

2 Entomopathogenic fungi

Entomopathogenic fungi (EPF) are mostly isolated from insect cadavers but are natural inhabitants of the soil (Litwin, Nowak, & Ro, 2020). Naturally, these microorganisms are endowed with the role of controlling the insect population in the ecosystem. However, the constant use of chemical fungicides, herbicides, and pesticides has resulted in a huge decline in the microbial population (Gilbert & Gill, 2010). This happened due to growth suppression, alterations in their metabolic trail, and damages in the cell structures due to the nonylphenols and other toxic surfactants used in the synthetic chemicals (Różalska, Glińska, & Długoński, 2014). Eventually, as a result of such adverse effects, these microbes cannot play their role in the ecosystem. Nevertheless, some EPF have developed survival strategies to overcome these pollutants' adverse effects (Szewczyk, Kuśmierska, & Bernat, 2018). EPF have several advantages, such as being biologically safe and can be cultivated on a large scale (Barta et al., 2019). These fungi can attack insects through tissues, not necessarily through the mouth (like bacteria or viruses). Additionally, they can infect insects' entire life cycle (Srinivasan, Sevgan, Ekesi, & Tamò, 2019). Moreover, most EPF overwinters and persevere on the soil for several seasons as they continue to attack their target insects (Rajula, Rahman, & Krutmuang, 2020; Tanada & Kaya, 2012). Further studies have shown that EPF can induce systemic resistance in plants that can help fight against a number of biotic pressures, including, but not limited to, pathogens such as phytoparasitic nematodes and abiotic stresses such as salinity (Bamisile, Dash, Akutse, Keppanan, & Wang, 2018).

3 Some of the current commercialized entomopathogenic fungi-based biopesticides

Biopesticides are mass-produced agents manufactured from living microorganisms or natural extracts to control plant pests (Chandler et al., 2011). Currently, more than 1000 species of EPF that are sourced from more than 100 genera are known and being explored for their potentiality of commercialization as biopesticides. These infect a variety of insect orders, and their utilization in biocontrol has improved in recent years (Barra-bucarei, Iglesias, & Torres, 2020; Shah et al., 2009). It has also been reported that most EPF species belong to the group Hypocreales (Ascomycota) and Entomophthoromycota. Even though they are dissimilar in some characteristics, both groups produce conidia or other asexual spores, which are the infectious elements (Barra-bucarei et al., 2020).

4 Entomopathogenic fungi on insect cadavers from the field and laboratory

Fig. 1A–D show various examples of entomopathogenic fungi on insect cadavers that were collected during a workshop on Survey, Isolation, and Taxonomy of Insect Fungi held between 17th and 20th September 2018 at Khao Yai National park, Nakhon Ratchasima, Thailand. The protocols for isolation and identification were followed by Thungrabeab, Blaeser, and Sengonca (2006), Luangsa-ard, Tasanathai, Mongkolsamrit, and Hywel-Jones (2007, 2008, 2010), Krutmuang et al. (2019), and Kepler et al. (2017)

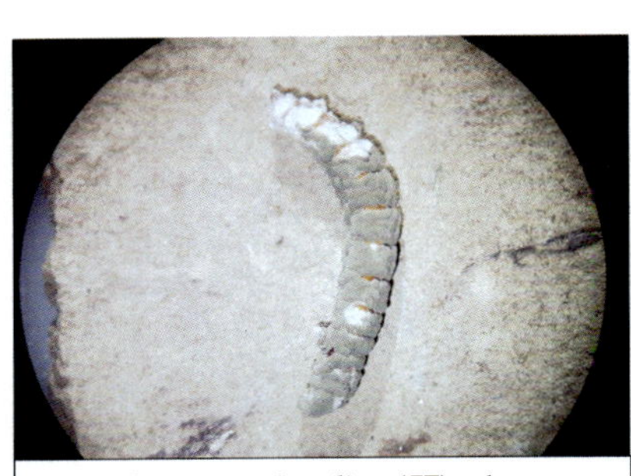

Metarhizium anisopliae **(57) v1 on mealworm: O. Coleoptera**

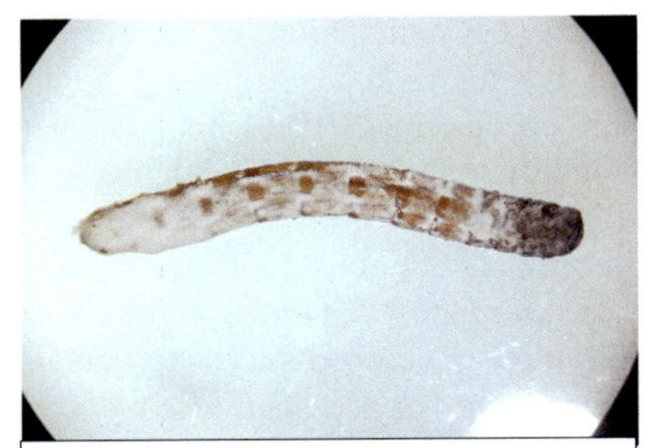

Beauveria bassiana **(35) v1 on mealworm: O. Coleoptera**

Metarhizium anisopliae **on cat flea: O. Siphonaptera**

Metarhizium cf flavoviride **(59) v1 on forest cockroach: O. Blattodea**

Beauveria bassiana **on termite: O. Isoptera**

Beauveria bassiana **on fruit fly: O. Diptera**

FIG. 1 (A–D) Various entomopathogenic fungi on various insect cadavers that were collected during a workshop on Survey, Isolation, and Taxonomy of Insect Fungi held between 17th and 20th September 2018 at Khao Yai National park, Nakhon Ratchasima, Thailand. *(Photo credit: Pittarate.)*

(Continued)

***Cordyceps loeiensis* (29) v4 on fly: O. Diptera**

***Cordyceps tuberculata* (35) on moth: O. Lepidoptera**

***Isaria javanica* (53) v1 on ant: O. Hymenoptera**

***Beauveria bassiana* on ant: O. Hymenoptera**

***Gillula pulchra* (39) v1 on spider**

***Ophiocordyceps pseudolloydii* (63) v2 on ant: O. Hymenoptera**

FIG. 1, CONT'D

(Continued)

Hypocrella discoidea (43) v1 on scale insect **O.** Hemiptera

Cordyceps gryllotalpidicola (33) v4 on beetle worm **O.** Coleoptera

Ophiocordyceps dipterigena (59) v2 on fly **O.** Diptera

Gibellula mirabilis (41) v4 on spider

Ophiocordyceps unilateralis (69) v1 on ant: **O.** Hymenoptera

Akanthomyces pistillariiformis (19) v2 on Caterpillar **O.** Lepidotera

FIG. 1, CONT'D

(Continued)

Isaria farinosa (51) v1 on Caterpillar O. Lepidoptera

Ophiocordyceps pseudolloydii (63) v2 On Ant O. Hymenoptera

Gibellula unica (43) v3 on Spider

Nomuraea atypicola (61) v1 on Spider

Metarhizium anisopliae on Tiger Centipede

Aschersonia marginata on Scale insect O. Hemiptera

FIG. 1—CONT'D

5 The most utilized entomopathogenic fungi as biopesticides

Although the number of entomopathogenic fungi that are characterized and identified is quite enormous, not all have been utilized efficiently to control the insect pests. This is because several experimental validations are required for the development of biopesticides. In this regard, two EPF are at the top of the list. The two EPF that have been used effectively are *Beauveria bassiana* and *Metarhizium anisopliae*. However, with the increased campaign toward green technologies in controlling pests and diseases, researchers are now focusing on the developments of other EPF-based biopesticides that are already known to have pest control activities (de Faria and Wraight, 2007; Mcguire and Northfield, 2020).

5.1 *Beauveria bassiana*

In the 19th century, the Italian scientist Agustino Bassi Bombex identified the *Beauveria bassiana* (Bals.-Criv) Vuill, nearly 30 years of research on white muscardine disease in *Bombyx mori* (Längle, 2006). Currently, *Beauveria bassiana* and *Metarhizium anisopliae* are the most utilized entomopathogenic fungi in the biopesticide industry. The advantage is that *Beauveria bassiana* can be applied like chemical pesticides and works very effectively. Several studies have proved that it can reduce various insect populations by 35%–75%, which goes a long way in managing insect pests in an ecofriendly manner (Brust, Wakil, & Qayyum, 2018). *Beauveria bassiana* is one of the efficient EPF that has been used to control a lot of insects worldwide. This fungus has been used to control over 60 insects, mostly from the forested areas, crop pests, and postharvest pests such as *Artona funeralis* (bamboo zygaenid), *Pyrrhalta aenescens* (green elm leaf beetle), *Lymantria xylina* (coast oak tussock moth), *Nephotettix cincticeps* (common rice leafhopper) *Nilaparvata lugens* (brown planthopper), and *Carposina nipponensis* (small peach borer) among others. Additionally, records show that the malaria vector is also prone to infection by this fungus. Although adult mosquitoes were used extensively in these studies, recent works also show the potential to control the larval stages (Bukhari, Takken, & Koenraadt, 2011). Studies are still ongoing on more insects that are susceptible to the white muscardine disease (Li et al., 2010). Interestingly, this fungus is found in most soils in most parts of the world and most climatic regions (Dannon et al., 2020; Zimmermann, 2007). Besides, *Beauveria bassiana* has been found to tolerate higher temperatures, which allows them to be effective in warm temperate regions. In Syria, some isolates isolated from the warm areas were able to grow in temperatures between 20°C and 35°C and were found to induce between 31% and 100% mortality to insects. In that study, two isolates that were characterized through molecular studies as *Beauveria bassiana* caused insect death by over 50% within 2 days (Alali et al., 2019; Song et al., 2019). Some studies have reported that consumption of *Beauveria bassiana* positively impacts the immune system in humans (Dannon et al., 2020).

5.1.1 *Mode of action of* Beauveria bassiana

This fungus is known to strike its target percutaneously like any other EPF. Firstly, it attaches to the cuticle and then germination ensues. The process proceeds to the penetration through the cuticle by extracellularly manufactured enzymes such as chitinases and proteases that break down components of the cuticle. The fungus then invades the host hemolymph and produces toxins and metabolites that help to get rid of the host immune system. Finally, after proliferation and subduing the host, sporulation occurs, and mycelial growth can be observed by the white muscardine disease on the host cadaver (Gabarty, Salem, Fouda, Abas, & Ibrahim, 2014; Keswani, Singh, & Singh, 2013).

5.1.2 *Mass production of* Beauveria bassiana

Ideally, to use any EPF effectively in controlling insect pests in agriculture and forestry, we need to ensure its mass production. Additionally, this mass-produced fungus should be cost-effective (Kassa et al., 2008). Fortunately, *Beauveria bassiana* being a facultative fungus can grow on disparate substrates such as molasses, potato, rice, millet, whey, and sorghum. Notably, the fungus produces a significant number of spores in most of these substrates. Also, most of these products are readily available or can be acquired easily at a lower cost (Abraham, Easwaramoorthy, & Santhalakshmi, 2003; Kassa et al., 2008). Several technologies have been developed to ensure that large numbers of this fungus are produced in the right formulations that are easy to use, to meet the increasing demand. In China, some companies developed large liquid and solid fermenters for this purpose and delivered large volumes required in the agricultural industry. These technologies and mechanization of the production process have been adopted in various parts of the world (Li et al., 2010). Some of the mass-produced products of *Beauveria bassiana* are enumerated in Table 1.

5.2 *Metarhizium anisopliae*

The fungus *Metarhizium anisopliae* was first identified and named by E. Metschnikoff as *Entomophthora anisopliae* in 1879. However, during his studies, Sorokin named the green muscardine disease Metarhizium, therefore, changing the genus name from Entomophthora to Metarhizium, *Metarhizium anisopliae* (Metsch.) Sorokin. After successful experiments on beetroot weevils and white grubs of leaf chafer between the years 1884 and 1888, the fungus was recognized as the introductory biological control agent (Zimmermann, 1993). The narrow host range and environmentally friendly nature, easy attachment to host insects, combined with simple mass production techniques, have made *Metarhizium anisopliae* an excellent EPF worldwide (Aw & Hue, 2017). *Metarhizium anisopliae* occurs worldwide and has been isolated from all kinds of soils and insect cadavers. In fact, it is known to be part of the soil flora in almost every

TABLE 1 **Table of biopesticides formulated from entomopathogenic fungi and their target insects.**

Fungi	Commercial product/ biopesticide	Target insect	Country formulating/ utilizing
Verticillium lecanii	Vertalec	Aphids	The Netherlands
Metarhizium anisopliae	Biogreen	Scarab larvae on pasture	Australia
V. lecanii	Mycotal	Whitefly and thrips	The Netherlands
Metarhizium anisopliae	Cobican	Sugarcane spittlebug	Venezuela
Beauveria bassiana	Conidia	Coffee-berry borer	Colombia
Beauveria bassiana	Ostrinil	Corn-borer	France
Beauveria bassiana	CornGuard	European corn-borer	United States
Beauveria bassiana 447	Baits Motel Stay-awhile	Ants	United States
Beauveria bassiana ATCC 74040	Naturalis L	Various insects	United States
Beauveria bassiana GHA	Mycotrol ES Mycotrol O BotaniGard 22WP BotaniGard ES	Various insects	United States
Metarhizium anisopliae	Bio-Path	Cockroaches	United States

Metarhizium anisopliae	Bio-Blast	Termites	United States
Beauveria bassiana HF23	balEnce	House fly	United States
Metarhizium anisopliae F52	Tick-Ex	Ticks and grubs	United States
Isaria fumosorosea Apopka 97	PFR-97	Whitefly and thrips	United States
Beauveria bassiana	Mycotrol GH	Grasshoppers, locusts	United States
Beauveria bassiana	Mycotrol WP and BotaniGard	Whitefly, aphids, thrips	United States
Beauveria bassiana	Naturalis-L	Cotton pests including bollworms	United States
I. fumosorosea	PFR-97	Whitefly	United States
Beauveria bassiana	Proecol	Army worm	Venezuela
Beauveria bassiana	Boverin	Colorado beetle	Former USSR
Beauveria bassiana	Boverol	Colorado beetle	Czechoslovakia
Beauveria bassiana	Boverosil	Colorado beetle	Czechoslovakia
Beauveria brongniartii	Engerlingspilz	Cockchafers	Switzerland
Beauveria brongniartii	Schweizer	Cockchafers	Switzerland
Beauveria brongniartii	Melocont	Cockchafers	Austria
Metarhizium flavoviride	Green Muscle	Locusts, grasshoppers	United Kingdom
Beauveria bassiana ATCC 74040	Naturalis L	Thrips, whitefly, mites	United Kingdom

Continued

TABLE 1 Table of biopesticides formulated from entomopathogenic fungi and their target insects—cont'd

Fungi	Commercial product/ biopesticide	Target insect	Country formulating/ utilizing
Beauveria bassiana GHA	Botanigard	Whiteflies, aphids, thrips	United Kingdom
Lecanicillium muscarium (Ve6) former *(Verticillium lecanii)*	Mycotal Vertalec	Whiteflies, thrips, aphids (except the Chrysanthemum aphid: *Macrosiphoniella sanborni*)	United Kingdom
I. fumosorosea Apopka 97	Preferal WG	Greenhouse whiteflies *(Trialeurodes vaporariorum)*	United Kingdom
I. fumosorosea Fe9901	Nofly	Whiteflies	United Kingdom
I. fumosorosea	Pae-Sin	Whitefly	Mexico
Beauveriabassiana	Conidia	Coffee berry borer	Germany
	BroadBrand	*Plutella xylostella, Thaumatotibia leucotret, Aonidiella aurantii, Tetranychus urticae, Phthorimaea opercullela*, stinkbug, thrips, whiteflies	
	Ostrinil	*Paysandisia archon* and *Ostrinia nubilali*	France
	Ballvéria	*Bemisia tabaci* biotype B	
	Nagestra	Root borers, caterpillars, sucking pests, locusts, Colorado potato beetles	India
	Beauvitech-WP	Whiteflies, thrips, and aphids	Kenya
	Bb-Protec	Whiteflies, herbivorous mites, wireworms	Switzerland

	Racer	Rice leaf folder, *Helicoverpa armigera*, *Spodoptera litura*, loopers, leaf-eating caterpillars, mealy bugs, coffee berry borers, fruit borers, cotton bollworm, root grubs, surface living larvae and nymph	India
	Daman	Root borers, caterpillars, sucking pests, locusts, Colorado potato beetle	India
Beauveria brongniartii	Bas-Eco	*Helicoverpa armigera,* berry borer, root grubs	India
Hirsutella thompsonii	No-Mite	Spider mites	India
	Mycohit	Acari	India
I. fumosorosea	Nofly	Whitefly	Spain
Metarhizium anisopliae	ICIPE 69[a]	Western flower thrips and bean flower thrips	Kenya
Metarhizium anisopliae	ICIPE 20[a]	Pea leaf miner	Kenya
Metarhizium anisopliae	ICIPE 78[a]	Spider mites	Kenya
Metarhizium anisopliae	ICIPE 62[a]	*Aphis gossypii; Brevicoryne brassicae; Lipaphis pseudobrassicae*	Kenya
Beauveria bassiana	CPD 9[a]	Pod sucking bug and pod borer	Nigeria
Metarhizium anisopliae	CPD 5[a] and CPD 12[a]	Pod sucking bug and pod borer	Nigeria
Beauveria bassiana	CPD 11[a]	Cowpea aphid	Nigeria
Metarhizium anisopliae	CPD 4[a] and CDP 5[a]		Nigeria
Beauveria bassiana	Myco Jaal	*Plutella xylostella* and *Pieris brassicae*	India
Beauveria bassiana	Bba5653[a]	*Plutella xylostella*	Benin

Continued

TABLE 1 Table of biopesticides formulated from entomopathogenic fungi and their target insects—cont'd

Fungi	Commercial product/ biopesticide	Target insect	Country formulating/ utilizing
Beauveria bassiana	Bb Plus	Thrips	South Africa
	Bb weevil	Weevil	South Africa
	Sparticus	Whiteflies	South Africa
Metarhizium anisopliae subsp. *acridum* IMI 330189	Green Muscle	Locust	South Africa
Beauveria bassiana GHA	Biopower	Aphid, diamond black moth	Kenya
	BotaniGard	Sucking insect pests	Kenya
Beauveria bassiana		*Monochamus alternatus* and *Dendrolimus punctatus*	China
Metarhizium anisopliae		Cockroaches, locusts, and grasshoppers	China
Conidiobolus thromboides	Vektor 25SL	Aphids	China
Beauveria bassiana	Biosoft ATEC Beauveria Larvo-Guard Biorin Biolarvex Biogrubex Biowonder Veera Phalada 101B Bioguard Bio-power	Coffee berry borer, diamondback moth, thrips, grasshoppers, whiteflies, aphids, codling moth	India

Metarhizium anisopliae	ABTEC Verticillium Meta-Guard Biomet Biomagic Meta Biomet Sun Agro Meta Bio-Magic	Coleoptera and Lepidoptera, termites, mosquitoes, leafhoppers, beetles, grubs	India
I. fumosorosea	Priority	Whitefly	India
I. lilacinus	Yorker ABTEC Paceilomyces Paecil Pacihit ROM biomite Bio-Nematon	Whitefly	India
Verticillium lecanii	Verisoft ABTEC Verticillium Vert-Guard Bioline Biosappex Versitile Ecocil Phalada 107 V Biovert Rich ROM Verlac ROM Gurbkill	Whitefly, coffee, green bug, homipteran pests	India

Continued

TABLE 1 Table of biopesticides formulated from entomopathogenic fungi and their target insects—cont'd

Fungi	Commercial product/ biopesticide	Target insect	Country formulating/ utilizing
	Sun Agro Verti Bio-Catch		
Beauveria bassiana	BotaniGard (imported) Ceremoni (imported)	Thrips, greenhouse whitefly, two-spotted spider mite	South Korea
I. fumosorosea	Bangsili	Two-spotted spider mite, greenhouse whitefly	South Korea
Beauveria bassiana fungus	Boverin	Insect pests, larvae of Colorado potato beetle	Russia
Beauveria bassiana GHA and Bt	Bitoxibacillin	Colorado potato beetle	Russia
Beauveria bassiana	Boverin	Colorado potato beetle, thrips, and various insects	Ukraine
Entomophthora spp. (spores and toxins)	Mycoaphidin	Pea aphid, peach aphids, and other aphids	Ukrainc
Metarhizium sp.	Metarilin	Soil stages of hard wings insects	Ukraine
Paecilomyces spp.	Pecilomin	Larvae of various insects	Ukraine
Verticillium lecanii	Verticilin	Flying insects	Ukraine
Verticillium lecani	*Verticilina granulara*-BL	*Trialeurodes vaporariorum*	Maldova

Beauveria bassiana	Biagro Bb-Vinchuca (imported) Biagro Bb-mosca	*Triatoma infestans, Musca domestica*	Argentina
Beauveria bassiana	Boveril PL 63	Coleoptera (Curculionidae), Acari (Tetranychidae)	Brazil
Metarhizium anisopliae	Biotech Metarril E9 Metarril 1037 Metarriz Methavida	Hemiptera (Cercopidae), Acari (Ixodidae)	Brazil
Sporothrix insectorum	No commercial name	Hemiptera (Tingidae	Brazil
Metarhizium anisopliae	Metaquino	Spittlebugs	Brazil
Beauveria bassiana MB-1 (INISAV 182-04)	Bibisav-2	*Atta insularis, Acromyrmex octospinosus, Attamyces bromatificus* (Formicidae)	Cuba
Beauveria bassiana LBB-1 (INISAV 180-04)	Basisav-1	*Cosmopolites sordidus* (banana weevil), *Cylas formicarius* (weevil), *Lissorhoptrus brevirostris* (aquatic weevil), *Pachnaeus* spp. (Curculionidae), *Thrips palmi, Diatraea saccharalis, Hypothenemus hampei, Diabrotica balteata, Pseudacysta perseae, Lagochirus dezayasi, Corythucha gossypii, Tipophorus nigritus, Phyllophaga* spp.	Cuba
Metarhizium anisopliae LBM-11 (INISAV 178-05)	Metasav-11	*Lissorhoptrus brevirostris* (Aquatic weevil), *Mocis* spp. (Lepidopteran larvae), *Prosapia bicincta, Cosmopolites sordidus* (banana weevil), *Tagosodes oryzicola, Oebalus insularis, Spodoptera* spp., *Spodoptera* spp., *Pachnaeus litus, Thrips palmi, Plutella xylostella, Hypothenemus hampei, Diabrotica balteata* (Curculionidae)	Cuba
Verticillium lecanii Y-57 (INISAV 179-05 and 180-05)	Vertisav-57	*Bemisia tabaci* (white fly), *Bemisia argentifolia, Frankliniella* spp., *Aleurotracholus tracheoides, Aphis gossypii, Myzus persicae, Lipaphis erizini, Brevicoryne brassicae, Thrips palmi, Bophilus microplus*	Cuba

Continued

TABLE 1 Table of biopesticides formulated from entomopathogenic fungi and their target insects—cont'd

Fungi	Commercial product/ biopesticide	Target insect	Country formulating/ utilizing
Beauveria bassiana HF23	Balance	House flies	Canada
Beauveria bassiana GHA	BotaniGard	Aphid, leafhoppers and planthoppers, mealybug, psyllid, scarab, beetle, thrips, weevils, whitefly	Canada
Metarhizium anisopliae F52	Met52	Black vine weevil	Canada
Metarhizium anisopliae	BioCane Granules	Grey-backed cane grub (scarabs)	Australia
Metarhizium anisopliae subsp. acridum	Green Guard	Locusts and grasshoppers	Australia
Metarhizium flavoviride	Chafer Guard	Redheaded pasture cockchafer	Australia

[a]*Isolates that have been tested but not yet designated for commercial application.*
Courtesy: Butt, T. M., Jackson, C., & Magan, N. (2001). Fungal biological control agents—Appraisal and recommendations. In *Fungi as biocontrol agents* (2001) (pp. 377–384); Ruiu, L. (2018). Microbial biopesticides in agroecosystems. *Agronomy*, *8*(11), 235. doi:10.3390/agronomy8110235; Srinivasan, R., Sevgan, S., Ekesi, S., & Tamò, M. (2019). Biopesticide based sustainable pest management for safer production of vegetable legumes and brassicas in Asia and Africa. *Pest Management Science*, (April). doi:10.1002/ps.5480; Kabaluk, J. T., Svircev, A. M., Goettel, M. S., & Woo, S. G. (2010). The use and regulation of microbial pesticides in representative jurisdictions worldwide. International Organization for Biological Control of Noxious Animals and Plants (IOBC); Maina, U. M., Galadima, I. B., Gambo, F. M., & Zakaria, D. (2018). A review on the use of entomopathogenic fungi in the management of insect pests of field crops. *Journal of Entomology and Zoology Studies*, *6*(1), 27–32.; Mascarin, G. M., & Jaronski, S. T. (2016). The production and uses of Beauveria bassiana as a microbial insecticide the production and uses of Beauveria bassiana as a microbial insecticide. *World Journal of Microbiology and Biotechnology*, (November). doi:10.1007/s11274-016-2131-3.

climatic condition or region (Bischoff et al., 2009). Faria and Wraight reported the wide use of *Metarhizium anisopliae* in the manufacture of mycoinsecticides and mycoacaricides (de Faria and Wraight, 2007; Wraight, Ugine, Ramos, & Sanderson, 2016). *Metarhizium anisopliae* has been the solution to spittlebugs in Brazil's sugarcane plantations. The sugarcane mills and private companies have mass-produced formulations of this fungus for the control of this pest that had been disastrous to the industry (Iwanicki et al., 2019).

5.2.1 *Mode of action of* Metarhizium anisopliae

The blend of cuticle-degrading enzymes and the mechanical pressure developed by *Metarhizium anisopliae* plays a significant role in gaining entry into the host through the cuticle (Barra-bucarei, Vergara, & Cortes, 2016). *Metarhizium anisopliae* is known to possess the *Mad1* and *Mad2* genes, which aides in the adhering to the cuticle of the targeted host and thereafter forming a germ tube as instigated by the external sources of nitrogen and carbon from the insect. During the germination period, the enzyme trehalase would be observed. The spores would then bulge to differentiate into appressoria, where the Mpl1 and ODC1 genes have been recognized to be responsible for the formation of appressorium. After the appressoria formation, the fungus proceeds to the penetration stage, where it produces several proteins such as carboxypeptidases, subtilisins, chymotrypsins, and trypsins. These proteins break down the procuticle of the target host. Once inside the host, the fungus colonizes the host and produces destruxins that suppress the insect's immune system. The host hemocytes would then cover the spores to escape any attack aimed at decimating them. Finally, sporulation occurs, and extrusion of the hyphae is observed when the green mycelium is formed on the carcass of the insect (Aw & Hue, 2017; Hubbard, Hynes, Erlandson, & Bailey, 2014).

5.2.2 *Mass production of* Metarhizium anisopliae

The species and substrate greatly impact the preparation of large amounts of fungal formulation (Anitha et al., 2015). The mass production of *Metarhizium anisopliae* is cost-effective, and the propagules can stay longer on the shelves. This fungus has been demonstrated to proliferate on various substrates, such as chickpea, sorghum, corn, pigeon pea, groundnut, rice, and green gram etc. Moreover, it grows easily on semisynthetic media such as carrot malt agar, Potato Dextrose Agar (PDA), and Sabouraud Maltose Agar with yeast extract (SMAY) (Agale et al., 2018). Specifically, *Metarhizium anisopliae* needs carbon and nitrogen to sporulate; hence, it would easily germinate and grow if the two nutrients are available. It has been observed that the optimized fermentation process can harvest massive amounts of conidia on rice substrate (Anitha et al., 2015; Prakash, Padmaja, & Kiran, 2008). Also, the temperature is very key to the production of conidia in rice substrates. It was found that higher numbers of conidia formation are observed at an average temperature of 25°C compared to a lower temperature of 20°C and below (Prakash et al., 2008). Different types of mycoinsecticides formulation exist, including contact powder, granules, wettable powder, suspensions, oil dispersion, and oil miscible flowable concentrates (de Faria and Wraight, 2007). In order to maintain the minimum production cost, this fungus's multiplication has been carried out on solid substrates that are easy and freely developed using available and cheap organic matter. Table 1 provides a number of specific formulations of *Metarhizium anisopliae* that have been developed around the world for a variety of insect pests.

6 The future of entomopathogenic fungi-based biopesticides

In the future, the entomopathogenic fungus should be used more widely to control pathogenic insects that generally destroy crops and plants. It has been documented that insects can develop resistance toward chemicals that are employed for their control. They have developed the resistance using various approaches, such as remodeling their cuticle over time to prevent the chemical from easily penetrating the cell, modifying their metabolic system to rapidly activate detoxifying enzymes such as cytochrome P450 esterase and monooxygenase, and performing mutational changes that have conferred knockdown resistance (Dang, Doggett, Singham, & Lee, 2017). Biopesticides use multipronged and all-inclusive modes of action, which carry the benefits of deferring the development of resistance by pests. The increased demand for organic farm products, mostly for export, is a great instigator for developing biopesticides (Gupta & Dikshit, 2010). Hubbard et al. (2014) reported that the use of biopesticides had been steadily increasing by about 10% annually (Hubbard et al., 2014). Moreover, research on the newly isolated entomopathogenic fungi and their potential for inclusion in the biopesticide industry is expanding day by day. Collected insect cadavers revealed the presence of disparate fungal pathogens that can be further explored for their efficacy and commercialization.

7 Studies on the compatibility of entomopathogenic fungi with other insecticides for IPM

IPM is the containment of crop pests and diseases through employing minimal use of chemicals but encouraging the use of natural/biological interventions (Bajwa & Kogan, 2002). Entomopathogenic fungi are naturally occurring microorganisms that do not pose any threat to the ecosystem and are considered ideal candidates for the IPM strategies. Therefore, in order to incorporate EPF in IPM, it is necessary to study their compatibility with other strategies involved. Notably, several studies have been conducted to enhance the efficiency of EPF against target hosts by combining it with plant extracts. Such studies are important as they will involve naturally occurring organisms. This is advantageous because they are locally available, hence the cost would be lower (Fernández-Grandon, Harte, Ewany, Bray, & Stevenson, 2020). The combination of *Beauveria bassiana* and *Metarhizium anisopliae* with plant extracts (neem and eucalyptus) has been reported to increase aphid mortality.

Specifically, eucalyptus combined with *Beauveria bassiana* caused 87% mortality to the aphids after 5 days of application. Interestingly, the prolificacy was also negatively affected when single and combined treatments of the plant extracts were used (Ali et al., 2018). Krutmuang et al. (2019) reported on the possibility of combining *Beauveria bassiana* with four fungicides in Thailand, and the results were impressive (Krutmuang et al., 2019). When Neves, Hirose, Tchujo, and Moino (2001) embarked on a compatibility study of three EPF, namely *Beauveria bassiana*, *Metarhizium anisopliae*, and *Isaria* sp. and neonicotinoid insecticides, and revealed that there was no effect on the germination of conidia unless higher concentrations were used. It was also evident that the use of the recommended dosage of the insecticides under the study did not negatively affect the conidia germination, production, and vegetative growth of the three EPF and were eventually recommended for IPM (Neves et al., 2001). These studies are a prerequisite before employing any kind of entomopathogenic fungi in integrated pest management. However, further research is necessary to explore the compatibility of EPF with natural products such as plant extracts. These will ensure the achievement of environmentally friendly products that the biopesticides seek to solve. In the future, therefore, the compatibility of EPF with other pesticides for IPM-related strategies should be studied in detail so that farmers and other stakeholders can be advised accordingly.

8 Some of the newly described entomopathogenic fungi

Entomologists, mycologists, taxonomists, and biologists are always on the lookout for newer and more potent entomopathogenic fungi around the world. New species of EPF are often reported. As a result, countless EPF always bring hope to the world. For a long time, *Beauveria bassiana*- and *Metarhizium anisopliae*-based biopesticides have been widely used against various pests (Chandler et al., 2011; Tupe, Pathan, & Deshpande, 2017). However, with the ongoing efforts in search of newer species of EPF, there is an optimism of getting more rigorous and potent EPF. Qu, Yu, Zhang, Han, and Zou (2018) recently discovered a new species *Ophiocordyceps ponerus* sp. nov. on soldier ant in China (Qu et al., 2018). Sanjuan et al. (2015) conducted molecular phylogenetic analysis on 35 samples from Colombia and Ecuador and described five new species from insect cadavers, namely, *Ophiocordyceps blattarioides*, *Ophiocordyceps tiputini*, *Ophiocordyceps araracuarensis*, *Ophiocordyceps fulgoromorphila*, and *Ophiocordyceps evansii*. It was also noted that the host and its habitat are fundamental components of identifying any new EPF (Sanjuan et al., 2015). Again, Xiao et al., in 2019, reported two new species from the family Ophiocordycipitaceae in Thailand. The species are *Ophiocordyceps globiceps* and *Ophiocordyceps sporangifera*, both of which are pathogenic to Coleoptera's larvae (Xiao, Hongsanan, Hyde, & Brooks, 2019). There are disparate species in this family and needs to be explored further for their possible formulations and use as biopesticides.

9 Mass production of entomopathogenic fungi-based biopesticides

For commercialization of EPF, we must ensure sustainable mass production of wettable powder, dry and liquid formulations. A study on formulated and unformulated potency of *Zoophthora radicans* against *Plutella xylostella* larvae noted that the formulated suspension was more effective than the unformulated one. The formulated fungus preserved the viability of the conidia for a longer period of time compared to the unformulated conidia. This encourages the formulation of EPF as they would stay longer without losing their efficacy (Batta, Rahman, Powis, Baker, & Schmidt, 2011).

Aerial conidia have proved to be the most effective propagules for EPF under normal environments. In this regard, most entomopathogenic hyphomycete fungi produce high numbers of small hydrophobic conidia in thick masses (Wraight, Jackson, & de Kock, 2001). Notably, the strong hydrophobic walls of many conidia of EPF help in conferring environmental stability that eventually increase their chances of staying longer on the shelves. However, research should be more focused on maintaining a high level of effectiveness of shelved formulations as opposed to the past reports that indicated low infection rates (Wraight et al., 2001). It is recommended to formulate EPF concentrations of 10^{13} spores. Several countries have produced aerial conidia of *Beauveria bassiana* using automated solid substrate technologies (Feng, Poprawski, & Khachatourians, 1994). Such technologies should be used more frequently in the future to formulate EPF that prefer solid substrates for commercialization purposes. However, it is known that some of these fungi can thrive more on liquid media. Although aerial conidia are preferred as mentioned earlier, hyphal bodies and submerged conidia have been reported to be just as virulent or even to a greater extent than the aerial conidia (Lacey, 1997; Lacey et al., 1999). It has become difficult to handle powdered formulations of fungi such as *Beauveria bassiana* and *Isaria fumosorosea* during wetting before spraying in the field. Fortunately, research has found both vegetable- and petroleum-derived oils to be intrinsically compatible with these fungi. This has made these products easier to handle. Now and in the future, more work is being done on UV persistent biopesticides so that when faced with the radiation in the field, they would not lose their efficacy (Wraight et al., 2001).

The recommendations that are required for yeast in determining their ability for commercialization are as follows:

(a) High and consistent efficacy
(b) Genetic stability
(c) Resistance to standard fungicides and other insecticides

(d) Nonproduction of secondary metabolites that may endanger human health
(e) Effectiveness against a wide range of pests
(f) Ability to endure adverse conditions
(g) Compatibility with other chemicals and physical treatments employed to the specific commodity
(h) The ability to grow on the affordable medium
(i) The stability of the formulation during storage

It is also important to note that formulations should have a fast growth rate, be unicellular with minimal size variation, osmotolerant, and temperature-stable (Mascarin & Jaronski, 2016).

10 Application of molecular technology in EPF-based biopesticides

With the recent advancements in science and molecular studies, it is clear that the effective production and utilization of EPF can be adapted to increase efficiency and sustainability in agriculture. Recombinant DNA techniques have been used to enhance the efficacy of fungal pathogens by inducing physical and chemical stimulation. It alters the habit of saprobic growth and induces the formation of appressorium, the infection structure that infects the host. Such advances are being further explored to increase the effectiveness of EPF so that they can be better alternatives to synthetic pesticides in all aspects (Leger & Screen, 2001). Essentially, molecular technologies should seek to enhance the following issues in EPF-based biopesticides:

(a) Decreasing the inoculum rate.
(b) Changing persistence.
(c) Improving mortality rate to the host.
(d) Restricting/expanding the specificity of the targeted host.
(e) Improving tolerance to the adverse environment.
(f) Enhancing resistance to fungicides.

11 Conclusion

Biopesticides based on entomopathogenic fungi are environmentally friendly and advantageous in many ways for controlling insect pests. Currently, there is a lot of work conducted on the identification and characterization of the entomopathogenic fungi. However, it is crucial to look for new and more potential EPF strains and produce commercial products from already known strains that have already proven effective against various insects. It is also encouraging to note that biotechnology is being applied in this effort; however, it should be more rigorous. The future of EPF-based biopesticides depends on the combined efforts of various scientists and other stakeholders.

References

Abraham, T., Easwaramoorthy, S., & Santhalakshmi, G. (2003). Mass production of *Beauveria bassiana* isolated from sugarcane root borer, *Emmalocera depresella* Swinhoe. *Sugar Tech, 5*(4), 225–229.

Agale, S. V., Gopalakrishnan, S., Ambhure, K. G., Chandravanshi, H., Gupta, R., & Wani, S. P. (2018). Mass production of entomopathogenic fungi *Metarhizium anisopliae* using different grains as a substrate. *International Journal of Current Microbiology and Applied Sciences, 7* (1), 2227–2232.

Aktar, W., Sengupta, D., & Chowdhury, A. (2009). Impact of pesticides use in agriculture: Their benefits and hazards. *Interdisciplinary Toxicology, 2*(1), 1–12. https://doi.org/10.2478/v10102-009-0001-7.

Alali, S., Mereghetti, V., Faoro, F., Bocchi, S., Al Azmeh, F., & Montagna, M. (2019). Thermotolerant isolates of *Beauveria bassiana* as potential control agent of insect pest in subtropical climates. *PLoS One, 14*(2), 1–13.

Ali, S., Farooqi, M. A., Sajjad, A., Ullah, M. I., Qureshi, A. K., Siddique, B., … Asghar, A. (2018). Compatibility of entomopathogenic fungi and botanical extracts against the wheat aphid, *Sitobion avenae* (Fab.) (Hemiptera: Aphididae). *Egyptian Journal of Biological Pest Control, 28* (1), 4–9. https://doi.org/10.1186/s41938-018-0101-9.

Anitha, S., Das, S. S. M., & Thivya, S. (2015). Mass production of *Metarhizium anisopliae* (Mets.) dust formulation using Jack fruit seed powder and testing its pathogenecity against *Corcyra cephalonica* Stainton. *European Journal of Biotechnology and Bioscience, 3*(10), 1–4.

Aw, K. M. S., & Hue, S. M. (2017). Mode of infection of *Metarhizium* spp. Fungus and their potential as biological control agents. *Journal of Fungi, 3*(2). https://doi.org/10.3390/jof3020030.

Bajwa, W. I., & Kogan, M. (2002). *Compendium of IPM definitions (CID)-What is IPM and how is it defined in the worldwide literature?* (pp. 1–14). IPPC Publication. 998.

Bamisile, B. S., Dash, C. K., Akutse, K. S., Keppanan, R., & Wang, L. (2018). Fungal endophytes: Beyond herbivore management. *Frontiers in Microbiology, 9*(Mar), 1–11. https://doi.org/10.3389/fmicb.2018.00544.

Barra-bucarei, L., Iglesias, A. F., & Torres, C. P. (2020). *Entomopathogenic fungi.* December 2019. https://doi.org/10.1007/978-3-030-24733-1.

Barra-bucarei, L., Vergara, P., & Cortes, A. (2016). Conditions to optimize mass production of *Metarhizium anisopliae* (Mets.) Sorokin 1883 in different substrates. *Chilean. Journal of Agricultural Research, 76*(December). https://doi.org/10.4067/S0718-58392016000400008.

Barta, M., Lalík, M., Rell, S., Kunca, A., Horáková, M. K., Mudrončeková, S., & Galko, J. (2019). Hypocrealean fungi associated with *Hylobius abietis* in Slovakia, their virulence against weevil adults and effect on feeding damage in laboratory. *Forests, 10*(8), 634. https://doi.org/10.3390/f10080634.

Batta, Y. A., Rahman, M., Powis, K., Baker, G., & Schmidt, O. (2011). Formulation and application of the entomopathogenic fungus : *Zoophthora radicans* (Brefeld) Batko (Zygomycetes : Entomophthorales). *Journal of Applied Microbiology,* 831–839. https://doi.org/10.1111/j.1365-2672.2011.04939.x.

Bischoff, J. F., Rehner, S. A., & Humber, R. A. (2009). A multilocus phylogeny of the *Metarhizium anisopliae* lineage. *Mycologia.* https://doi.org/10.3852/07-202. March.

Brust, G. E., Wakil, W., & Qayyum, M. A. (2018). Sustainable management of arthropod pests of tomato: Minor pests. In *Sustainable management of arthropod pests of tomato.* https://doi.org/10.1016/B978-0-12-802441-6.00008-5.

Bukhari, T., Takken, W., & Koenraadt, C. J. M. (2011). Development of *Metarhizium anisopliae* and *Beauveria bassiana* formulations for control of malaria mosquito larvae. *Parasites & Vectors*, *4*(1), 1–14.

Chandler, D., Bailey, A. S., Tatchell, G. M., Davidson, G., Greaves, J., & Grant, W. P. (2011). The development, regulation, and use of biopesticides for integrated pest management. *Philosophical Transactions of the Royal Society B: Biological Sciences*, *366*(1573), 1987–1998. https://doi.org/10.1098/rstb.2010.0390.

Dang, K., Doggett, S. L., Singham, G. V., & Lee, C. (2017). Insecticide resistance and resistance mechanisms in bed bugs, *Cimex* spp. (Hemiptera : Cimicidae). *Parasites & Vectors*, *10*(1), 1–31. https://doi.org/10.1186/s13071-017-2232-3.

Dannon, H. F., Dannon, A. E., Douro-Kpindou, O. K., Zinsou, A. V., Houndete, A. T., Toffa-Mehinto, J., ... Manuele, T. A. M.Ò. (2020). Toward the efficient use of *Beauveria bassiana* in integrated cotton insect pest management. *Journal of Cotton Research*, *3*(1), 1–21.

de Faria, M. R., & Wraight, S. P. (2007). Mycoinsecticides and Mycoacaricides: A comprehensive list with worldwide coverage and international classification of formulation types. *Biological Control*, *43*(3), 237–256. https://doi.org/10.1016/j.biocontrol.2007.08.001.

Feng, M. G., Poprawski, T. J., & Khachatourians, G. G. (1994). Production, formulation, and application of the entomopathogenic fungus *Beauveria bassiana* for insect control : Current status. *Biocontrol Science and Technology*, *4*(1), 3–34.

Fernández-Grandon, G. M., Harte, S. J., Ewany, J., Bray, D., & Stevenson, P. C. (2020). Additive effect of botanical insecticide and entomopathogenic fungi on pest mortality and the behavioral response of its natural enemy. *Plants*, *9*(2), 173.

Gabarty, A., Salem, H. M., Fouda, M. A., Abas, A. A., & Ibrahim, A. A. (2014). Pathogencity induced by the entomopathogenic fungi *Beauveria bassiana* and *Metarhizium anisopliae* in *Agrotis ipsilon (Hufn.). Journal of Radiation Research and Applied Sciences*, *7*(1), 95–100. https://doi.org/10.1016/j.jrras.2013.12.004.

Gilbert, L. I., & Gill, S. S. (2010). In L. I. Gilbert, & S. S. Gill (Eds.), *Insect control: Biological and synthetic agents* Academic Press.

Gupta, S., & Dikshit, A. K. (2010). Biopesticides : An ecofriendly approach for pest control. *Journal of Biopesticides*, *3*(Special Issue), 186–188.

Hubbard, M., Hynes, R. K., Erlandson, M., & Bailey, K. L. (2014). The biochemistry behind biopesticide efficacy. *Sustainable Chemical Processes*, *2*(1), 1–8.

Iwanicki, N. S., Pereira, A. A., Botelho, A. B. R. Z., Rezende, J. M., de Andrade Moral, R., Zucchi, M. I., & Júnior, I. D. (2019). Monitoring of the field application of *Metarhizium anisopliae* in Brazil revealed high molecular diversity of *Metarhizium* spp. in insects, soil, and sugarcane roots. *Scientific Reports*, 1–12. https://doi.org/10.1038/s41598-019-38594-8. July 2018.

Kassa, A., Brownbridge, M., Parker, B. L., Skinner, M., Gouli, V., Gouli, S., ... Hata, T. (2008). Whey for mass production of *Beauveria bassiana* and *Metarhizium anisopliae*. *Mycological Research*, *112*, 583–591. https://doi.org/10.1016/j.mycres.2007.12.004.

Kepler, R. M., Luangsa-Ard, J. J., Hywel-Jones, N. L., Quandt, C. A., Sung, G. H., Rehner, S. A., ... Chen, M. (2017). A phylogenetically-based nomenclature for *Cordycipitaceae* (*Hypocreales*). *IMA Fungus*, *8*(2), 335–353. https://doi.org/10.5598/imafungus.2017.08.02.08.

Keswani, C., Singh, S. P., & Singh, H. B. (2013). *Beauveria bassiana*: Status, mode of action, applications, and safety issues. *Biotech Today*. https://doi.org/10.5958/j.2322-0996.3.1.002. January.

Krutmuang, P., Pittarate, S., Rahman, A., Rajula, J., & Thungrabeab, M. (2019). Compatibility of ten isolates of *Beauveria bassiana* (Balsamo) Vuillemin with four commonly used fungicides in Thailand. *Fundamental and Applied Agriculture*, *4*, 1. https://doi.org/10.5455/faa.70566.

Kumar, S. (2012). Biopesticides: A need for food and environmental safety. *Journal of Biofertilizers & Biopesticides*, *3*(4), 4–6. https://doi.org/10.4172/2155-6202.1000e107.

Lacey, L. A. (1997). *Manual of techniques in insect pathology*. San Diego, CA: Academic Press.

Lacey, L. A., Kirk, A. A., Millar, L., Mercadier, G., Vidal, C., Kirk, A. A., ... Vidal, C. (1999). Ovicidal and Larvicidal activity of conidia and Blastospores of *Paecilomyces fumosoroseus* (Deuteromycotina : Hyphomycetes) against *Bemisia argentifolii* (Homoptera : Aleyrodidae) with a description of a bioassay system allowing prolonged survival of control insects. *Biocontrol Science and Technology*, *9*(1), 9–18. https://doi.org/10.1080/09583159929866.

Längle, T. (2006). *Beauveria bassiana* (Bals.-Criv.) Vuill.–A biocontrol agent with more than 100 years of history of safe use. In *Workshop on current risk assessment and regulation practice Salzau, Germany* (pp. 8–10).

Leger, R. S., & Screen, S. (2001). Prospects for strain improvement of fungal pathogens of insects and weeds. In *Fungi as biocontrol agents* (pp. 219–238). Oxford: CABI (Council of Applied Biology International).

Li, Z., Alves, S. B., Roberts, D. W., Fan, M., Delalibera, I., Jr., Tang, J., ... Rangel, D. E. (2010). Biocontrol science and technology biological control of insects in Brazil and China : History, current programs, and reasons for their successes using entomopathogenic fungi. *Biocontrol Science and Technology*. https://doi.org/10.1080/09583150903431665.

Litwin, A., Nowak, M., & Ro, S. (2020). Entomopathogenic fungi : Unconventional applications. *Reviews in Environmental Science and Bio/Technology*, *1*, 23–42. https://doi.org/10.1007/s11157-020-09525-1.

Luangsa-ard, J. J., Tasanathai, K., Mongkolsamrit, S., & Hywel-Jones, N. L. (2007). *Atlas of invertebrate-pathogenic fungi of Thailand. Vol. 1*. Pathum Thani, Thailand: National Center for Genetic Engineering and Biotechnology, National Science and Technology Development Agency.

Luangsa-ard, J. J., Tasanathai, K., Mongkolsamrit, S., & Hywel-Jones, N. L. (2008). *Atlas of invertebrate-pathogenic fungi of Thailand*. Pathum Thani, Thailand: National Center for Genetic Engineering and Biotechnology, National Science and Technology Development Agency.

Luangsa-ard, J. J., Tasanathai, K., Mongkolsamrit, S., & Hywel-Jones, N. L. (2010). *Atlas of invertebrate-pathogenic fungi of Thailand*. Thailand: National Center for Genetic Engineering and Biotechnology.

Mascarin, G. M., & Jaronski, S. T. (2016). The production and uses of *Beauveria bassiana* as a microbial insecticide the production and uses of *Beauveria bassiana* as a microbial insecticide. *World Journal of Microbiology and Biotechnology*. https://doi.org/10.1007/s11274-016-2131-3. November.

Moura, G., Biaggioni, R., Delalibera, Í., Kort, É., Fernandes, K., Luz, C., & Faria, M. (2019). Current status and perspectives of fungal entomopathogens used for microbial control of arthropod pests in Brazil. *Journal of Invertebrate Pathology*, *165*(December 2017), 46–53. https://doi.org/10.1016/j.jip.2018.01.001.

Neves, P. M. O. J., Hirose, E., Tchujo, P. T., & Moino, A. (2001). Compatibility of entomopathogenic fungi with neonicotinoid insecticides. *Neotropical Entomology*, *30*(2), 263–268. https://doi.org/10.1590/s1519-566x2001000200009.

Mcguire, A. V., & Northfield, T. D. (2020). Tropical occurrence and agricultural importance of *Beauveria bassiana* and *Metarhizium anisopliae*. *Frontiers in Sustainable Food Systems*, *4*(January), 6. https://doi.org/10.3389/fsufs.2020.00006.

Prakash, G. B., Padmaja, V., & Kiran, R. S. (2008). Statistical optimization of process variables for the large-scale production of *Metarhizium anisopliae* conidiospores in solid-state fermentation. *Bioresource Technology*, *99*, 1530–1537. https://doi.org/10.1016/j.biortech.2007.04.031.

Qu, J., Yu, L., Zhang, J., Han, Y., & Zou, X. (2018). A new entomopathogenic fungus, *Ophiocordyceps ponerus* sp. nov., from China. *Phytotaxa*, *343*(2), 116–126.

Rajula, J., Rahman, A., & Krutmuang, P. (2020). Entomopathogenic fungi in Southeast Asia and Africa and their possible adoption in biological control. *Biological Control*. https://doi.org/10.1016/j.biocontrol.2020.104399.

Różalska, S., Glińska, S., & Długoński, J. (2014). *Metarhizium robertsii* morphological flexibility during nonylphenol removal. *International Biodeterioration & Biodegradation*, *95*, 285–293. https://doi.org/10.1016/j.ibiod.2014.08.002.

Sanjuan, T. I., Franco-Molano, A. E., Kepler, R. M., Spatafora, J. W., Tabima, J., Vasco-Palacios, A. M., & Restrepo, S. (2015). Five new species of entomopathogenic fungi from the Amazon and evolution of neotropical Ophiocordyceps. *Fungal Biology*, *9*(1), 901–916. https://doi.org/10.1016/j.funbio.2015.06.010.

Shah, F. A., Ansari, M. A., Watkins, J., Phelps, Z., Cross, J., & Butt, T. M. (2009). Biocontrol science and technology influence of commercial fungicides on the germination, growth, and virulence of four species of entomopathogenic fungi. *Biocontrol Science and Technology*, *19*(7), 743–753. https://doi.org/10.1080/09583150903100807.

Skinner, M., Parker, B. L., & Kim, J. S. (2014). *Integrated pest management*. https://doi.org/10.1016/B978-0-12-398529-3.00011-7.

Song, M. H., Yu, J. S., Kim, S., Lee, S. J., Kim, J. C., Nai, Y. S., … Kim, J. S. (2019). Downstream processing of *Beauveria bassiana* and *Metarhizium anisopliae* -based fungal biopesticides against *Riptortus pedestris*: Solid culture and delivery of conidia. *Biocontrol Science and Technology*, *29*(6), 514–532. https://doi.org/10.1080/09583157.2019.1566951.

Srinivasan, R., Sevgan, S., Ekesi, S., & Tamò, M. (2019). Biopesticide based sustainable pest management for safer production of vegetable legumes and brassicas in Asia and Africa. *Pest Management Science*. https://doi.org/10.1002/ps.5480. April.

Szewczyk, R., Kuśmierska, A., & Bernat, P. (2018). Ametryn removal by *Metarhizium brunneum*: Biodegradation pathway proposal and metabolic background revealed. *Chemosphere*, *190*, 174–183. https://doi.org/10.1016/j.chemosphere.2017.10.011.

Tanada, Y., & Kaya, H. K. (2012). *Insect pathology*. Academic Press.

Thungrabeab, M., Blaeser, P., & Sengonca, C. (2006). Effect of temperature and host plant on the efficacy of different entomopathogenic fungi from Thailand against *Frankliniella occidentalis* (Pergande) and *Thrips tabaci* Lindeman (Thysanoptera: Thripidae) in the laboratory. *Journal of Plant Diseases and Protection*, *113*(4), 181–187.

Tupe, S. G., Pathan, E. K., & Deshpande, M. V. (2017). Development of *Metarhizium anisopliae* as a Mycoinsecticide: From isolation to field performance. *JoVE (Journal of Visualized Experiments)*, (125), e55272. 1–8 https://doi.org/10.3791/55272.

Wraight, S. P., Jackson, M. A., & de Kock, S. L. (2001). Production, stabilization, and formulation of fungal biocontrol agents. In *Fungi as biocontrol agents* (pp. 253–287). Oxford: CABI (Council of Applied Biology International).

Wraight, S. P., Ugine, T. A., Ramos, M. E., & Sanderson, J. P. (2016). Efficacy of spray applications of entomopathogenic fungi against western flower thrips infesting greenhouse impatiens under variable moisture conditions. *Biological Control*, *97*, 31–47. https://doi.org/10.1016/j.biocontrol.2016.02.016.

Xiao, Y., Hongsanan, S., Hyde, K. D., & Brooks, S. (2019). Two new entomopathogenic species of Ophiocordyceps in Thailand. *MycoKeys*, (47), 53–74. https://doi.org/10.3897/mycokeys.47.29898.

Zimmermann, G. (1993). The Entomopathogenic fungus *Metarhizium anisopliae* and it's potential as a biocontrol agent. *Pesticide Science*, *37*(4), 375–379.

Zimmermann, G. (2007). Review on safety of the entomopathogenic fungi *Beauveria bassiana* and *Beauveria brongniartii*. *Biocontrol Science and Technology*, *17*(6), 553–596. https://doi.org/10.1080/09583150701309006.

Chapter 5

Microbial fortification during vermicomposting: A brief review

Linee Goswami, Pralay Shankar Gorai, and Narayan Chandra Mandal
Mycology and Plant Pathology Laboratory, Department of Botany, Visva-Bharati, Santiniketan, West Bengal, India

Chapter outline

1 Introduction

Vermicomposting is an earthworm-mediated mesophilic degradation process. It results in an odorless, nutrient-rich, detox substrate. This process usually accelerates the decomposition of slowly degrading organic substrates (Goswami et al., 2014). Earthworms ingest organic matter along with some portion of the soil and excrete out almost equal amount as vermicast. These vermicasts bear the load of beneficial microorganisms, mostly released from the earthworm gut. Earthworms are used to fragment the substrate to stimulate the microbial population and fasten the biochemical degradation (Sen & Chandra, 2009). The vermicast composition varies depending on the substrate composition and the earthworm species. Both of these factors largely govern microbial community structures

Recent Advancement in Microbial Biotechnology. https://doi.org/10.1016/B978-0-12-822098-6.00011-2

(Hussain et al., 2016). Apart from these two factors, abiotic factors such as pH, aeration, water retention, moisture content, particle size, granulation, etc. also play significant roles in determining the microbial activity and their fortification.

Vermicomposting involves a complicated food web that begins with the systematic recycling and slow degradation of carbon-rich substances. Just like any other food web, vermicomposting also has its trophic levels. Microbes are an essential part of these food chains. The initial process starts with diverse groups of microorganisms as primary consumers, followed by microbe-feeders (secondary consumer), detritivores, microbial detritivores (tertiary consumer), and carnivores. Primary consumers include bacteria, fungi, and other ciliates to start the mineralization process (Domínguez, Aira, & Gómez-Brandón, 2010). Detritivores like microarthropods, nematodes, and earthworms are present as secondary consumers. They act as storehouses for different groups of microorganisms.

Microbes play an indispensable role in nutrient cycling because of their diversity and abundance in the natural environment. Accelerated mineralization results in slow and sustained release of nutrients in bioavailable forms. Plant roots absorb water-soluble nutrients from rhizospheric soils. The root apoplast absorbs these nutrients to reach in the above-ground biomass through xylem loading. In the above-ground biomass, the nutrients get accumulated and converted into essential metabolites through several biochemical processes. Therefore, these microbes can be considered as the catalyst to fasten the whole process. Thus, it maintains the soil-plant-animal continuum in the ecosystem services.

Presence of wide varieties of decomposers determine the fertility status of the soil. To maintain soil health, functional groups of decomposers viz. bacteria and fungi work in close association with other members of the soil fauna. Such interactions are often termed as competition, mutualism, predation, etc. These biotic associations are extremely important as they determine the functional microbial diversity and future substrate quality (Sampedro & Domínguez, 2008). Even though soil fauna like earthworms accelerate the process of vermicomposting, the energy source in the entire process remains unknown (Aira, Monroy, & Domínguez, 2007). Several hypotheses juggle between nonselective feedstock to inert carbon sources (Goswami et al., 2013; Sampedro, Jeannotte, & Whalen, 2006). This flow of energy is crucial for a continuous cycle. When organic matter and some portion of soil pass through the long cylindrical alimentary canal of the earthworm, they undergo several biochemical reactions. Studies showed presence of nitrogen fixers and phosphate solubilizers in earthworm vermicast (Hussain et al., 2016, 2018; Singh et al., 2015). N-fixers tend to facilitate N availability through the combined action of earthworm body fluids, mucous, and nitrogenous excretory products. Whereas, P solubilizer accelerates the P mineralization process with the help of phosphatase enzyme activity in the vermicast (Devi et al., 2020). A group called *endo*-symbiotic microbe, residing inside the gizzard, releases extracellular enzymes that dissolve cellulose and phenol-rich compounds in the substrate. Apart from these, physical activities like burrowing, tunneling contribute to a better environment for microbial respiration. It also improves soil aeration, moisture content, and nutrient distribution. Such activities of decomposers also affect soil health and compost quality.

In this review, we made an effort to understand how earthworm ecology, population, reproduction, and food habits can influence the microbial diversity of any vermicompost or a vermi-amended surface or earthworm habitats. The critical question is to understand how certain microbes groups dominate, and their interdependence can alter nutrient cycling patterns, effects soil fertility, and plant growth. How is it correlated with lithospheric activities? We reviewed recent works on high throughput sequencing and metagenomics data to understand microbial diversity patterns for detailed investigations.

2 Influence of vermicomposting and aerobic composting processes on microbial dominance

Vermicomposting and composting are the two most economically viable bioconversion processes. During these processes, substrate composition and species variation mainly govern their microbial profile, which finally manipulates the whole process. There are marked differences in the community structure during aerobic composting and vermicomposting.

2.1 Impact on bacterial profile

The aerobic composting is thermophilic in nature, so it is home for thermophilic bacteria. Whereas vermicomposting is mesophilic, therefore it is home for mesophilic bacteria. Due to this fundamental difference between the two systems, microbes behave differently and play different roles in substrate decomposition . The study showed that *Bacillus*, *Flavobacterium*, *Pseudomonas*, and *Cellulomonas* are predominant under aerobic compost. Except for *Pseudomonas*, all other genera remain inactive under vermicomposting (Cai, Gong, Sun, Li, & Yu, 2018). Hussain et al. (2018) reported an alteration in the microbial community structure in lignocellulosic biowaste under vermicomposting and composting. This study reported a sharp increase in the nitrogen-fixing population in *Eudrillus eugeniea*, *Eisenia fetida*, and *Perionyx excavatus* mediated vermicomposting system as compared to conventional single species vermicomposting and aerobic composting. It was evident from the study that the relative abundance of bacteria was highest under vermicomposting treatment combinations than aerobic composting treatments. Wang et al. (2017) reported five major phyla present in *Eisenia fetida* vermicompost, such as *Proteobacteria*, *Tenericutes*, *Bacteriodetes*, *Firmicutes*, and *Verrumicrobia.* Castillo, Romero, and Nogales (2013) reported a sharp temporal decrease in α and γ proteobacteria population in winery and olive mill waste. Chen, Chang, Chen, Zhang, and Yu (2017) identified 65 bacterial genera belonging to 33 phyla and 21 fungal genera belonging to 5 different fungal phyla under vermicompost and compost. Among the bacterial phyla, the dominant ones were *Proteobacteria*, *Bacteriodetes*, *Firmicutes*, and *Actinobacteria.* They reported Bacteriodetes and Firmicutes dominance under composting treatments; whereas, under vermicomposting treatments, γ-Proteobacteria and Actinobacteria showed higher abundance occupying about 84% of the total population. In the same substrate, among the fungal population, the dominant phyla found were Ascomycota, Basidiomycota, and Zygomycota, contributing to 76% of the total population.

Factors that affect bacterial growth rates are physicochemical properties, moisture content, and temperature of the substrate. Several workers found temporal variations of bacterial and fungal dominance under both the bioconversion systems while working with a wide variety of feed mixtures (Budroni et al., 2020; Huang et al., 2018; Pathma & Sakthivel, 2013; Wang et al., 2017). Huang et al. (2018) reported an increase in Proteobacteria and a decrease in Firmicutes, when *Eisenia fetida* was inoculated for 20 days under three systems containing cow dung, fruit, and vegetable wastes, and sewage sludge, respectively. Domínguez et al. (2010) reported a change in bacterial α and β diversity under fresh dead scotch broom vermicomposting, when incubated with *Eisenia andrei* and analyzed at an interval of 14, 42, and 91 DAI. They observed a distinct change in bacterial core community structure on 0 and 14 DAI, whereas no changes were recorded at 42 and 91 DAI of vermicomposting. Huang, Li, Wei, Chen, and Fu (2013) reported the dominance of Actinobacteria and Bacteriodetes in vegetable waste vermicompost. Similarly, Yasir et al. (2009) demonstrated presence of four major phyla, i.e., *Proteobacteria*, *Bacteriodetes*, *Firmicutes*, and *Actinobacteria* under paper mill sludge vermicompost. Pathma and Sakthivel (2013) isolated a set of 193 OTUs of bacteria from a straw and goat manure-based system incubated with *Eisenia fetida*. Later, with the help of molecular phylogenetic identification via 16 s rRNA sequence homology, they recorded a 57% dominance of *Bacillus* sp. in the substrates, followed by 15% of *Pseudomonas* and 12% of *Microbacterium*. Apart from that, groups such as Acinetobacter (5%), Chryseobacterium (3%), Arthobacter, Pseudoxanthomonas, Stentrophomonas, Paenibacillus, Rhodococcus, Enterobacter, Rheinheimer, Cellulomonas, etc. were also present. Out of 193 identified OTUs of bacteria, 96 numbers (49%) showed potential antagonism against plant phytopathogens. Whereas, 51% showed biofertilizer potential with high siderophore activity, ACC deaminase production, and phosphate solubilization.

deAngelis et al. (2011) reported that Firmicutes and Actinobacteria taxa decompose lignocellulosic substrates, whereas Proteobacteria and Actinobacteria are more active during lignin decomposition. Similarly, Bacteriodetes participate in cellulose and chitin degradation (Grady, MacDonald, Liu, Richman, & Yuan, 2016; Kirby, 2005; Zhang et al., 2014). Their activities are mostly regulated by the extracellular enzymes. All these enzymes escalate the process of complex macromolecular breakdown and release secondary metabolites (Lim, Chiam, & Wang, 2014). Studies showed that Bacteriodetes like *Bacteroides cellulosolvens*, Proteobacteria like *Pseudomonas* sp. CL3, and *Cellvibrio japonicus* tend to initiate the degradation of polysaccharides like cellulose, xyloglucan, mannan, arbanan, pectin, etc. The presence of earthworm contributes to the whole process via activation of a group of cellulase enzymes i.e., *endo*-β-1, 4-D-glucanase, *exo*-β-1, 4-D-glucanase, and β-1, 4-D-glucosidase that further fasten the process of polysaccharide degradation (Cheng & Chang, 2011; Gardner, 2016; Van Dyk & Pletschke, 2012). Therefore, apart from time and temperature, substrate composition is a crucial factor to regulate the selective growth of the microbial population.

Blomstrom, Lalander, Komakech, Vinnerås, and Boqvist (2016) carried out a metagenomics study of vermicomposted substrates containing an 80:20 ratio of cattle manure: food waste. They reported that the relative abundance of Proteobacteria was the highest in the vermicompost substrate. Among the identified OTUs, six predominant classes were zetaproteobacteria, alphaproteobacteria, gammaproteobacteria, rhizobiales, aeromonades, and burkholderiales. These bacterial groups are specifically responsible for nitrogen metabolism and fixation. They also recorded presence of pathogens like *Brucella* spp., *Salmonella* sp., *Escherichia coli*, *Enterococcus* spp., and *Clostridium* sp. in the substrate. However, their presence was found very low (below 10%) in the vermicomposted substrates. The third-largest genera identified were *Actinobacteria*, out of which 93% belonged to *Microbacterium*, *Mycobacterium*, *Corynebacterium*, *Streptomyces*, and *Cellulomonas* (Blomstrom et al., 2016). Two-third of the bacterial population was identified as phylum Firmicutes and class *Clostridia*. Similar results were also reported by the previous researchers (Hill & Baldwin, 2012; Romero-Tepal et al., 2014; Yasir et al., 2009). *Clostridium* sp., an anaerobic sulfite reducing bacteria, actively associated with cellulose degradation, was found to be present high in number under aerobic composting processes (Haagsma, 1991). Aira, Olcina, Pérez-Losada, and Domínguez (2016) reported that vermicomposting changed bacterial diversity in fresh cow dung substrate when exposed to *Eisenia fetida*. The substrate showed higher β diversity and higher nutrient availability. Therefore, it was concluded that the microbial population's presence and fortification were directly correlated with the earthworm population.

Pathma and Sakthivel (2013) observed that vermicomposting bacteria show antagonism against plant and human pathogens. Studies carried out by Eastman et al. (2001) and Yadav, Tare, and Ahammed (2010) with human fecal slurry showed a sharp reduction in the load of *Salmonella* sp., *Escherichia coli*, and helminthic ova under vermicompost. Aira, Gómez-Brandón, González-Porto, and Domínguez (2011) showed that vermicomposting could reduce pathogen load in cow manure and could be used for industrial-scale production. Cai et al. (2018) gave a detailed analysis of bacterial profile under vermicomposting and aerobic composting surfaces at an interval of 30 days, as shown in Fig. 1.

2.2 Impact on fungal growth

Fig. 2 gives a detailed account of change in fungal populations under vermicomposting and aerobic composting substrates at a 30-day interval up to 150 days.

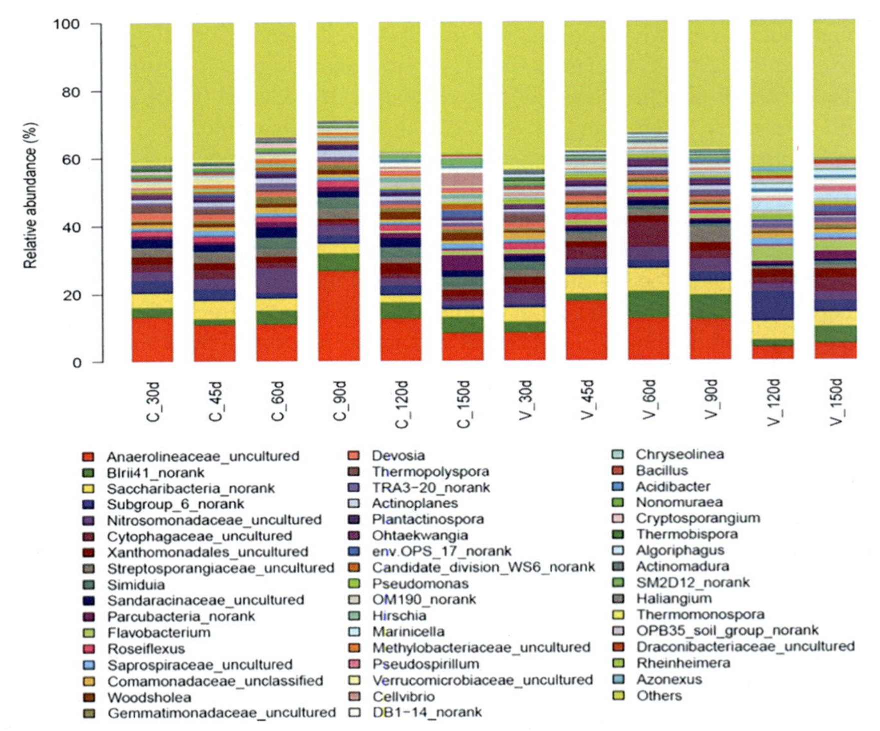

FIG. 1 Temporal variations of relative abundance of bacterial taxa under vermicomposting and aerobic composting substrates. Samples collected at an interval of 15 days, from the 30th day of incubation period. Substrate composition was green wastes (A mixture of *Robinia pseudoacacia* Linn., *Cotinus coggygria* Scop., and *Ilex chinensis Sims* litters). *(Courtesy Cai, L., Gong, X., Sun, X., Li, S., Yu, X. (2018). Comparison of chemical and microbiological changes during the aerobic composting and vermicomposting of green waste.* PLoS One, 13 *(11), e0207494. https://doi.org/10.1371/journal.pone.0207494.)*

Among the fungal population, *Basidiomycota* grows profusely under vermicomposting, whereas A*scomycota* are prominent under aerobic composting (Langarica-Fuentes et al., 2014). Varma, Nashine, Sastri, and Kalamdhad (2017) reported that when there is a spike in *Ascomycota* under the aerobic composting system, it led to lignin peroxidase enzyme activation. Chigineva, Aleksandrova, Marhan, Kandeler, and Tiunov (2011) reported that fungal hyphae networks work in coordination to release hydrolytic enzymes for lignocellulosic degradation. This process can translocate N-rich nutrients to substrates with poor N-availability. The growth of fungal mycelium promotes earthworm colony formation and microbial proliferation by providing nutrition to the earthworm hatchlings. Chen et al. (2017) revealed that vermicomposting changes microbial profile within a week of earthworm incubation. They found altered dominance of *Aspergillus*, *Trichosporon*, and *Nectriaceae*_unclassified under vermicomposting, whereas *Anaerolineaceae*_uncultured, *Mortierella*,

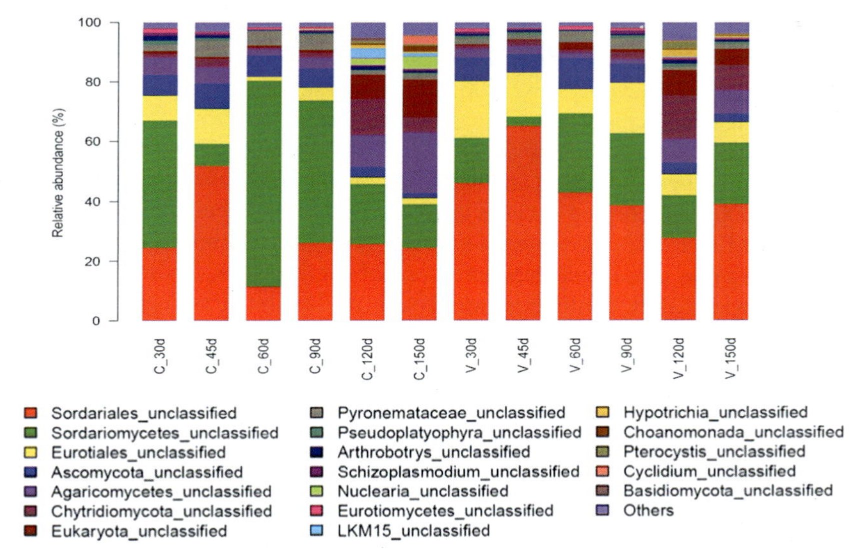

FIG. 2 Comparative assessment of temporal variations in fungal diversity under vermicompost and aerobic compost. *(Courtesy Cai, L., Gong, X., Sun, X., Li, S., Yu, X. (2018). Comparison of chemical and microbiological changes during the aerobic composting and vermicomposting of green waste.* PLoS One, 13 *(11), e0207494. https://doi.org/10.1371/journal.pone.0207494.)*

and *Trichosporon* were dominant under aerobic composting. When analyzed, the initial prestabilized substrate showed relative abundance in the order of *Mortierella* (44%) > *Trichosporon* (12%) > *Emericella* (10%) > *Aspergillus* (8%). At the same time, 1 week of earthworm inoculation showed the change in the order of *Aspergillus* > *Trichosporon* > *Nectriaceae*_unclassified. They could cause cellulose hydrolysis and showed tolerance against lignocellulose-derived inhibitors (Wang, Gao, & Bao, 2016; Zhou, Selvam, & Wong, 2016). Post 3 weeks of incubation, further changes were detected in the bacterial and fungal populations under both the systems. However, dominant phyla in both systems remains unaltered. Fascinatingly, Castillo et al. (2013) observed that a few actinobacteria from earthworm gut release chitinase enzyme in olive mill waste substrates. Because the chitinase enzyme dissolves chitin from the fungal cell wall, fungal growth was arrested in the substrate. Schmidt-Dannert (2016) observed that Basidiomycota release necessary extracellular enzymes via oxidative biotransformation in a postvermicomposting surface, leading to lignin decomposition.

3 Influence of earthworm ecological categories on microbial dominance and their relative abundance

Earthworms belong to phylum annelids and family oligochete. They are clitellated hermaphroditic soil-dwelling organisms. The earthworm population is one of the most important factors for soil fertility management. They are often termed as "farmer's best friend" because of the role in nutrient cycling (Bhattacharya & Chattopadhyay, 2002). Because they share major biomass load under tropical and temperate soil ecosystems, their activities can directly affect soil formation, composition, and diversity. They also play a significant role in organic matter decomposition and balancing the nutrient turnover rate (Lavelle & Spain, 2001). According to Gates (1972), around 4000 species of earthworms were identified to date. However, we are still far away from understanding their whole life cycle processes. Depending on habitat selection, feeding behavior, and burrowing habits, earthworms are broadly divided into three ecological categories (Bouche, 1977). They are:

(1) Epigeic earthworms: Epigeic earthworms are the uppermost soil-dwelling nematodes. They feed on decomposed organic matters. They are the most voracious feeder of litters and survives on the surface and organic horizon of soil. For example, *Eisenia fetida*, *Eudrillus eugeniea*, *Perionyx excavatus*, *Dendrobaena octaedra*, *Dendrobaena attemsi*, *Dendrodrilus rubidus*, *Eiseniella tetraedra*, *Heliodrilus oculatus*, *Lumbricus castaneus*, *Lumbricus festivus*, *Eisenia andrei*, etc. Another subgroup possesses qualities of both epigeic and endogeic earthworms, often termed as epi-endogeic earthworms. For example, *Lumbricus rubellus*.

(2) Anecics earthworms: Anecics are commonly defined as burrowers. They feed on a mixture of soil and organic matter. For example, *Lampito mauritii*, *Aporrectodea longa*, *Aporrectodea nocturna*, and *Lumbricus terrestris*.

(3) Endogeic earthworms: Endogeic earthworms are known as soil feeders and mostly live in deep burrows. For example, *Metaphire posthuma*, *Allolobophora chlorotica*, *Apporectodea caliginosa*, *Apporectodea icterica*, and *Apporectodea rosea*.

Several abiotic factors govern the earthworm population and their survival under the natural environment. Such factors and different environmental conditions are listed below.

(1) Physiological endurance
(2) Reproductive rate
(3) Tolerance to external shock
(4) Life cycle duration and adaptation
(5) Resistance to external handling
(6) Coping mechanism

Epigeic earthworms show these behaviors very prominently. Interestingly, few anecics and endogeic species also bear these traits. Hence, those species can be considered as bioagents for proper vermicomposting.

Earthworms ingest coarse particulate organic matter along with some amount of soil; release faeces in shape on holorganic pellets (Lavelle &

Spain, 2001). These pellets specifically provide a better surface to volume ratio than the decomposed litter surface, thus fastening the decomposition ratio. Earthworms contribute towards the growth of beneficial microbe communities in the soil surface. When exposed to organic-rich substrates, depending on the substrate composition, they tend to proliferate microbial growth and facilitate the decomposition process. They have high efficacy in converting recalcitrant carbon forms like lignocellulosic substrates, chitin, cellulose, hemicellulose, etc. into labile forms. Thus, they can provide nourishment, and microbes can easily reproduce. Studies showed that epigeic earthworms like *Eisenia andrei*, *Eisenia fetida*, *Perionyx excavatus*, and *Eudrillus eugeniea* (Domínguez, 2004; Goswami, Mukhopadhyay, Bhattacharya, Das, & Goswami, 2018; Paul, Goswami, Pegu, & Bhattacharya, 2020); anecics earthworms like *Lampito mauritii* (Goswami et al., 2014; Maity, Bhattacharya, & Chaudhury, 2009); and endogeic earthworms like *Metaphire posthuma* (Sahariah, Goswami, Kim, Bhattacharyya, & Bhattacharya, 2015) show excellent vermicomposting potential. The use of monoculture or double species culture or even multiple species consortium is quite beneficial and fast. These vermireactors are termed as smart vermiconversion systems (Devi et al., 2020; Hussain et al., 2018). These vermireactors can provide a faster recycling process and superior quality product for commercial-scale production. It takes almost 40–50 DAI for lignocellulosic substrates and spent mushroom substrate, respectively. Fig. 3 shows how the microbial community structure gets affected by earthworm ecological categories during vermicomposting.

Singh et al. (2015) reported Proteobacteria and Bacteriodetes as dominant phyla under *Eisenia fetida* vermicomposting. Braga et al. (2016) reported that γ-proteobacteria could use labile carbons from the substrate and grow. The presence of earthworms fastens this process. Gong et al. (2018) reported that Proteobacter to Actinobacter ratio relatively alters under vermicomposting. Devi et al. (2020) documented an increase in the beneficial microbe and nutrient availability in spent mushroom substrate vermicompost under *Perionyx excavatus*. Similar findings were reported by Sampedro and Whalen (2007) that showed remarkable changes under PLFA-profiles for relative concentrations of aerobic bacteria, microeukaryotes, fungi, etc. Advanced scientific methods, including high-throughput techniques and the next-generation sequencing techniques, confirmed findings like how earthworm species could affect the microbial species dominance with time. Several workers reported that the existence of endogeic earthworms like *Aporrectodea trapezoides*, *Metaphire guillelmi*, and *Pontoscolex corethrurus* tend to increase loads of family Flavobacteriaceae and order Sphingobacteriales from phyla Bacteroidete; Rhodocyclaceae from β-Proteobacteria; Paenibacillaceae, Verrucomicrobia, and *Nitrosovibrio* from Firmicutes (de Menezes et al., 2018; Gong et al., 2018; Huang et al., 2020). Groups of Chitinophagaceae, Cytophagaceae from phyla Bacteroidete, Neisseriaceae from Proteobacteria, and Microbacteriaceae from Actinobacteria were found to grow profusely when *Aporrectodea trapezoids* and *Pontoscolex corethrurus*

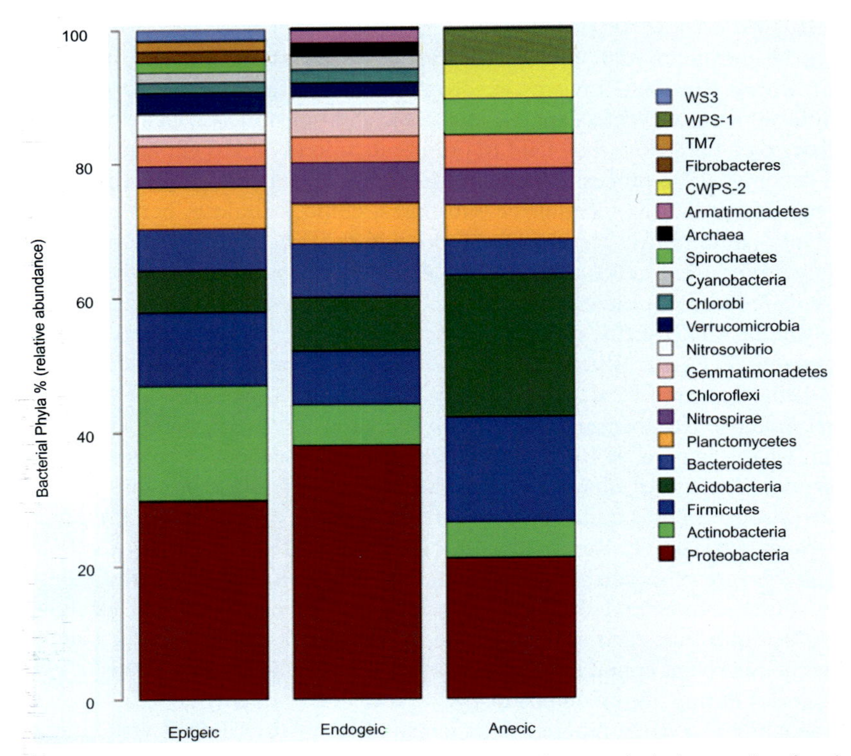

FIG. 3 Variations of bacterial and fungal diversity under different ecological categories of earthworms. The data shown here was collected from the peer-reviewed works published in Web of Science within the span of 2009–18. *(Courtesy Medina-Sauza, R. M., Álvarez-Jiménez, M., Delhal, A., Reverchon, F., Blouin, M., Guerrero-Analco, J. A., ... Barois, I. (2019). Earthworms building up soil microbiota, a review.* Frontiers in Environmental Science, *7, 81. https://doi.org/10.3389/fenvs.2019.00081.)*

were used for vermicomposting. These groups were found to release some chitinolytic enzymes that can dissolve chitin and initiate the decomposition process for chitin rich substrates (Bernard et al., 2012). Gong et al. (2018) reported reduced load of Chloroflexi and Fibrobacters under anecics earthworm *Metaphire guillelmi* when mulching is done for rice cultivation. However, the same microbial load increased, when straw was used instead of mulch. Therefore, it was understood that organic amendments and their composition directly affect microbial species dominance, nutrient fortification, and soil fertility management. Sometimes, earthworms exert an inverse effect on bacterial growth, whereas no effect recorded on the fungal population after 30 days of the incubation period (Aira, Sampedro, Monroy, & Domínguez, 2008).

Among epigeic earthworms, *Eisenia fetida*, *Perionyx excavatus*, and *Eudrillus eugeniea* have been widely used for vermicomposting throughout the world.

Their gut microbial diversity showed the presence of three major phyla, including *Proteobacteria*, *Actinobacteria*, and *Firmicutes* (Medina-Sauza et al., 2019) (Fig. 3). Phyla like *Verrucomicrobia*, a methanotroph, and *Chloroflexi*, a phototroph, present in the gut of *Eisenia fetida*, whereas *Spirochaetes*, i.e., chemoheterotrophs found in the gut fluid of *Perionyx excavatus* (Singh et al., 2015). Apart from that, *Aeromonas*, *Bacillus*, *Clostridium*, *Paenibacillus*, *Propionibacterium*, and *Staphyloccoccus* were also found in an ample amount inside *Eisenia fetida* gut. A low oxygen environment creates anoxic conditions inside the hollow tube of earthworm gut. This can alter microbial community dominance and overall structure (Koubova, Chroňáková, Pižl, Sánchez-Monedero, & Elhottová, 2015). Among all gut microbes isolated from such conditions, *Bacillus* and *Paenibacillus* are the most important ones, as they degrade complex aromatic compounds under oxygen-deficit conditions and provide food (Konig, 2006). Another group of earthworms, i.e., *Dendrobaena veneta* could increase the load of gammaproteobacteria like *Kluyvera cryocrescens* with *Pseudomonas putida* (Fjøsne, Myromslien, Wilson, & Rudi, 2018). The study showed that epigeic earthworms mostly trigger diversity, abundance, and specificity of the microbe population. However, an overall network profiling and metagenomics confirmed that earthworms in soil primarily promote eight major bacterial phyla. They are, namely, *Proteobacteria*, *Actinobacteria*, *Firmicutes*, *Acidobacteria*, *Planctomycetes*, *Bacteroidetes*, *Nitrospirae*, and *Chloroflexi* (Medina-Sauza et al., 2019). They also play an active role in soil nutrient availability and dynamics. Hence, such microbial profile documentation and analysis can be a report card for soil fertility status.

4 Influence of microbial structural change and temporal dominance on nutrient availability

Soil macro- and micro-faunal distribution and their activity control the process of nutrient solubilization and mobilization. Bacteria, along with fungi and other soil fauna, determine nutrient mineralization rate and transfer of nutrients and essential elements via trophic levels in a food web. Many researchers used different earthworm categories in the last decade to explore the possibilities and advancement of vermicomposting processes. They showed that earthworms react differently to different substrates. Hence, the quality of vermicast also tends to be different. Earthworms can directly or indirectly affect nutrient mineralization through physiological activity, metabolism, cell proliferation, and biosynthesis pathways (Braga et al., 2016).

As described in the earlier Section 3, earthworms tend to alter microbial load in the substrate. How do they do that? It begins with litter feeding. Epigeic earthworms most abundantly feed on slowly decomposing organic substances on the soil surface. These substances, when passing through the earthworm gut, get mixed with different enzymes. These enzymes break down complex macromolecules present in the substrate. These, in turn, increase the surface area, and gradually, the surface becomes microbe inhabitable (Hussain et al., 2016). The ecological variation of earthworms imparts varied effect on the nutrient mineralization process. During vermicomposting, due to consistent burrowing activity, the substrates tend to have higher porosity. This creates an aerobic environment in the substrate. It inhibits the growth of anaerobic bacteria. Thus, it can determine the nature of the microbial profile and substrate quality. Recent studies showed that endogeic earthworms like *Apporectodea caliginosa* and *Metaphire posthuma* can be excellent carbon fixers (Abail, Sampedro, & Whalen, 2017; Sahariah et al., 2015). During these studies, they showed remarkable responses towards carbon mineralization processes. Whereas anecics earthworms like *Lampito mauritii* and *Lumbricus rubellus* are excellent metal detoxifiers (Goswami et al., 2014; Sturzenbaum, Winters, Galay, Morgan, & Kille, 2001). Bernard et al. (2012) reported that fast catabolic actions and growth rate of r-strategist gut microflora present in earthworm's hollow cylindrical gut are the major driving force behind the accelerated substrate decomposition process. Braga et al. (2016) stated that the enrichment of fast-growing γ-proteobacteria-like species in vermicast might increase soil organic matter (SOM) mineralization.

Earthworms require a very low amount of carbon and nitrogen for their daily activity. Owing to such a low-lying life, earthworms contribute to a negligible amount of respiratory CO_2 to the surrounding environment. However, they contribute the highest to the conversion of labile carbon forms to stable carbons during assimilation and degradation of recalcitrant forms to labile forms during the process of mineralization (Abail et al., 2017). Budroni et al. (2020) showed that pH and total organic carbon (TOC) content in the substrate could influence bacterial profiles. They reported that 22.3% (*w*/w) TOC% in fermented cow manure (FCM) and 34.8% (w/w) in fermented brewer's spent grains (FBSG) at pH level >4.5 tend to support bacterial taxa with lignocellulosic degradation potential. They used Illumina Mi-sequencing of 16s rRNA amplicons to produce a data set of 75,863 and 84,060 raw sequences per sample. This study confirmed that the bacterial population found in both FBSG and FCM substrates (As shown in Fig. 4) are very diverse. After analyzing using higher sequence clustering programs, it showed significantly high numbers of operational taxonomic units (OTUs) under the FBSG substrate compared to the FCM substrate. This could be correlated positively with TOC% in both substrates. Hence, the workers concluded that TOC% of the substrate could influence the community structure temporally.

The relative abundance of *Paenibacillaceae*, *Comamonadaceae*, was very high in the FBSG and FCM substrates. These bacterial taxa could initiate a complex carbohydrate degradation process to convert them into simpler molecules through the slow release of chitinolytic enzymes, lignin modifying enzymes, cellulase, and hemicellulase enzymes. The presence of the Chitinophagaceae groups of bacteria tends to release β-glucosidase enzymes associated with chitin and cellulose degradation (Bailey, Fansler, Stegen, & McCue, 2013).

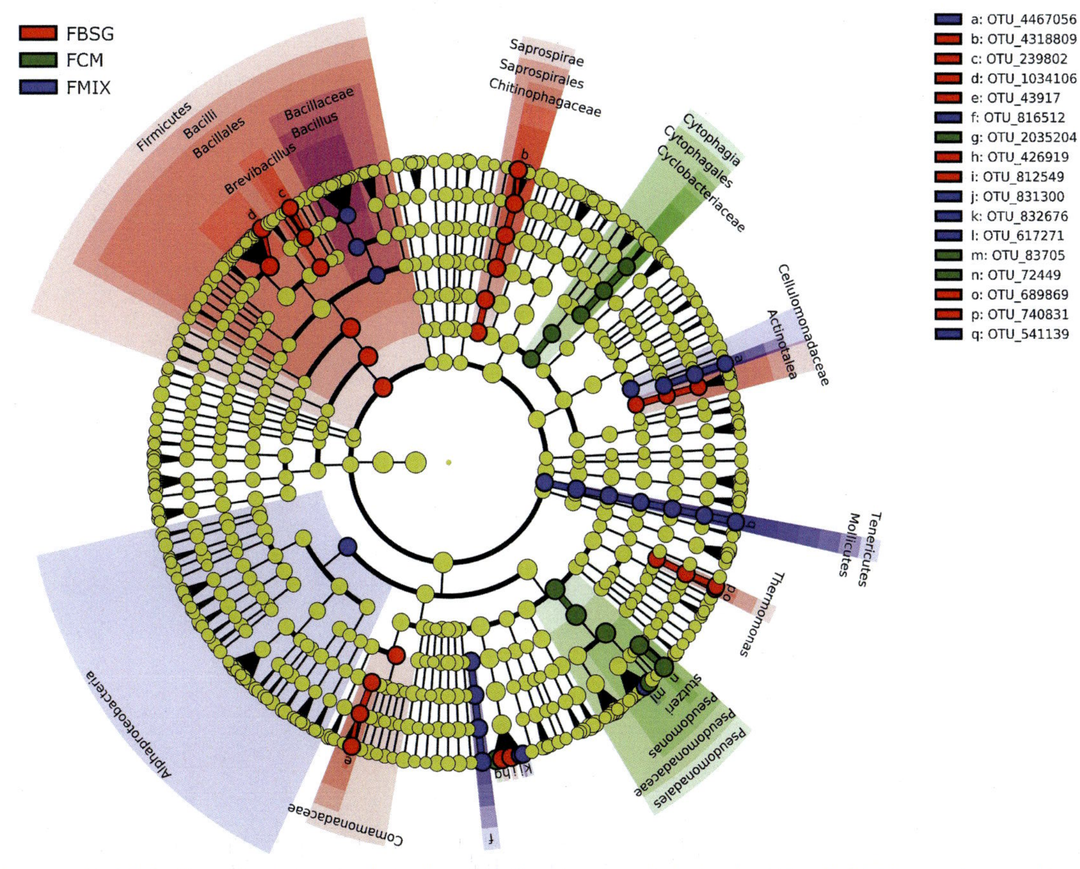

FIG. 4 A cladogram using LEfSe (linear discriminant analysis effect size) algorithms from the isolated and identified OTUs from fermented brewer's spent grains (FBSG), fermented cow manure (FCM) and a mixture of FBSG: FCM (50:50 *v*/v) under *Eisenia fetida*. *(Courtesy Budroni, M., Mannazzu, I., Zara, S., Saba, S., Pais, A., & Zara, G. (2020). Composition and functional profiling of the microbiota in the casts of Eisenia fetida during vermicomposting of brewers' spent grains.* Biotechnology Repor*ts,* 25, *e00439.)*

Their presence was also recorded higher under FBSG during the study. Therefore, the presence of earthworms stimulated the growth of microorganisms in the substrate.

FBSG substrates with low pH and highest TOC%, incubated with *Eisenia fetida*, showed a higher abundance of bacterial taxa and high C mineralization and N assimilation rate, resulting in nutrient-rich vermicast. Aira et al. (2016) reported similar findings in the order of *Eudrillus eugeniea* > *Eisenia andrei* = *Eisenia fetida* for their influence on carbon mineralization, N assimilation, and P-solubilization rate. All these activated enzymes showed positive correlations with sustained nutrient release. Aira et al. (2016) reported a surge in alkaline phosphatase activity in the order *Eisenia andrei* > *Eudrillus eugeniea* > *Eisenia fetida* for 60 days. This increased P-availability in the pig manure substrates. Hussain et al. (2018) reported similar findings with N and P availability, highest recorded under *Eisenia fetida* + *Eudrillus eugeniea* + *Perionyx excavatus* consortium. A change in microbial community structure in soil alters the normal functioning of the surrounding environment. Apart from the C cycle, P and K cycles are also being regulated by earthworm categorization and species selection.

Earthworms play significant roles in the nitrogen (N) cycling processes in soil. Living organisms cannot use direct atmospheric nitrogen; hence, its conversion from inorganic form to dissolved and organic forms is crucial for sustaining life on Earth. Lightning and thunders directly transfer atmospheric N to soil. Symbiotic bacteria like rhizobium present in the root nodules of leguminous plants can fix these nitrogen. The fixed N gets converted into soluble forms with the help of Nitrobacter and Nitrosomonas groups of bacteria. Soluble NO_3^- or NO_2^- ions are readily absorbed by plants and other rhizospheric microbes. Thus, it enters the food web and gets stored in the form of different proteins. When living organisms die, their body decomposes and release protein-rich components to the soil surface. Now, detritivores, including earthworms along with denitrifying bacteria, start their process. With the help of selected gut microbes and protease enzymes, nitrogen-rich organic compounds break down into soluble forms inside the earthworm gut (Aira et al., 2008; Aira & Domínguez, 2008; Lazcano, Gómez-Brandón, & Domínguez, 2008). Denitrifying microbes rapidly convert nitrate nitrogen into gaseous nitrogen. This initiates the nitrogen mineralization process. Earthworms, on the other hand, release half-digested substrates mixed with N-rich mucus, excretory products. This renders a platform for these microbes to grow at a faster pace and reproduce. Aira et al. (2016) reported that such an environment often favors the growth of *Actinobacteria*, *Proteobacteria*, *Firmicutes*, etc. in the decomposing substrates. Hussain et al. (2018) documented these changes in nitrogen-fixing and phosphate-solubilizing bacteria (PSB) profiles in the order of *Eisenia fetida* + *Eudrillus eugeniea* + *Perionyx excavatus* > *Eisenia fetida* + *Eudrillus eugeniea* > *Eisenia fetida* + *Perionyx excavatus* > Compost ($P = .000$; LSD = 0.02); PSB count: *Eisenia fetida* + *Eudrillus eugeniea* + *Perionyx excavatus* > *Eisenia fetida* + *Eudrillus eugeniea* > *Eisenia fetida* + *Perionyx excavatus* > Compost ($P = .000$; LSD = 0.02). Fig. 5A shows clearly how nutrient profiles alter under different consortium-mediated vermicomposting and composting processes. It was evident that NFB and PSB bacteria's presence attributed to an increase in nutrient availability and enzymes like urease, phosphatase activity towards 40–50 days of incubation. These processes increased the microbial biomass carbon and reduced metabolic quotients in the substrate upon maturity.

4.1 Alteration of microbial respiration and biomass: Its impact on soil fertility

Soil organic carbon is one of the most important profiles that are responsible for maintaining soil health. Microbes use carbon for various metabolic activities like microbial biomass production, excretion, and respiration. Therefore, any sudden change in soil carbon can alter the microbial community structure. It can ultimately jeopardize the soil homeostasis. However, there is a dearth of works to establish a direct link among the parameters to understand how microbes affect nutrient stabilization and solubilization processes. Therefore, an *in-depth* study is required to establish the correlation between substrate composition, nutrient fortification, microbial growth, and earthworm speciation.

Apart from earthworm gut fluids, microbial biomass carbon is another critical factor that balances nutrient turnover rate and mineralization during vermicomposting. Bardgett (2005) observed that soil bears the load of grazers like microbe-feeding protozoans under the natural environment. These microbe feeder protozoans have a very low assimilation rate. Therefore, they tend to release high volumes of nitrogen-rich nutrients as excretory products and total microbial biomass. These microbial biomass stock up in the soil over the years and build the fertile soil organic layer. Clarholm (1994) coined the term "microbial loop" to describe such a solution of labile and remobilized nutrients. Initially, bacteria grow profusely due to a higher amount of labile carbons released from complex polysaccharides in the substrates. However, with time, as the decomposition process proceeds, both earthworms and microbes compete for food. During this time, carbon-based substrate behaves as a limiting factor for bacterial growth and earthworm health (Tiunov & Scheu, 2004). Therefore, there are many possibilities that the microbial load reduces in the matured substrate. Their metabolic quotient also gradually decreases with compost maturity. Hence, both microbial biomass carbon and metabolic quotient can adequately account for microbial health in soil or substrates (Six, Frey, Thiet, & Batten, 2006). Pathma and Sakthivel (2013) observed that microbial biomass shows temporal variations. This may be attributed to the higher bacterial and fungal diversity and activity in vermicompost samples.

Few workers reported reduced fungal diversity in the 30-day vermicast than the initial substrate (Frostegård & Bååth, 1996; Zelles, 1997). Goswami et al. (2018) reported that during 60 days of tannery sludge vermicomposting using

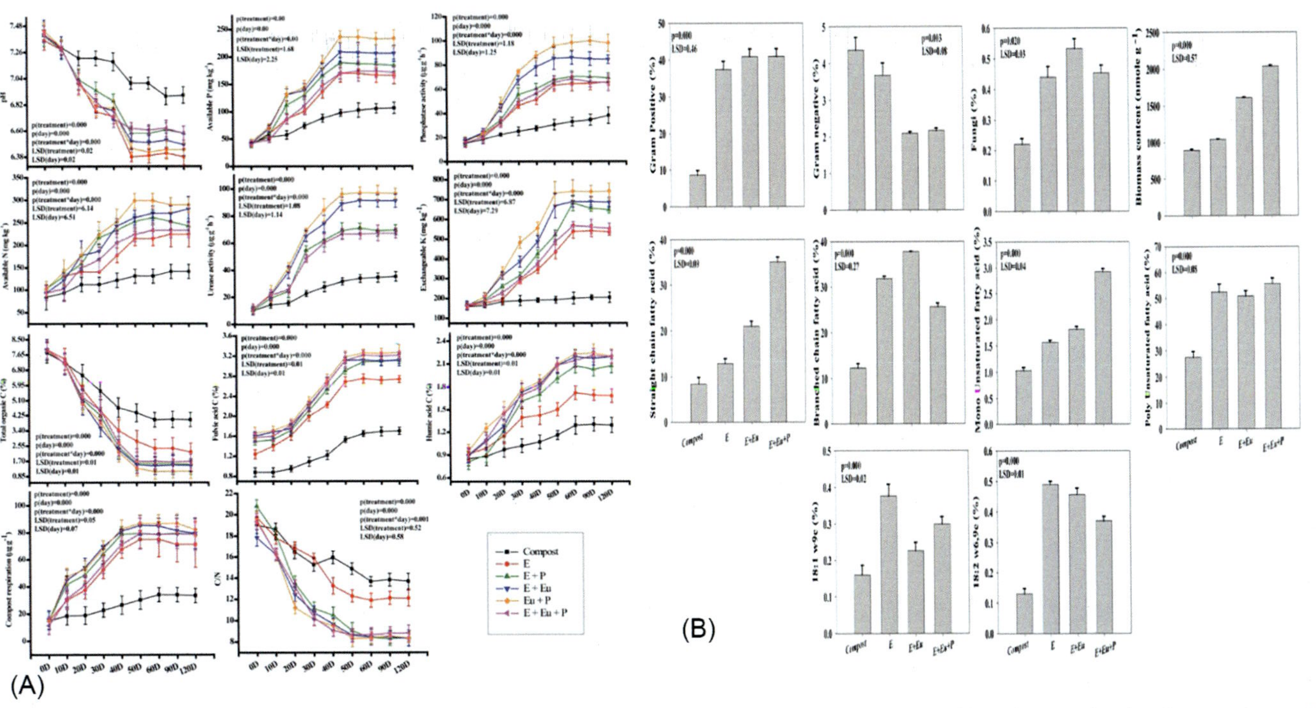

FIG. 5 (A): Physicochemical alterations and nutrient availability under vermicomposting (consortium mediated, double and monoculture) and composting mediated substrates collected temporally at an interval of 10 DAI. (B) PLFA analysis of the substrates during the experimentation under both vermicomposting and composting treatments. *(Courtesy Hussain, N., Das, S., Goswami, L., Das, P., Sahariah, B., & Bhattacharya, S. S. (2018). Intensification of vermitechnology for kitchen vegetable waste and paddy straw employing earthworm consortium: Assessment of maturity time, microbial community structure, and economic benefit.* Journal of Cleaner Production, 182, *414–426.)*

Eudrillus eugeniea, metabolic quotient gets reduced temporally upon compost maturity. Similar findings were reported by Paul, Das, Raul, and Bhattacharya (2018) when the textile industry sludge was used as a substrate for vermicomposting. Such activities and responses may be attributed to the metabolic activities of the earthworm gut-microflora. Abail et al. (2017) said that the presence of both endogeic and anecics earthworms could result in 70% nutrient solubilization; but, epigeic earthworms alone could increase only up to 43%. There have also been reports of functional differences in each category. For example, epigeic earthworms tend to increase P solubilization and mobilization, but endogeic earthworms are more effective towards the N-assimilation process. A higher metabolic quotient tends to represent contaminated surfaces or stress. Phospholipid fatty acid assay is a signature technique to analyse the community structure and microbial biomass in any substrate. Hussain et al. (2018) used PLFA techniques (Fig. 5B) and reported a higher percentage of Gram-positive and -negative bacteria in the consortium-mediated vermicompost. The presence of lower concentrations of branched-chain fatty acids under consortium-mediated vermicompost treatments confirmed these claims. The study also confirmed that vermicomposting could reduce the microbial biomass load by four to five times compared to the aerobic composting process.

5 Microbial gene expression as a functional biomarker of dominance under vermicomposting systems

The microbial gene expression study can be a functional biomarker for particular group of microorganisms present in any substrate. Studies showed that bacterial genes behave explicitly under certain environment. This leads to very specific physicochemical changes in the soil or even vermicomposting and composting surfaces (Bai et al., 2015; Ribbons et al., 2018). For example, earthworms are excellent denitrifying agents under natural condition. Therefore, their presence influence the denitrifying bacterial diversity in the surrounding soil.

Studies carried out by different workers showed that both epigeic and endogeic earthworms can manipulate the soil denitrifying population. They tend to upregulate denitrifying genes like nitrite reductase (*nirK*; *nirS*) and few nitrous oxide reductase genes like *nosZ*, etc. in the substrate. Braker, Zhou, Wu, Devol, and Tiedje (2000) observed the upregulated expression of two nitrite reductase genes (*nirK* and *nirS*) present in a denitrifying soil environment. Upregulation of denitrifying gene expression corresponds to N_2O emissions. However, the burrowing activities of earthworms create aerobic environment that tend to reduce N_2O emission in the long run.

Chapuis-Lardy et al. (2010) recorded the expression of two denitrifying genes *nirK* (nitrite reductase) and *nosZ* (Nitrous oxide reductase) present in the rhizospheric soil samples collected from different experimental sites. They described that studying the expression of *nirK* and *nirS* isolated from different surfaces, could give a clear idea about the N-mineralization status of the surface. Therefore, these genes' expression profile can be used as functional markers, and it varies under different surfaces. The expression of functional markers is also species-specific. Braga et al. (2016) reported that endogeic earthworms like *Pontoscolex corethrurus* upregulate *nirK* and *nosZ* gene expression in the rhizospheric soil. Nebert, Bloem, Lubbers, and van Groenigen (2011) observed that epi-endogeic *Lumbricus rubellus* shows an upregulated *nosZ* gene expression in soil. However, endogeic *Apporectodea caliginosa* showed no effect when incubated under similar environment and constant time period. Under the similar experimental conditions, anecics earthworm *Maoridrilus transalpinus* showed reduced N_2O emission (Kim, Robinson, Lee, Boyer, & Dickinson, 2017) indicating downregulation of denitrification process.

Because of such specificity, these gene expressions can broadly be considered as soil fertility indicators. They could also be used as functional genetic markers to identify the effect of earthworm species selection upon the substrate and feed mixture quality. Hence, it could be summarized that any modification in the selective bacterial populations could be assessed by monitoring the changes in the functional biomarkers in the surrounding environment.

6 Effect on bioremediation

Microbes play important role in the bioremediation process. During the bioconversion process, both aerobic composting and vermicomposting tend to decrease the substrate volume via degradation. Aerobic composting is a time-consuming process where the thermophilic bacteria play the major role.

However, under vermicomposting, mesophilic bacteria perform the conversion process with the help of earthworms. It fastens the process and reduces the toxic load in the substrate by immobilizing the bioavailable metals. There are several bacteria reported to be tolerant to toxic metals. The presence and specificity of heavy metals can influence the growth, metabolism, and community structure of microbial populations under the natural environment (Zampieri, Pinto, Schultz, de Oliveira, & de Oliveira, 2016). Several workers have isolated, identified heavy metal tolerant *Bacillus*, *Arthrobacter*, *Microbacterium*, and *Actinobacteria* from the metal-polluted environment (Piotrowska-Seget, Cycoń, & Kozdroj, 2005). The mechanism of metal remediation by microbes involves different cascade pathways and several environmental factors. They have been found to sequester and precipitate metals via phase distribution and speciation. These behaviors arrest heavy metal mobility. Such traits of selective metal-tolerant bacterial strains were found in 106 genera to date. They are extremely useful in environmental pollution remediation (Qu et al., 2018). Most of these strains are isolated from polluted soils like mine sites, industries like tannery industries, municipal sewage plants, roots and rhizospheric soil of hyperaccumulators, and water resources like the sea. Wang et al. (2017) reported a total of 12 genera of heavy metal-resistant/tolerant bacteria collected from animal excreta. Their minimum inhibitory concentrations and tolerance capacity require further investigation. Further, in vitro and *in-depth* study of such bacterial cultures are a must go to understand their feasibility under the natural environment and their efficacy in metal reduction.

7 Conclusion

Since the very onset, the interwinding relationships between microbes and other organisms play an inevitable role in the life process continuum. Microbes can not only thrive in extreme environment but also facilitate modifications in the external environment. Their activites largely govern the structural changes on the Earth's surface. From the lithosphere to the atmosphere, they have their well-defined line of work and contribute mainly to the maintainence of the homeostasis. Microbes regulate nutrient solubilization, assimilation, mineralization, etc. During continuous life processes, an enormous amount of organic waste is generated on a daily basis. Bioconversion processes, like vermicomposting and aerobic composting, can recycle any organic debris, minimize their waste load, and expedite detoxification rate. Therefore, in this review, we have documented and analyzed how the diversity and abundance of different microbes present on the vermicomposting and aerobic composting surfaces can influence the recycling process and their fate in the long run. The presence of specific species can either fortify or curb specific microbial populations that release extracellular enzymes to facilitate certain biochemical reactions. These biochemical reactions control the nutrient mineralization process. When microbes thrive in a specific environment, it can be assessed by observing the changes in functional biomarker expression in the surrounding environment. Pollutants like toxic metals, hydrocarbons, biogenic pollutants, parasites, pesticides, fungicides, etc. can be regulated by inoculation and exposure of identified beneficial microbial species under an optimized external environment. But there is a dearth of scientific experimentation in this regard. The current study sheds light on the future scopes for in-depth studies to establish the microbial influence on nutrient solubilization and detoxification process. The areas associated with metal remediation and organic amendments are the current burning topics for environmental research and sustainable development.

Acknowledgment

First author, L.G., would like to acknowledge the financial assistance received from University Grants Commission, India in the form of Dr. DS Kothari Postdoctoral fellowship for the years 2019–22 (BL/18-19/0215).

Conflict of interest

None.

References

Abail, Z., Sampedro, L., & Whalen, J. K. (2017). Short-term carbon mineralization from endogeic earthworm casts as influenced by properties of the ingested soil material. *Applied Soil Ecology*, *116*, 79–86.

Aira, M., & Domínguez, J. (2008). Optimizing vermicomposting of animal wastes: Effects of rate of manure application on carbon loss and microbial stabilization. *Journal of Environmental Management*, *88*(4), 1525–1529.

Aira, M., Gómez-Brandón, M., González-Porto, P., & Domínguez, J. (2011). Selective reduction of the pathogenic load of cow manure in an industrial-scale continuous-feeding vermireactor. *Bioresource Technology*, *102*(20), 9633–9637.

Aira, M., Monroy, F., & Domínguez, J. (2007). Earthworms strongly modify microbial biomass and activity triggering enzymatic activities during vermicomposting independently of the application rates of pig slurry. *Science of the Total Environment*, *385*(1–3), 252–261.

Aira, M., Olcina, J., Pérez-Losada, M., & Domínguez, J. (2016). Characterization of the bacterial communities of casts from *Eisenia andrei* fed with different substrates. *Applied Soil Ecology*, *98*, 103–111.

Aira, M., Sampedro, L., Monroy, F., & Domínguez, J. (2008). Detritivorous earthworms directly modify the structure, thus altering the functioning of a microdecomposer food web. *Soil Biology and Biochemistry*, *40*(10), 2511–2516.

Bai, S. H., Reverchon, F., Xu, C. Y., Xu, Z., Blumfield, T. J., Zhao, H., ... Wallace, H. M. (2015). Wood biochar increases nitrogen retention in field settings mainly through abiotic processes. *Soil Biology and Biochemistry*, *90*, 232–240.

Bailey, V. L., Fansler, S. J., Stegen, J. C., & McCue, L. A. (2013). Linking microbial community structure to β-glucosidic function in soil aggregates. *The ISME Journal*, *7*(10), 2044–2053.

Bardgett, R. (2005). *The biology of soil: A community and ecosystem approach*. Oxford University Press.

Bernard, L., Chapuis-Lardy, L., Razafimbelo, T., Razafindrakoto, M., Pablo, A. L., Legname, E., ... Chotte, J. L. (2012). Endogeic earthworms shape bacterial functional communities and affect organic matter mineralization in a tropical soil. *The ISME Journal*, *6*(1), 213–222.

Bhattacharya, S. S., & Chattopadhyay, G. N. (2002). Increasing bioavailability of phosphorus from fly ash through vermicomposting. *Journal of Environmental Quality*, *31*(6), 2116–2119.

Blomstrom, A. L., Lalander, C., Komakech, A. J., Vinnerås, B., & Boqvist, S. (2016). A metagenomic analysis displays the diverse microbial community of a vermicomposting system in Uganda. *Infection Ecology & Epidemiology*, *6*(1), 32453.

Bouche, M. B. (1977). Strategies lombriciennes. *Ecological Bulletins*, 122–132.

Braga, L. P., Yoshiura, C. A., Borges, C. D., Horn, M. A., Brown, G. G., Drake, H. L., & Tsai, S. M. (2016). Disentangling the influence of earthworms in sugarcane rhizosphere. *Scientific Reports*, *6*, 38923.

Braker, G., Zhou, J., Wu, L., Devol, A. H., & Tiedje, J. M. (2000). Nitrite reductase genes (nirK and nirS) as functional markers to investigate diversity of denitrifying bacteria in Pacific Northwest marine sediment communities. *Applied and Environmental Microbiology*, *66*(5), 2096–2104.

Budroni, M., Mannazzu, I., Zara, S., Saba, S., Pais, A., & Zara, G. (2020). Composition and functional profiling of the microbiota in the casts of *Eisenia fetida* during vermicomposting of brewers' spent grains. *Biotechnology Reports*, *25*, e00439.

Cai, L., Gong, X., Sun, X., Li, S., & Yu, X. (2018). Comparison of chemical and microbiological changes during the aerobic composting and vermicomposting of green waste. *PLoS One*, *13* (11), e0207494. https://doi.org/10.1371/journal. pone.0207494.

Castillo, J. M., Romero, E., & Nogales, R. (2013). Dynamics of microbial communities related to biochemical parameters during vermicomposting and maturation of agroindustrial lignocellulose wastes. *Bioresource Technology*, *146*, 345–354.

Chapuis-Lardy, L., Brauman, A., Bernard, L., Pablo, A. L., Toucet, J., Mano, M. J., … Blanchart, E. (2010). Effect of the endogeic earthworm *Pontoscolex corethrurus* on the microbial structure and activity related to CO_2 and N_2O fluxes from a tropical soil (Madagascar). *Applied Soil Ecology*, *45*(3), 201–208.

Chen, Y., Chang, S. K., Chen, J., Zhang, Q., & Yu, H. (2017). Characterization of microbial community succession during vermicomposting of medicinal herbal residues. *Bioresource Technology*, *249*, 542–549.

Cheng, C. L., & Chang, J. S. (2011). Hydrolysis of lignocellulosic feedstock by novel cellulases originating from *Pseudomonas* sp. CL3 for fermentative hydrogen production. *Bioresource Technology*, *102*(18), 8628–8634.

Chigineva, N. I., Aleksandrova, A. V., Marhan, S., Kandeler, E., & Tiunov, A. V. (2011). The importance of mycelial connection at the soil–litter interface for nutrient translocation, enzyme activity and litter decomposition. *Applied Soil Ecology*, *51*, 35–41.

Clarholm, C. (1994). *The microbial loop in the soil*. Uppsala, Sweden: Division for Soil Biology, University of Agricultural Sciences, John Wiley and Sons Ltd.

de Menezes, A. B., Prendergast-Miller, M. T., Macdonald, L. M., Toscas, P., Baker, G., Farrell, M., … Thrall, P. H. (2018). Earthworm-induced shifts in microbial diversity in soils with rare versus established invasive earthworm populations. *FEMS Microbiology Ecology*, *94*(5), fiy051. https://doi.org/10.1093/femsec/fiy051.

deAngelis, K. M., Allgaier, M., Chavarria, Y., Fortney, J. L., Hugenholtz, P., Simmons, B., … Hazen, T. C. (2011). Characterization of trapped lignin-degrading microbes in tropical forest soil. *PLoS One*, *6*(4), e19306.

Devi, J., Deb, U., Barman, S., Das, S., Bhattacharya, S. S., Tsang, Y. F., … Kim, K.-H. (2020). Appraisal of lignocellusoic biomass degrading potential of three earthworm species using vermireactor mediated with spent mushroom substrate: Compost quality, crystallinity, and microbial community structural analysis. *Science of the Total Environment*, *716*, 135215. https://doi.org/10.1016/j.scitotenv.2019.135215.

Domínguez, J. (2004). In C. A. Edwards (Ed.), *Earthworm ecology* (pp. 401–424). CRC Press.

Domínguez, J., Aira, M., & Gómez-Brandón, M. (2010). Vermicomposting: Earthworms enhance the work of microbes. In H. Insam, et al. (Eds.), *Microbes at Work*. Berlin, Heidelberg: Springer.

Eastman, B. R., Kane, P. N., Edwards, C. A., Trytek, L., Gunadi, B., Stermer, A. L., & Mobley, J. R. (2001). The effectiveness of vermiculture in human pathogen reduction for USEPA biosolids stabilization. *Compost Science & Utilization*, *9*, 38–49.

Fjøsne, T., Myromslien, F. D., Wilson, R. C., & Rudi, K. (2018). Earthworms are associated with subpopulations of Gammaproteobacteria irrespective of the total soil microbiota composition and stability. *FEMS Microbiology Letters*, *365*(9), fny071.

Frostegård, A., & Bååth, E. (1996). The use of phospholipid fatty acid analysis to estimate bacterial and fungal biomass in soil. *Biology and Fertility of Soils*, *22*(1–2), 59–65.

Gardner, J. G. (2016). Polysaccharide degradation systems of the saprophytic bacterium *Cellvibrio japonicus*. *World Journal of Microbiology and Biotechnology*, *32*(7), 121.

Gates, G. E. (1972). Burmese earthworms: An introduction to the systematics and biology of megadrile oligochaetes with special reference to Southeast Asia. *Transactions of the American Philosophical Society*, *62*(7), 1–326.

Gong, X., Jiang, Y., Zheng, Y., Chen, X., Li, H., Hu, F., … Scheu, S. (2018). Earthworms differentially modify the microbiome of arable soils varying in residue management. *Soil Biology and Biochemistry*, *121*, 120–129.

Goswami, L., Mukhopadhyay, R., Bhattacharya, S. S., Das, P., & Goswami, R. (2018). Detoxification of chromium-rich tannery industry sludge by Eudrillus eugeniae: Insight on compost quality fortification and microbial enrichment. *Bioresource Technology*, *266*, 472–481.

Goswami, L., Patel, A. K., Dutta, G., Bhattacharyya, P., Gogoi, N., & Bhattacharya, S. S. (2013). Hazard remediation and recycling of tea industry and paper mill bottom ash through vermiconversion. *Chemosphere*, *92*(6), 708–713.

Goswami, L., Sarkar, S., Mukherjee, S., Das, S., Barman, S., Raul, P., ... Bhattacharya, S. S. (2014). Vermicomposting of tea factory coal ash: Metal accumulation and metallothionein response in *Eisenia fetida* (Savigny) and Lampito mauritii (Kinberg). *Bioresource Technology*, *166*, 96–102.

Grady, E. N., MacDonald, J., Liu, L., Richman, A., & Yuan, Z.-C. (2016). Current knowledge and perspectives of Paenibacillus: A review. *Microbial Cell Factories*, *15*, 203. https://doi.org/10.1186/s12934-016-0603-7.

Haagsma, J. (1991). Pathogenic anaerobic bacteria and the environment. *Revue Scientifique et Technique*, *10*(3), 749–764.

Hill, G. B., & Baldwin, S. A. (2012). Vermicomposting toilets, an alternative to latrine style microbial composting toilets, prove far superior in mass reduction, pathogen destruction, compost quality, and operational cost. *Waste Management*, *32*(10), 1811–1820.

Huang, K., Li, F., Wei, Y., Chen, X., & Fu, X. (2013). Changes of bacterial and fungal community compositions during vermicomposting of vegetable wastes by *Eisenia foetida*. *Bioresource Technology*, *150*, 235–241.

Huang, K., Xia, H., Wu, Y., Chen, J., Cui, G., Li, F., ... Wu, N. (2018). Effects of earthworms on the fate of tetracycline and fluoroquinolone resistance genes of sewage sludge during vermicomposting. *Bioresource Technology*, *259*, 32–39.

Huang, K., Xia, H., Zhang, Y., Li, J., Cui, G., Li, F., ... Wu, N. (2020). Elimination of antibiotic resistance genes and human pathogenic bacteria by earthworms during vermicomposting of dewatered sludge by metagenomic analysis. *Bioresource Technology*, *297*, 122451.

Hussain, N., Das, S., Goswami, L., Das, P., Sahariah, B., & Bhattacharya, S. S. (2018). Intensification of vermitechnology for kitchen vegetable waste and paddy straw employing earthworm consortium: Assessment of maturity time, microbial community structure, and economic benefit. *Journal of Cleaner Production*, *182*, 414–426.

Hussain, N., Singh, A., Saha, S., Kumar, M. V. S., Bhattacharyya, P., & Bhattacharya, S. S. (2016). Excellent N-fixing and P-solubilizing traits in earthworm gut-isolated bacteria: A vermicompost based assessment with vegetable market waste and rice straw feed mixtures. *Bioresource Technology*, *222*, 165–174.

Kim, Y. N., Robinson, B., Lee, K. A., Boyer, S., & Dickinson, N. (2017). Interactions between earthworm burrowing, growth of a leguminous shrub and nitrogen cycling in a former agricultural soil. *Applied Soil Ecology*, *110*, 79–87.

Kirby, R. (2005). Actinomycetes and lignin degradation. *Advances in Applied Microbiology*, *58*, 125–168.

Konig, H. (2006). *Bacillus* species in the intestine of termites and other soil invertebrates. *Journal of Applied Microbiology*, *101*(3), 620–627.

Koubova, A., Chroňáková, A., Pižl, V., Sánchez-Monedero, M. A., & Elhottová, D. (2015). The effects of earthworms *Eisenia* spp. on microbial community are habitat dependent. *European Journal of Soil Biology*, *68*, 42–55.

Langarica-Fuentes, A., Zafar, U., Heyworth, A., Brown, T., Fox, G., & Robson, G. D. (2014). Fungal succession in an in-vessel composting system characterized using 454 pyrosequencing. *FEMS Microbiology Ecology*, *88*(2), 296–308.

Lavelle, P., & Spain, A. V. (2001). *Soil ecology*. Springer Science & Business Media.

Lazcano, C., Gómez-Brandón, M., & Domínguez, J. (2008). Comparison of the effectiveness of composting and vermicomposting for the biological stabilization of cattle manure. *Chemosphere*, *72*(7), 1013–1019.

Lim, J. W., Chiam, J. A., & Wang, J. Y. (2014). Microbial community structure reveals how microaeration improves fermentation during anaerobic co-digestion of brown water and food waste. *Bioresource Technology*, *171*, 132–138.

Maity, S., Bhattacharya, S., & Chaudhury, S. (2009). Metallothionein response in earthworms Lampito mauritii (Kinberg) exposed to fly ash. *Chemosphere*, *77*, 319–324.

Medina-Sauza, R. M., Álvarez-Jiménez, M., Delhal, A., Reverchon, F., Blouin, M., Guerrero-Analco, J. A., … Barois, I. (2019). Earthworms building up soil microbiota, a review. *Frontiers in Environmental Science*, *7*, 81.

Nebert, L. D., Bloem, J., Lubbers, I. M., & van Groenigen, J. W. (2011). Association of earthworm-denitrifier interactions with increased emission of nitrous oxide from soil mesocosms amended with crop residue. *Applied and Environmental Microbiology*, *77*(12), 4097–4104.

Pathma, J., & Sakthivel, N. (2013). Molecular and functional characterization of bacteria isolated from straw and goat manure based vermicompost. *Applied Soil Ecology*, *70*, 33–47.

Paul, S., Das, S., Raul, P., & Bhattacharya, S. S. (2018). Vermi-sanitization of toxic silk industry waste employing Eisenia fetida and Eudrilus eugeniae: Substrate compatibility, nutrient enrichment and metal accumulation dynamics. *Bioresource Technology*, *266*, 267–274. https://doi.org/10.1016/j.biortech.2018.06.092.

Paul, S., Goswami, L., Pegu, R., & Bhattacharya, S. S. (2020). Vermiremediation of cotton textile sludge by Eudrillus eugeniae: Insight into metal budgeting, chromium speciation, and humic substance interactions. *Bioresource Technology*, *314*, 123753. https://doi.org/10.1016/j.biortech.2020.123753.

Piotrowska-Seget, Z., Cycoń, M., & Kozdroj, J. (2005). Metal-tolerant bacteria occurring in heavily polluted soil and mine spoil. *Applied Soil Ecology*, *28*(3), 237–246.

Qu, M., Chen, J., Huang, Q., Chen, J., Xu, Y., Luo, J., … Zheng, Y. (2018). Bioremediation of hexavalent chromium contaminated soil by a bioleaching system with weak magnetic fields. *International Biodeterioration & Biodegradation*, *128*, 41–47.

Ribbons, R. R., Kepfer-Rojas, S., Kosawang, C., Hansen, O. K., Ambus, P., McDonald, M., … Vesterdal, L. (2018). Context-dependent tree species effects on soil nitrogen transformations and related microbial functional genes. *Biogeochemistry*, *140*(2), 145–160.

Romero-Tepal, E. M., Contreras-Blancas, E., Navarro-Noya, Y. E., Ruíz-Valdiviezo, V. M., Luna-Guido, M., Gutiérrez-Miceli, F. A., & Dendooven, L. (2014). Changes in the bacterial community structure in stored wormbed leachate. *Journal of Molecular Microbiology and Biotechnology*, *24*(2), 105–113.

Sahariah, B., Goswami, L., Kim, K. H., Bhattacharyya, P., & Bhattacharya, S. S. (2015). Metal remediation and biodegradation potential of earthworm species on municipal solid waste: A parallel analysis between *Metaphire posthuma* and *Eisenia fetida*. *Bioresource Technology*, *180*, 230–236.

Sampedro, L., & Domínguez, J. (2008). Stable isotope natural abundances (δ13C and δ15N) of the earthworm *Eisenia fetida* and other soil fauna living in two different vermicomposting environments. *Applied Soil Ecology*, *38*(2), 91–99.

Sampedro, L., Jeannotte, R., & Whalen, J. K. (2006). Trophic transfer of fatty acids from gut microbiota to the earthworm *Lumbricus terrestris* L. *Soil Biology and Biochemistry*, *38*(8), 2188–2198.

Sampedro, L., & Whalen, J. K. (2007). Changes in the fatty acid profiles through the digestive tract of the earthworm *Lumbricus terrestris* L. *Applied Soil Ecology*, *35*(1), 226–236.

Schmidt-Dannert, C. (2016). Biocatalytic portfolio of Basidiomycota. *Current Opinion in Chemical Biology*, *31*, 40–49.

Sen, B., & Chandra, T. S. (2009). Do earthworms affect dynamics of functional response and genetic structure of microbial community in a lab-scale composting system? *Bioresource Technology*, *100*(2), 804–811.

Singh, A., Singh, D. P., Tiwari, R., Kumar, K., Singh, R. V., Singh, S., … Nain, L. (2015). Taxonomic and functional annotation of gut bacterial communities of *Eisenia foetida* and *Perionyx excavatus*. *Microbiological Research*, *175*, 48–56. https://doi.org/10.1016/j.micres.2015.03.003.

Six, J., Frey, S. D., Thiet, R. K., & Batten, K. M. (2006). Bacterial and fungal contributions to carbon sequestration in agroecosystems. *Soil Science Society of America Journal*, *70*(2), 555–569.

Sturzenbaum, S. R., Winters, C., Galay, M., Morgan, A. J., & Kille, P. (2001). Metal ion trafficking in earthworms—Identification of a cadmium specific metallothionein. *Journal of Biological Chemistry*, *276*(36), 34013–34018.

Tiunov, A. V., & Scheu, S. (2004). Carbon availability controls the growth of detritivores (Lumbricidae) and their effect on nitrogen mineralization. *Oecologia*, *138*(1), 83–90.

Van Dyk, J. S., & Pletschke, B. I. (2012). A review of lignocellulose bioconversion using enzymatic hydrolysis and synergistic cooperation between enzymes—Factors affecting enzymes, conversion and synergy. *Biotechnology Advances*, *30*(6), 1458–1480.

Varma, V. S., Nashine, S., Sastri, C. V., & Kalamdhad, A. S. (2017). Influence of carbide sludge on microbial diversity and degradation of lignocellulose during in-vessel composting of agricultural waste. *Ecological Engineering*, *101*, 155–161.

Wang, J., Gao, Q., & Bao, J. (2016). Genome sequence of *Trichosporon cutaneum* ACCC 20271: An oleaginous yeast with excellent lignocellulose derived inhibitor tolerance. *Journal of Biotechnology*, *228*, 50–51.

Wang, Y., Han, W., Wang, X., Chen, H., Zhu, F., Wang, X., & Lei, C. (2017). Speciation of heavy metals and bacteria in cow dung after vermicomposting by the earthworm, *Eisenia fetida*. *Bioresource Technology*, *245*, 411–418.

Yadav, K. D., Tare, V., & Ahammed, M. M. (2010). Vermicomposting of source-separated human faeces for nutrient recycling. *Waste Management*, *30*, 50–56.

Yasir, M., Aslam, Z., Kim, S. W., Lee, S. W., Jeon, C. O., & Chung, Y. R. (2009). Bacterial community composition and chitinase gene diversity of vermicompost with antifungal activity. *Bioresource Technology*, *100*(19), 4396–4403.

Zampieri, B. D. B., Pinto, A. B., Schultz, L., de Oliveira, M. A., & de Oliveira, A. J. F. C. (2016). Diversity and distribution of heavy metal-resistant bacteria in polluted sediments of the Araça Bay, São Sebastião (SP), and the relationship between heavy metals and organic matter concentrations. *Microbial Ecology*, *72*(3), 582–594.

Zelles, L. (1997). Phospholipid fatty acid profiles in selected members of soil microbial communities. *Chemosphere*, *35*(1–2), 275–294.

Zhang, H., Tan, S. N., Wong, W. S., Ng, C. Y. L., Teo, C. H., Ge, L., … Yong, J. W. H. (2014). Mass spectrometric evidence for the occurrence of plant growth promoting cytokinins in vermicompost tea. *Biology and Fertility of Soils*, *50*(2), 401–403.

Zhou, Y., Selvam, A., & Wong, J. W. (2016). Effect of Chinese medicinal herbal residues on microbial community succession and anti-pathogenic properties during co-composting with food waste. *Bioresource Technology*, *217*, 190–199.

Chapter 6

Potential of compost for sustainable crop production and soil health

Sharjeel Ahmad[a,b], Rabia Khalid[b], Saira Abbas[c], Rifat Hayat[b], and Iftikhar Ahmed[a]

[a]National Microbial Culture Collection of Pakistan (NCCP), Bio-resource Conservation Institute (BCI), National Agriculture Research Center (NARC), Islamabad, Pakistan, [b]PirMehr Ali Shah Arid Agriculture University, Rawalpindi, Pakistan, [c]Department of Zoology, University of Science and Technology, Bannu, Pakistan

Chapter outline

1 Introduction

Composting is the disintegration of waste materials such as leaves, shredded twigs, and kitchen scraps from plants in the controlled environmental conditions to get valuable organic substance (Smith, Brown, Ogilvie, Rushton, & Bates, 2001). Composting is a solid biodegradable phase, self-heating process, of organic wastes (Finstein & Morris, 1975). Nowadays, composting becomes a major and crucial factor in sustainable agriculture (Ryckeboer et al., 2003). The main purposes of composting are (i) to add plant nutrients in their available form, (ii) to improve the soil quality, (iii) to employ the organic wastes, and (iv) to save money (Kaushal & Bharti, 2015).

Recent Advancement in Microbial Biotechnology. https://doi.org/10.1016/B978-0-12-822098-6.00005-7

The process of composting is useful and convenient and is a good substitute for land filling. Natural recycling (composting) occurs by the disintegration of organic wastes by microbes, being consumed by invertebrates, and returns to the soil to provide nourishment to aid plant growth. Composting helps in nutrients recycling in the soil; continuous use results in improving soil fertility and controlling soil erosion by enhancing structural stability. Composting is observed as an environmentally friendly waste treatment technology (Tweib, Rahman, & Kalil, 2011). This method is very valuable in devastating pathogens and enhancing nitrification process (Fauziah & Agamuthu, 2009). It can assist to the carbon sequestration and can relatively switch the use of peat and fertilizers. Composting is generally based on the presence and working of microbes. There is diverse genetic heterogeneity in microbes that are present in the soil environment (Sait, Hugenholtz, & Janssen, 2002).

Composting becomes an economically beneficial process that is easily adopted by farmers in order to utilize their field organic waste and to use an alternative of fertilizers. The whole composting process has three basic phases: (i) microflora activities, (ii) mesophilic stage (temperature booster stage), and (iii) thermophilic stage (heat-loving phase) (Hoitink & Boehm, 1999). By passing through these phases, organic wastes are eventually converted into humus that is a fine dark dry organic matter (Büyüksönmez, Rynk, Hess, & Bechinski, 2000). Composting is a substitute for the solid waste management system.

It not only increases organic carbon content in soil, but it is also used to rectify soil structure, water holding capacity, water infiltration rate, and soil tilth. The volume of the waste material is decreased by the use of this process resulting in nutrient rich and stable material as compared to the initial material (Barral, Paradelo, Moldes, Domínguez, & Díaz-Fierros, 2009). The compost characteristics mostly depend on the quality of initial material and the technique by which composting is carried out (Bernal, Alburquerque, & Moral, 2009). The compositions of some compostable wastes are given in Table 1.

It involves several factors that are interconnected at the microbial level i.e., the moisture content, oxidative input, thermal relationship, and available nutrients (Ryckeboer et al., 2003). In recent composting techniques, there are different additives or substrates that are employed in the wastes during the composting process that enhances the quality of compost that can be organic (Gabhane et al., 2012; Zhang & Sun, 2017), mineral (Himanen & Hänninen, 2009; Wang et al., 2017), biological (Jurado et al., 2015; Wang et al., 2017), or a mixed substrate (Hayawin et al., 2014; Wang et al., 2017) (which are known as bulking agents, having both beneficial or detrimental effects or act as additives) like urea (Villasenor, Rodriguez, & Fernandez, 2011). This chapter covers the different stages of composting, its types, its biochemistry, and compost impact on crop and soil quality.

TABLE 1 The composition of some important compostable dry organic waste materials (Berkeley, 1950; Galier & Partridge, 1969; Glathe, 1959).

Components	Garbage (%)	Plant material (%)	Sewage sludge (% activated)
Volatile matter	85–90	–	–
Moisture	60–70	–	4.2
Total sugar	–	–	–
Total carbohydrate	32	–	–
Cellulose	10	15–60	2.6
Starch	–	–	–
Protein	12–18	5–40	37.0
Lignin	–	5–30	6.9
Water soluble	–	5–30	1.8–2.8
Phosphorous	0.5–0.6	0.05–1.5	2.5
Potassium	0.7–1.8	0.3–0.6	0.4
Lipids	9–15	5–15	4.7
Crude fiber	4–18	–	–

2 Composting, types, and phases

2.1 Process of composting

Composting is a self-contained process that utilizes microbes in preferably thermophilic temperature conditions (40–65°C). Composting is a core and economically favorable process involving a wide variety of microbes. The main steps of composting processes is to select the category of compost bin either you want to use open pile or a compost bin, then select the composter location if we want to use flat, well-drained, or sunny location, then initiate with a coating of bristly materials (like twigs) to permit drainage and aeration, and cover this coating with leaves or simple covering among layers of green materials (nitrogen-rich materials) and browns (carbon-rich materials), add the kitchen as well as yard wastes, keep on adding layers till the trunk is fully loaded. Place the compost in a suitable place to collect when it is fully prepared (Agnew & Leonard, 2003).

For piles, which are mostly brown colored material (dead leaves), the application of commercial fertilizers will aid to supply nitrogen and accelerate the composting process. If the pile is too wet or too dry, then add water or dry material according to the condition (Büyüksönmez et al., 2000; Büyüksönmez, Rynk, Hess, & Bechinski, 1999). The optimum temperature for composting is 30–80°C (Albrecht, Joffre, Petit, Terrom, & Périssol, 2009). The end products are considered as stable and are without any phytotoxic agents and pathogens (Wichuk & McCartney, 2007). Composting involves the progressions of microbial communities at different stages (Insam & De Bertoldi, 2007; Mehta, Palni, Franke-Whittle, & Sharma, 2014).

The main parameters controlling the rate of composting process are the water content, nutrient contents, pH level, oxygen demand, the temperature required, time period, and numerous types of mesophilic and thermophilic microbes present (Mohee & Mudhoo, 2005). The most suitable composting materials are sewage sludge, different firms (industrial) trashes (e.g., food, pulp, and paper), yard and garden scraps, municipal solid wastes (up to 70% organic matter by weight), soft pruning, clippings and leaves, kitchen wastes like fruit, peelings, teabag sand eggshell, and torn paper, mixed with grass cuttings and used carefully (Chang & Hsu, 2008; Diener, Collins, Martin, & Bryan, 1993).

Composting process can be done in many ways ranging from using minor, indoor worm bins that requires a few pounds of food per week to the huge commercial operations that requires many tons of organic wastes in long, outdoor piles called windrows (Trautmann & Krasny, 1998). Some common use methods of composting are static pile, in-vessel, bin composter, windrow, and vermicomposting (Fauziah & Agamuthu, 2009).

2.2 Types of composting

There are numerous forms of composting resulted in a product that not only improves soil physical and chemical properties but also reduces the cost of fertilizers, which is briefly demonstrated in Fig. 1.

2.2.1 Aerobic composting

The aerobic composting is the decomposition of organic waste matters in the presence of air (oxygen) (Hepperly, Lotter, Ulsh, Seidel, & Reider, 2009). In this type, the moisture content is approximately 55%–75%, and the carbon to nitrogen ratio (C/N) is 30 (Tweib et al., 2011). The aerobic composting forms transitional compounds like organic acids and aerobic microorganisms (*Nocardia*, *Pseudomonas* spp., *Bacillus*, and *Mycobacterium* spp.) disintegrate them further. The final compost, with its quite volatile form of organic matter, has the least chances of phytotoxicity. This method is considered very beneficial and productive than the anaerobic composting for agricultural production. During aerobic composting, the microbial population is very high and varies with the stage of disintegration of organic material and type of organic wastes (Atkinson, Jones, & Gauthier, 1996), as shown in Table 2.

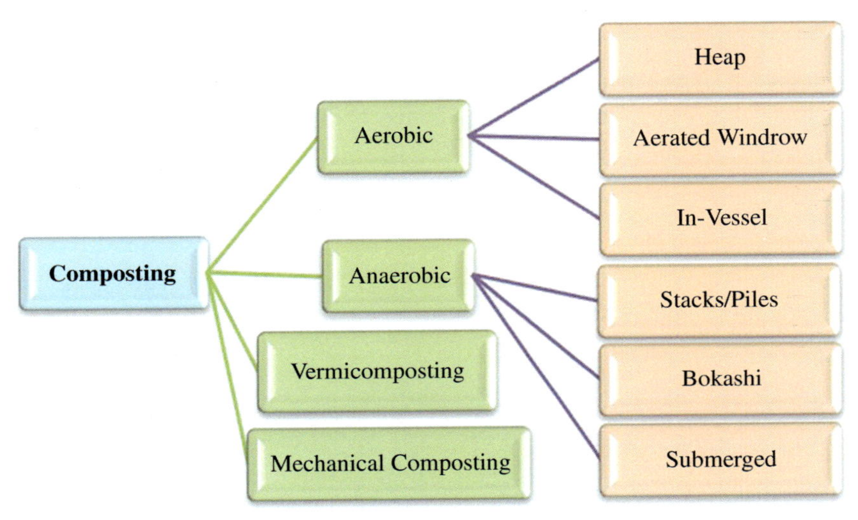

FIG. 1 Different types of composting.

The prokaryotic diversity during the process of aerobic composting is identified by many researchers (Beffa et al., 1998; Strom, 1985). The fungal dominance in municipal wastes decomposition was also reported (Lim, Chin, AnizaYusof, Yahya, & Tee, 2019; Rajeswari, Padmavathy, & Aanand, 2018; von Klopotek, 1969). The whole process of aerobic composting is the natural process by which organic matter is decomposed and forms a humus-like substance (Misra, Roy, & Hiraoka, 2003).

2.2.1.1 Heap method

The most conventional and common method of aerobic composting is the heap method, where organic matter is split into three different kinds and to be set in a heap and coated with a lean layer of soil or dry leaves. The heap method needs to mix every week, and it requires about 3 weeks giving the resultant material. This method is the same as the pit method (Inckel, de Smet, Tersmette, & Veldkamp, 2005).

2.2.1.2 Aerated windrow composting

The next method is aerated windrow composting, in which (organic) waste material is shaped in the lines (rows) of long piles called "windrows" and oxygenated by rotating the pile periodically by either mechanically or manually. Windrow composting often requires a large scale of land, equipment, labors, and patience to generate this experiment. By this method, the resultant material is prepared in 3–9 weeks, depending on environmental conditions and management (Zafar, 2017). The ideal pile height and width area are about 4–8′ and

TABLE 2 Different microorganism presence in numerous type of waste material at various temperature phases during composting (Chang & Hsu, 2008; Hassen et al., 2001; Möller & Stinner, 2009; Rivard et al., 1995; Ryckeboer et al., 2003; Schulze, 1962).

Sr. No.	Organisms	Source material	Temperature phase
1	*Achromobacter* sp.	Waste paper	Mesophilic, Thermophilic
2	*Absidia* sp.	Municipal solid waste	Mesophilic
3	*Absidia corymbifera*	Municipal solid waste, garden waste, wheat straw, barley straw and straw compost and mushroom compost	Mesophilic, Thermophilic
4	*Absidia orchidis*	Mushroom compost	Mesophilic
5	*Absidia ramosa*	Garden and kitchen waste and sewage sludge and mushroom compost	Thermophilic
6	*Achromobacter xylosoxidans*	Vegetables, fruits and garden waste	Mesophilic
7	*Acidovorax facilis*	Vegetables, fruits and garden waste	Mesophilic
8	*Acidovorax* sp.	Waste paper	Mesophilic
9	*Acinetobacter* sp.	Vegetables, fruits and garden waste	Mesophilic, Thermophilic
10	*Acremoniella* sp.	Lawn cuttings and barley straw	Thermophilic
11	*Acremonium atrogriseum*	Vegetables, fruits and garden waste	Mesophilic
12	*Acremonium breve*	Vermicompost	Mesophilic
13	*Acremonium butyric*	Mushroom compost	Mesophilic
14	*Acremonium charticola*	Garden waste	Mesophilic
15	*Acremonium chrysogenum*	Vegetables, fruits and garden waste, garden waste, 70% green waste, 10% kitchen waste and 20% shredded wood and vermicompost	Mesophilic

TABLE 2 Different microorganism presence in numerous type of waste material at various temperature phases during composting (Chang & Hsu, 2008; Hassen et al., 2001; Möller & Stinner, 2009; Rivard et al., 1995; Ryckeboer et al., 2003; Schulze, 1962)—cont'd

Sr. No.	Organisms	Source material	Temperature phase
16	*Acremonium furcatum*	Vegetables, fruits and garden waste	Mesophilic
17	*Acremonium kiliense*	Vegetables, fruits and garden waste	Mesophilic
18	*Acremonium murorum*	Garden and kitchen waste and sewage sludge	Mesophilic
19	*Acremonium sclerotigenum*	Garden waste and vermicompost	Mesophilic
20	*Acremonium* sp.	Vegetables, fruits and garden waste, garden waste, mushroom compost and vermicompost	Mesophilic
21	*Acremonium strictum*	Vegetables, fruits and garden waste, garden waste, lawn cuttings and barley straw and vermicompost	Mesophilic
22	*Acremonium thermophilum*	Vegetables, fruits and garden waste	Mesophilic, Thermophilic
23	*Acremonium verruculosum*	Vermicompost	Mesophilic
24	*Acrodontium griseum*	Garden waste	Mesophilic
25	*Acrophialophora fusispora*	Garden waste and vermicompost	Mesophilic
26	*Actinomucor elegans*	Vegetables, fruits and garden waste, garden waste	Mesophilic
27	*Actinomucor* sp.	Municipal solid waste	Mesophilic
28	*Actinomyces* sp.	Manure, horse manure and waste paper	Thermophilic
29	*Agaricus bisporus*	Mushroom compost	Mesophilic
30	*Alcaligenes faecalis*	Waste paper	Mesophilic, Thermophilic

Continued

TABLE 2 Different microorganism presence in numerous type of waste material at various temperature phases during composting (Chang & Hsu, 2008; Hassen et al., 2001; Möller & Stinner, 2009; Rivard et al., 1995; Ryckeboer et al., 2003; Schulze, 1962)—cont'd

Sr. No.	Organisms	Source material	Temperature phase
31	*Alcaligenes* sp.	Manure, horse manure	Mesophilic
32	*Aleurisma* sp.	Municipal solid waste	Mesophilic
33	*Alternaria alternate*	Vegetables, fruits and garden waste, Municipal solid waste, garden and kitchen waste and sewage sludge, grass (lawn cuttings), garden waste, wheat straw, barley straw and straw compost and vermicompost	Mesophilic
34	*Alternaria* sp.	Manure and horse manure	Mesophilic
35	*Amphibacillus xylanus*	Vermicompost	Mesophilic
36	*Aphanoas custerreus*	Vermicompost	Mesophilic
37	*Apiospora montagnei*	Vermicompost	Mesophilic
38	*Armillaria mellea*	Garden and kitchen waste and sewage sludge	Mesophilic
39	*Arthrinium phaeospermum*	Garden waste	Mesophilic
40	*Arthrinium phaeospermum* spp.	Garden waste	Mesophilic
41	*Arthrobacter ilicis*	Waste paper	Mesophilic
42	*Arthrobacter* sp.	Vegetables, fruits and garden waste	Mesophilic
43	*Arthrobotrys* sp.	Municipal solid waste, garden waste	Mesophilic
44	*Arthrobotrys* sp.	Municipal solid waste, garden waste	Mesophilic
45	*Arthrobotrys amerospora*	Vegetables, fruits and garden waste	Mesophilic

TABLE 2 Different microorganism presence in numerous type of waste material at various temperature phases during composting (Chang & Hsu, 2008; Hassen et al., 2001; Möller & Stinner, 2009; Rivard et al., 1995; Ryckeboer et al., 2003; Schulze, 1962)—cont'd

Sr. No.	Organisms	Source material	Temperature phase
46	*Arthrobotrys amerospora* spp.	Vegetables, fruits and garden waste	Mesophilic
47	*Arthrobotrys oligospora*	Vegetables, fruits and garden waste, garden and kitchen waste and sewage sludge, garden waste and mushroom compost	Mesophilic
48	*Arthrobotrys oligospora* spp.	Vegetables, fruits and garden waste, garden and kitchen waste and sewage sludge, garden waste and mushroom compost	Mesophilic
49	*Ascodesmis microscopic*	Garden waste	Mesophilic
50	*Ascodesmis microscopic*	Garden waste	Mesophilic
51	*Aspergillus niger*	Municipal solid waste, grass (lawn cuttings), garden waste, mushroom compost and vermicompost	Mesophilic
52	*Aspergillus orchraceous*	Garden waste, mushroom compost and vermicompost	Mesophilic
53	*Aspergillus oryzae*	Garden waste	Mesophilic
54	*Aspergillus parasiticus*	Mushroom compost	Mesophilic
55	*Aspergillus puniceus*	Vermicompost	Mesophilic
56	*Aspergillus ruber*	Mushroom compost	Mesophilic
57	*Aspergillus* sp.	Vegetables, fruits and garden waste, Municipal solid waste, garden and kitchen waste and sewage sludge, grass (lawn cuttings), garden waste,	Mesophilic, Thermophilic

Continued

TABLE 2 Different microorganism presence in numerous type of waste material at various temperature phases during composting (Chang & Hsu, 2008; Hassen et al., 2001; Möller & Stinner, 2009; Rivard et al., 1995; Ryckeboer et al., 2003; Schulze, 1962)—cont'd

Sr. No.	Organisms	Source material	Temperature phase
		tree bark, hard-wood bark; bark compost, peanut shells, manure, horse manure, vermicompost	
58	*Aspergillus sulphureus*	Cattle manure and forestry wastes	Mesophilic
59	*Aspergillus sydowii*	Mushroom compost	Mesophilic
60	*Aspergillus terreus*	Grass (lawn cuttings), wheat straw, barley straw and straw compost and mushroom compost	Mesophilic, Thermophilic
61	*Aspergillus terreus*	Garden waste	Mesophilic
62	*Aspergillus terreus*	Garden waste, vermicompost	Mesophilic
63	*Aspergillus versicolor*	Vegetables, fruits and garden waste, kitchen waste and shredded newspapers, garden waste and tree bark, hard-wood bark; bark compost	Mesophilic
64	*Aspergillus wentii*	Cattle manure and forestry wastes	Mesophilic
65	*Aureobasidium pullulans*	Grass (lawn cuttings), garden waste and mushroom compost	Mesophilic
66	*Aureobasidium pullulans*	Cattle manure and forestry wastes	Mesophilic
67	*Aureobasidium* sp.	Wheat straw, barley straw and straw compost	Mesophilic
68	*Azotobacter* sp.	Cattle manure and forestry wastes	Mesophilic

TABLE 2 Different microorganism presence in numerous type of waste material at various temperature phases during composting (Chang & Hsu, 2008; Hassen et al., 2001; Möller & Stinner, 2009; Rivard et al., 1995; Ryckeboer et al., 2003; Schulze, 1962)—cont'd

Sr. No.	Organisms	Source material	Temperature phase
69	*Bacillus amyloliquefaciens* *Bacillus badius* *Bacillus cereus*	Vegetables, fruits and garden waste, tree bark, hard-wood bark; bark compost, manure, horse manure	Mesophilic, Thermophilic
70	*Bacillus circulans*	Cattle manure and forestry wastes	Mesophilic
71	*Bacillus coagulans*	Vegetables, fruits and garden waste, kitchen waste and shredded newspapers and mushroom compost	Thermophilic
72	*Bacillus denitrifican*	Mushroom compost	Thermophilic
73	*Bacillus licheniformis*	Vegetables, fruits and garden waste, kitchen waste and shredded newspapers, garden waste and mushroom compost	Mesophilic, Thermophilic
74	*Bacillus megaterium*	Waste paper	Mesophilic, Thermophilic
75	*Bacillus mycoides*	Tree bark, hard-wood bark; bark compost	Mesophilic
76	*Bacillus oleronius*	Vegetables, fruits and garden waste	Mesophilic
77	*Bacillus pallidus*	Waste paper	Thermophilic
78	*Bacillus pumilus*	Vegetables, fruits and garden waste	Mesophilic, Thermophilic
79	*Bacillus schlegelii*	Vegetables, fruits and garden waste	Thermophilic
80	*Bacillus smithii*	Vegetables, fruits and garden waste and mushroom compost	Thermophilic

Continued

TABLE 2 Different microorganism presence in numerous type of waste material at various temperature phases during composting (Chang & Hsu, 2008; Hassen et al., 2001; Möller & Stinner, 2009; Rivard et al., 1995; Ryckeboer et al., 2003; Schulze, 1962)—cont'd

Sr. No.	Organisms	Source material	Temperature phase
81	*Bacillus* sp.	Vegetables, fruits and garden waste, Food waste and sawdust, garden waste, tree bark, hard-wood bark; bark compost, waste paper and cattle manure and forestry wastes	Mesophilic, Thermophilic
82	*Bacillus sphaericus*	Vegetables, fruits and garden waste, kitchen waste and shredded newspapers, peanut shells	Mesophilic
83	*Bacillus subtilis*	Vegetables, fruits and garden waste, kitchen waste and shredded newspapers, tree bark, hard-wood bark; bark compost, peanut shells, mushroom compost, waste paper	Mesophilic, Thermophilic
84	*Bacillus thuringiensis*	Vegetables, fruits and garden waste	Mesophilic
85	*Bacteroides* sp.	Cattle manure and forestry wastes	Mesophilic
86	*Bradyrhizobium* sp.	Waste paper	Mesophilic
87	*Brevibacillus brevis*	Kitchen waste and shredded newspapers	Thermophilic
88	*Brevibacillus agri*	Vegetables, fruits and garden waste	Mesophilic
89	*Brevibacillus laterosporus*	Waste paper	Mesophilic
90	*Brevibacillus thermoruber*	Mushroom compost	Thermophilic
91	*Brevundimonas* sp.	Vermicompost	Mesophilic
92	*Brevundimonas diminuta*	Vegetables, fruits and garden waste	Mesophilic

TABLE 2 Different microorganism presence in numerous type of waste material at various temperature phases during composting (Chang & Hsu, 2008; Hassen et al., 2001; Möller & Stinner, 2009; Rivard et al., 1995; Ryckeboer et al., 2003; Schulze, 1962)—cont'd

Sr. No.	Organisms	Source material	Temperature phase
93	*Chaetomium* sp.	Garden and kitchen waste and sewage sludge, garden waste and cattle manure and forestry wastes	Mesophilic
94	*Chaetomium thermophile*	Mushroom compost	Thermophilic
95	*Chaetomium thermophile*	Mushroom compost	Thermophilic
96	*Chaetomium thermophilum*	Municipal solid waste, garden and kitchen waste and sewage sludge, wheat straw, barley straw and straw compost and mushroom compost	Thermophilic
97	*Chrysosporium indicum*	Vermicompost	Mesophilic
98	*Chrysosporium merdarium*	Vermicompost	Mesophilic
99	*Chrysosporium queenslandicum*	Vermicompost	Mesophilic
100	*Chrysosporium tropicum*	Vermicompost	Mesophilic
101	*Circinella umbellate*	Municipal solid waste	Mesophilic
102	*Cunninghamella elegans*	Vermicompost	Mesophilic
103	*Curvularia harveyi*	Vermicompost	Mesophilic
104	*Curvularia pallescens*	Mushroom compost	Mesophilic
105	*Cylindrocarpon* sp.	Garden waste	Mesophilic
106	*Cylindrocarpon destructans*	Vermicompost	Mesophilic
107	*Cylindrocarpon lichenicola*	Vegetables, fruits and garden waste	Mesophilic
108	*Dactylaria* sp.	Vegetables, fruits and garden waste	Mesophilic
109	*Desulfotomaculum thermosapovorans*	Waste paper	Thermophilic

Continued

TABLE 2 Different microorganism presence in numerous type of waste material at various temperature phases during composting (Chang & Hsu, 2008; Hassen et al., 2001; Möller & Stinner, 2009; Rivard et al., 1995; Ryckeboer et al., 2003; Schulze, 1962)—cont'd

Sr. No.	Organisms	Source material	Temperature phase
110	*Doratomyces microspores*	Garden waste, mushroom compost and manure, horse manure	Mesophilic
111	*Doratomyces* sp.	Garden and kitchen waste and sewage sludge	Mesophilic
112	*Doratomyces medius*	Vegetables, fruits and garden waste	Mesophilic
113	*Doratomyces purpureofuscus*	Municipal solid waste, mushroom compost and vermicompost	Mesophilic
114	*Doratomyces stemonitis*	Vegetables, fruits and garden waste, municipal solid waste, garden waste and wheat straw, barley straw and straw compost	Mesophilic
115	*Emericella nidulans*	Garden and kitchen waste and sewage sludge, garden waste, wheat straw, barley straw and straw compost and mushroom compost	Mesophilic, Thermophilic
116	*Emericella nidulans*	Vermicompost	Mesophilic
117	*Enterobacter cloacae*	Tree bark, hard-wood bark; bark compost	Thermophilic
118	*Flavimonas oxyzihabitans*	Vegetables, fruits and garden waste	Mesophilic
119	*Flavobacterium johnsoniae*	Waste paper	Mesophilic
120	*Flavobacterium mizutaii*	Vegetables, fruits and garden waste	Mesophilic
121	*Flavobacterium* sp.	Mushroom compost	Mesophilic

TABLE 2 Different microorganism presence in numerous type of waste material at various temperature phases during composting (Chang & Hsu, 2008; Hassen et al., 2001; Möller & Stinner, 2009; Rivard et al., 1995; Ryckeboer et al., 2003; Schulze, 1962)—cont'd

Sr. No.	Organisms	Source material	Temperature phase
122	*Geobacillus stearothermophilus*	Vegetables, fruits and garden waste, kitchen waste and shredded newspapers, mushroom compost	Thermophilic
123	*Gliocladium* sp.	Garden and kitchen waste and sewage sludge, garden waste	Mesophilic
124	*Gliocladium deliquescens* Spp.	Mushroom compost	Mesophilic
125	*Gliocladium penicillioides*	Grass (lawn cuttings)	Mesophilic
126	*Gliocladium roseum*	Vegetables, fruits and garden waste, Municipal solid waste	Mesophilic
127	*Gliocladium viride*	Municipal solid waste	Mesophilic
128	*Gliomastix* sp.	Municipal solid waste	Mesophilic
129	*Gliomastix atrogriseum*	Vegetables, fruits and garden waste	Mesophilic
130	*Gloeophyllum trabeum*	Garden and kitchen waste and sewage sludge	Mesophilic
131	*Graphium* sp.	Municipal solid waste	Mesophilic
132	*Graphium putredinis*	Vegetables, fruits and garden waste and vermicompost	Mesophilic
133	*Gymnoascacea* sp.	Garden waste	Mesophilic
134	*Harpographium* sp.	Garden and kitchen waste and sewage sludge	Mesophilic
135	*Heterosporium* sp.	Municipal solid waste	Mesophilic
136	*Hormiscium* sp.	Municipal solid waste	Mesophilic, Thermophilic
137	*Humicola fuscoatra*	Vegetables, fruits and garden waste	Mesophilic

Continued

TABLE 2 Different microorganism presence in numerous type of waste material at various temperature phases during composting (Chang & Hsu, 2008; Hassen et al., 2001; Möller & Stinner, 2009; Rivard et al., 1995; Ryckeboer et al., 2003; Schulze, 1962)—cont'd

Sr. No.	Organisms	Source material	Temperature phase
138	*Humicola fuscoatra*	Garden waste and vermicompost	Mesophilic
139	*Humicola grisea*	Mushroom compost and vermicompost	Thermophilic
140	*Humicola grisead*	Municipal solid waste and mushroom compost	Mesophilic, Thermophilic
141	*Humicola insolensa*	Garden and kitchen waste and sewage sludge, wheat straw, barley straw and straw compost and mushroom compost	Mesophilic, Thermophilic
142	*Hydrogenobacter sp.*	Vegetables, fruits and garden waste	Thermophilic
143	*Janthinobacterium lividum*	Tree bark, hard-wood bark; bark compost	Mesophilic
144	*Mortierella alpine*	Garden waste and vermicompost	Mesophilic
145	*Mortierella humilis*	Garden waste	Mesophilic
146	*Mortierella reticulate*	Vegetables, fruits and garden waste	Mesophilic
147	*Mortierella* sp.	Municipal solid waste, garden waste and vermicompost	Mesophilic
148	*Mortierella echinosphaera*	Garden waste and vermicompost	Mesophilic
149	*Mortierella exigua*	Garden waste	Mesophilic
150	*Mortierella globalpina*	Garden waste	Mesophilic
151	*Mortierella hyalina*	Vegetables, fruits and garden waste, garden waste	Mesophilic
152	*Mortierella indohii*	Garden waste	Mesophilic
153	*Mortierella polycephala*	Garden waste	Mesophilic

TABLE 2 Different microorganism presence in numerous type of waste material at various temperature phases during composting (Chang & Hsu, 2008; Hassen et al., 2001; Möller & Stinner, 2009; Rivard et al., 1995; Ryckeboer et al., 2003; Schulze, 1962)—cont'd

Sr. No.	Organisms	Source material	Temperature phase
154	*Mortierella stylospora*	Vegetables, fruits and garden waste	Mesophilic
155	*Mortierella turficola*	Vegetables, fruits and garden waste	Mesophilic
156	*Mortierella vinacea*	Vegetables, fruits and garden waste	Mesophilic
157	*Mucor abundans*	Vermicompost	Mesophilic
158	*Mucor caninus*	Mushroom compost	Mesophilic
159	*Mucor circinelloides*	Vegetables, fruits and garden waste	Mesophilic
160	*Mucor circinelloides*	Municipal solid waste, garden waste and manure, horse manure	Mesophilic
161	*Mucor circinelloides*	Vermicompost	Mesophilic
162	*Mucor circinelloides*	Municipal solid waste	Mesophilic
163	*Mucor fragilis*	Mushroom compost	Mesophilic
164	*Mucor hiemalis*	Vegetables, fruits and garden waste	Mesophilic
165	*Mucor microspores*	Vegetables, fruits and garden waste	Mesophilic
166	*Mucor miehei*	Mushroom compost	Thermophilic
167	*Mucor plumbeus*	Vermicompost	Mesophilic
168	*Mucor pusillus*	Garden and kitchen waste and sewage sludge and mushroom compost	Thermophilic
169	*Mucor racemosus*	Vegetables, fruits and garden waste and mushroom compost	Mesophilic
170	*Mucor racemosus*	Vermicompost	Mesophilic

Continued

TABLE 2 Different microorganism presence in numerous type of waste material at various temperature phases during composting (Chang & Hsu, 2008; Hassen et al., 2001; Möller & Stinner, 2009; Rivard et al., 1995; Ryckeboer et al., 2003; Schulze, 1962)—cont'd

Sr. No.	Organisms	Source material	Temperature phase
171	*Mucor* sp.	Vegetables, fruits and garden waste, Municipal solid waste, tree bark, hard-wood bark; bark compost, mushroom compost and waste paper	Mesophilic, Thermophilic
172	*Myceliophthora* sp.	Vermicompost	Mesophilic
173	*Myceliophthora thermophila*	Wheat straw, barley straw and straw compost and vermicompost	Thermophilic
174	*Mycena sp.*	Municipal solid waste	Mesophilic
175	*Mycogone nigra*	Municipal solid waste	Mesophilic
176	*Myrothecium* sp.	Mushroom compost	Mesophilic
177	*Nectria* sp.	Garden waste	Mesophilic
178	*Nectria inventa*	Garden waste	Mesophilic
179	*Nectria ventricosa*	Garden waste	Mesophilic
180	*Neosartorya spinosa*	Vermicompost	Mesophilic
181	*Neosartorya fischeri*	Vermicompost	Mesophilic
182	*Nigrospora* sp.	Vermicompost	Mesophilic
183	*Oedocephalumglomerulosum*	Municipal solid waste and mushroom compost	Mesophilic
184	*Penicillium dupontii*	Municipal solid waste, garden and kitchen waste and sewage sludge	Thermophilic
185	*Penicillium chrysogenum*	Vegetables, fruits and garden waste, garden waste, mushroom compost and vermicompost	Mesophilic
186	*Penicillium citrinum*	Garden waste and vermicompost	Mesophilic

TABLE 2 Different microorganism presence in numerous type of waste material at various temperature phases during composting (Chang & Hsu, 2008; Hassen et al., 2001; Möller & Stinner, 2009; Rivard et al., 1995; Ryckeboer et al., 2003; Schulze, 1962)—cont'd

Sr. No.	Organisms	Source material	Temperature phase
187	*Penicillium commune*	Vegetables, fruits and garden waste	Mesophilic
188	*Penicillium corylophilum*	Vegetables, fruits and garden waste, garden waste	Mesophilic
189	*Penicillium corymbiferum*	Vegetables, fruits and garden waste	Mesophilic
190	*Penicillium crustosum*	Vegetables, fruits and garden waste	Mesophilic
191	*Penicillium cyclopium*	Vegetables, fruits and garden waste	Mesophilic
192	*Penicillium dierckxii*	Garden waste	Mesophilic
193	*Penicillium digitatum*	Municipal solid waste, garden waste and vermicompost	Thermophilic
194	*Penicillium diversum*	Garden waste	Mesophilic
195	*Penicillium echinulatum*	Garden waste and vermicompost	Mesophilic
196	*Penicillium expansum*	Vegetables, fruits and garden waste, garden waste	Mesophilic
197	*Penicillium glabrum*	Garden waste	Mesophilic
198	*Penicillium glandicola*	Vermicompost	Mesophilic
199	*Penicillium herquei*	Garden waste and vermicompost	Mesophilic
200	*Penicillium implicatum*	Garden waste and vermicompost	Mesophilic
201	*Penicillium islandicum*	Mushroom compost and vermicompost	Mesophilic
202	*Penicillium italicum*	Vermicompost	Mesophilic
203	*Penicillium spinulosum*	Vegetables, fruits and garden waste and vermicompost	Mesophilic

Continued

TABLE 2 Different microorganism presence in numerous type of waste material at various temperature phases during composting (Chang & Hsu, 2008; Hassen et al., 2001; Möller & Stinner, 2009; Rivard et al., 1995; Ryckeboer et al., 2003; Schulze, 1962)—cont'd

Sr. No.	Organisms	Source material	Temperature phase
204	*Penicillium stoloniferum*	Vegetables, fruits and garden waste	Mesophilic
205	*Penicillium velutinum*	Vegetables, fruits and garden waste	Mesophilic
206	*Penicillium verrucosum*	Garden waste	Mesophilic
207	*Penicillium viridicatum*	Vegetables, fruits and garden waste	Mesophilic
208	*Penicillium waksmanii*	Vegetables, fruits and garden waste and garden waste	Mesophilic
209	*Peziza ostracoderma*	Garden waste	Mesophilic
210	*Peziza vesiculosa*	Mushroom compost	Mesophilic
211	*Phialemonium obovatum*	Garden waste	Mesophilic
212	*Phialophora cyclaminis*	Vermicompost	Mesophilic
213	*Phoma* sp.	Municipal solid waste and vermicompost	Mesophilic
214	*Phomaexigua*	Vermicompost	Mesophilic
215	*Phoma glomerata alternariacea*	Municipal solid waste	Mesophilic
216	*Phoma herbarum*	Garden waste	Mesophilic
217	*Phomopsis* sp.	Garden waste and vermicompost	Mesophilic
218	*Pichia* sp.	Cattle manure and forestry wastes	Mesophilic
219	*Piptocephalis* sp.	Municipal solid waste	Mesophilic
220	*Pithomyces* sp.	Vermicompost	Mesophilic
221	*Plectosporium tabacinum*	Vegetables, fruits and garden waste	Mesophilic
222	*Pleurotus ostreatus*	Garden and kitchen waste and sewage sludge	Mesophilic

TABLE 2 Different microorganism presence in numerous type of waste material at various temperature phases during composting (Chang & Hsu, 2008; Hassen et al., 2001; Möller & Stinner, 2009; Rivard et al., 1995; Ryckeboer et al., 2003; Schulze, 1962)—cont'd

Sr. No.	Organisms	Source material	Temperature phase
223	*Preussia* sp.	Vermicompost	Mesophilic
224	*Preussiafleischhakii*	Garden waste	Mesophilic
225	*Pseudallescheria boydii*	Municipal solid waste, garden waste and vermicompost	Mesophilic
226	*Pseudeurotium zonatum*	Garden waste	Mesophilic
227	*Pseudo gymnoascus*	Municipal solid waste	Mesophilic
228	*Pullularia* sp.	Municipal solid waste	Mesophilic
229	*Pythium irregular*	Vegetables, fruits and garden waste	Mesophilic
230	*Pythium oligandrum*	Vegetables, fruits and garden waste	Mesophilic
231	*Rhinocladiella atrovirens*	Garden and kitchen waste and sewage sludge	Mesophilic
232	*Rhizomucor* sp.	Tree bark, hard-wood bark; bark compost	Thermophilic
233	*Rhizomucor pusillus*	Municipal solid waste, garden waste, wheat straw, barley straw and straw compost, mushroom compost and manure, horse manure	Mesophilic, Thermophilic
234	*Rhizopus arrhizus*	Vegetables, fruits and garden waste	Mesophilic
235	*Rhizopus chinensis*	Mushroom compost	Thermophilic
236	*Rhizopus microspores*	Mushroom compost	Thermophilic
237	*Rhizopus oryzae*	Garden waste	Mesophilic
238	*Rhizopus stolonifer*	Municipal solid waste, garden waste and mushroom compost	Mesophilic

Continued

TABLE 2 Different microorganism presence in numerous type of waste material at various temperature phases during composting (Chang & Hsu, 2008; Hassen et al., 2001; Möller & Stinner, 2009; Rivard et al., 1995; Ryckeboer et al., 2003; Schulze, 1962)—cont'd

Sr. No.	Organisms	Source material	Temperature phase
239	*Rhodotorula* sp.	Peanut shells and cattle manure and forestry wastes	Mesophilic
240	*Sepedonium* sp.	Municipal solid waste and manure, horse manure	Mesophilic, Thermophilic
241	*Sepedonium niveum*	Mushroom compost	Mesophilic
242	*Sistotremabrink mannii*	Municipal solid waste	Mesophilic
243	*Sordaria fimicola*	Garden waste	Mesophilic
244	*Spicaria* sp.	Vermicompost	Mesophilic
245	*Sporothrix schenckii*	Vegetables, fruits and garden waste	Mesophilic
246	*Sporotrichum* sp.	Manure, horse manure	Thermophilic
247	*Sporotrichum thermophile*	Garden and kitchen waste and sewage sludge and mushroom compost	Mesophilic, Thermophilic
248	*Stachybotrys* sp.	Garden and kitchen waste and sewage sludge and waste paper	Mesophilic
249	*Stachybotrys chartarum*	Municipal solid waste, garden waste and vermicompost	Mesophilic
250	*Staphylotrichum coccosporum*	Garden waste and vermicompost	Mesophilic
251	*Stemphylium* sp.	Municipal solid waste	Mesophilic
252	*Stereum* sp.	Garden waste	Mesophilic
253	*Stereum hirsutum*	Garden waste	Mesophilic
254	*Stibellathermophilab*	Mushroom compost	Thermophilic
255	*Stylopage* sp.	Municipal solid waste	Mesophilic
256	*Syncephalastrum racemosum*	Vermicompost	Mesophilic
257	*Syncephalis* sp.	Vegetables, fruits and garden waste	Mesophilic

TABLE 2 Different microorganism presence in numerous type of waste material at various temperature phases during composting (Chang & Hsu, 2008; Hassen et al., 2001; Möller & Stinner, 2009; Rivard et al., 1995; Ryckeboer et al., 2003; Schulze, 1962)—cont'd

Sr. No.	Organisms	Source material	Temperature phase
258	*Talaromyces emersonii*	Garden and kitchen waste and sewage sludge, wheat straw, barley straw and straw compost, and mushroom compost	Thermophilic
259	*Terrabacter* sp.	Cattle manure and forestry wastes	Mesophilic
260	*Trichurusterrophilus*	Municipal solid waste	Mesophilic
261	*Ulocladium* sp.	Peanut shells	Mesophilic
262	*Ulocladium alternariae*	Garden waste	Mesophilic
263	*Ulocladium consortiale*	Vegetables, fruits and garden waste, municipal solid waste	Mesophilic
264	*Verticillium lecanii*	Vegetables, fruits and garden waste, garden waste, lawn cuttings and barley straw	Mesophilic
265	*Verticillium leptobactrum*	Vegetables, fruits and garden waste	Mesophilic
266	*Verticillium nigrescens*	Vermi-compost	Mesophilic
267	*Verticillium* sp.	Vegetables, fruits and garden waste, municipal solid waste, garden waste	Mesophilic
268	*Volutella ciliata*	Garden waste	Mesophilic
269	*Westerdykella dispersa*	Vermi-compost	Mesophilic
270	*Zygorhynchus heterogamous*	Municipal solid waste and grape marc	Mesophilic
271	*Zygorhynchus* sp.	Manure, horse manure	Mesophilic
272	*Zygorhynchus moelleri*	Vegetables, fruits and garden waste and wheat straw, barley straw and straw compost	Mesophilic

14–16′. Large volumes of different wastes like grease, organic liquids, and other animal byproducts are being decomposed through this method (Chen, de Haro, Moore, & Falen, 2011).

2.2.1.3 In-vessel compositing

In this type of composting, organic (waste) materials are filled into the mechanical equipment's like drum, silo or concrete-lined trench, etc. where self-controlled conditions are applied. This process can utilize substantial amounts of waste and accommodate any kind of organic waste and works aerially hence generates very little odor. It requires 2–6 weeks to give the resultant material. The size of the vessel or container is varied. This method can be employed in cold weather areas (Chen et al., 2011).

2.2.2 Vermicomposting

Vermicomposting is a kind of composting, certain species of earthworms are employed to make compost. It is basically a mesophilic process that employs microbes and earthworms. The earthworms feed on organic waste material and excreted them out in a granular form (cocoons). Earthworms reduce the volume of organic matter by 40%–60%. Earthworm needs moisture content of about 32%–66% and pH 7. After 60–90 days, the worms will double themselves. These extra worms can be employed as fish food and sold to another farmer. The main food of worms is food scrape and animal manures (Zafar, 2017). It requires temperature of about 25–37°C that would depend on the species of worms. There are usually four species of worms that are extensively applied in vermicomposting, which includes two tropical species of worms i.e., *Eudriluseugeniae* (Hayawin et al., 2014; Reinecke, Viljoen, & Saayman, 1992) and *Perionyx excavatus* (Alavi et al., 2017; Malińska et al., 2017). These are also classified as epigenic worms. There are some other worms that play a vital role in high decomposition, and they are *Eisenia Andrei* and *Eisenia fetida* (Dominguez & Aira, 2010).

2.2.3 Anaerobic composting

The anaerobic composting is the decomposition of organic (waste) material in the absence of air (Zafar, 2017). The absence of air (O_2) encourages the anaerobic bacteria (*Clostridium*, *Bacteroides*, *Actinomyces*, *Fusobacterium*, *Veillonella*, etc.) to break down food scraps and other natural waste (Jeffreys, 1959). It has many types (Lim et al., 2019); some are as follows:

2.2.3.1 Stacks or piles

A simple pile of food scraps and other green organic matter can work best in the anaerobic style of composting. This is usually done as kitchen composting. The anaerobic microbes decomposed the organic wastes into the final product within 6–12 months (Claassen & Carey, 2004).

2.2.3.2 Bokashi composting

The bokashi method is a special anaerobic composting in which fermentation is carried out in a closed container. The microorganisms breakdown the food, producing the fermented material that is further treated with an aerobic compost pile (Inckel et al., 2005). The basic objectives of bokashi composting are its affordability, easiness, less time consumption, bad odors, and improving soil quality (Jørgensen, 2019).

2.2.3.3 Submerged composting

Keeping the composting material submerged in water in a way that the unwanted odor is prevented. This method can be used with an open container or a closed system. In this process, the organic wastes are put in an enclosed container of a suitable size and covered with water, which will lead to slow break down of organic wastes because water act as a cooling agent, which slows down the metabolism of the microbes (Miller, 1993).

2.2.4 Mechanical composting (composting equipment)

Composting equipment is utilized to prepare compost, which is a dark and rich, easily soluble soil conditioner (humus). The product formed through the biological decomposition of organic (waste) materials are animal waste, food byproducts, municipal sludge, and animal carcasses. The organic composting process can be done in a composter resulting in nutritionally rich compost. There are many types of composting equipment, such as compost bins, compost tumblers, compost containers, and compost turners. There are further different kinds of compost bins, such as the covered bridged organic hot bin, the coated mesh wired bin, and the earth engine double bin (Sweeten, 2008). Compost tumbler can prepare compost in a very short time. A compost tumbler is capable of preparing compost in just 3 weeks. Compost turners are not used for preparing a large amount of compost (used for less than 10,000 tons of compost per season) (Yan-mei, 2008). A composter is also a type of compost equipment that produces large qualities of compost in a short while and can reach to the temperature above 65.5°C (Williams, Ziegenfuss, & Sisk, 1992). The diagrammatic structure of the vermicompost is shown in Fig. 2.

2.3 Phases of composting

There are four stages of compositing: (i) mesophilic phase, (ii) thermophilic phase, (iii) cooling phase, and (iv) curing phase.

2.3.1 Mesophilic phase

It is also known as primary phase. The mesophiles usually survive at moderate temperatures, 20–45°C (68–113°F), including human pathogens. The thermophiles grow above 45°C (113°F), and some survive in this temperature, or even above.

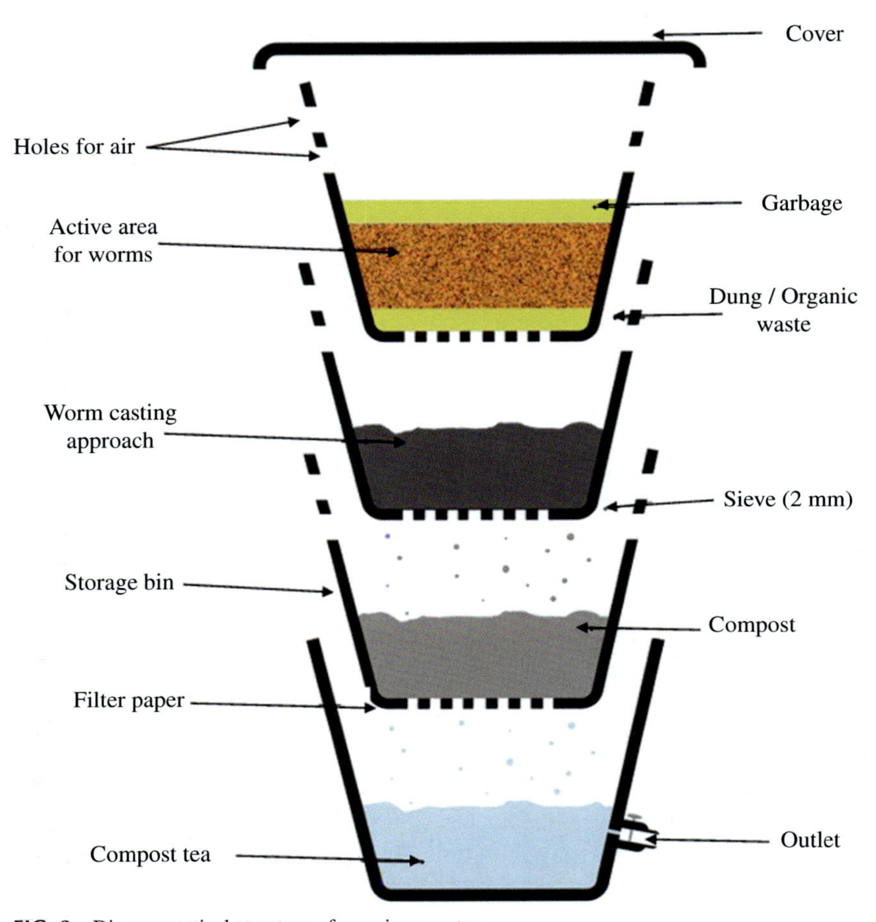

FIG. 2 Diagrammatical structure of vermicomposter.

Different communities of microorganisms are balanced during these composting phases. The primary breakdown is done by mesophilic microorganisms, which quickly disintegrate or decompose the soluble, readily disruptive (waste) compounds. The heat that they evolved induced the compost temperature to enhance rapidly (Linch & Poole, 1979). The main goals achieved by this phase are the temperature rising (for best working of thermophilic bacteria), odor dislodgement, and making environment unfavorable for vectors (Hassen et al., 2001).

2.3.2 Thermophilic phase

When temperature rises to almost 40°C, the activity of mesophilic microorganisms decreases, and they are exchanged by thermophilic or "heat-loving" (Trautmann & Krasny, 1998). At the temperatures of 55°C and even above, many microbes that are human and plant pathogens are demolished. Because the temperature over about 65°C slay many sets of microbes that limit the disintegration rate, compost managers utilize the aeration and integration for retaining the temperature falls below this point (Linch & Poole, 1979).

2.3.3 Cooling and curing phase

During the thermophilic phase, raised temperature get a move on the decomposition of structural molecules of plant wastes like proteins, fats, and complex carbohydrates like cellulose. As the content of these high-energy compounds in the wastes becomes less, the compost temperature started decreasing gradually (cooling phase), and then again, mesophilic microorganisms take over for the final phase of "curing" or "maturation of the remaining organic matter" (Linch & Poole, 1979).

3 Biochemistry of composting

Different microorganisms are involved in the disintegration of organic (waste) materials. Initially, they start breaking down organic waste aerobically under favorable conditions. During this period, the microorganisms use organic material to fulfill their metabolic and growth activities. As the temperature begins to increase during the biological oxidation, the organic material act as insulator and heat is reserved in the compost pile, but when the temperature reaches to about 44–52°C (111–125.6°F), the decomposition starts and then the temperature starts decreases and finally reaches to the ambient temperature. In this method, the pH initially becomes acidic and then alkaline, and at last, it reaches near neutral (Gray, Sherman, & Biddlestone, 1971). Initially, the composting mass (material) is at room temperature, and when the decomposition starts, there is a steady increase in the temperature due to the reproduction of microorganisms. When the temperature reaches above 40°C (104°F), then the mesophilic stage exchanged by the thermophilic stage (it needs 2–3 days to reach this stage) (Teensma, 1961). During the composting process, there is a continuous alter in the microflora population and communities that are as listed in Table 2. Different microbial communities predominate according to their suitability with environmental conditions, which mainly includes bacteria, fungi, algae, yeast, actinomycetes, and molds (Bagstam, 1978; Crawford, 1978), some of which are primary, secondary, or tertiary decomposers (Middleton, Van Der Valk, Williams, Mason, & Davis, 1992; Ryckeboer et al., 2003).

3.1 Composting and microorganisms

3.1.1 Bacteria

The bacteria are microscopic living organisms and survive in abundance in compost. They make up 80%–90% of the total means of microbes and have naturally existed in a gram of compost. Bacteria, accountable for most of the breakdown and heat formation in compost, are nutritionally the most erratic kinds of compost making organisms, utilizing a wide nature of enzymes to chemically decompose a variety of organic (waste) materials (Trautmann & Krasny, 1998). At the beginning of this process, mesophilic bacteria dominate. Most of these are those kinds that can also be present in topsoil. As the compost heats up over 40°C, thermophilic bacteria activate. The microbial populations during this phase are dominated by members of the genus Bacillus. The population of Bacillus species is high at temperatures between 50°C and 55°C but decreases vividly at 60°C or above. When conditions become unfavorable, bacilli live by creating spores that are highly resistant to heat, cold, or dryness; they found everywhere in nature and become active whenever favorable conditions have appeared. At the highest compost temperatures, the *Thermus* genus of the bacteria has become active (often wonder how microbes progressed in nature that can resist the high temperatures found in active compost) (Forsyth & Webley, 1948).

3.1.2 Actinomyces

The actinomyces are the microorganisms that cause the characterized earthy smell. Like other bacteria, they have no nuclei, but they grow multicellular filaments like fungi and are facultative anaerobic. Actinomyces work best on moderate temperature. In composting, they play a vital role in the breakdown of convoluted and resistant organic material such as cellulose, lignin, chitin, and proteins. Actinomyces are found in large clusters, and mostly its working has seen at the last stage of decomposition (Stutzenberger, Kaufman, & Lossin, 1970; Waksman, Cordon, & Hulpoi, 1939).

3.1.3 Fungi

The fungi include molds and yeasts and are accountable for the disintegration of many complicated plant compounds in soil and compost. In compost, fungi are core organism because they decompose the tough raw wastes, permit the bacteria to initiate the breakdown process once most of the cellulose has been converted into the simple form (Moubasher, Abdel-Hafez, Abdel-Fattah, & Moharram, 1982). They spread and grow by forming many cells or filaments, and they are able to decompose organic residues that are too acidic or less nitrogenous material for further bacterial action. Most of the fungi are highly saprophytes; they can live on dead material and obtain energy by decomposing organic (waste) material in deceased plants and animals. They are plentiful during both mesophilic and thermophilic phases of composting. They usually grow in the outermost region of compost where temperatures are usually high (Eastwood, 1952).

3.1.4 Worms

Worms (Aristotle called worms the "intestine of the earth") are heterotrophic organisms that can obtain energy by decomposing the plants and animal parts as well as the organic fecal matter. They are different from other decomposers (fungi, bacteria, and protists), which are not able to ingest the discrete lumps of matter (Hayawin, Astimar, Ibrahim, et al., 2014; Malińska et al., 2017).

3.1.5 Rotifers

Rotifers are the microscopic multicellular microbes found in compost, which help in the disintegration of organic wastes and also ingest bacteria and fungi (Ogello, Wullur, Sakakura, & Hagiwara, 2018). Rotifers are usually having one or two groups of vibrating cilia on the head. Their body shape is round and comprise three parts: a head, trunk, and tail. The rotifers in compost are present diversely in the water where they adhere to plant substances and feed microorganisms (Graves, Gwendolyn, Donald, James, & Dana, 2000).

3.2 parameters

Several parameters that are kept in mind during composting and the effect of these parameters on the soil are briefly explained in the following sections, also given in Tables 3 and 4.

3.2.1 Aeration

About 5–9 cubic feet of air per pound of volatile matter per day is required (Schulze, 1962), 7.7 cubic feet of air per pound of volatile matter per day for municipal compost is required (Kosobucki, Chmarzynski, & Buszewski, 2000), and 9–23 cubic feet of air was consumed per day per pound of volatile matter. By employing windrows method, turning the windrow at 15-in. depth will increased the oxygen percentage of about 18.6%. Oxygen was usually as low as 1%–2% at a 24-in. depth (Donay, Fernandes, Lagrange, & Herrmann, 2007). Oxygen is also essential for the aerobic microbes during the disintegration of organic (waste) material, which is easily circulated in the grinded tissues (Trautmann & Krasny, 1998).

3.2.2 C:N ratio

During this process, most of the organic (waste) material is transformed into CO_2 and H_2O (Wei, Fan, & Wang, 2001), then the odor of ammonia is observed because of carbon, which is unavailable due to the presence of a lot of newspapers or the very low C/N ratio. If this situation happens, then some of the nitrogen is evolved into air in the form of ammonia instead of by used microorganisms (Diaz & De Bertoldi, 2007). The ideal carbon-to-nitrogen (C:N) ratio for composting is generally considered to be around 30:1. C:N ratios below 30:1 allow the fast microbial growth and speedy decomposition, lead to the release of

TABLE 3 The properties of mature compost and its effect on soil (Strom, 1985).

Property	Value	Effect on soil
pH	6–8.4	Alkaline compost can raise pH in acidic soil
C:N ratio	<12:1	At >25:1, *N* immobilization occur
Organic matter	30%–70%	Affect application rate that will vary widely
Moisture content	40%–50%	Higher moisture will increase the transportation and handling costs
Soluble salts	0–10 mmhos/cm	Phytotoxic, if SSP (soluble salts percentage) enhances
Nutrient content	2%–5%	If greater than 2%, additional fertilizer may be needed

TABLE 4 The C/N ratio and nitrogenous content percentage of different organic wastes (Amlinger, Götz, Dreher, Geszti, & Weissteiner, 2003; Golueke, 1972).

Sr. No.	Organic materials (wastes)	C/N ratio	Nitrogen (percent of dry weight)
1	Urine	0.8	15–18
2	Cow manure	18	1.7
3	Blood	3	10–14
4	Sheep manure	–	3.8
5	Pig manure	–	3.8
6	Activated sludge	6	5
7	Non-legume vegetable waste	11–12	2.5–4
8	Saw dust	200–500	0.1
9	Straw, wheat	128–150	0.3–0.5
10	Grass clippings	12–15	3–6
11	Potato tops	25	1.5
12	Straw, oats	48	1.1
13	Horse manure	25	2.3
14	Poultry manure	15	6.3

excess amount of nitrogen, and loss of the nutrient as the ammonia gas having undesirable odors. C:N ratios above 30:1 do not provide sufficient amount of nitrogen for optimum growth of microbes. This may cause the compost cooled that leads to slow degradation (Trautmann & Krasny, 1998). Total carbon was calculated by the dynamic flash combustion, and the total content of nitrogen was analyzed by protein analysis (Hassen et al., 2001). The C:N ratios of some wastes are given in Table 4.

3.2.3 pH

During composting, the pH generally varies between 5.5 and 8.5, which largely depends on the composition of the ingredients (Trautmann & Krasny, 1998). The pH of the compost was determined by distilled water: compost (*w*/*v*) with a 1:10 (Bong et al., 2017).

3.2.4 Moisture content

The optimum moisture content needed for composting is 50%–60% by weight because it provides optimum water to balance microbial activity but not so much that airflow is blocked (high moisture content causes microbial growth inhibition) (Trautmann & Krasny, 1998). The moisture content was calculated by weighing after heating in an oven at 105°C. Changing in the temperature was recorded by using a mercury thermometer (Donay et al., 2007).

3.2.5 Microbial population

The prior grinding of the material used for composting increases the surface area for microbes to work efficiently on it (Berkeley, 1950; Gotaas and World Health Organization, 1956), which enhances the release of CO_2 twice as compared to ungrounded material (Gray et al., 1971). The bacteria, fungi, and actinomyces numbers were recorded by the dilute plate method on tryptic soy agar, Rose Bengal medium, and actinomycete isolation agar, respectively. The inoculation temperature of mesophilic and thermophilic microbes are 35°C and 50°C, respectively (Okereke & Kanu, 2004). The growth rates of microbes during composting are given in Table 5.

3.2.6 Temperature

It is one of the key factors of the composting process. It is produced as a byproduct during the microbial degradation of organic material (Nakasaki, Shoda, & Kubota, 1985; Venglovsky et al., 2005). According to Margesin, Cimadom, and Schinner (2006), this production is usually observed in two phases of composting i.e., mesophilic phase (lower than 45°C) and thermophilic phase (greater than 45°C). This temperature fluctuation affects the microbial activity during composting (Epstein, 2011).

TABLE 5 Growth rate of microbial communities (approximately) during composting process (Galitskaya et al., 2017; Miller, 1993).

Organisms	Population rate
Bacteria (during mesophilic stage)	10^9–10^{13} g^{-1} substrate
Bacteria (during thermophilic stage)	10^8–10^{12} g^{-1} substrate
Bacteria (average population)	$(3.0 \pm 0.2) \times 10^6$ g^{-1} substrate
Actinomycetes (during mesophilic stage)	10^7–10^9 g^{-1} substrate
Actinomycetes (during thermophilic stage)	10^8–10^{12} g^{-1} substrate
Actinomycetes (average population)	10^6–10^{12} g^{-1} substrate
Fungi (average population)	10^5–10^8 g^{-1} substrate

3.2.7 *Enzymatic activity*

Enzyme production was investigated by solid-state fermentation. In this type of fermentation, each substrate was cut into 1 cm pieces and oven-dried, after that 70 g of each substrate was mixed in 150 mL of berg's mineral salts medium, and the moisture content was rectified to 70% by adding the distilled water and then autoclaved before use. Bacteria were grown in the berg's mineral salt medium at 35°C for 24 h. Cell suspensions were adjusted to the 0.5 McFarland standards, and 1 mL of culture was inoculated into the substrate at 35°C for 13 days. Then cellulose and xylanase were assayed daily (Donay et al., 2007) using different agricultural wastes, including rice straws, etc. (Bong et al., 2017).

3.3 Chemical reactions in the composting process

Composting material varies from highly heterogeneous to virtually homogeneous wastes in municipal refuse and sludge. The biochemical breakdown of such complex wastes is a complex process resulting in many intermediates wastes, yielding the final product at the end (Polprasert & Koottatep, 2017), which are given in Fig. 3.

3.3.1 *Nitrification*

The breakdown of complex hydrocarbons is supported by different bacteria. The nitrification is done by two groups nitrifying bacteria that are Nitrosomonas and Nitrobacter carrying out the reaction as follows:

$$NH_4 + {}^3/_4O_2 \xrightarrow{NitrosomonaS} NO_2^- + 2H^+ + H_2O$$

$$NO_2^- + {}^1/_2O_2 \xrightarrow{Nitrobacter} NO_3^-$$

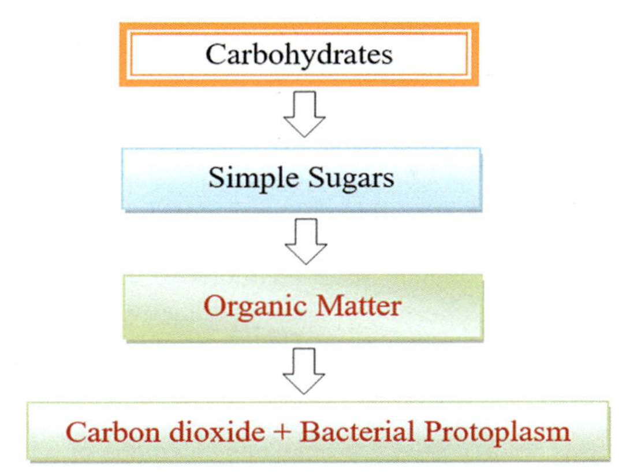

FIG. 3 The process of breakdown of the carbohydrates are as follow (Polprasert & Koottatep, 2017).

The Nitrosomonas bacteria are responsible for transforming NH_4^+ to NO_2^- and the Nitrobacter bacteria are able to convert NO_2^- to NO_3^-. Plants need NO_3^- for their growth. It also helps in the maturation phase to produce good quality compost (which is employed as a fertilizer or the soil conditioner) (Silverman & Ehrlich, 1964). The nitrifying bacteria require low temperature (below 40°C) and have a slow growth rate, and they are inactive at high temperatures (thermophilic phase), and they become active when the reaction is completed. The overall oxidation nitrification is:

$$NH_4{}^+ + 2O_2 \rightarrow NO_{3^-} + 2H^+ + H_2O$$

The reaction of NH_4^+ synthesis in cell tissues is:

$$NH_4{}^+ + 4CO_2 + HCO_3{}^- + H_2O \rightarrow C_5H_7O_2N + 5O_2$$

The overall nitrification reaction is:

$$22NH_4{}^- + 37O_2 + 4CO_2 + HCO_3{}^- \rightarrow 21NO_3{}^- + C_5H_7O_5N + 20H_2O + 42H^+$$

The last reaction illustrates the overall conversion of NH_4^+ into the final simplified products (Finstein & Morris, 1975).

4 Composting and sustainable environment

Composting plays a vital role in cleaning and maintaining the environment. It is useful in improving soil quality, reducing waste in landfills, and minimizing the utilization of the chemical in the field (Chaney & Ryan, 1994; Helmke, Robarge, Korotev, & Schomberg, 1979; Own, 2010; Smith et al., 2001). It is estimated that about 70% of household wastes are the organic material that has given the least attention with respect to pollution control and disease reduction. It plays an integral role in soil water retention, and it acts as sponge when mixed with the soil that not only results in maximum water retention but also improves root growth, leaching of essential nutrients from the soil, and the rate of erosion by increasing the adhesive properties of soil and by improving soil structure. It is an effective way to reduce greenhouse gases like methane because methane-producing microbes are not active in the presence of air (O_2) (Doorn & Barlaz, 1995). It helps the soil to sequester carbon dioxide. Composting is an effective waste management technique as it is utilizing waste in an efficient way by employing microorganisms, making it safer for soil and atmosphere as compared to chemical fertilizers (El-Shafei, Yehia, & El-Naqib, 2008; Khan & Faisal, 2008). Applying compost will aid in controlling greenhouse gas emissions, pesticide use, less fuel consumption, and sequester carbon dioxide (Eureka, 2008). The degree of nutritional enhancement of soil will depend on the material and technology (its relationship with the environment) used in the composting (Krebs, Gupta, Furrer, & Schulin, 1998).

4.1 Composting and bioremediation

The critical issues like environmental pollution and, in particular, contaminants source like polycyclic aromatic hydrocarbons (PAHs), which could be composted and degraded by the composting approach (Antizar-Ladislao, Lopez-Real, & Beck, 2004). The wastewater treatment from olive mills performed successfully with wheat straw by the technique: forced aerated composting (Tomati, Galli, Pasetti, & Volterra, 1995). Composting plays a crucial role in the remediation of soil and sediments polluted with hydrocarbons (Williams & Keehan, 1993). It is estimated that about 73% of PAHs were degraded by the composting process (Beaudin et al., 1996).

Different explosive, toxic, contaminated, and health hazard substances are easily disintegrated with the help of composting like a wood containing PAH after composting for 61 days had significantly reduced PAH concentration (Barker & Bryson, 2002; Briggs, Riley, & Whitwell, 1998; Chaney & Ryan, 1994). Composting and bioremediation are integrated to disintegrate toxic metallic and organic contaminants and their byproduct to ensure biosafety (Barker, 1997; Chaney et al., 2001; Stratton, Barker, & Rechcigl, 1995; Stratton & Rechcigl, 1998). The remediation process with composting is the same as the biological processes in the soil; however, composting accelerates the destruction process (Büyüksönmez et al., 1999; Büyüksönmez et al., 2000; Rao, Grethlein, & Reddy, 1995; Rao, Grethlein, & Reddy, 1996; Williams & Keehan, 1993). The concentration of numerous water soluble and acid hydrolyzable compounds fractions is declined, which leads to the production of complexes with humus substances; therefore, limit their mobility and availability for plant usage (Heyes, Moore, & Rudd, 1998; Paré, Dinel, & Schnitzer, 1999). The copper, manganese, and zinc leaching are being declined due to their association with organic compounds during composting in compost hog manure (Hsu & Lo, 2000). McMahon et al. (2008) showed the quality of timber composting by using mixed board wood pieces from constructional and demolition wastes with the poultry manure, ecobiowaste, and green wastes (acts as a nutritional supplement). The composting process can be employed in diesel oil degradation by adding sewage sludge and compost as a catalyst to enhance degradation (Namkoong, Hwang, Park, & Choi, 2002). As Jørgensen, Puustinen, and Suortti (2000) described how the composting of contaminated soils in biopiles be done rapidly by adding organic material.

5 Composting and sustainable soil health

Compost is nowadays the most efficient alternative of chemical fertilizers for farmers. The compost is used as a natural and ecological beneficial means of improving crop production by enhancing soil fertility (Ouédraogo, Mando, & Zombré, 2001). Compost is not only valuable for plant growth but also beneficial for the microflora present in the soil. The addition of compost improves the activity of microflora because compost acts as a source of energy. The decomposed organic material (compost) contains macro and micronutrients, which are essential for plant growth. Composting also abolishes the weed plants and plant pathogens during crop production by the fluctuation of temperature (Neher, 2019). The effect of compost on soil are given in Table 3 and Fig. 4.

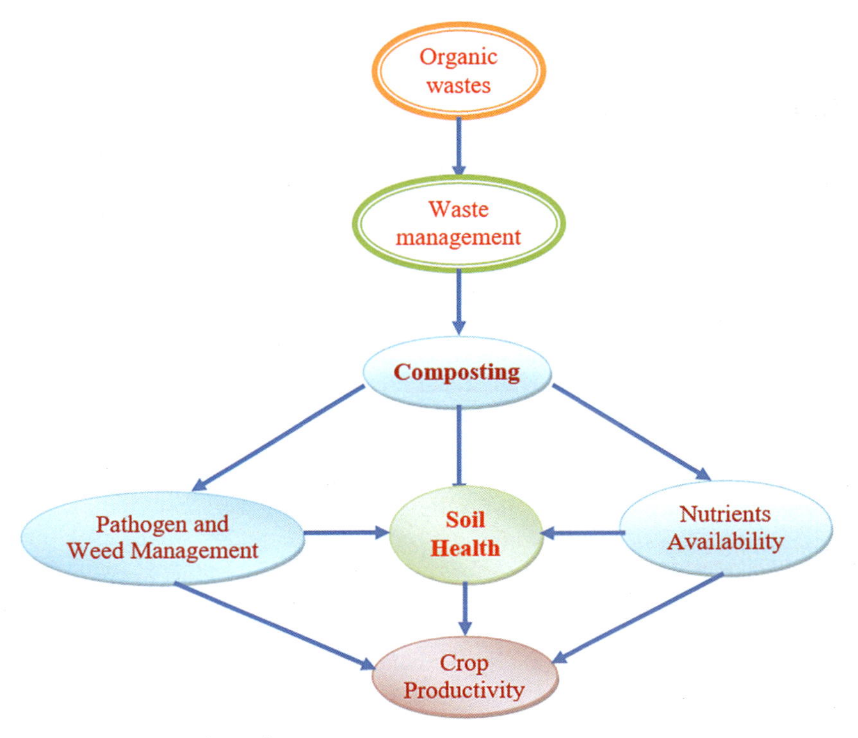

FIG. 4 Advantages of organic wastes.

It was believed that the compost amendment field gives a 32% higher yield than the unfertilized area (Chew et al., 2019). The compost amendment field gives equal yield as the chemical fertilized field. The crop yield, height, root, and dry weight of shoots are increased about 18%, 19.5%, 19%, and 5.3%, respectively, by the addition of compost. The phosphorous availability increased with the amendment of the organic wastes in the weathered soil (Andriamananjara, Rabeharisoa, Prud'homme, & Morel, 2016). The microbial activity mediates composting. There is abundant genetic heterogeneity in microbes that are present in our environment (Sait et al., 2002). Composting becomes a very economically beneficial process that is easily adopted by farmers to utilize their field organic waste and to use an alternative of fertilizers. The amendments of compost in soil play a vital role in maintaining its chemical, physical and biological properties, and also in the plant growth (Cogger, Hummel, Hart, & Bary, 2008).

Compost contains nutrients that are beneficial for the optimal growth of plants (such as nitrogen, phosphorus, and potassium) and is also an excellent supplier of macronutrients that are needed by plants. Nutrients are available to the plant according to their needs and give the soil a porous layer due to the working of different microflora, which helps them to conserve moisture and nutrients and provide nutrients to plants in their available forms (Niggli, Schmid, & Fliessbach, 2008). It also acts as inoculants to soils, adding microflora and large creatures such as earthworms, etc. and neutralizes various soil toxins and metals by bonding them and act as pH buffer, so the plant is less affected by changes in soil pH. It can also be used as mulch, soil amendment, as a potting material, compost tea (which is the core for the hydroponic system), and as a growing media (Aslam, Horwath, & Vander Gheynst, 2008; Aslam, Vander Gheynst, & Rumsey, 2008).

The use of compost is nowadays becoming a natural and ecological way to improve not only soil fertility but also soil conditions and crop production (Ouédraogo et al., 2001). Its application, in addition to enhancing crop production, also reduces the plant diseases and acts as a slow long-term release fertilizer (Hepperly et al., 2009; Saison et al., 2006). Compost is not only used for crops but also applied to vegetables and orchid and even in the tree growing. The application of compost increases the yield of the crop. It increases the sorghum yield by 45% than nonamended production (Ouédraogo et al., 2001). Vegetables like radish, which need very careful and highly fertile soils, are easily managed and highly yielded by the application of compost (Imthiyas & Seran, 2014). It also has a great effect on plant root length, stems, girth, dry mass, and fresh weight (Sarker, Kashem, & Osman, 2012). Compost application is considered as an economical and environmentally favorable method to improve the soil quality and crop productivity as well as the sustainability of production by reproducing soil organic matter and supplying micro and macronutrients. Compost also helps to enhance macro aggregate stability. Composting improves soil structure and enhances the water holding capacity

(Duong, 2013). It also improves soil porosity, reduces bulk density, increases gas exchange and water permeability, and improves the aggregation capacity of soil by improving the root zone environment. It also helps in controlling wind and water erosion by cementing soil particle with each other or with the plants (Lichtfouse, 2014). Taking the microscopic view, another beneficial characteristic of compost is evident, which is due to the presence of different worms, centipedes, sowbugs, and others, which shows that the compost is a healthy living material. The presence of these decomposer organisms proves that there is still some organic material being slowly decomposed with time and releasing nutrients (they are also indicators of balanced soil ecology, which includes organisms that keep diseases and pests in check). Many experiments have shown that the rich soil life in compost helps to control diseases and pests that might otherwise infest a more sterilized soil lacking natural checks against their spread (Stratton et al., 1995).

6 Compost and sustainable crop production

Organic farming is also known as a sustainable agriculture system that is ecosystem friendly. The organic farming system does not allow the use of fertilizer and pesticide but depends upon the natural environment that promotes plant nutrition and maintaining soil health, quality, and fertility (which are the chief aim of organic farming) (Lichtfouse, 2014; Magdoff & Weil, 2004; Council Regulation, 1999). Organic farming is all things that will encourage soil fertility and health and increase its productivity without any artificial amendments (Lichtfouse, 2014). Composting is a vital part of organic farming due to degraded soil health. The foremost importance of compost use is the increase in soil organic matter (Magdoff & Weil, 2004). Fig. 4 shows the beneficial aspects of composting on crop productivity.

Mamo, Halbach, and Rosen (2002) used the four municipal solid waste composts coded as: PRAIRIELAND (Truman, MN), Recomp (St. Cloud, MN), Swift (Benson, MN), and Wright (Buffalo, MN) on sandy loam soils to measure their effect on heavy metals in soil. They concluded that the addition of these types of compost into soil increased some macro- and micronutrients. Trace metals (lead, cadmium, mercury, and arsenic, which were not essential to plant growth) did not accumulate. Alike the previous example, Sikora, Tester, Taylor, and Parr (1982) grew crop plants with sewage sludge compost in two different soils and observed that phosphorus content in these plants of compost amended loamy sand soil thatwas approximately double from that compost amended silt loam soil. However, Evanylo et al. (2008) observed that the available P increases about 225% after the application of compost @144 $t\,ha^{-1}$ (dry weight). The elemental composition of green compost and vermicompost are given in Table 6.

Compost is able to provide plants the nutrients in their available form (Lichtfouse, 2014). Humus substances (compost) have high water holding capacity. Humus has an important effect on soil physical properties, such as

TABLE 6 The elemental composition of garden compost and vermicompost (Nagavallemma et al., 2004).

Sr. No.	Nutrient elements	Garden compost (%)	Vermicompost (%)
1	Nitrogen	0.8	0.51–1.61
2	Organic Carbon	12.2	9.8–13.4
3	Potassium	0.48	0.15–0.73
4	Phosphorous	0.35	1.19–1.02
5	Zinc	0.0012	0.0042–1.110
6	Manganese	0.0414	0.0105–0.2038
7	Sodium	<0.01	0.058–0.158
8	Copper	0.0017	0.0026–0.0048
9	Calcium	2.27	1.18–7.61
10	Iron	1.169	0.2050–1.3313
11	Magnesium	0.57	0.093–1.568

by stabilizing aggregate structure, which will result in conserving water and maintaining aeration in the soil (Magdoff & Weil, 2004). Lynch, Voroney, and Warman (2005) and Erhart and Hartl (2010) described the application of compost that enhances the soil water contents.

7 Composting and biogas

Biogas is anaerobic digestion (AD) or fermentation of different organic wastes like animal wastes, municipal wastes, and agricultural wastes, etc. (Klass, 2004). AD or anaerobic composting is the conversion of an organic substance into methane and carbon dioxide (Zhu, Yang, & Li, 2015). This process works in four stages i.e., hydrolysis of complex substances like lipids, proteins and carbohydrates, acidogenesis of byproducts of hydrolysis stage, acetogenesis of fatty acids, and at last methanogenesis (Abatzoglou & Boivin, 2009; Demirbas & Balat, 2009; Kao et al., 2012; Ramaraj & Dussadee, 2015; Salminen & Rintala, 2002). The AD of organic wastes will reduce the pathogenicity, odor formation and is an alternative to commercial fertilizer; therefore, it plays an essential role in environment safety as a source of biogas production by which CH_4 and CO_2 emission (Frascarelli, 2011; Holm Nielsen, Al Seadi, & Oleskowicz-Popiel, 2009; Igliński et al., 2012). Torquati, Venanzi, Ciani, Diotallevi, and Tamburi (2014) investigated that biogas is the driving tool to reduce the need for burning fossil fuel.

The biogas used as an agricultural tool is a relatively new approach to improve quality and productivity (Odlare, Pell, & Svensson, 2008). There are numerous kinds of wastes that are employed on the bases of which its quality will be evaluated: (Monnet, 2003) (i) solid wastes: organic wastes, combustible, noncombustible, ash, dead animals, constructional, and hazardous wastes; (ii) liquid wastes: human wastes, industrial water-wastes, sewage sludge, and organic liquids; (iii) municipal wastes: rubbish, trash, kitchen, paper, plastic, glass, and metal wastes; (iv) industrial wastes: scrap items, oils and chemicals, etc.; (v) medical wastes: syringes, surgical instruments, discarded medicines, etc.; and (vi) hazardous wastes: plastics, dyes, paints, etc. (Ahmed et al., 2020). It is the final remains (Gerardi, 2003) that may contain different persistent organic chemicals like phthalates (Angelidaki, Mogensen, & Ahring, 2000; Hartmann & Ahring, 2003; Nilsson, 2000; Nilsson, Kylin, & Sundin, 2000), phenolic compounds (Angelidaki et al., 2000; Levén, Nyberg, Korkea-Aho, & Schnürer, 2006; Leven & Schnürer, 2005), pesticides (Nilsson, 2000), polychlorinated and poly hydrocarbons (Angelidaki et al., 2000; Nilsson, 2000; Nilsson, Waldebäck, Liljegren, Kylin, & Markides, 2001), and dioxin-like compounds (Engwall & Schnürer, 2002; Olsman et al., 2002).

It prevents different soil-borne diseases and has direct (inhibition of diseases in plants and microbial resistance) (Hoitink & Boehm, 1999; Yu et al., 2006) or indirect effects (stimulation of biological activities) (Odlare et al., 2008). Numerous results were reported on the effect of biogas on soil health (Debosz, Petersen, Kure, & Ambus, 2002; Ernst, Müller, Göhler, & Emmerling, 2008; Jedidi, Hassen, Van Cleemput, & M'Hiri, 2004; Leifeld, Siebert, & Kögel-Knabner, 2002; Marinari, Masciandaro, Ceccanti, & Grego, 2000; Odlare et al., 2008) and crop productivity (Båth & Rämert, 1999; Garg, Pathak, Das, & Tomar, 2005; Marchaim, 1992; Odlare et al., 2008; Rivard et al., 1995; Svensson, Odlare, & Pell, 2004; Tiwari, Tiwari, & Upadhyay, 2000). It causes a decrease in the formation of greenhouse gases and its emission by reducing the carbon content of biogas residues (Amon, Moitzi, Schimpl, Kryvoruchko, & Wagner-Alt, 2002; Drury et al., 2006; Möller & Stinner, 2009). Its production is not limited to geographical areas and environmental conditions like wind, solar, and other energies (Pöschl, Ward, & Owende, 2010).

8 Conclusion

Composting is the core technology that is economically and environmentally beneficial. The major benefits include cost saving, save environmental pollution, and avoid different diseases. The rise in the population leads to the excess production of municipal solid waste in all over the world; therefore, this technology helps us to utilize these wastes. Garbage or organic wastes are produced nowadays in every home, which results in environmental and soil pollution and causes health hazards. Composting, which is an environmentally friendly technique, is implemented in this way that such wastes are converted into wealth. Composting is a sustainable technique for waste disposal. Composting saves water by helping the soil to conserve moisture and reduces the water run-off and helping in the recycling of the organic resources. It is also utilized to avoid a health hazard, which is caused by the use of other fertilizers. Composting is the microbial-mediated process that can be utilized to produce the compost on a diverse level ranging from home to a large commercial scale. Nowadays, a lot of researchers are working in it, and it is becoming popular day by day among farmers due to its easy adaptability and cost.

References

Abatzoglou, N., & Boivin, S. (2009). A review of biogas purification processes. *Biofuels, Bioproducts and Biorefining*, *3*, 42–71.

Agnew, J. M., & Leonard, J. J. (2003). The physical properties of compost. *Compost Science & Utilization*, *11*, 238–264.

Ahmed, M., et al. (2020). Wastes to be the source of nutrients and energy to mitigate climate change and ensure future sustainability: Options and strategies. *Journal of Plant Nutrition*, 1–25.

Alavi, N., Daneshpajou, M., Shirmardi, M., Goudarzi, G., Neisi, A., & Babaei, A. A. (2017). Investigating the efficiency of co-composting and vermicomposting of vinasse with the mixture of cow manure wastes, bagasse, and natural zeolite. *Waste Management*, *69*, 117–126.

Albrecht, R., Joffre, R., Petit, J. L., Terrom, G., & Périssol, C. (2009). Calibration of chemical and biological changes in cocomposting of biowastes using near-infrared spectroscopy. *Environmental Science & Technology*, *43*, 804–811.

Amlinger, F., Götz, B., Dreher, P., Geszti, J., & Weissteiner, C. (2003). Nitrogen in biowaste and yard waste compost: Dynamics of mobilisation and availability—A review. *European Journal of Soil Biology*, *39*(3), 107–116.

Amon, B., Moitzi, G., Schimpl, M., Kryvoruchko, V., & Wagner-Alt, C. (2002). *Methane, nitrous oxide and ammonia emissions from management of liquid manures, final report 2002*. On behalf of "federal ministry of agriculture, forestry, environmental and water management" and "federal ministry of education, science and culture" research project.

Andriamananjara, A., Rabeharisoa, L., Prud'homme, L., & Morel, C. (2016). Drivers of plant-availability of phosphorus from thermally conditioned sewage sludge as assessed by isotopic labeling. *Frontiers in Nutrition*, *3*, 19.

Angelidaki, I., Mogensen, A. S., & Ahring, B. K. (2000). Degradation of organic contaminants found in organic waste. *Biodegradation*, *11*, 377–383.

Antizar-Ladislao, B., Lopez-Real, J., & Beck, A. (2004). Bioremediation of polycyclic aromatic hydrocarbon (PAH)-contaminated waste using composting approaches. *Critical Reviews in Environmental Science and Technology*, *34*, 249–289.

Aslam, D. N., Horwath, W., & Vander Gheynst, J. S. (2008). Comparison of several maturity indicators for estimating phytotoxicity in compost-amended soil. *Waste Management*, *28*, 2070–2076.

Aslam, D. N., Vander Gheynst, J. S., & Rumsey, T. R. (2008). Development of models for predicting carbon mineralization and associated phytotoxicity in compost-amended soil. *Bioresource Technology*, *99*, 8735–8741.

Atkinson, C. F., Jones, D. D., & Gauthier, J. J. (1996). Biodegradability and microbial activities during composting of poultry litter. *Poultry Science*, *75*, 608–617.

Bagstam, G. (1978). Population changes in microorganisms during composting of spruce bark. I. Influence of temperature control. *European Journal of Applied Microbiology and Biotechnology*.

Barker, A. V. (1997). *Composition and uses of compost*. ACS Publications.

Barker, A. V., & Bryson, G. M. (2002). Bioremediation of heavy metals and organic toxicants by composting. *The Scientific World Journal*, *2*, 407–420.

Barral, M. T., Paradelo, R., Moldes, A. B., Domínguez, M., & Díaz-Fierros, F. (2009). Utilization of MSW compost for organic matter conservation in agricultural soils of NW Spain. *Resources, Conservation and Recycling*, *53*, 529–534.

Båth, B., & Rämert, B. (1999). Organic household wastes as a nitrogen source in leek production. *Acta Agriculturae Scandinavica, Section B-Plant Soil Science*, *49*, 201–208.

Beaudin, N., Caron, R. F., Legros, R., Ramsay, J., Lawlor, L., & Ramsay, B. (1996). Cocomposting of weathered hydrocarbon-contaminated soil. *Compost Science & Utilization*, *4*, 37–45.

Beffa, T., et al. (1998). Mycological control and surveillance of biological waste and compost. *Medical Mycology*, *36*, 137–145.

Berkeley. (1950). *Composting for disposal of organic refuse*. University of California, Institute of Engieering Research, Sanitary Engineering Research, Laboratory.

Bernal, M. P., Alburquerque, J. A., & Moral, R. (2009). Composting of animal manures and chemical criteria for compost maturity assessment. A review. *Bioresource Technology*, *100*, 5444–5453.

Bong, C. P.-C., et al. (2017). Towards low carbon society in Iskandar Malaysia: Implementation and feasibility of community organic waste composting. *Journal of Environmental Management*, *203*, 679–687.

Briggs, J. A., Riley, M. B., & Whitwell, T. (1998). Quantification and remediation of pesticides in runoff water from containerized plant production. *Journal of Environmental Quality*, *27*, 814–820.

Büyüksönmez, F., Rynk, R., Hess, T. F., & Bechinski, E. (1999). Occurrence, degradation and fate of pesticides during composting: Part I: Composting, pesticides, and pesticide degradation. *Compost Science & Utilization*, *7*, 66–82.

Büyüksönmez, F., Rynk, R., Hess, T. F., & Bechinski, E. (2000). Literature review: Occurrence, degradation and fate of pesticides during composting: Part II: Occurrence and fate of pesticides in compost and composting systems. *Compost Science & Utilization*, *8*, 61–81.

Chaney, R. L., & Ryan, J. A. (1994). *Heavy metals and toxic organic pollutants in MSW-composts: Research results on phytoavailability, bioavailability, fate, etc*. US Environmental Protection Agency.

Chaney, R. L., et al. (2001). *Heavy metal aspects of compost use*. Boca Raton, FL: Lewis Publishers.

Chang, J. I., & Hsu, T.-E. (2008). Effects of compositions on food waste composting. *Bioresource Technology*, *99*, 8068–8074.

Chen, L., de Haro, M. M., Moore, A., & Falen, C. (2011). The composting process. *Dairy Manure Compost Production and Use in Idaho*, *2*, 513–532.

Chew, K. W., Chia, S. R., Yen, H.-W., Nomanbhay, S., Ho, Y.-C., & Show, P. L. (2019). Transformation of biomass waste into sustainable organic fertilizers. *Sustainability*, *11*, 2266.

Claassen, V. P., & Carey, J. L. (2004). Regeneration of nitrogen fertility in disturbed soils using composts. *Compost Science & Utilization*, *12*, 145–152.

Cogger, C., Hummel, R., Hart, J., & Bary, A. (2008). Soil and redosier dogwood response to incorporated and surface-applied compost. *HortScience*, *43*, 2143–2150.

Council Regulation. (1999). Council Regulation (EC) no 1804/1999 of 19 July 1999 supplementing Regulation (EEC) no 2092/91 on organic production of agricultural products and indications referring thereto on agricultural products and foodstuffs to include livestock production. *Official Journal of the European Communities*, *222*, 08.

Crawford, D. L. (1978). Lignocellulose decomposition by selected streptomyces strains. *Applied and Environmental Microbiology*, *35*, 1041–1045.

Debosz, K., Petersen, S. O., Kure, L. K., & Ambus, P. (2002). Evaluating effects of sewage sludge and household compost on soil physical, chemical and microbiological properties. *Applied Soil Ecology*, *19*, 237–248.

Demirbas, M. F., & Balat, M. (2009). Progress and recent trends in biogas processing. *International Journal of Green Energy*, *6*, 117–142.

Diaz, L. F., & De Bertoldi, M. (2007). History of composting. In *Waste management series* (pp. 7–24). Elsevier.

Diener, R. G., Collins, A. R., Martin, J. H., & Bryan, W. B. (1993). Composting of source-separated municipal solid waste for agricultural utilization—A conceptual approach for closing the loop. *Applied Engineering in Agriculture*, *9*, 427–436.

Dominguez, J., & Aira, M. (2010). New developments and insights on vermicomposting in Spain. In *Vermiculture technology: Earthworms, organic wastes, and environmental management* (p. 409).

Donay, J. L., Fernandes, P., Lagrange, P. H., & Herrmann, J. L. (2007). Evaluation of the inoculation procedure using a 0.25 McFarland standard for the BD Phoenix automated microbiology system. *Journal of Clinical Microbiology*, *45*, 4088–4089.

Doorn, M. R. J., & Barlaz, M. A. (1995). *Estimate of global methane emissions from landfills and open dumps*. Final report, January 1992-September 1994 Durham, NC (United States): Pechan (EH) and Associates, Inc.

Drury, C. F., Reynolds, W. D., Tan, C. S., Welacky, T. W., Calder, W., & McLaughlin, N. B. (2006). Emissions of nitrous oxide and carbon dioxide. *Soil Science Society of America Journal*, *70*, 570–581.

Duong, T. T. T. (2013). *Compost effects on soil properties and plant growth*. University of Adelaide.

Eastwood, D. J. (1952). The fungus flora of composts. *Transactions of the British Mycological Society*, *35*, 215–220.

El-Shafei, A., Yehia, M., & El-Naqib, F. (2008). Impact of effective microorganisms compost on soil fertility and rice productivity and quality. *MISR Journal of Agricultural Engineering*, *25*, 1067–1093.

Engwall, M., & Schnürer, A. (2002). Fate of Ah-receptor agonists in organic household waste during anaerobic degradation—Estimation of levels using EROD induction in organ cultures of chick embryo livers. *Science of the Total Environment*, *297*, 105–108.

Epstein, E. (2011). *Industrial composting: Environmental engineering and facilities management*. CRC Press.

Erhart, E., & Hartl, W. (2010). Compost use in organic farming. In *Genetic engineering, Biofertilisation, Soil Quality and Organic Farming* (pp. 311–345). Springer.

Ernst, G., Müller, A., Göhler, H., & Emmerling, C. (2008). C and N turnover of fermented residues from biogas plants in soil in the presence of three different earthworm species (Lumbricus terrestris, Aporrectodea longa, Aporrectodea caliginosa). *Soil Biology and Biochemistry*, *40*, 1413–1420.

Eureka. (2008). *Recycling, composting and greenhouse gas reductions in Minnesota*. (Make Dirt Not Waste).

Evanylo, G., Sherony, C., Spargo, J., Starner, D., Brosius, M., & Haering, K. (2008). Soil and water environmental effects of fertilizer-, manure-, and compost-based fertility practices in an organic vegetable cropping system. *Agriculture, Ecosystems & Environment*, *127*, 50–58.

Fauziah, S. H., & Agamuthu, P. (2009). Sustainable household organic waste management via vermicomposting. *Malaysian Journal of Science*, *28*, 135–142.

Finstein, M. S., & Morris, M. L. (1975). Microbiology of municipal solid waste composting. In *Advances in applied microbiology* (pp. 113–151). Elsevier.

Forsyth, W. G. C., & Webley, D. M. (1948). *The microbiology of composting II. A study of the aerobic themophilic bacterial flora developing in grass composts* (1st ed., pp. 34–39). Wiley Online Library.

Frascarelli, A. (2011). Le energie rinnovabili in agricoltura. *Agriregionieuropa anno*, *7*.

Gabhane, J., et al. (2012). Additives aided composting of green waste: Effects on organic matter degradation, compost maturity, and quality of the finished compost. *Bioresource Technology*, *114*, 382–388.

Galier, W. S., & Partridge, J. R. (1969). Physical and chemical analysis of domestic municipal refuse from Raleigh, North Carolina. *Compost Science*, *10*(3), 12–15.

Galitskaya, P., Biktasheva, L., Saveliev, A., Grigoryeva, T., Boulygina, E., & Selivanovskaya, S. (2017). Fungal and bacterial successions in the process of co-composting of organic wastes as revealed by 454 pyrosequencing. *PLoS One*, *12*(10), e0186051.

Garg, R. N., Pathak, H., Das, D. K., & Tomar, R. K. (2005). Use of flyash and biogas slurry for improving wheat yield and physical properties of soil. *Environmental Monitoring and Assessment*, *107*, 1–9.

Gerardi, M. H. (2003). *The microbiology of anaerobic digesters*. John Wiley & Sons.

Glathe, H. (1959). Biological processes in the composting of refuse. *International Research Group on Refuse Disposal, Bulletin*, (7), 8–10.

Golueke, C. G. (1972). *Composting: A study of the process and its principles* (6th ed.). Emmaus, PA: Rodale Press.

Gotaas, H. B., & World Health Organization. (1956). *Composting: Sanitary disposal and reclamation of organic wastes/Harold B. Gotaas*. World Health Organization.

Graves, E. R., Gwendolyn, M. H., Donald, S., James, N. K., & Dana, C. (2000). *Chapter 2. Composting. Part 637* (pp. 1–88).

Gray, K. R., Sherman, K., & Biddlestone, A. J. (1971). Review of composting. 2. Practical process. *Process Biochemistry*, *6*, 22.

Hartmann, H., & Ahring, B. K. (2003). Phthalic acid esters found in municipal organic waste: Enhanced anaerobic degradation under hyper-thermophilic conditions. *Water Science and Technology*, *48*, 175–183.

Hassen, A., Belguith, K., Jedidi, N., Cherif, A., Cherif, M., & Boudabous, A. (2001). Microbial characterization during composting of municipal solid waste. *Bioresource Technology*, *80*, 217–225.

Hayawin, Z. N., Astimar, A. A., Ibrahim, M. H., Abdul Khalil, H. P. S., Syirat, Z. B., & Menon, N. R. (2014). The growth and reproduction of Eisenia fetida and Eudrilus eugeniae in mixtures of empty fruit bunch and palm oil mill effluent. *Compost Science & Utilization*, *22*, 40–46.

Hayawin, Z. N., Astimar, A. A., Ridzuan, R., Syirat, Z. B., Ravi Menon, N., & Jalani, N. F. (2014). *Study on the effect of adding zeolite and other bio-adsorbance in enhancing the quality of the palm based vermicompost* (pp. 926–934). Trans Tech Publication.

Helmke, P. A., Robarge, W. P., Korotev, R. L., & Schomberg, P. J. (1979). Effects of soil-applied sewage sludge on concentrations of elements in earthworms. *Journal of Environmental Quality*, *8*, 322–327.

Hepperly, P., Lotter, D., Ulsh, C. Z., Seidel, R., & Reider, C. (2009). Compost, manure and synthetic fertilizer influences crop yields, soil properties, nitrate leaching and crop nutrient content. *Compost Science & Utilization*, *17*, 117–126.

Heyes, A., Moore, T. R., & Rudd, J. W. M. (1998). Mercury and methylmercury in decomposing vegetation of a pristine and impounded wetland. *Journal of Environmental Quality*, *27*, 591–599.

Himanen, M., & Hänninen, K. (2009). Effect of commercial mineral-based additives on composting and compost quality. *Waste Management*, *29*, 2265–2273.

Hoitink, H. A. J., & Boehm, M. J. (1999). Biocontrol within the context of soil microbial communities: A substrate-dependent phenomenon. *Annual Review of Phytopathology*, *37*, 427–446.

Holm Nielsen, J. B., Al Seadi, T., & Oleskowicz-Popiel, P. (2009). The future of anaerobic digestion and biogas utilization. *Bioresource Technology*, *100*, 5478–5484.

Hsu, J.-H., & Lo, S.-L. (2000). Characterization and extractability of copper, manganese, and zinc in swine manure composts. *Journal of Environmental Quality*, *29*, 447–453.

Igliński, B., Buczkowski, R., Iglińska, A., Cichosz, M., Piechota, G., & Kujawski, W. (2012). Agricultural biogas plants in Poland: Investment process, economical and environmental aspects, biogas potential. *Renewable and Sustainable Energy Reviews*, *16*, 4890–4900.

Imthiyas, M. S. M., & Seran, T. H. (2014). Influence of compost with reduced level of chemical fertilizers on the accumulation of dry matter in leaves of radish (Raphanus sativus L.). *J. Agric. Sci. Eng.*, *1*.

Inckel, M., de Smet, P., Tersmette, T., & Veldkamp, T. (2005). *The preparation and use of compost* (7th ed.). Agromisa Foundations Wage-Ningen.

Insam, H., & De Bertoldi, M. (2007). Microbiology of the composting process. In *Waste management series* (pp. 25–48). Elsevier.

Jedidi, N., Hassen, A., Van Cleemput, O., & M'Hiri, A. (2004). Microbial biomass in a soil amended with different types of organic wastes. *Waste Management & Research*, *22*, 93–99.

Jeffreys, G. A. (1959). *Simultaneous aerobic and anaerobic composting process*. (Google Patents).

Jørgensen, F. A. (2019). *Recycling*. MIT Press.

Jørgensen, K. S., Puustinen, J., & Suortti, A. M. (2000). Bioremediation of petroleum hydrocarbon-contaminated soil by composting in biopiles. *Environmental Pollution*, *107*, 245–254.

Jurado, M. M., Suárez-Estrella, F., López, M. J., Vargas-García, M. C., López-González, J. A., & Moreno, J. (2015). Enhanced turnover of organic matter fractions by microbial stimulation during lignocellulosic waste composting. *Bioresource Technology*, *186*, 15–24.

Kao, C.-Y., et al. (2012). A mutant strain of microalga Chlorella sp. for the carbon dioxide capture from biogas. *Biomass and Bioenergy*, *36*, 132–140.

Kaushal, A., & Bharti, U. (2015). Microbial kitchen waste composting: Effective environmentally sound alternative of solid waste management. *International Journal of Recent Scientific Research*, *6*, 7651–7654.

Khan, S., & Faisal, M. N. (2008). An analytic network process model for municipal solid waste disposal options. *Waste Management*, *28*, 1500–1508.

Klass, D. L. (2004). Biomass for renewable energy and fuels. *Encyclopedia of Energy*, *1*, 193–212.

Kosobucki, P., Chmarzynski, A., & Buszewski, B. (2000). Sewage sludge composting. *Polish Journal of Environmental Studies*, *9*, 243–248.

Krebs, R., Gupta, S. K., Furrer, G., & Schulin, R. (1998). Solubility and plant uptake of metals with and without liming of sludge-amended soils. *Journal of Environmental Quality*, *27*, 18–23.

Leifeld, J., Siebert, S., & Kögel-Knabner, I. (2002). Biological activity and organic matter mineralization of soils amended with biowaste composts. *Journal of Plant Nutrition and Soil Science*, *165*, 151–159.

Levén, L., Nyberg, K., Korkea-Aho, L., & Schnürer, A. (2006). Phenols in anaerobic digestion processes and inhibition of ammonia oxidising bacteria (AOB) in soil. *Science of the Total Environment*, *364*, 229–238.

Leven, L., & Schnürer, A. (2005). Effects of temperature on biological degradation of phenols, benzoates and phthalates under methanogenic conditions. *International Biodeterioration & Biodegradation*, *55*, 153–160.

Lichtfouse, E. (2014). *Sustainable agriculture reviews*. Springer.

Lim, W. J., Chin, N. L., AnizaYusof, Y., Yahya, A., & Tee, T. P. (2019). Modelling of pilot-scale anaerobic food wastes composting process with dry leaves or cow manure. *Pertanika Journal of Science & Technology*, *27*.

Linch, J. M., & Poole, N. J. (1979). *Microbial ecology: A conceptual approach*. John Wiley & Sons.

Lynch, D. H., Voroney, R. P., & Warman, P. R. (2005). Soil physical properties and organic matter fractions under forages receiving composts, manure or fertilizer. *Compost Science & Utilization*, *13*, 252–261.

Magdoff, F., & Weil, R. R. (2004). Soil organic matter management strategies. *Soil Organic Matter in Sustainable Agriculture*, 45–65.

Malińska, K., Golańska, M., Caceres, R., Rorat, A., Weisser, P., & Ślęzak, E. (2017). Biochar amendment for integrated composting and vermicomposting of sewage sludge—The effect of biochar on the activity of Eisenia fetida and the obtained vermicompost. *Bioresource Technology*, *225*, 206–214.

Mamo, M., Halbach, T. R., & Rosen, C. J. (2002). *Utilization of muncipal solid waste compost for crop production. Website reference; file. E:\Utilization% 20Of% 20Municipal% 20Solid% 20Waste% 20Comp ost% 20For% 20Cr*. USA: Department of Soil, Water and Climate, University of Minnesota.

Marchaim, U. (1992). *Biogas processes for sustainable development*. Food and Agriculture Organization.

Margesin, R., Cimadom, J., & Schinner, F. (2006). Biological activity during composting of sewage sludge at low temperatures. *International Biodeterioration & Biodegradation*, *57*, 88–92.

Marinari, S., Masciandaro, G., Ceccanti, B., & Grego, S. (2000). Influence of organic and mineral fertilisers on soil biological and physical properties. *Bioresource Technology*, *72*, 9–17.

McMahon, V., Garg, A., Aldred, D., Hobbs, G., Smith, R., & Tothill, I. E. (2008). Composting and bioremediation process evaluation of wood waste materials generated from the construction and demolition industry. *Chemosphere*, *71*, 1617–1628.

Mehta, C. M., Palni, U., Franke-Whittle, I. H., & Sharma, A. K. (2014). Compost: Its role, mechanism and impact on reducing soil-borne plant diseases. *Waste Management*, *34*, 607–622.

Middleton, B. A., Van Der Valk, A. G., Williams, R. L., Mason, D. H., & Davis, C. B. (1992). Litter decomposition in an Indian monsoonal wetland overgrown withPaspalum distichum. *Wetlands*, *12*, 37–44.

Miller, F. C. (1993). *Ecological process control of composting. Soil microbial ecology* (pp. 529–536). New York: Marcel Dekker.

Misra, R. V., Roy, R. N., & Hiraoka, H. (2003). *On-farm composting methods*. Rome, Italy: UN-FAO.

Mohee, R., & Mudhoo, A. (2005). Analysis of the physical properties of an in-vessel composting matrix. *Powder Technology*, *155*, 92–99.

Möller, K., & Stinner, W. (2009). Effects of different manuring systems with and without biogas digestion on soil mineral nitrogen content and on gaseous nitrogen losses (ammonia, nitrous oxides). *European Journal of Agronomy*, *30*, 1–16.

Monnet, F. (2003). *An introduction to anaerobic digestion of organic wastes*. A report by Remade Scotland.

Moubasher, A. H., Abdel-Hafez, S. I. I., Abdel-Fattah, H. M., & Moharram, A. M. (1982). Fungi of wheat and broad-bean straw composts. *Mycopathologia*, *78*, 161–168.

Nagavallemma, K. P., Wani, S. P., Lacroix, S., Padmaja, V. V., Vineela, C., Rao, M. B., & Sahrawat, K. L. (2004). Vermicomposting: Recycling wastes into valuable organic fertilizer. Global Theme on Agroecosystems Report no. 8. http://oar.icrisat.org/id/eprint/3677.

Nakasaki, K., Shoda, M., & Kubota, H. (1985). Effect of temperature on composting of sewage sludge. *Applied and Environmental Microbiology*, *50*, 1526–1530.

Namkoong, W., Hwang, E.-Y., Park, J.-S., & Choi, J.-Y. (2002). Bioremediation of diesel-contaminated soil with composting. *Environmental Pollution*, *119*, 23–31.

Neher, D. (2019). Compost and plant disease suppression. *Biocycle*, *60*, 22.

Niggli, U., Schmid, H., & Fliessbach, A. (2008). *Organic farming and climate change*. Geneva: International Trade Centre (ITC).

Nilsson, M.-L. (2000). *Occurrence and fate of organic contaminants in wastes*. University of Agricultural Sciences.

Nilsson, M.-L., Kylin, H., & Sundin, P. (2000). Major extractable organic compounds in the biologically degradable fraction of fresh, composted and anaerobically digested household waste. *Acta Agriculturae Scandinavica, Section B-Plant Soil Science*, *50*, 57–65.

Nilsson, M.-L., Waldebäck, M., Liljegren, G., Kylin, H., & Markides, K. E. (2001). Pressurized-fluid extraction (PFE) of chlorinated paraffins from the biodegradable fraction of source-separated household waste. *Fresenius' Journal of Analytical Chemistry*, *370*, 913–918.

Odlare, M., Pell, M., & Svensson, K. (2008). Changes in soil chemical and microbiological properties during 4 years of application of various organic residues. *Waste Management*, *28*, 1246–1253.

Ogello, E. O., Wullur, S., Sakakura, Y., & Hagiwara, A. (2018). Composting fishwastes as low-cost and stable diet for culturing Brachionus rotundiformis Tschugunoff (Rotifera): Influence on water quality and microbiota. *Aquaculture*, *486*, 232–239.

Okereke, H. C., & Kanu, I. J. (2004). In A. Onyeagba (Ed.), *Identification and characterization of microorganisms. Laboratory guide for microbiology* (pp. 95–110). Okigwe: Crystal Publishers.

Olsman, H., BjÖRnfoth, H., Van Bavel, B., LindstrÖM, G., SchnüRer, A., & Engwall, M. (2002). Characterisation of dioxin-like compounds in anaerobically digested organic material by bioassay-directed fractionation. *Organohalogen Compounds*, *58*, 345–348.

Ouédraogo, E., Mando, A., & Zombré, N. P. (2001). Use of compost to improve soil properties and crop productivity under low input agricultural system in West Africa. *Agriculture, Ecosystems & Environment*, *84*, 259–266.

Own, G. Y. (2010). *Landscape and gardening*. Bridgeton, MO: St. Louis Community College Continuing Education.

Paré, T., Dinel, H., & Schnitzer, M. (1999). Extractability of trace metals during co-composting of biosolids and municipal solid wastes. *Biology and Fertility of Soils*, *29*, 31–37.

Polprasert, C., & Koottatep, T. (2017). *Organic waste recycling: Technology, management and sustainability*. IWA publishing.

Pöschl, M., Ward, S., & Owende, P. (2010). Evaluation of energy efficiency of various biogas production and utilization pathways. *Applied Energy*, *87*, 3305–3321.

Rajeswari, C., Padmavathy, P., & Aanand, S. (2018). Composting of fish waste: A review. *International Journal of Applied Research*, *4*, 242–249.

Ramaraj, R., & Dussadee, N. (2015). Biological purification processes for biogas using algae cultures: A review. *International Journal of Sustainable and Green Energy. Special Issue: Renewable Energy Applications in the Agricultural Field and Natural Resource Technology*, *4*, 20–32.

Rao, N., Grethlein, H. E., & Reddy, C. A. (1995). Mineralization of atrazine during composting with untreated and pretreated lignocellulosic substrates. *Compost Science & Utilization*, *3*, 38–46.

Rao, N., Grethlein, H. E., & Reddy, C. A. (1996). Effect of temperature on composting of atrazine-amended lignocellulosic substrates. *Compost Science & Utilization*, *4*, 83–88.

Reinecke, A. J., Viljoen, S. A., & Saayman, R. J. (1992). The suitability of Eudrilus eugeniae, Perionyx excavatus and Eisenia fetida (Oligochaeta) for vermicomposting in southern Africa in terms of their temperature requirements. *Soil Biology and Biochemistry*, *24*, 1295–1307.

Rivard, C. J., et al. (1995). Anaerobic digestion of municipal solid waste. *Applied Biochemistry and Biotechnology*, *51*, 125.

Ryckeboer, J., et al. (2003). A survey of bacteria and fungi occurring during composting and self-heating processes. *Annals of Microbiology*, *53*, 349–410.

Saison, C., et al. (2006). Alteration and resilience of the soil microbial community following compost amendment: Effects of compost level and compost-borne microbial community. *Environmental Microbiology*, *8*, 247–257.

Sait, M., Hugenholtz, P., & Janssen, P. H. (2002). Cultivation of globally distributed soil bacteria from phylogenetic lineages previously only detected in cultivation-independent surveys. *Environmental Microbiology*, *4*, 654–666.

Salminen, E., & Rintala, J. (2002). Anaerobic digestion of organic solid poultry slaughterhouse waste—A review. *Bioresource Technology*, *83*, 13–26.

Sarker, A., Kashem, A., & Osman, K. T. (2012). Influence of city finished compost and nitrogen, phosphorus and potassium (NPK) fertilizer on yield, nutrient uptake and nutrient use efficiency of radish (Raphanus sativus L.) in an acid soil. *International Journal of Agricultural Sciences*, *2*, 315–321.

Schulze, K. L. (1962). Continuous thermophilic composting. *Applied and Environmental Microbiology*, *10*, 108–122.

Sikora, L. J., Tester, C. F., Taylor, J. M., & Parr, J. F. (1982). Phosphorus uptake by fescue from soils amended with sewage sludge compost 1. *Agronomy Journal*, *74*, 27–33.

Silverman, M. P., & Ehrlich, H. L. (1964). Microbial formation and degradation of minerals. In *Advances in applied microbiology* (pp. 153–206). Elseier.

Smith, A., Brown, K., Ogilvie, S., Rushton, K., & Bates, J. (2001). *Waste management options and climate change: Final report* (pp. 137–150). DG Environment: European Commission.

Stratton, M. L., Barker, A. V., & Rechcigl, J. E. (1995). Compost. In *Soil amendments and environmental quality* (pp. 249–309).

Stratton, M. L., & Rechcigl, J. E. (1998). Agronomic benefits of agricultural, municipal, and industrial by-products and their co-utilization: An overview. In *Beneficial co-utilization of agricultural, municipal and industrial by-products* (pp. 9–34). Springer.

Strom, P. F. (1985). Effect of temperature on bacterial species diversity in thermophilic solid-waste composting. *Applied and Environmental Microbiology*, *50*, 899–905.

Stutzenberger, F. J., Kaufman, A. J., & Lossin, R. D. (1970). Cellulolytic activity in municipal solid waste composting. *Canadian Journal of Microbiology*, *16*, 553–560.

Svensson, K., Odlare, M., & Pell, M. (2004). The fertilizing effect of compost and biogas residues from source separated household waste. *The Journal of Agricultural Science*, *142*, 461–467.

Sweeten, J. M. (2008). *Composting manure and sludge*. (Texas FARMER Collection).

Teensma, B. (1961). Composting city refuse in the Netherlands. *Compost Science*, *1*, 11–17.

Tiwari, V. N., Tiwari, K. N., & Upadhyay, R. M. (2000). Effect of crop residues and biogas slurry incorporation in wheat on yield and soil fertility. *Journal of the Indian Society of Soil Science*, *48*, 515–520.

Tomati, U., Galli, E., Pasetti, L., & Volterra, E. (1995). Bioremediation of olive-mill wastewaters by composting. *Waste Management & Research*, *13*, 509–518.

Torquati, B., Venanzi, S., Ciani, A., Diotallevi, F., & Tamburi, V. (2014). Environmental sustainability and economic benefits of dairy farm biogas energy production: A case study in Umbria. *Sustainability*, *6*, 6696–6713.

Trautmann, N. M., & Krasny, M. E. (1998). *Composting in the classroom: Scientific inquiry for high school students*. Kendall/Hunt Publishing Company.

Tweib, S. A., Rahman, R. A., & Kalil, M. S. (2011). *A literature review on the composting*. University Kebangsaan.

Venglovsky, J., et al. (2005). Evolution of temperature and chemical parameters during composting of the pig slurry solid fraction amended with natural zeolite. *Bioresource Technology*, *96*, 181–189.

Villasenor, J., Rodriguez, L., & Fernandez, F. J. (2011). Composting domestic sewage sludge with natural zeolites in a rotary drum reactor. *Bioresource Technology*, *102*, 1447–1454.

von Klopotek, A. (1969). Enzymatic reduction of Iron oxide by Fungi. *Applied Lvlicrobiol*, *41*, 1.

Waksman, S. A., Cordon, T. C., & Hulpoi, N. (1939). Influence of temperature upon the microbiological population and decomposition processes in composts of stable manure. *Soil Science*, *47*, 83–114.

Wang, Q., et al. (2017). Improvement of pig manure compost lignocellulose degradation, organic matter humification and compost quality with medical stone. *Bioresource Technology*, *243*, 771–777.

Wei, Y.-S., Fan, Y.-B., & Wang, M.-J. (2001). A cost analysis of sewage sludge composting for small and mid-scale municipal wastewater treatment plants. *Resources, Conservation and Recycling*, *33*, 203–216.

Wichuk, K. M., & McCartney, D. (2007). A review of the effectiveness of current time–temperature regulations on pathogen inactivation during composting. *Journal of Environmental Engineering and Science*, *6*, 573–586.

Williams, R. T., & Keehan, K. R. (1993). *Hazardous and industrial waste composting* (pp. 363–382). Worthington, OH: Renaissance Press.

Williams, R. T., Ziegenfuss, P. S., & Sisk, W. E. (1992). Composting of explosives and propellant contaminated soils under thermophilic and mesophilic conditions. *Journal of Industrial Microbiology*, *9*, 137–144.

Yan-mei, Y. (2008). The design of the organic solid wastes compost equipment. *Journal of Chongqing Jiaotong University (Natural Science)*, *6*.

Yu, F., et al. (2006). Application of biogas fermentation residue in Ziziphus jujuba cultivation. *Ying Yong Sheng Tai Xue Bao [The Journal of Applied Ecology]*, *17*, 345–347.

Zafar, S. (2017). *Solid waste management in Bahrain*.

Zhang, L., & Sun, X. (2017). Addition of fish pond sediment and rock phosphate enhances the composting of green waste. *Bioresource Technology*, *233*, 116–126.

Zhu, J., Yang, L., & Li, Y. (2015). Comparison of premixing methods for solid-state anaerobic digestion of corn Stover. *Bioresource Technology*, *175*, 430–435.

Chapter 7

Fungal bioprocessing of lignocellulosic materials for biorefinery

Oscar Fernando Vázquez-Vuelvas[a], Jose Antonio Cervantes-Chávez[b], Francisco Javier Delgado-Virgen[c], Laura Leticia Valdez-Velázquez[d], and Rosa Jazmin Osuna-Cisneros[c]

[a]Biochemical Engineering and Bioprocessing Laboratory, Chemical Science Faculty, University of Colima, Colima, México, [b]Queretaro Autonomous University, Basic and Applied Microbiology Unit, Natural Science Faculty, Santiago de Querétaro, Mexico, [c]Plant and Microbial Biotechnology Laboratory, Mexico's National Technologic, Colima Institute of Technology, Colima, México, [d]Biological Products Laboratory, Chemical Engineering Laboratory, Chemical Science Faculty, University of Colima, Colima, México

Chapter outline

Recent Advancement in Microbial Biotechnology. https://doi.org/10.1016/B978-0-12-822098-6.00009-4

1 Introduction

In recent years, lignocellulosic biomass has gained an important technologic and economic value. The development of environmentally friendly processes to transform lignocellulose into commodities of industrial importance has transformed the way of how modern societies perceive future economic organization. Although the use of carbon as source of energy has been important through the mankind evolution, in recent decades, excessive contaminating agents emitted by industrial and domestic activities have caused high increments of normal atmospheric pollutant, as a result of demanding modern lifestyles.

From the beginning of first human societies, carbon sources, now identifiable as biomass, were used to access to heat and light, to process food and other activities. During the industrial revolution, different sources of carbon become essential for the high production of energy. Intrinsically, the generation of carbon dioxide was incremented with the exploitation of petroleum to refine fuels for energy and other numerous chemicals as a source of raw materials. The excessive demand for natural resources, mainly from the nonrenewable origin, to elaborate many commodities, has resulted in an uncountable generation of by-products and waste materials with low or no reintegration to the natural carbon cycle. Each consumer good represents several times the quantity of carbon requirements to get over the user, interpreted by the packing materials, fuels for manufacturing and transportation energy, among others.

Different consequences of atmospheric contamination, such as global warming or stratospheric ozone depletion, are attributed to the emission of carbon dioxide and chlorofluorocarbons. These carbon compound derivatives are still emitted without restriction policies in developing countries due to linear and petroleum-based economies. Consequently, resource consumption and no recycling practices are reflected in this high dependence on petroleum. Curiously, the technological development to process lignocellulose materials was growing with energy requirements during the industrial revolution. A high amount of energy is required for various industrial processes, including textile, pulp, paper, and metallurgical industries. During the last century, wood and cotton were almost the unique sources of lignocellulose as raw material for producing commodities. However, during the last quarter of the twentieth century, a larger variety of uses of lignocellulose as a feedstock started to be implemented, and the use of microorganisms to process it were deeply researched. Greener chemical and engineering processes were designed to use potential agricultural residues as raw materials to meet the requirements of fuel, biopolymers, and chemicals. At this point in the technological revolution, the fungi kingdom has a key role in producing numerous important metabolites to manufacture many biomaterials, contributing with food additives and feed products, enzymes, platform chemical building blocks and biofuels. With the arrival of biological engineering solutions such as genetic engineering, proteomics, metabolomics and also the progress in advanced systems to control bioprocessing equipment, many operating difficulties have been diminished, even the control of culture parameters has gotten improved to reach and expand a great diversity of produced metabolites using fungi.

2 Lignocelullosic biomass and its chain value

2.1 Economy of biomaterials

Residual biomass is a bioorganic material that may be classified as waste, from a linear economy point of view (Kapoor et al., 2020; Nizami et al., 2017; Sharma, Gaur, Kim, & Pandey, 2020; Vea, Romeo, & Thomsen, 2018). Plant biomass is constituted in a major proportion by lignocellulose, which is formed by a biological process from the atmospheric carbon dioxide via the photosynthetic pathway, and following the carbon cycle, this is a renewable material that is permanently regenerated by the plant kingdom (Ma et al., 2019). Therein, lignocellulose is a biodegradable and environmentally friendly bioresource, available from several reservoirs, and represents a huge and promising carbon source for the transition from a petroleum-based to a biomass-based economy (Chojnacka, Moustakas, & Witek-Krowiak, 2020; Liao et al., 2020). The lignocellulosic biomass includes agricultural and forestry sources as a harvested and primary raw material for elaborating different products, and wasted biomass and industrial residues as usable materials for second-generation bio-based factories (Kapoor et al., 2020).

2.2 Knowledge-based bioeconomy for biorefineries

Since lignocellulose is present in every agricultural crop and represents a non-digestible substrate for the human organism, lignocellulosic material becomes a common by-product or even residual material with the potential use as feedstock for biofactories, usually and currently named biorefineries. The biorefinery concept is employed to efficiently produce high-value metabolites from diverse materials such as lignocellulosic biomass, algal biomass, industrial and municipal food waste, microbial treated waste, and manure (Ubando,

Felix, & Chen, 2020; Vea et al., 2018). Therein, the use of residual or wasted lignocellulose biomass as raw material makes it economically attractive because this inexpensive commodity acquires value through the application of technical transformations described by the concept of knowledge-based bioeconomy, which has been defined as "the method of changing life science knowledge into innovative, self-renovating and competitive products" by the European Commission. The term "bioeconomy" comprehends all industries and economic sectors that produce, manage, and otherwise utilize biological resources (such as agriculture, forestry, fisheries, and other bioresource industries) and related services (supply and consumer industries) (Aguilar, Bochereau, & Matthiessen, 2009; Ahmad et al., 2020; Hu et al., 2013; Jiménez-González & Woodley, 2010; Kapoor et al., 2020; Sharma et al., 2020).

2.3 Circular bioeconomy

As distinguished from the beginning of the industrial revolution, and so far, most of the global economies developed a linear system and have depended on taking resources to produce, make and use for consumption, and dispose of them (Kapoor et al., 2020). However, the concept of knowledge-based bioeconomy has promoted the evolution of factories toward the migration or creation of biorefineries. In the same context, the notion of reuse wasted lignocellulose instead of disposing of it as spoilage or other low-value uses has also acquired an associated concept of biorefineries. Consequently, the reutilization outlook for obtaining benefits from waste as a potential resource of feedstock led to the thought of the circular economy (Kapoor et al., 2020; Ubando et al., 2020; Vea et al., 2018). Circular economy, projected as a bio-based ideology, biocircular economy or circular bioeconomy is perceived as an improving element of utilization of lignocellulosic biomass and the recovery of its residue for reuse, recycle, and remanufacture, through the development of bioprocesses applied on biorefineries (Chojnacka et al., 2020; Ubando et al., 2020). Circular bioeconomy redesigns the life cycle of a conventional product, minimizing merchandise usage and waste through the incorporation of terms such as "reuse" and "remanufacture" to the linear chain of traditional economic systems to develop a closed-loop technical and bioeconomic cycle (Fig. 1). All this conceptualization has only one purpose, the equilibrium of the social, economic, technical, and environmental factors to the progress of the global population (Dragone et al., 2020; Kapoor et al., 2020; Sperandio & Ferreira Filho, 2019; Ubando et al., 2020; Vea et al., 2018).

2.4 Valorization of lignocellulosic biomass

The concept of circular bioeconomy implicates the reuse of wasted products after their consumption stage. From an economic point of view, the use of wastes enhances the value of them as raw material, and consequently, materials

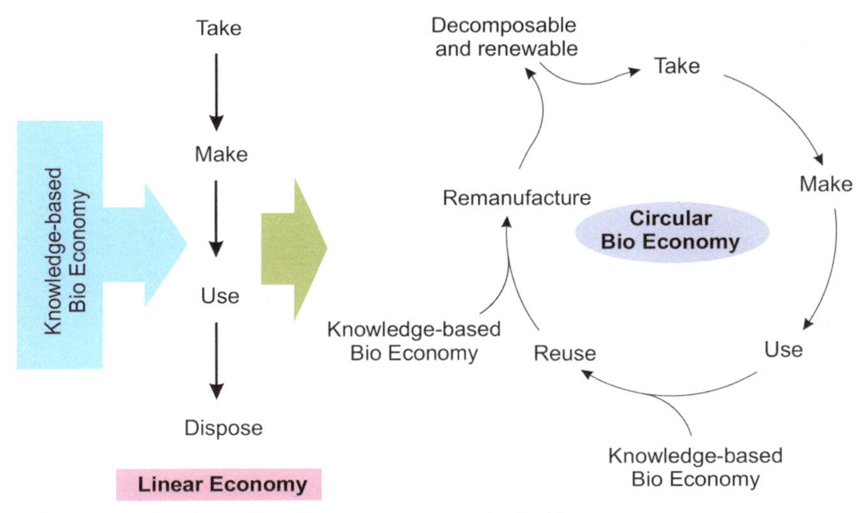

FIG. 1 Evolutive change from linear economy to a circular bioeconomy concept.

as lignocellulosic biomass become valuable. Therefore, the valorization concept incorporates to the economical ecosystem and is assigned to lignocellulolytic feedstock; its appreciation is intrinsically related to the product elaborated by biotechnological processing (Meghana & Shastri, 2020). In this regard, circular bioeconomy represents an advantageous manner of holding the biomaterial and bioproduct value for a longer duration (Kapoor et al., 2020). Then, biomass valorization is defined as the enhancement of the appreciation of feedstocks to biofactories to refine different materials. Biorefineries act as a value-increasing platform by combining the conversion technologies and the applicability to suitable lignocellulolytic material (Fig. 2) (Ubando et al., 2020).

In order to develop an increasing value to lignocellulosic biomass, the biorefinery concept is further classified into the type of feedstock (Takkellapati, Li, & Gonzalez, 2018):

- Biorefinery type I. It uses only lignocellulosic material, has optimized production capability, and elaborates only one product. For example, a factory that produces biodiesel from seed oil or ethanol from corn grain.
- Biorefinery type II. This is similar to biorefinery type I; it uses only lignocellulose material but is competent for producing several commodities. For example, the elaboration of a variety of chemicals, monosaccharides as building blocks, and ethanol from cereals.
- Biorefinery type III. This is an advanced biorefinery and can exploit several kinds of biomaterials and transformation techniques to elaborate a wide variety of products. This type is subdivided into four classes: (1) whole-crop biorefinery, (2) green biorefinery, (3) lignocellulosic biorefinery, and (4) two-platform concept biorefinery.

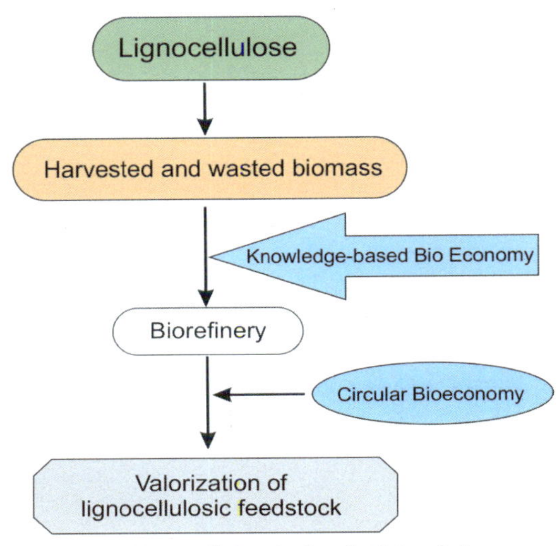

FIG. 2 Valorization process of lignocellulose feedstock. Value chain structure in a bioeconomy.

3 Benefits of lignocellulosic materials for biorefineries

3.1 Availability of lignocellulose

Lignocellulosic feedstock represents a neutral, easily available, and renewable carbon source for biorefineries, and it has been considered as a promising source to obtain bioprocessed materials (Kapoor et al., 2020). Lignocellulosic biomass is present in agricultural and forestry residues, municipal wastes, and agro-industrial residues. Agricultural residues, considered as a primary source of lignocellulosic biomass, are mainly originated as subproducts during the harvesting process (Kapoor et al., 2020). Predominantly, this biomass is recovered from residues from cereal crops such as barley, rice, wheat, oat, soybean brans and straws, corn stover, and sorghum stalks; seeds such as tea, tobacco, tomato seeds, and linseed; and other vegetable sources such as foliage waste, bay laurel leaves, palm trees, date seeds, wood, among many others (Kapoor et al., 2020; Soccol et al., 2017). Domestic waste, constituted by the organic fraction of municipal solid Waste (OFMSW), and agro-industrial residues are considered as secondary source of lignocellulose biomass and are constituted mainly by residues such as sugarcane, cassava, and agave bagasses; coffee pulp, coffee, and coconut husk; citrus or fruit peel; and pulp waste from diverse agroindustrial processes. All these materials are composed mainly of lignocellulose, besides starch, pectin, and other fibers (Nizami et al., 2017; Soccol et al., 2017; Ubando et al., 2020; Vea et al., 2018). Moreover, about 1300 million tons/year of lignocellulosic material are generated globally. Consequently, lignocellulosic material belongs to the biomass sector that does not represent a problem with human feeding crops and can be utilized widely as feedstock to biorefineries (Ubando et al., 2020).

3.2 Advantages of lignocellulosic feedstock for biorefineries

With the purpose of substituting the present petroleum-based world economy, a biorefinery becomes an entity that projects sustainable development of value-added bioproducts. Several convenient reasons exist to consider lignocellulosic biomass as the primary raw material for biorefineries. These reasons may be grouped as technoenvironmental and socioeconomical aspects, which involve the following details:

3.2.1 Technical and environmental advantages

Renewable carbon source raw material. Abundance of the lignocellulosic feedstock make it the most attractive material for bioprocessing because permanent regeneration is almost guaranteed (Ferreira & Taherzadeh, 2020; Liao, de Beeck, et al., 2020).

Biodegradable and neutral product. Closely related to the previous aspect, lignocellulose is a neutral carbon source and raw material for producing biomaterials, biofuels, or chemicals, which are biodegradable and/or biotransformed to be regenerated as lignocellulose, making this cycle an environmentally friendly and sustainable process (Aguilar et al., 2009).

Biological resources. Microorganisms are ubiquitous and their utilization to develop bioprocessing to transform lignocellulosic material acquires relevance through the isolation, exploitation, and genetic engineering on wild microbiota.

Knowledge and Technology. Bioconversion fundamentals and experience involves the joint work of enterprises and academia; this is to say, scientific knowledge and technical experience may be developed collaboratively to generate local adaptation of technology (Kapoor et al., 2020).

3.2.2 Social and economic aspects

No competition with food crops. Lignocellulosic materials, as the second generation of feedstock, do not conflict with human food crops; in fact, they represent complementary exploitation of the bioresources (de Jong & Jungmeier, 2015; Ferreira & Taherzadeh, 2020; Takkellapati et al., 2018).

Low-value feedstock. Since lignocellulose is permanently available, this residual biomaterial can be considered as low-price biomass, which bioprocessed can become medium- to high-value commodities or products (Kocabas, Ogel, & Bakir, 2014).

Regionally and locally based economy. The evolution of a circular bioeconomy may be established at local and regional levels, which may eventually evolve to create self-providing biofactories to become integral (waste-free) biorefineries (Kapoor et al., 2020).

4 Lignocellulosic materials, structure, and characteristics

Currently, the use of biomass has increased enormously and particularly that derived from agricultural or agroindustrial residues for combustible production or chemical products; this has become a key point to sustainable development, given that residues are renewable in contrast to the raw material of fossil origin (Gallezot, 2007; Meents, Watanabe, & Samuels, 2018). Lignocellulose represents a material localized ubiquitously in the plant cell wall of grasses, trees, and food crops. Its complex structure is constituted by three major building blocks that together create a recalcitrant structure to protects plant cells from microbial degradation (Frommhagen et al., 2017; Liao, de Beeck, et al., 2020; Liu, Olson, Wu, Broberg, & Sandgren, 2017; Meents et al., 2018). Due to its complexity, lignocellulosic biomass has to be pretreated and fractionated into its main constituent; subsequently, the individual fractions can be selectively converted into a set of value-added end-products through biological, chemical, and/or thermal approaches (Liao et al., 2020). The main component of lignocellulose is cellulose, which represents 25%–55%, 24%–50% of hemicellulose, and 10%–35% of lignin, among other components as pectins, proteins, waxes, terpenes, fats, and inorganic matter (Liao, de Beeck, et al., 2020).

4.1 Cellulose

Cellulose is a native and the most abundant on Earth linear homo biopolymer composed of β-(1–4)-linked glycosidic units in a twofold helical conformation. This linearity is explained by the presence of hydroxyl groups with different polarities that develop strong intramolecular and intermolecular hydrogen bonds; this makes cellulose a slightly stiff material (Kondo, 1998). Cellulose is ordered in cellobiose monomers, has a reducing and nonreducing sugar at the end of the polymeric chain, and generally contain from 500 to 15,000 glucose monomers, but this composition behavior depends on the plant species (Klemm, Heublein, Fink, & Bohn, 2005; Liao, de Beeck, et al., 2020). Most of the cellulose is present in the plant cell as microfibrils, crystalline areas of cellulose are crisscrossed with amorphous areas (Kondo, 1998; Kulasinski, Keten, Churakov, Derome, & Carmeliet, 2014; Liao, de Beeck, et al., 2020). Planar sheets constitute these crystalline zones as ribbons of hydrogen-bonded cellulose chains, and chemical and thermal processing of native cellulose (cellulose I) can affect the structure and provoke the formation of other noncrystalline frameworks of cellulose (cellulose II, III, and IV), with different crystal unit cells (Fig. 3) (Klemm et al., 2005; Kondo, 1998; Liao, de Beeck, et al., 2020; Zugenmaier, 2001). The amorphous region is characterized by being easily penetrable by solvents, enzymes, and reagents and is therefore easily hydrolyzable, while the crystalline region is highly resistant to penetration by solvents, enzymes, and reagents (Gorshkova, Kozlova, & Mikshina, 2013; Karimi & Taherzadeh, 2016; Meents et al., 2018).

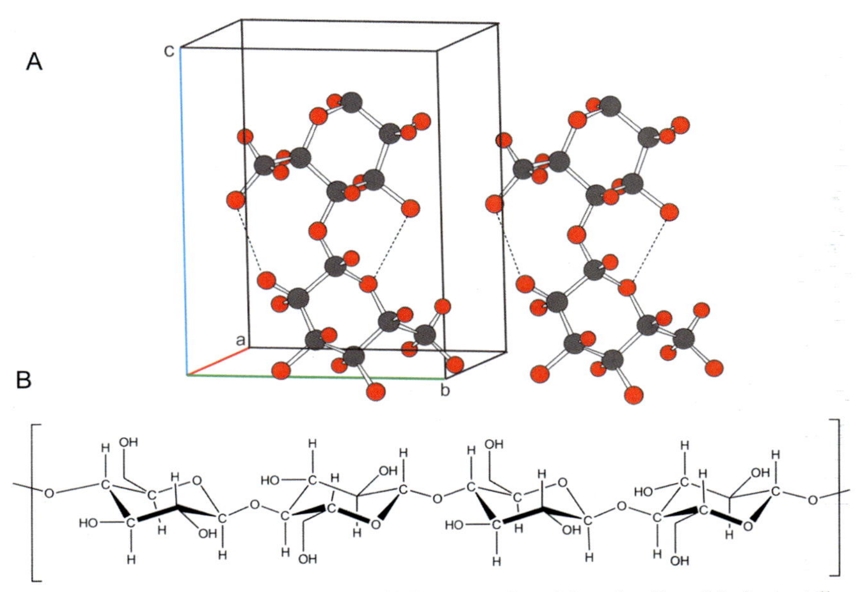

FIG. 3 Molecular structure of cellulose. (A) Representation of the unit cell model of a crystallographic structure of cellulose Iβ to demonstrate the intramolecular hydrogen bonding. (B) Chemical structure of cellulose.

4.2 Hemicellulose

Hemicellulose is almost as abundant as cellulose and is a component of lignocellulose. It represents a more heterogeneous group of biopolymers constituted by glycosidic-linked pentoses (xylose, arabinose), hexoses (glucose, mannose, galactose, fucose, rhamnose), sugar acids [(4-*O*-methyl)glucuronic acid], and ester-linked acetyl and feruloyl groups (Fig. 4) (Liao, de Beeck, et al., 2020; Scheller & Ulvskov, 2010). There are two common types of the major components of hemicellulose: xylans for hardwood and glucomannans for softwood (Zhang, Li, Xiong, Hong, & Chen, 2015). The hemicellulose structure is cross-linked with the cellulose matrix and varies according to different plant species and tissue types (Ma et al., 2019). Hemicellulose is not crystalline, is less recalcitrant, presents amorphous sections due to its branched framework, and has a smaller polymeric chain than cellulose (around 100 and 200) (Oh et al., 2015). Hemicellulose is bound to cellulose via hydrogen bonding and lignin by covalent bonding to constitute lignin-carbohydrate complexes, identified as "holocellulose" fraction of lignocellulose (Liao, de Beeck, et al., 2020; Liu et al., 2019).

4.3 Lignin

Lignin is a recalcitrant and complex mixture of insoluble phenolic macromolecules constituted of the three phenyl propanols, coniferyl (guaiacyl) alcohol or

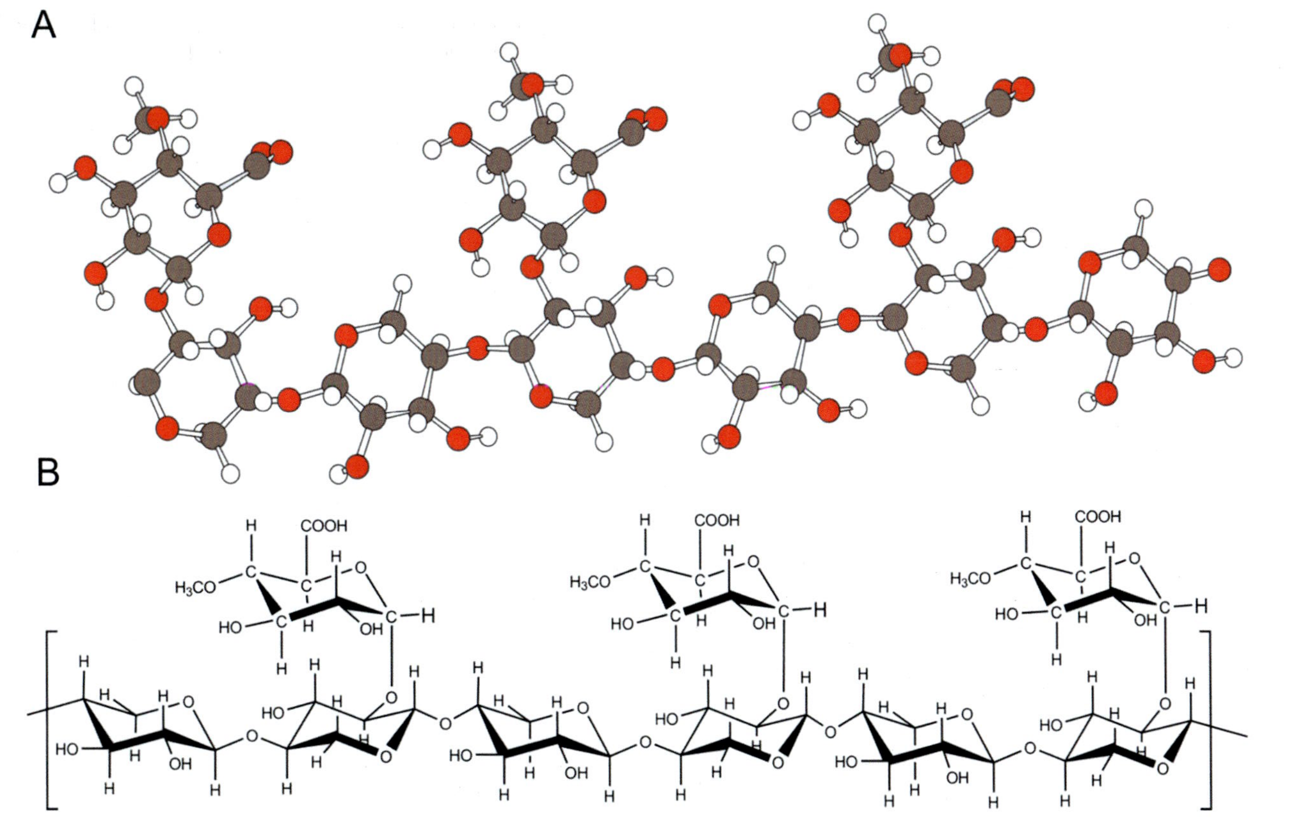

FIG. 4 Molecular structure of hemicellulose. (A) Representation of the atomic bonding of a 3D of 4-*O*-methylglucuronoxylan in hemicellulose. (B) Chemical structure of hemicellulose.

(G) unit, sinapyl (syringyl) alcohol or (S) unit, and p-coumaryl (p-hydroxyphenyl) alcohol or (H) unit (Cao et al., 2018; Gazi, 2019; Liao, de Beeck, et al., 2020). The amorphous heterogeneous polymeric structure is crosslinked to cellulose and hemicellulose via covalent bonding and hydrogen bonds. There are differences between lignin from softwoods and lignin from hardwoods; the first contains (G) units while the second contains, along with those, additional (S) units (Gazi, 2019). Lignin is a basic component that reinforces and provides rigidity and shape to plant tissues, is also considered as an important defense mechanism to microbial attack, and can prevent water destroys the polysaccharide-protein matrix of plant cells (Achyuthan et al., 2010; Cao et al., 2018; Gazi, 2019; Kawaguchi, Ogino, & Kondo, 2017; Liao, de Beeck, et al., 2020; Tolbert, Akinosho, Khunsupat, Naskar, & Ragauskas, 2014).

5 Fungi and their lignocellulose degrading abilities

For many decades, nonrenewable fossils resources like petroleum have been used to synthesize chemicals, chemical building blocks, or its derivatives. Nevertheless, the transition to the use of renewable carbon sources, such as lignocellulose, is a global tendency due to the importance of producing valuable chemical building blocks from different sources of biomass, with low costs and employing environmentally friendly transformations (Guo, Chang, & Lee, 2018). For a long time, lignocellulosic biomass has been considered as waste because of crystallinity of its chemical structure and recalcitrancy nature, that is why lignocellulose degradation constitutes a challenging task. Fortunately, the use of fungi during the microbial valorization process of plant biomass, waste material, and low-value industrial by-products into technologically functional materials is becoming a key target in bioeconomy to avoid using fossil resources. In the kingdom of microorganisms, fungi and yeasts are promising candidates to carry out a variety of bioprocesses, considering its enzymatic arsenal to metabolize several recalcitrant feedstocks to be degraded, like cellulose, hemicellulose, molasses, and lignin; as well as pectin and chitin present in general and fishery industry (Amoah, Kahar, Ogino, & Kondo, 2019; Ma et al., 2020; Park, Kim, Youn, & Choi, 2018).

Due to their metabolic machinery, only saprotrophic fungi (mainly basidiomycetes) and some bacteria can degrade lignocellulosic biomass. The wood-degrading species are divided into three different types within fungi, depending on the depolymerizing mechanisms, white fungi, brown fungi, and the soft-rot ones (Rouches, Herpoël-Gimbert, Steyer, & Carrere, 2016). The degradation mechanism of soft-rot fungi (ascomycetes and deuteromycetes) is not yet well understood. Brown-rot fungi (BRF) leave a brown-colored residue (hence their name) due to a nonenzymatic attack on wood that leads to lignin modification instead of degradation/mineralization and cellulose depolymerization. These organisms are particularly known by their high hydrolyzing capacity of all polysaccharides from wood without causing substantial lignin degradation (Arantes, Jellison, & Goodell, 2012). By contrast, white-rot fungi (WRF) leave a whitish residue on decaying wood, develop enzymatic machinery to degrade lignin, and hence, can use cellulose and hemicellulose as a source of carbon and energy. They were first investigated to evaluate their capacity to delignify for improving digestibility (Rouches et al., 2016). Their main involved enzymes in this process are metalloproteins, such as the lignin peroxidase enzyme (LiP), the manganese peroxidase enzyme (MnP), and other variety of versatile peroxidase enzymes, as well as the laccase enzyme (phenol oxidases, POX), the heme peroxidase enzymes, the hydrogen peroxide generating oxidase enzyme from the glucose/methanol/choline oxidase/dehydrogenase enzymes (GMC) and the copper radical oxidase enzyme superfamilies. Extracellular GMC enzymes involved in the lignocellulose degradation process are cellobiose dehydrogenase, pyranose 2-oxidase, aryl-alcohol oxidase (AAO), and methanol oxidase (Carro, Serrano, Ferreira, & Martínez, 2016). Most of these biocatalysts are absent in the genomes of BRF and some other poor wood rotters (Fernández-Fueyo et al., 2016). WRF are generally preferred as the most efficient in delignification. Their unique enzymatic system gives them the ability to attack phenolic structures and transform lignin into CO_2 (Rouches et al., 2016).

It is important to note that not all WRF produce all these enzymes. They can have either one or three of them, even different paired combinations. Other peroxidase enzymes have also been previously documented, such as the manganese independent peroxidase enzyme, the dye-decolorizing peroxidase enzyme, and the aromatic peroxygenase enzyme, to name a few. Several accessory oxidases are also involved; some induce the H_2O_2 production utilized by peroxidase enzymes, while others alter distinct types of substrates such as glyoxal, aromatic alcohols, etc. Mycelia can also affect nonenzymatic lignin denaturalization via the Fenton reaction, which induces the formation of free radicals that breakthrough cell walls and make it easier for the entry of lignin-degrading enzymes. Laccases and MnP enzymes act mainly on phenols, whereas LiP cleaves mainly nonphenolic (lignin) components (up to 90% of the macromolecule). LiP has been documented to react with phenols in the presence of veratryl alcohol. In contrast, the peroxidase enzymes can degrade phenolic and nonphenolic compounds (Rouches et al., 2016).

Lytic polysaccharide monooxygenases (LPMO) are important copper enzymes that carry out the oxidative breakage of glycosidic bonds, contributing to the degradation of recalcitrant natural polysaccharides such as lignocellulose (Johansen, 2016; Liu, Olson, Wu, Broberg, & Sandgren, 2017). They are found both in fungi and bacteria and catalyze the breakage of bonds by oxidation of glycan residues in polymeric carbohydrates utilizing external electron donors and O_2. Cu(I) is employed to switch on O_2 to form a Cu(II)-superoxyl middle reactant and afterward adds an oxygen atom on positions either C_1 or C_4 of the two proximate glycan residues at the scissile bond (Liu, Olson, Wu, Broberg, & Sandgren, 2017). Currently, LPMOs are divided into seven groups in the

Carbohydrate-Active enZymes database (Carbohydrate-Active enZYmes Database, 2020), including AA9–11 and AA13–16, with different oxidative regioselectivity, substrate specificity, and species origins (Johansen, 2016). AA9s, AA11s, AA13s, AA14s, and AA16s are mainly from eukaryote with cellulolytic, chitinase, starch-degrading, and xylanase activity, respectively. AA10s are from bacteria, eukaryote, viruses, or archaea with cellulolytic or chitinase activity; AA15s are from a eukaryote, viruses, or archaea also with cellulase or chitinase activity (Zhou & Zhu, 2020). *Lentinula edodes* (Shiitake mushrooms), for example, presents the capacity of utilization with agricultural straw resources, with one hundred and one lignocellulolytic enzymes identified, similar to other WRF (Chen et al., 2016). Relevant factors contribute to the diversity of this enzymatic machinery, many of them may have an affinity to electron donor, some other indicate preferences to minimize undesired oxidation reactions or physiochemical conditions. Diverse pH optimal values and mineral salt tolerances are examples of adaptions, which are consequence of continuous battles between pathogens and their respective hosts (Johansen, 2016).

LPMO need electrons to switch on their active site. These electrons may be given by reducing entities like ascorbic or gallic acid, flavocytochrome-dependent cellobiose dehydrogenases, light-induced pigments, light-driven chemical oxidation of water, or diphenol-regenerating GMC-oxidoreductases. Monophenols are not the best electron donors for the activity of LPMO, due to their comparatively high redox potential. Molecules with a 1,2-benzenediol or a 1,2,3-benzenetriol moiety possess a lower redox potential compared to monophenols. This enables them to reduce the copper ion in the LPMO active site and enhance the LPMO activity (Frommhagen et al., 2017).

The efficiency of delignification depends on the fungal strain, the substrate, and the cultivation conditions. Solid cultures, better known under solid-state fermentation (SoSF), are an optimal option for cultivating fungi since a greater amount of substrate per volume unit can be pretreated in a solid process than in liquid culture. Some of the most relevant aspects to optimize in SoSF are the amount of inoculum, moisture content of around 70%–80%, a temperature between 25°C and 30°C, adequate aeration (which will depend on the substrate particle size, packing behavior, and its porosity), the use of supplements, and decontamination (Rouches et al., 2016). For most of the WRF, delignification is promoted by controlling starvation with nitrogen addition, which can restrain delignification while triggering fungal growth and substrate consumption. On the other hand, some reports have published a positive effect of nitrogen addition on the lignin-degrading enzyme production, occasionally controlled by the pH as well as the nitrogen source. Moderately low pH values usually favor ligninolytic enzyme production. The addition of cofactors such as manganese, copper, and iron can also have positive effects on delignification (Rouches et al., 2016). For example, it has been found that with increasing the Cu^{2+} concentrations in the growing media using copper sulfate, *Ganoderma lucidum* MDU-7 exhibited higher secretion of a lignocellulolytic consortium constituted by laccase, cellulolytic, and xylanolytic enzymes, along with an increment of the production of phenolic and antioxidant compounds (Jain et al., 2020).

6 Genetic engineering to clear fungi the way to use alternative feedstocks

For a long time, research and industry efforts were conducted to isolate and characterize novel microorganisms. This selection was made by direct screening between a population to identify those strains with the most robust metabolite of interest or high enzymatic activity. Now, with the implementation of genomics, genetic engineering, and metabolomics, the design of the desired strain with specific traits is possible, improving the production of valued metabolites starting with waste as raw material, bringing the new industry named biofactory and biorefinery (Champreda et al., 2019; Dragone et al., 2020).

6.1 Genetic manipulation of microorganisms

It is worth mentioning that, to engineer a microorganism, it is essential to count on a robust toolbox to manipulate its genome. These tools include a well-annotated genome sequence, availability of constitutive, inducible, and strong promoters, an established genetic transformation system and the most recently developed CRISPR/Cas9 genome editing (Fig. 5) (Larroude, Trabelsi, Nicaud, & Rossignol, 2020; Wang et al., 2020; Zhou et al., 2020). Then, to exemplify the importance of genetic engineering in the biorefinery processes, it is necessary to describe an interesting compound due to its applications and economic importance, citric acid. This organic acid is broadly employed in the industries of cosmetics, food, and pharmaceutics, with an average annual production calculated around 2 million tons and the necessities in the industry are constantly increasing (Steiger, Mattanovich, & Sauer, 2017). Traditionally, this acid is produced using a wild type strain of *Aspergillus niger*, but recently its production was

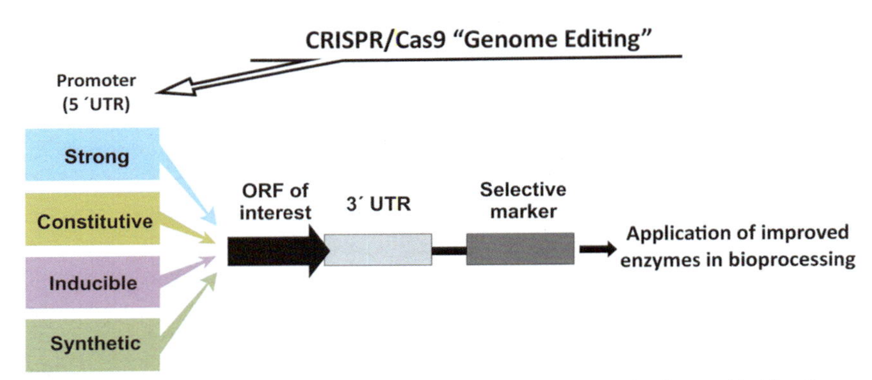

FIG. 5 Application of genetic engineering to optimize the enzyme production used in biorefinery.

improved by engineering an *A. niger* strain in order to meet the needs of the industry (Steiger, Rassinger, Mattanovich, & Sauer, 2019). This modification was conducted by manipulation of the genetic expression of Aspni7|1,165,828, putatively involved in the citric acid secretion. This gene was identified in the genome database of *A. niger* using an itaconic acid transporter from *Ustilago maydis*. By CrisperCas technology, the participation of this gene was demonstrated in the production of citric acid in *A. niger* and was named citrate export A (CexA) (Steiger, Rassinger, Mattanovich, & Sauer, 2019). Then, *A. niger* was engineered to regulate the expression of this gene, using a constitutive and inducible expression systems. With the constitutive promoter *pmbfA*, the citric acid secretion was increased by 37% (33 g/L) compared with the wild type strain. In contrast, the inducible expression system based on the doxycycline system was also evaluated to regulate the expression of CexA. When this system was turned on with 1 μg/mL of doxycycline, fivefold (109 g/L) production of citric acid was obtained compared with the uninduced control condition (Steiger, Rassinger, Mattanovich, & Sauer, 2019).

6.2 Novel adaptations of microorganisms in the biorefinery

Recently, the search for new models to produced value-added metabolites using agricultural, forestry, or by-product wastes turned the sight toward the dimorphic fungus *U. maydis*. This fungus is mainly studied to conduct research on the plant-pathogen interaction field (Matei & Doehlemann, 2016). Then, as a plant pathogen, it possesses enzymes to depolymerize biomass components, *U. maydis* converts hemicellulose into fungal biomass, and this fungus is amenable to be engineered to enhance the segregation of xylanases and celullases (Geiser et al., 2016). Besides, it produces prized secondary metabolic products in a natural manner, such as malic acid, itaconic acid, and glycolipids. It is worth to mention the fact of *U. maydis* is tolerant to the impurities presented in the bioreactor when using biomass previously chemo-hydrolyzed since no pure feedstock material will be used in biorefinery (Regestein et al., 2018), highlighting the importance for the search for new candidates to be established for biorefinery.

Nevertheless, some issues must be addressed before its implementation in a biorefinery process since the hydrolytic enzymes are only produced when the fungus is playing the plant pathogen role, which is characterized by hyphal growth (Matei & Doehlemann, 2016). This condition is vital to be assumed regarding the fermentation process in stirred tanks; a problem to solve is the mechanical damage suffered by the hyphae cells, an issue mainly observed in filamentous fungi since some of them grow in this form as a reaction to stress conditions generated during the fermentation (Geiser et al., 2016). As mentioned earlier, *U. maydis* is a dimorphic fungus, so it can grow in the form of a yeast avoiding the problem imposed by the hyphal form in relation to the fermentation; nevertheless, in response to acid pH, this fungus grows in the hyphal manner (Ruiz-Herrera, Leon, Guevara-Oivera, & Carabez-Trejo, 1995). Since genetic manipulation in *U. maydis* is easily carried out, the deletion of the *fuz7* gene-rendered mutants cannot affect the dimorphic switch (Banuett & Herskowitz, 1994). With all this in mind, this fungus has been genetically modified to fully metabolize "cheap" waste feedstocks like sugar beet pulp (123 US$/ton) obtained as a byproduct of the sugar industry, which is used by *U. maydis* to produce the added value itaconic acid (1500 US$/ton) (De Carvalho, Magalhães, & Soccol, 2018).

6.3 A successful strategy to implement fungal plant pathogens as itaconic acid producers

Itaconic acid is an unsaturated dicarbonic acid that was recently considered as one of the most promising platform chemicals to establish a bio-based building block derived from biomass for the synthesis of chemicals and polymers; its global trade is predicted to reach US$ 570 million this year (Hosseinpour Tehrani et al., 2019; Klement & Büchs, 2013). Itaconic acid for industrial utilization is usually produced by *A. terreus* (100 g/L); originally, the yield reached by *U. maydis* wild type strain is not even close to that obtained by *A. terreus*. Today genetically modified *U. maydis* strains are excellent microorganisms using cellulose and hemicellulose as a substrate and are able to produce 220 g/L of itaconic acid (Becker et al., 2020; Hosseinpour Tehrani, Becker, et al., 2019; Klement et al., 2012). Either in *U. maydis* and *A. terreus*, the biosynthetic genes of itaconic acid are clustered and are coregulated by the cluster-associated regulator *Ria1*; besides, their function is compartmentalized in the cytosol and in the mitochondria (Hosseinpour Tehrani et al., 2019; Hosseinpour Tehrani, Becker, et al., 2019). Generation of itaconic acid by *U. maydis* was increased by the heterologous expression of a gene involved in mitochondrial transport *mttA* from *A. terreus* under the genetic control of the strong promoter *eTEF* in the genetic background △*Um_mtt1* (Hosseinpour Tehrani, Becker, et al., 2019; Hosseinpour Tehrani, Geiser, et al., 2019). Similarly, in *A. terreus* and *A. niger*, overexpression of the itaconic acid transporter *mfsA* from *A. terreus* increased the itaconic production (Hossain et al., 2016), whereas the overexpression of the corresponding gene *itp1* in *U. maydis*, did not improve the production (Geiser et al., 2016). Taking into account that biorefinery uses are not pure substrates, *U. maydis* arises as an excellent organism due to its capacity to secrete itaconic acid from the plant leftovers as compared with *A. terreus,* whose growth is inhibited by impurities present in the feedstock; this is supported by the itaconic acid obtained using beech wood as a fermentative material (Regestein et al., 2018).

Another interesting fungus useful for itaconic acid production is *U. cynodontis*, pathogen of *Cynodon dactylon*. This species was implemented considering its ability to tolerate acid conditions (Wierckx et al., 2020); nevertheless this condition triggers the hyphal growth in *U. cynodontis,* which represents a bottleneck during the fermentation process since the yeast morphology is a more preferable condition than hyphal morphology (Klement et al., 2012). With this in mind, *U. cynodontis* NBRC9727 strain was genetically modified to stably grow in the yeast-like morphology; this was achieved by deletion of *ras2*, *fuz7*, and *ubc3* genes, as previously reported in *U. maydis* (Klose, De Sá, & Kronstad, 2004). The deletion of these genes in *U. cynodontis* did not affect the cell fitness and pH tolerance but increased the yield of itaconic acid (Hosseinpour Tehrani, Tharmasothirajan, Track, Blank, & Wierckx, 2019). Further experiments using *Δfuz7* mutant were conducted to additionally improved the itaconic acid production; this was achieved by deletion of *cyp3* gene (P450 monooxygenase); to avoid the synthesis of (*S*)-2-hydroxyparaconate and the heterologous overexpression of the transporter *mttA* from *A. terreus* driven by the control of *Etef* promoter and the multicopy expression of the regulator *ria1* integrated into the genome driven by its own promoter, rendered the engineered strain $\Delta fuz7^{r}$ $\Delta cyp3^{r}$ $P_{etefmttA}$ $P_{ria1ria1}$ with an increased 6.5-fold itaconic acid production as compared with the wild type strain (Hosseinpour Tehrani, Saur, Tharmasothirajan, Blank, & Wierckx, 2019; Hosseinpour Tehrani, Tharmasothirajan, Track, Blank, & Wierckx, 2019).

Another useful basidiomycete implemented to increase the repertoire of engineered strains to improve itaconic acid production was *Ustilago vetiveriae* TZ1, by overexpression of *ria* and *mtt* genes from *U. maydis,* rendered independent strains producing 2- and 1.5-fold itaconic acid respectively (Zambanini et al., 2017).

7 From recalcitrant biomass to a more accessible feedstock

In order to make accessible the monosaccharide molecules present in lignocellulose, the industry has implemented several processes, either mechanical, chemical, and, more recently, the biological or enzymatic treatment (de Paula et al., 2019; Tolbert et al., 2014). Glucose, cellobiose and celloligosaccharides are obtained from lignocellulose by the action of celullases and oxidoreductases (Brethauer & Studer, 2014; Houfani, Anders, Spiess, Baldrian, & Benallaoua, 2020). Considering the intrinsic recalcitrant structure of lignocellulose to be degraded, in nature, at least 115 cellulases are involved in the processing of decayed plants contributing to the carbon cycle (Devendran et al., 2016).

The WRF belong to the basidiomycete group. Even though they can depolymerize all the elements present in the plant cell walls, their implementation in the biorefinery is challenging due to the lack of a versatile biotechnological toolbox, neither genetic transformation systems, genetic markers or genomes data are available. That is why the efforts have been directed to other cellulolytic fungi equipped with genetic and molecular tools like *Trichoderma reesei* or *Aspergillus niger* (Qin et al., 2012; Rantasalo et al., 2018, 2019; Tong, Zheng, Tong, Shi, & Sun, 2019; Xu et al., 2019).

T. reesei strains are widely used to the biotechnological production of enzymes to degrade lignocellulose; this fungus is amenable to be genetically engineered since some strains efficiently carried out the genetic recombination, expression of heterologous proteins, and several technologies have been implemented in this species (He et al., 2020; Wu et al., 2020; Zhang, Wu, Wang, Wang, & Wei, 2018). *T. reesei* is a saprotrophic fungus, it secretes endo and exoglucanases; β-glucosidases and hemicellulases (Bischof, Ramoni, & Seiboth, 2016). The main enzymes involved in the plant cell wall degradation are cellobiohydrolases; β-endo xylanases, β-glucosidase, and β-xylosidase (Derntl, MacH, & MacH-Aigner, 2019). The overexpression in *T. reesei* of the endoglucanase *eg2* and cellobiohydrolase *bgl1* genes on the strain QEB4 rendered high cellulolytic activity, converting 94.2% of the cellulose presented in corncob residues (Gao, Qian, Wang, Qu, & Zhong, 2017; Qian et al., 2017). Another genetic manipulation to improve the saccharification process by *T. reesei* was the heterologous expression of the cellobiohydrolase *bglA* gene from *A. niger* in controlled conditions by the *T. reesei cbh1* promoter; the resulting strain SCB18 showed 51.3-fold BGL activity as compared with the parental strain (Gao, Qian, Wang, Qu, & Zhong, 2017).

Even though *T. reesei* successfully secretes cellulolytic enzymes, it simultaneously secretes proteases as well. This fact represents a challenge and a opportunity of improving *T. reesei* for industrial purposes, in order to synthesize specific proteases under same conditions to produce cellulases (Qian et al., 2019). Then, *T. reesei* was engineered to produce the strain ΔP70 to eliminate the function of three proteases *tre81070*, *tre120998*, and *tre123234*; the resulting triple mutant strain showed reduced protease activity whereas the high activity of *endo*-β-1,4-glucanase, β-glucosidase, xylanase, cellobiohydrolase was measured as compared with the control strain (Qian et al., 2019). Another parameter to be controlled to enhance the manufacture of celullases is the pH (Li et al., 2013). In order to keep the intracellular pH homeostasis, the deletion of the H^+ATPase *tre76238* was carried out (Liu et al., 2019). The mutant Δ76238 showed increased cellulase production; this result was attributed to the high intracellular gradient of generated H^+, which promotes the import of Ca^{++} ions, activating the calcineurin pathway through the *crz1* transcription factor (Liu et al., 2019). Also, it is possible that the proton accumulation in Δ76238 mutant activates the transglycosylation activity of the β-glucosidases and hence, the production of sophorose, a well-known cellulase inducer that enhances the synthesis of cellulolytic enzymes during the fermentation; this study clearly opens the possibilities to conduct more research related with the field of regulation of cellulases production (Li et al., 2013; Liu et al., 2019).

The transcription factor Xyr1 is a master regulator of cellulose depolymerization (Druzhinina & Kubicek, 2017), its overexpression by the strong tef promoter provoked an overproducing-cellulases strain since the endoxylanolytic activity was increased to 7.5-fold on xylan and 1.5-fold on CMC as substrates, and higher β-xylosidase activity fourfold on xylan compared to the wild-type like Δ*tmus53* (Derntl, MacH, & MacH-Aigner, 2019; Druzhinina & Kubicek, 2017). Moreover, the creation of a new transcription factor produced by fusing the DNA binding domain of Xyr1 with Ypr1, the transactivation domain Ypr1 (the regulator involved in the synthesis of sorbicillinoid), produced the new strain TXY1(Xyr1-Ypr1), which showed upregulation of the cell wall degrading enzymes encoding genes: *cbh1*, *cbh2*, *egl1*, *bgl1*, *xyn1*, *xyn2*, *bxl1*, and *xyl1* (Derntl, MacH, & MacH-Aigner, 2019). Contrary, during the degradation of lignocellulose by *T. reesei,* the participation of transcriptional repressors Cre1, Ace1, and Rce1 were demonstrated (Druzhinina & Kubicek, 2017; Shida, Furukawa, & Ogasawara, 2016). Interestingly, the protein Cft1 was identified as a novel repressor of transcription factors involved in the transcriptional activation of cellulases encoding genes; then, its deletion rendered strains with a higher capacity to hydrolyze cellulose due to the enhanced activity of several cellulases (Meng, Zhang, Liu, Zhao, & Bai, 2020).

To expand the range of microorganisms necessary to benefit from cellulose, the search and engineering for new members are imperative. In this context, *Myceliophthora thermophila*, efficiently grows on cellulose material as a carbon source due to its ability to secrete hydrolytic enzymes to depolymerize cellulose producing cellobiose, making this fungus an attractive organism for its implementation in biorefinery (Singh, 2014). Recently, a strain for efficient use of cellobiose and cellulose was constructed by overexpression of the malate transporter (*mae*) and the pyruvate carboxylase (*pyc*) genes (Li et al., 2019). In this fungus, the intracellular catabolism of cellobiose proceeds through two mechanisms, firstly by its hydrolysis producing two glucose molecules, and the phosphorolytic pathway that breaks down the cellobiose to glucose and glucose-1-phosphate through the action of cellobiose phosphorylase (Lynd, Weimer, Van Zyl, & Isak, 2002).

In order to improve the uptake of cellobiose in *M. thermophila*, the *cdt-1* transporter gene from *Neurospora crassa* was expressed in controlled conditions by the strong constitutive promoter *eif* (elongation initiation factor) in the strain JG207. The modified strain JG207cdt harbors 6 copies of the *cdt-1* gene, its expression increased the uptake of cellobiose by 51%, (Li et al., 2019). In the same way, the expression of the cellobiose phosphorylase (*Mtcpp*) gene was upregulated under the growth on Avicel® or cellulose (Li, Gu, et al., 2019). To improve this fungus' capacity to be utilized in a biofactory, its native *Mtcpp* gene and the corresponding homologue *Ctcpp* gene from *Clostridium thermocellum* were overexpressed under the control of the strong *eif* promoter (Li, Gu, et al., 2019). These two phosphorylase genes increased the production of malate by 9.5% and 5.2% respectively compared with the JG207cdn strain overexpressing the cellobiose transporter (Li, Gu, et al., 2019).

Yarrowia lipolytica is a nonconventional yeast widely used in the industry due to its generally recognized as safe status and its natural capacity to accumulate up to 50% lipids of its dry-weight. Being genetically amenable to be modified, a robust toolbox has been developed, turning it attractive to use it as a platform for the manufacture of biocombustibles (Blazeck et al., 2014; Schwartz, Curtis, Löbs, & Wheeldon, 2018). This yeast is able to easily use six-carbon sugars, and fortunately, its growth on cheaper substrates has been improved through the use of genetic engineering.

Even though the versatility displayed by *Yarrowia lipolytica,* this yeast is unable to use cellulose or cellobiose as carbon source, although the analysis of its genome revealed six sequences putatively related with a family of GH3 β-glucosidases (Guo et al., 2015; Michely, Gaillardin, Nicaud, & Neuvéglise, 2013). After several years of hard engineering work, there are now *Y. lipolytica* strains that are able to grow on a variety of cheap substrates like cellulose, cellobiose, xylose, starch, and xylan being possible to use renewable feedstocks (Ganesan et al., 2019). Recently, *Y. lipolytica* strains with the capacity of using cellobiose, have been constructed by the heterologous expression of the cellodextrin transporter and β-glucosidase from *N. crassa*, cdt-1 and gh1–1, respectively, in controlled conditions by a strong constitutive promoter (Lane, Zhang, Wei, Rao, & Jin, 2015). The strain Po1f-BC produced lipid bodies and citric acid (5.1 g/L) using cellobiose as a sole carbon source in contrast to the control strain Po1f-INT03; also, the modified strain showed uptake of 2.36 g cellobiose/gcell/h (Lane, Zhang, Wei, Rao, & Jin, 2015). By the same time, the constitutive expression of BGL1 and BGL2 genes in *Y. lipolytica* produced the strains ZetaB1 and ZetaB2, respectively, which showed the ability to grow on cellobiose as a sole carbon source (Guo et al., 2015).

Lately, a novel strategy to activate the transcriptional silent β-glucosidases BGL1 and BGL2 genes was applied. Utilizing a CRISPR system, the transcription of these genes was activated, and the corresponding strains showed an enhanced capacity to grow on cellobiose (Schwartz, Curtis, Löbs, & Wheeldon, 2018). To further improve the utilization of cellobiose, BGL1, and BGL2 were expressed in the previously mentioned strains ZetaB1 and ZetaB2, showing an increased capacity to use cellobiose (Guo et al., 2015). To consolidate a bioprocess, *Y. lipolytica* was transformed with the exocellobiohydrolases produced by *T. reesei Tr*EG I and rh*Tr*EG II, *Pf*CBH I from *Penicillium funiculosum* and *Nc*CBH I from *N. crassa,* in all cases using the TEF promoter and the *Y. lipolytica* lipase 2 pre-pro region, different combinations were prepared and some strains showed a high capacity to degrade Avicel, CMC and PASC (Guo et al., 2018). The YCL3 strain was proficient to grow on media CIM-V cellulose consuming 40% of it, whereas the strain YCL6 used 60% of the substrate (Guo, Robin, et al., 2018). Finally, after several rounds of genetic modification, it was feasible to consolidate a bioprocess by removing the genetic markers of the above strains and producing the pair of new cellulolytic CYLp and CYLx strains. Then, CYLp strain was engineered to enhance lipid synthesis by overexpression of SCD1 and DGA genes to produce CYLpO the obese strain (Guo, Robin, et al., 2018; Qiao et al., 2015). This strain produced sevenfold fatty acid methyl esters (FAME) in contrast to the control strain, it processes 50 g/L of cellulose and segregates 5.0 g/L of FAME in

96h. Afterward, the cellulolytic strain CYLx was transformed with the Δ12 hydroxylase *FAH12* gene from *Claviceps purpurea,* creating the strain CYLxR to yield ricinoleic acid (RA), a impotant exploited commodity in the petrochemical industry, using cellulose as feedstock. This strain produced 2.2g RA/L in the decane (organic) phase hydrocarbon in 96h, consuming 11g/L of cellulose (Guo, Robin, et al., 2018). Finally, the CYLp strain was transformed with the native lipase 2 genes to produce active lipase resulting in the strain CYLpL producing 592U lipase/g cellulose (Guo, Robin, et al., 2018); altogether these genetic modifications constituted a successfully consolidated bioprocess.

8 Agroindustrial fruit pulp-rich peel and fishery residual biomasses

8.1 Complementing the ability to degrade fruit peel pectin-rich residual biomass

Wasted fruit peels are rich in pectin, mainly composed of polygalacturonic acid. A full set of the enzymes needed for the metabolism of glucuronic acid are present in the genome of *U. maydis*: D-galacturonic acid reductase, L-galacturonate dehydratase, 2-Keto-3-deoxy-L-galactonato aldolase, and L-glyceraldehyde reductase. Nonetheless, not all the enzymes needed to produce glucuronic acid from pectin are present, or if they are, their expression is induced only *in planta* (Doehlemann et al., 2008). In order to produce a strain able to fully metabolize pectin to glucuronic acid, *U. maydis* was transformed with the following genes: endopolygalacturonase from *U. maydis* (UmPgu1), the exopolygalacturonase from *Klebsiella* sp. (KpPguB), the endopolygalacturonase from *Pectobacterium corotovora* (PcPen1), exopolygalacturonase from *A. tubingensis* (AtPgaX), and *A. aculeatus* (AaPg1), the expression of all of them was controlled by the strong pOMA promoter (Stoffels et al., 2020).

8.2 Chitin, from a protective shell to a valued product

Another important waste-valuable product to be considered is chitin. This is a polymer composed of *N*-acetyl-glucosamine joined by β (1–4) glycosidic linkages; its chemical structure is similar to lignocellulose but possesses an acetylamino group instead of the hydroxyl group at the C_2 position (Cauchie, 2002). This is the second most abundant and renewable biopolymer after cellulose, with the advantage of having nitrogen in its chemical structure. Chitin biomass includes crustacean shells, exoskeletons of arthropods, the fungal cell wall of fungi, and insects' intestinal mucosa (Luo et al., 2020). These raw materials are considered as waste and simultaneously as an ecological problem, mainly due to crustacean shells, since an estimated 6–8 tons are disposed of annually; nevertheless, it is a very cheap feedstock, around 100 US dollars per ton (Ma et al., 2020). Chitin is also effective in producing a variety of valuable carbon and nitrogenous compounds used in medicine, cosmetics, agriculture, and renewable fuels. In spite of the benefits obtained by its valorization, this is still mainly conducted by chemical and physical procedures with the corresponding chemical residues production (Zhou et al., 2020). One of the valued compounds obtained from chitin is chitosan, which is the deacetylated form of chitin by the action of deacetylases. Its industrial importance is related to its remarkable properties since it is a non-toxic antioxidant, biocompatible, anticancer, and biodegradable material (Darbasi, Askari, Kiani, & Khodaiyan, 2017). The fungal chitin deacetylases are widely used to produce chitosan; the corresponding genes from *Colletotricum lindemutianum*, *Aspergillus nidulans*, and *Podospora anserina* were expressed in heterologous systems to improve its activity and efficiently obtain chitosan from chitin (Hoßbach et al., 2018; Kang, Chen, Zhai, & Ma, 2012; Liu et al., 2017).

Chitosan is enzymatically hydrolyzed by chitosanases (glycosylhydrolases), producing water-soluble bioactive chitooligosaccharides (COS). In order to supply the industrial demand of these COS, it is a prerequisite to count with chitosanases able to maintain their activity under fermentation conditions (Luo et al., 2020). The bottle neck to be solved, is the small amount of chitosanases secreted by microorganism in their natural way. Then, *Bacillus amyloliquefaciens* was explored by its catalytic activity toward chitosan. The gene BaCsn46B was identified as a chitosanase belonging to family 46. The yeast *Pichia pastoris* was engineered to overexpress this gene in line with the pAOX promoter induced by methanol and cloned in the plasmid pPIC9K-BaCsn46B. The recombinant protein was successfully produced, and it was able to produce different COS in a 5L fermenter using chitosan as a feedstock (Luo et al., 2020).

9 Fungal bioprocessing to produce metabolites on biorefineries

9.1 Biorefinery processing

A biorefinery can be described as a process, industrial plant, or a cluster of factories, oriented to carry out the transformation of feedstock to obtain marketable products like food, feed, materials and chemicals, and energy sources as fuels to power or heat (Takkellapati et al., 2018; Ubando et al., 2020). From the point of view of resemblance between petrochemical and bio-based refineries, the former is specialized in making fine separations of the crude oil hydrocarbons by physicochemical unit processes to get pure components. The latter, a biorefinery, effects conversions utilizing renewable feedstock from several sources and makes use of pretreatments, which are necessary for preparing biomass to subsequent transformations and separations for reaching an eventually refined or purified bioproduct (de la Torre et al., 2019).

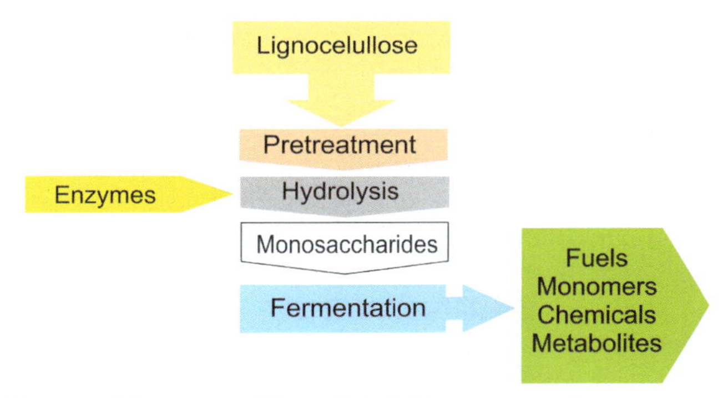

FIG. 6 Main steps of the process of lignocellulosic biomass conversion.

9.2 Pretreatment of lignocellulosic biomass

The utilization of lignocellulosic material requires, as mentioned before, necessarily pretreatment steps aimed to isolate the cellulose component from hemicellulose and lignin (Takkellapati et al., 2018). The common biomass pretreatments focus to alter the particle dimension and the framework of the lignocellulose structure to separate the phenolic macromolecule of lignin from cellulose and hemicellulose polysaccharides. Afterward, these substrates can be bioprocessed enzymatically to obtain different biomolecules, such as monosaccharides, as shown by the scheme of Fig. 6. Eventually, monosugars can constitute the platform to be bioprocessed to produce biofuels, chemicals and metabolites that can be considered as monomers to polymerize and produce biomaterials (Liguori, Ventorino, Pepe, & Faraco, 2016; Takkellapati et al., 2018).

The pretreatment consists of physical, chemical, thermochemical, and biological processing or a combination of them. Physical processing are milling and grinding; chemical processing includes acid or alkali pretreatment, oxidative, and organosolv methods; hydrothermal processing consists of steam explosion or ammonia fiber explosion; and biological processing is represented basically by fermentation (Takkellapati et al., 2018). In the biorefinery context, the contribution and importance of fungi as a tool in the biological pretreatments, has been described above. Fungi are widely employed to prepare lignocellulosic biomass for fermentative purposes of depolymerization. Undoubtedly, monosaccharide production using the enzymatic hydrolysis assisted by microorganisms is an environmental friendly process; nonetheless, the beneficial attributes that represent the supply of cellulases to this industry constitute a bottleneck, mainly due to the high economic cost of the refining process to obtain efficient catalysts (Chandel, Garlapati, Singh, Antunes, & da Silva, 2018).

Naturally, the most skilled organisms to degrade lignocellulose are fungi, considering their ability to grow on agriculture and forestry residues as a sole carbon source and at low water activities (Brethauer & Studer, 2014; Houfani et al., 2020). From fungi, ascomycete, basidiomycete, and zygomycete divisions bioprocess substrates on aerobic conditions to obtain products at laboratory, bench-scale and even larger levels, because their hydrolytic enzymes are mainly produced extracellularly (Soccol et al., 2017; Srivastava, Singh, & Saini, 2020). Although some anaerobic fungi have been reported as good depolymerizing enzyme producers, it's difficult to recover them because enzymes are often membrane-bound as cellosomes (Ferreira, Mahboubi, Lennartsson, & Taherzadeh, 2016; Ivarsson, Schnürer, Bengtson, & Neubeck, 2016).

9.3 Bioprocessing of lignocellulosic feedstock

After lignocellulosic pretreatment, the fungal bioprocessing usually employs two technologies, submerged or liquid-state fermentation (LSF) and solid state fermentation (SoSF).

9.3.1 LSF bioreactors for bioprocessing lignocellulose

LSF is carried out by models as bubble column, airlift, and classical stirred tank bioreactors (Niglio, Procentese, Russo, Piscitelli, & Marzocchella, 2019). These systems are suitable for the required aerobic conditions by fungi to metabolize substrates. Some advantages of LSF are the use of instrumentation and control of important operative parameters as pH, dissolved oxygen and temperature, as well as the capability of biomass separation after the fermentation, mixing, aeration, and scaling up. Nonetheless the aqueous media, the key factor is to outspread oxygen to the filamentous fungi; these systems need a shearing environment proportioned by using air sparger in bubbles with impellers in stirred tanksor airlift column bioreactors. However, excessive hydrodynamic stress can lead to lower productivities of enzymes and even cell damage or disruption (Esperança et al., 2020). In addition, the liquid culture usually requires substrate and nutrients in relatively low concentration to avoid viscous conditions that inhibit oxygen mass transferring from air to the aqueous system; which, in combination with the size of pellet or mycelial fungal biomass, generates systems that need to be studied for the production of bioproducts. These LSF systems are usually adequate for, but not limited to, soluble substrates.

9.3.2 SoSF bioreactors for bioprocessing lignocellulose

SoSF is a bioprocess where fungi grow in an environment free of water or with exceptionally low content of free water. This lower water activity makes SoSF less susceptible to bacterial contamination, substrate inhibitions, and favors higher enzymatic productivity and then higher final concentration of products. This type of system is similar to those natural conditions where fungi mainly grow, so fermentations utilizing lignocellulose as substrates are effective. Moreover, from the economical approach, SoSF needs lower capital because of the operating costs are lower due to reduced stirring and downstream processes (Soccol et al., 2017).

The SoSF bioreactors, these days, are designed in several conformations. Depending on the mixing system, they are classified into stirred or static bioreactors. Stirred bioreactors are subdivided into a stirred drum or horizontal drum, while static bioreactors are subdivided into fixed bed and perforated trays. Finally, by the aeration type, they are subdivided into systems that employees forced ventilation and those that do not (Mahmoodi, Najafpour, & Mohammadi, 2019; Ruiz-Leza, Rodriguez-Jasso, Rodriguez-Herrera, Contreras-Esquivel, & Aguilar, 2007; Soccol et al., 2017).

9.4 Bioprocessing types of lignocellulose

The main bioprocessing configuration systems dedicated to the conversion of crops and residual biomass into sugars and then, to fuel, material, or chemical bioproducts (de la Torre et al., 2019; Mondala, 2015; Oh et al., 2015; Takkellapati et al., 2018), are listed below and illustrated in Fig. 7.

- Two-step depolymerization and fermentation (TD&F).
- Simultaneous depolymerization and fermentation (SD&F).
- Multiple bioprocessing transformation (MBT).

9.5 Production of fungal bioprocessed metabolites

Numerous reviews and reports are published describing protocols about the use of second generation feedstock such as agriculture residues, agro-industrial waste, and OFMSW that have been fungal bioprocessed to produce different value-added metabolites. For example, organic acids constitute an environmentally friendly option to be used as platform chemicals; nonetheless, many of them are produced by chemical synthesis, a great diversity of filamentous fungi have the metabolic capacity to produce them starting from lignocellulosic feedstock. Table 1 lists a set of fungal strains and their metabolites obtained utilizing lignocellulose as a raw material.

Itaconic acid was originally produced using *Aspergillus itaconicus*, although nowadays is elaborated by *Aspergillus terreus* utilizing hydrolysates

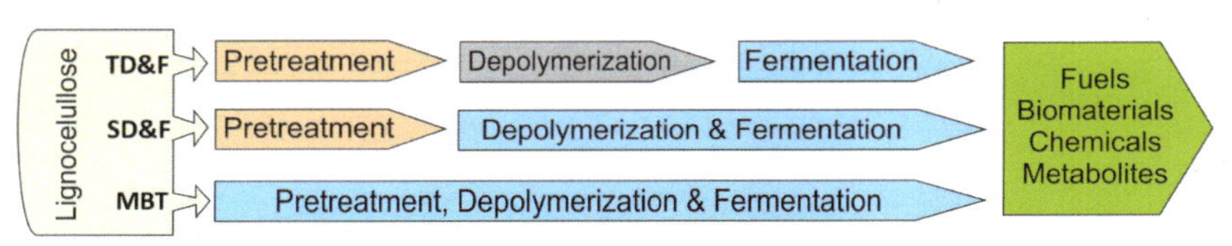

FIG. 7 Processes of the transformation of bioprocessed products by TD&F, SD&F, and MBT.

TABLE 1 Fungal strains and their bioprocessing metabolites obtained from lignocellulosic biomass.

Fungal strain	Fermentation	Substrate	Metabolite	Productivity	Reference
Aspergillus niger	SSF	Corn stover	Citric acid	136.6 g/L	Hou and Bao (2018)
Aspergillus niger	LSF	Corn stover	Gluconic acid	76.77 g/L	Zhang, Zhang, and Bao (2016)
Aspergillus terreus	LSF	Wheat bran	Itaconic acid	49.65 g/L	Liu et al. (2017)
Rhizopus oryzae	LSF	Wheat bran	Fumaric acid	20.2 g/L	Wang, Huang, Li, Wen, and Jia (2015)
Trichoderma reesei	LSF	Wheat bran	Erythritol	5 mg/L	Jovanović, Mach, and Mach-Aigner (2014)
Echinodontium taxodii	SoSF	Moso bamboo	Manganese peroxidase	46.52 U/gds	Kong et al. (2016)
Mucor indicus	LSF	OFMSW	Ethanol	194 g/kg of OFMSW	Mahmoodi, Karimi, and Taherzadeh (2018)
Aspergillus fumigatus	SoSF	Rice straw	β-xylosidase	150.25 U/mL	Jin, Song, Ma, and Liu (2020)
Aspergillus niger	SoSF	Orange peel pomace and sugar cane bagasse	Exopectinase Endopectinase	75 U/gds 16 U/gds	Mahmoodi et al. (2019)
Trichoderma reesei	LSF	Sugar cane bagasse, wheat bran and orange peel pomace	Total Cellulase Endoglucanase Exoglucanase Xylanase	1.76 U/mL 4.28 U/mL 1.25 U/mL 36.85 U/mL	Silva, Hergesel, Campioni, Carvalho, and Oliva-Neto (2018)
Aspergillus niger	SoSF	Brewer's spent grain	Xylanase Cellulase Cellobiase Feruloyl esterase	61.58 U/mL 0.22 U/mL 28.22 U/mL 2.61 U/mL	Paz, Outeiriño, Pérez Guerra, and Domínguez (2019)

TABLE 1 Fungal strains and their bioprocessing metabolites obtained from lignocellulosic biomass—cont'd

Fungal strain	Fermentation	Substrate	Metabolite	Productivity	Reference
Penicillium oxalicum	LSF	Wheat bran	Carboxy Methyl Cellulase Filter Paperase	36.49 U/mL 13.63 U/mL	Han, Liu, Song, and Qu (2018)
Trichoderma viride	SoSF	Corn cob Eucalyptus sawdust Filter paper	Total Cellulase	26.87 U/mL 3.692 U/mL 15.771 U/mL	Sartori et al. (2015)
Trichoderma longibrachiatum	SoSF	Sugar cane bagasse Soybean hull	Exoglucanase β-glucosidase Xylanase	24.9 U/gds 144.5 U/gds 384.4 U/gds	Chang and Webb (2017)
Myceliophtora thermophila	SoSF	Sugarcane bagasse Wheat bran	Endoglucanase Xylanase	148 U/gds 900 U/gds	Perez, Casciatori, and Thoméo (2019)
Aspergillus niger	SoSF	Semi exhausted olive	Xylanase Carboxy Methyl Cellulase β-glucosidase	26.8 U/gds 30.5 U/gds 13.8 U/gds	Filipe et al. (2020)

U/mL, Units of enzymatic activity. A unit of the respective enzymatic activity represents the amount of enzyme that releases 1 μmol of product per minute of reaction. *U/gds*, Units of enzymatic activity per gram of dried solid substrate. *LSF*, liquid-state fermentation; *OFMSW*, Organic fraction of municipal solid waste; *SoSF*, solid-state fermentation.

from various biomaterials, such as wheat bran. *Aspergillus terreus* CICC40205 mutant was inoculated to wheat bran (32.5% soluble substance, 44.2% hemicelluloses, 10% cellulose, and 13.3% lignin) hydrolysate by diluted sulfuric acid hydrothermal pretreatment. The LSF of *A. terreus* CICC40205 with a substrate of hydrolyzed glucose and xylose from wheat bran reached a titer of 49.65 g/L of itaconic acid, which compared to the productivity of 86 g/L produced by *A. terreus* DSM-23081 starting solely from glucose substrate, is a good result, considering the source of substrate (Liu, Deng, et al., 2017).

Fumaric acid was reported to be produced in LSF employing hydrolysates of lignocellulosic residue wheat bran, pretreated only hydrothermally, and bioprocessed by *Rhizopus oryzae* wild 1.22. The culture broth resulted with the presence of fumaric acid with a titer of 20.2 g/L (Wang, Huang, Li, Wen, & Jia, 2015).

For the manufacturing of cellulolytic and hemicellulolytic enzymes, *Aspergillus tubingensis* was used as an enzymatic cocktail for saccharification of sugar cane bagasse, to produce 2G bioethanol production. Biomaterial such as copra meal, wheat bran, palm kernel cake, groundnut shell, rice husk, rice straw, and jatropha seed cake were bioprocessed by *Aspergillus tubingensis* NKBP-5 with a SSF system to obtain an enzymatic extract that was partially purified. Cellulolytic (FPase, Endo-β-glucanase, β-glucosidase) and hemicellulolytic (Endo-β-mannasase, α-galactosidase, Endo-β-xylanase, β-xylosidase) enzymes were detected at different activities for all the lignocellulosic materials. This semipurified enzymatic extract was employed to hydrolyze sugarcane bagasse for saccharification and use it for alcoholic fermentation ethanol production via *Candida shehatae* NCIM 3501 (Prajapati, Jana, Suryawanshi, & Kango, 2020).

10 Conclusions

Lignocellulose represents a vital feedstock for the production of numerous metabolites used to obtain important consumer products in modern life. This is a renewable biomaterial composed of reusable nutrients that can be metabolized and separated into various compounds by fungi. Since lignocellulose is derived from agricultural byproducts, agroindustrial, and domestic residues, this feedstock presents variability in components and their respective quantities. Nonetheless, the heterogeneity of the lignocellulosic structure, fungi metabolize it through their enzymatic machinery and biotransform it into chemical building blocks and organic acids for biomaterials, food additives, feed, and biofuels.

Genetic engineering has targeted fungi to procure the expression of degrading enzymes in higher levels than wild strains in order to increment productivity in the bioprocessing of lignocellulosic materials. These engineering improvements to industrially important fungi, such as *Aspergillus niger* or *Trichoderma reesei*, are focused on tolerating different anomalous operating conditions during different stages of biological pretreatments of lignocellulose feedstock. In the same context, solid and liquid fermentations have been researched, not only to improve the productivity of metabolites but also to change conditions of pretreated substrates and cultures. These efforts are the pathway to develop better protocols to maximize production during the bioprocessing of lignocellulose.

The research and development of lignocellulose bioprocessing by fungi have contributed to the generation of important technological solutions in accordance with the concept of a biorefinery. The contributions to this knowledge field has transformed the perception of the value for lignocellulose and related wastes by the modern society, and specifically by regional economies. All economies that bet to be organized in agreement with the knowledge-based economy concept and invest in renewable feedstock bioprocessing, will undoubtedly contribute to valorization processes of locally based biofactories, and also to a sustainable future life style for humanity.

References

Achyuthan, K. E., Achyuthan, A. M., Adams, P. D., Dirk, S. M., Harper, J. C., Simmons, B. A., et al. (2010). Supramolecular self-assembled chaos: Polyphenolic lignin's barrier to cost-effective lignocellulosic biofuels. *Molecules*, *15*(12), 8641–8688. https://doi.org/10.3390/molecules15118641.

Aguilar, A., Bochereau, L., & Matthiessen, L. (2009). Biotechnology as the engine for the knowledge-based bio-economy. *Biotechnology and Genetic Engineering Reviews*, *26*(1), 371–388. https://doi.org/10.5661/bger-26-371.

Ahmad, B., Yadav, V., Yadav, A., Rahman, M. U., Yuan, W. Z., Li, Z., et al. (2020). Integrated biorefinery approach to valorize winery waste: A review from waste to energy perspectives. *Science of the Total Environment*, *719*, 137315. https://doi.org/10.1016/j.scitotenv.2020.137315.

Amoah, J., Kahar, P., Ogino, C., & Kondo, A. (2019). Bioenergy and biorefinery: Feedstock, biotechnological conversion and products. *Biotechnology Journal*, *14*(6). https://doi.org/10.1002/biot.201800494.

Arantes, V., Jellison, J., & Goodell, B. (2012). Peculiarities of brown-rot fungi and biochemical Fenton reaction with regard to their potential as a model for bioprocessing biomass. *Applied Microbiology and Biotechnology*, *94*(2), 323–338. https://doi.org/10.1007/s00253-012-3954-y.

Banuett, F., & Herskowitz, I. (1994). Morphological transitions in the life cycle of *Ustilago maydis* and their genetic control by de *a* and *b* loci. *Experimental Mycology*, *18*, 247–266.

Becker, J., Hosseinpour Tehrani, H., Gauert, M., Mampel, J., Blank, L., & Wierckx, N. (2020). An *Ustilago maydis* chassis for itaconic acid production without by-products. *Microbial Biotechnology*, *13*(2), 350–362. https://doi.org/10.1111/1751-7915.13525.

Bischof, R. H., Ramoni, J., & Seiboth, B. (2016). Cellulases and beyond: The first 70 years of the enzyme producer *Trichoderma reesei*. *Microbial Cell Factories*, *15*(1), 1–13. https://doi.org/10.1186/s12934-016-0507-6.

Blazeck, J., Hill, A., Liu, L., Knight, R., Miller, J., Pan, A., et al. (2014). Harnessing *Yarrowia lipolytica* lipogenesis to create a platform for lipid and biofuel production. *Nature Communications*, *5*. https://doi.org/10.1038/ncomms4131.

Brethauer, S., & Studer, M. H. (2014). Consolidated bioprocessing of lignocellulose by a microbial consortium. *Energy and Environmental Science*, *7*(4), 1446–1453. https://doi.org/10.1039/c3ee41753k.

Cao, L., Yu, I. K. M., Liu, Y., Ruan, X., Tsang, D. C. W., Hunt, A. J., et al. (2018). Lignin valorization for the production of renewable chemicals: State-of-the-art review and future prospects. *Bioresource Technology*, *269*(June), 465–475. https://doi.org/10.1016/j.biortech.2018.08.065.

Carbohydrate-Active enZYmes Database. (2020). http://www.cazy.org/.

Carro, J., Serrano, A., Ferreira, P., & Martínez, A. T. (2016). Fungal aryl-alcohol oxidase in lignocellulose degradation and bioconversion. In V. Gupta (Ed.), *Microbial enzymes in bioconversions of biomass (biofuel and biorefinery)* (pp. 301–322). Cham: Springer. https://doi.org/10.1007/978-3-319-43679-1_12.

Cauchie, H. M. (2002). Chitin production by arthropods in the hydrosphere. *Hydrobiologia*, *470*, 63–95. https://doi.org/10.1023/A:1015615819301.

Champreda, V., Mhuantong, W., Lekakarn, H., Bunterngsook, B., Kanokratana, P., Zhao, X. Q., et al. (2019). Designing cellulolytic enzyme systems for biorefinery: From nature to application. *Journal of Bioscience and Bioengineering*, *128*(6), 637–654. https://doi.org/10.1016/j.jbiosc.2019.05.007.

Chandel, A. K., Garlapati, V. K., Singh, A. K., Antunes, F. A., & da Silva, S. S. (2018). The path forward for lignocellulose biorefineries: Bottlenecks, solutions and perspective on commercialization. *Bioresource Technology*, *264*, 370–381. https://doi.org/10.1016/j.biortech.2018.06.004.

Chang, C. W., & Webb, C. (2017). Production of a generic microbial feedstock for lignocellulose biorefineries through sequential bioprocessing. *Bioresource Technology*, *227*, 35–43. https://doi.org/10.1016/j.biortech.2016.12.055.

Chen, L., Gong, Y., Cai, Y., Liu, W., Zhou, Y., Xiao, Y., et al. (2016). Genome sequence of the edible cultivated mushroom *Lentinula edodes* (shiitake) reveals insights into lignocellulose degradation. *PLoS One*, *11*(8), 1–20. https://doi.org/10.1371/journal.pone.0160336.

Chojnacka, K., Moustakas, K., & Witek-Krowiak, A. (2020). Bio-based fertilizers: A practical approach towards circular economy. *Bioresource Technology*, *295*(October 2019), 122223. https://doi.org/10.1016/j.biortech.2019.122223.

Darbasi, M., Askari, G., Kiani, H., & Khodaiyan, F. (2017). Development of chitosan based extended-release antioxidant films by control of fabrication variables. *International Journal of Biological Macromolecules*, *104*, 303–310. https://doi.org/10.1016/j.ijbiomac.2017.06.055.

De Carvalho, J. C., Magalhães, A. I., & Soccol, C. R. (2018). Biobased itaconic acid market and research trends – is it really a promising chemical? *Chimica Oggi/Chemistry Today*, *36*(4), 56–58.

de Jong, E., & Jungmeier, G. (2015). Chapter 1—Biorefinery concepts in comparison to petrochemical refineries. In A. Pandey, R. Höfer, M. Taherzadeh, K. M. Nampoothiri, & C. Larroche (Eds.), *Industrial biorefineries and white biotechnology* (pp. 3–33). Elsevier. https://doi.org/10.1016/B978-0-444-63453-5.00001-X.

de la Torre, I., Martin-Dominguez, V., Acedos, M. G., Esteban, J., Santos, V. E., & Ladero, M. (2019). Utilisation/upgrading of orange peel waste from a biological biorefinery perspective. *Applied Microbiology and Biotechnology*, *103*(15), 5975–5991. https://doi.org/10.1007/s00253-019-09929-2.

de Paula, R. G., Antoniêto, A. C. C., Ribeiro, L. F. C., Srivastava, N., O'Donovan, A., Mishra, P. K., et al. (2019). Engineered microbial host selection for value-added bioproducts from lignocellulose. *Biotechnology Advances*, *37*(6), 0–1. https://doi.org/10.1016/j.biotechadv.2019.02.003.

Derntl, C., MacH, R. L., & MacH-Aigner, A. R. (2019). Fusion transcription factors for strong, constitutive expression of cellulases and xylanases in *Trichoderma reesei*. *Biotechnology for Biofuels*, *12*(1), 1–18. https://doi.org/10.1186/s13068-019-1575-8.

Devendran, S., Abdel-Hamid, A. M., Evans, A. F., Iakiviak, M., Kwon, I. H., MacKie, R. I., et al. (2016). Multiple cellobiohydrolases and cellobiose phosphorylases cooperate in the ruminal bacterium *Ruminococcus albus* 8 to degrade cellooligosaccharides. *Scientific Reports*, *6*(July), 1–15. https://doi.org/10.1038/srep35342.

Doehlemann, G., Wahl, R., Vranes, M., de Vries, R. P., Kämper, J., & Kahmann, R. (2008). Establishment of compatibility in the *Ustilago maydis*/maize pathosystem. *Journal of Plant Physiology*, *165*(1), 29–40. https://doi.org/10.1016/j.jplph.2007.05.016.

Dragone, G., Kerssemakers, A. A. J., Driessen, J. L. S. P., Yamakawa, C. K., Brumano, L. P., & Mussatto, S. I. (2020). Innovation and strategic orientations for the development of advanced biorefineries. *Bioresource Technology*, *302*(December 2019), 122847. https://doi.org/10.1016/j.biortech.2020.122847.

Druzhinina, I. S., & Kubicek, C. P. (2017). Genetic engineering of *Trichoderma reesei* cellulases and their production. *Microbial Biotechnology*, *10*(6), 1485–1499. https://doi.org/10.1111/1751-7915.12726.

Esperança, M. N., Mendes, C. E., Rodriguez, G. Y., Cerri, M. O., Béttega, R., & Badino, A. C. (2020). Sparger design as key parameter to define shear conditions in pneumatic bioreactors.

Biochemical Engineering Journal, *157*(February), 107529. https://doi.org/10.1016/j.bej.2020.107529.

Fernández-Fueyo, E., Ruiz-Dueñas, F. J., López-Lucendo, M. F., Pérez-Boada, M., Rencoret, J., Gutiérrez, A., et al. (2016). A secretomic view of woody and nonwoody lignocellulose degradation by *Pleurotus ostreatus*. *Biotechnology for Biofuels*, *9*(1), 1–18. https://doi.org/10.1186/s13068-016-0462-9.

Ferreira, J. A., Mahboubi, A., Lennartsson, P. R., & Taherzadeh, M. J. (2016). Waste biorefineries using filamentous ascomycetes fungi: Present status and future prospects. *Bioresource Technology*, *215*, 334–345. https://doi.org/10.1016/j.biortech.2016.03.018.

Ferreira, J. A., & Taherzadeh, M. J. (2020). Improving the economy of lignocellulose-based biorefineries with organosolv pretreatment. *Bioresource Technology*, *299*(October 2019), 122695. https://doi.org/10.1016/j.biortech.2019.122695.

Filipe, D., Fernandes, H., Castro, C., Peres, H., Oliva-Teles, A., Belo, I., et al. (2020). Improved lignocellulolytic enzyme production and antioxidant extraction using solid-state fermentation of olive pomace mixed with winery waste. *Biofuels, Bioproducts and Biorefining*, *14*(1), 78–91. https://doi.org/10.1002/bbb.2073.

Frommhagen, M., Mutte, S. K., Westphal, A. H., Koetsier, M. J., Hinz, S. W. A., Visser, J., et al. (2017). Boosting LPMO-driven lignocellulose degradation by polyphenol oxidase-activated lignin building blocks. *Biotechnology for Biofuels*, *10*(1), 1–16. https://doi.org/10.1186/s13068-017-0810-4.

Gallezot, P. (2007). Catalytic routes from renewables to fine chemicals. *Catalysis Today*, *121*(1), 76–91. https://doi.org/10.1016/j.cattod.2006.11.019.

Ganesan, V., Spagnuolo, M., Agrawal, A., Smith, S., Gao, D., & Blenner, M. (2019). Advances and opportunities in gene editing and gene regulation technology for *Yarrowia lipolytica*. *Microbial Cell Factories*, *18*(1), 1–9. https://doi.org/10.1186/s12934-019-1259-x.

Gao, J., Qian, Y., Wang, Y., Qu, Y., & Zhong, Y. (2017). Production of the versatile cellulase for cellulose bioconversion and cellulase inducer synthesis by genetic improvement of *Trichoderma reesei*. *Biotechnology for Biofuels*, *10*(1), 1–16. https://doi.org/10.1186/s13068-017-0963-1.

Gazi, S. (2019). Valorization of wood biomass-lignin via selective bond scission: A minireview. *Applied Catalysis B: Environmental*, *257*(July), 117936. https://doi.org/10.1016/j.apcatb.2019.117936.

Geiser, E., Reindl, M., Blank, L. M., Feldbrügge, M., Wierckx, N., & Schipper, K. (2016). Activating intrinsic carbohydrate-active enzymes of the smut fungus *Ustilago maydis* for the degradation of plant cell wall components. *Applied and Environmental Microbiology*, *82*(17), 5174–5185. https://doi.org/10.1128/AEM.00713-16.

Gorshkova, T. A., Kozlova, L. V., & Mikshina, P. V. (2013). Spatial structure of plant cell wall polysaccharides and its functional significance. *Biochemistry (Moscow)*, *78*(7), 836–853. https://doi.org/10.1134/S0006297913070146.

Guo, H., Chang, Y., & Lee, D. J. (2018). Enzymatic saccharification of lignocellulosic biorefinery: Research focuses. *Bioresource Technology*, *252*(November 2017), 198–215. https://doi.org/10.1016/j.biortech.2017.12.062.

Guo, Z., Duquesne, S., Bozonnet, S., Cioci, G., Nicaud, J. M., Marty, A., et al. (2015). Development of cellobiose-degrading ability in *Yarrowia lipolytica* strain by overexpression of endogenous genes. *Biotechnology for Biofuels*, *8*(1), 1–16. https://doi.org/10.1186/s13068-015-0289-9.

Guo, Z. P., Robin, J., Duquesne, S., O'Donohue, M. J., Marty, A., & Bordes, F. (2018). Developing cellulolytic *Yarrowia lipolytica* as a platform for the production of valuable products in consolidated bioprocessing of cellulose. *Biotechnology for Biofuels*, *11*(1), 1–15. https://doi.org/10.1186/s13068-018-1144-6.

Han, X., Liu, G., Song, W., & Qu, Y. (2018). Production of sodium gluconate from delignified corn cob residue by on-site produced cellulase and co-immobilized glucose oxidase and catalase. *Bioresource Technology, 248*, 248–257. https://doi.org/10.1016/j.biortech.2017.06.109.

He, J., Liu, X., Xia, J., Xu, J., Xiong, P., & Qiu, Z. (2020). One-step utilization of non-detoxified pretreated lignocellulose for enhanced cellulolytic enzyme production using recombinant *Trichoderma reesei* RUT C30 carrying alcohol dehydrogenase and nicotinate phosphoribosyltransferase. *Bioresource Technology, 310*, 123458. https://doi.org/10.1016/j.biortech.2020.123458.

Hossain, A. H., Li, A., Brickwedde, A., Wilms, L., Caspers, M., Overkamp, K., et al. (2016). Rewiring a secondary metabolite pathway towards itaconic acid production in *Aspergillus niger*. *Microbial Cell Factories, 15*(1), 1–15. https://doi.org/10.1186/s12934-016-0527-2.

Hoßbach, J., Bußwinkel, F., Kranz, A., Wattjes, J., Cord-Landwehr, S., & Moerschbacher, B. M. (2018). A chitin deacetylase of *Podospora anserina* has two functional chitin binding domains and a unique mode of action. *Carbohydrate Polymers, 183*(July 2017), 1–10. https://doi.org/10.1016/j.carbpol.2017.11.015.

Hosseinpour Tehrani, H., Becker, J., Bator, I., Saur, K., Meyer, S., Rodrigues Lóia, A. C., et al. (2019). Integrated strain- and process design enable production of 220 g L-1 itaconic acid with *Ustilago maydis*. *Biotechnology for Biofuels, 12*(1), 1–11. https://doi.org/10.1186/s13068-019-1605-6.

Hosseinpour Tehrani, H., Geiser, E., Engel, M., Hartmann, S. K., Hossain, A. H., Punt, P. J., et al. (2019). The interplay between transport and metabolism in fungal itaconic acid production. *Fungal Genetics and Biology, 125*, 45–52. https://doi.org/10.1016/j.fgb.2019.01.011.

Hosseinpour Tehrani, H., Saur, K., Tharmasothirajan, A., Blank, L. M., & Wierckx, N. (2019). Process engineering of pH tolerant *Ustilago cynodontis* for efficient itaconic acid production. *Microbial Cell Factories, 18*(1), 213. https://doi.org/10.1186/s12934-019-1266-y.

Hosseinpour Tehrani, H., Tharmasothirajan, A., Track, E., Blank, L. M., & Wierckx, N. (2019). Engineering the morphology and metabolism of pH tolerant *Ustilago cynodontis* for efficient itaconic acid production. *Metabolic Engineering, 54*(May), 293–300. https://doi.org/10.1016/j.ymben.2019.05.004.

Hou, W., & Bao, J. (2018). Simultaneous saccharification and aerobic fermentation of high titer cellulosic citric acid by filamentous fungus *Aspergillus niger*. *Bioresource Technology, 253* (November 2017), 72–78. https://doi.org/10.1016/j.biortech.2018.01.011.

Houfani, A. A., Anders, N., Spiess, A. C., Baldrian, P., & Benallaoua, S. (2020). Insights from enzymatic degradation of cellulose and hemicellulose to fermentable sugars—A review. *Biomass and Bioenergy, 134*(January 2019). https://doi.org/10.1016/j.biombioe.2020.105481.

Hu, X., Wei, B., Zhang, B., Xu, X., Jin, Z., & Tian, Y. (2013). Synthesis and characterization of dextrin monosuccinate. *Carbohydrate Polymers, 97*(1), 111–115. https://doi.org/10.1016/j.carbpol.2013.04.054.

Ivarsson, M., Schnürer, A., Bengtson, S., & Neubeck, A. (2016). Anaerobic fungi: A potential source of biological H_2 in the oceanic crust. *Frontiers in Microbiology, 7*(May), 1–8. https://doi.org/10.3389/fmicb.2016.00674.

Jain, K. K., Kumar, A., Shankar, A., Pandey, D., Chaudhary, B., & Sharma, K. K. (2020). De novo transcriptome assembly and protein profiling of copper-induced lignocellulolytic fungus *Ganoderma lucidum* MDU-7 reveals genes involved in lignocellulose degradation and terpenoid biosynthetic pathways. *Genomics, 112*(1), 184–198. https://doi.org/10.1016/j.ygeno.2019.01.012.

Jiménez-González, C., & Woodley, J. M. (2010). Bioprocesses: Modeling needs for process evaluation and sustainability assessment. *Computers and Chemical Engineering, 34*(7), 1009–1017. https://doi.org/10.1016/j.compchemeng.2010.03.010.

Jin, X., Song, J., Ma, J., & Liu, G. Q. (2020). Thermostable β-xylosidase from *Aspergillus fumigatus*: Purification, characterization and potential application in lignocellulose bioethanol production. *Renewable Energy*, *155*, 1425–1431. https://doi.org/10.1016/j.renene.2020.04.054.

Johansen, K. S. (2016). Lytic polysaccharide monooxygenases: The microbial power tool for lignocellulose degradation. *Trends in Plant Science*, *21*(11), 926–936. https://doi.org/10.1016/j.tplants.2016.07.012.

Jovanović, B., Mach, R. L., & Mach-Aigner, A. R. (2014). Erythritol production on wheat straw using *Trichoderma reesei*. *AMB Express*, *4*(1), 1–12. https://doi.org/10.1186/s13568-014-0034-y.

Kang, L., Chen, X., Zhai, C., & Ma, L. (2012). Synthesis and high expression of chitin deacetylase from *Colletotrichum lindemuthianum* in *Pichia pastoris* GS115. *Journal of Microbiology and Biotechnology*, *22*(9), 1202–1207. https://doi.org/10.4014/jmb.1112.12026.

Kapoor, R., Ghosh, P., Kumar, M., Sengupta, S., Gupta, A., Kumar, S. S., et al. (2020). Valorization of agricultural waste for biogas based circular economy in India: A research outlook. *Bioresource Technology*, *304*(February), 123036. https://doi.org/10.1016/j.biortech.2020.123036.

Karimi, K., & Taherzadeh, M. J. (2016). A critical review on analysis in pretreatment of lignocelluloses: Degree of polymerization, adsorption/desorption, and accessibility. *Bioresource Technology*, *203*, 348–356. https://doi.org/10.1016/j.biortech.2015.12.035.

Kawaguchi, H., Ogino, C., & Kondo, A. (2017). Microbial conversion of biomass into bio-based polymers. *Bioresource Technology*, *245*, 1664–1673. https://doi.org/10.1016/j.biortech.2017.06.135.

Klement, T., & Büchs, J. (2013). Itaconic acid—A biotechnological process in change. *Bioresource Technology*, *135*, 422–431. https://doi.org/10.1016/j.biortech.2012.11.141.

Klement, T., Milker, S., Jäger, G., Grande, P. M., Domínguez de María, P., & Büchs, J. (2012). Biomass pretreatment affects *Ustilago maydis* in producing itaconic acid. *Microbial Cell Factories*, *11*(1), 43. https://doi.org/10.1186/1475-2859-11-43.

Klemm, D., Heublein, B., Fink, H. P., & Bohn, A. (2005). Cellulose: Fascinating biopolymer and sustainable raw material. *Angewandte Chemie, International Edition*, *44*(22), 3358–3393. https://doi.org/10.1002/anie.200460587.

Klose, J., De Sá, M. M., & Kronstad, J. W. (2004). Lipid-induced filamentous growth in *Ustilago maydis*. *Molecular Microbiology*, *52*(3), 823–835. https://doi.org/10.1111/j.1365-2958.2004.04019.x.

Kocabas, A., Ogel, Z. B., & Bakir, U. (2014). Xylanase and itaconic acid production by *Aspergillus terreus* NRRL 1960 within a biorefinery concept. *Annals of Microbiology*, *64*(1), 75–84. https://doi.org/10.1007/s13213-013-0634-9.

Kondo, T. (1998). Hydrogen bonds in cellulose and cellulose derivatives. In S. Dumitriu (Ed.), *Polisaccharides. Structural, diversity and functional versatility* (pp. 131–171). Marcel Decker.

Kong, W., Chen, H., Lyu, S., Ma, F., Yu, H., & Zhang, X. (2016). Characterization of a novel manganese peroxidase from white-rot fungus *Echinodontium taxodii* 2538, and its use for the degradation of lignin-related compounds. *Process Biochemistry*, *51*(11), 1776–1783. https://doi.org/10.1016/j.procbio.2016.01.007.

Kulasinski, K., Keten, S., Churakov, S. V., Derome, D., & Carmeliet, J. (2014). A comparative molecular dynamics study of crystalline, paracrystalline and amorphous states of cellulose. *Cellulose*, *21*(3), 1103–1116. https://doi.org/10.1007/s10570-014-0213-7.

Lane, S., Zhang, S., Wei, N., Rao, C., & Jin, Y. S. (2015). Development and physiological characterization of cellobiose-consuming *Yarrowia lipolytica*. *Biotechnology and Bioengineering*, *112*(5), 1012–1022. https://doi.org/10.1002/bit.25499.

Larroude, M., Trabelsi, H., Nicaud, J. M., & Rossignol, T. (2020). A set of *Yarrowia lipolytica* CRISPR/Cas9 vectors for exploiting wild-type strain diversity. *Biotechnology Letters*, *42*(5), 773–785. https://doi.org/10.1007/s10529-020-02805-4.

Li, J., Gu, S., Zhao, Z., Chen, B., Liu, Q., Sun, T., et al. (2019). Dissecting cellobiose metabolic pathway and its application in biorefinery through consolidated bioprocessing in *Myceliophthora thermophila*. *Fungal Biology and Biotechnology*, *6*(1), 1–12. https://doi.org/10.1186/s40694-019-0083-8.

Li, J., Lin, L., Sun, T., Xu, J., Ji, J., Liu, Q., et al. (2019). Direct production of commodity chemicals from lignocellulose using *Myceliophthora thermophila*. *Metabolic Engineering*, 1–11. https://doi.org/10.1016/j.ymben.2019.05.007.

Li, C., Yang, Z., He Can Zhang, R., Zhang, D., Chen, S., & Ma, L. (2013). Effect of pH on cellulase production and morphology of *Trichoderma reesei* and the application in cellulosic material hydrolysis. *Journal of Biotechnology*, *168*(4), 470–477. https://doi.org/10.1016/j.jbiotec.2013.10.003.

Liao, Y., de Beeck, B. O., Thielemans, K., Ennaert, T., Snelders, J., Dusselier, M., et al. (2020). The role of pretreatment in the catalytic valorization of cellulose. *Molecular Catalysis*, *487*(January), 110883. https://doi.org/10.1016/j.mcat.2020.110883.

Liao, Y., Koelewijn, S.-F., Van den Bossche, G., Van Aelst, J., Van den Bosch, S., Renders, T., et al. (2020). A sustainable wood biorefinery for low–carbon footprint chemicals production. *Science*, *367*(6484). https://doi.org/10.1126/science.aau1567. 1385 LP–1390.

Liguori, R., Ventorino, V., Pepe, O., & Faraco, V. (2016). Bioreactors for lignocellulose conversion into fermentable sugars for production of high added value products. *Applied Microbiology and Biotechnology*, *100*(2), 597–611. https://doi.org/10.1007/s00253-015-7125-9.

Liu, Z., Gay, L. M., Tuveng, T. R., Agger, J. W., Westereng, B., Mathiesen, G., et al. (2017). Structure and function of a broad-specificity chitin deacetylase from *Aspergillus nidulans* FGSC A4. *Scientific Reports*, *7*(1), 1–12. https://doi.org/10.1038/s41598-017-02043-1.

Liu, B., Olson, Å., Wu, M., Broberg, A., & Sandgren, M. (2017). Biochemical studies of two lytic polysaccharide monooxygenasesfrom the white-rot fungus *Heterobasidion irregulare* and their roles in lignocellulose degradation. *PLoS One*, *12*(12), 1–21. https://doi.org/10.1371/journal.pone.0189479.

Liu, P., Zhang, G., Chen, Y., Zhao, J., Wang, W., & Wei, D. (2019). Enhanced cellulase production by decreasing intercellular pH through H^+-ATPase gene deletion in *Trichoderma reesei* RUT-C30. *Biotechnology for Biofuels*, *12*(1), 1–15. https://doi.org/10.1186/s13068-019-1536-2.

Luo, S., Qin, Z., Chen, Q., Fan, L., Jiang, L., & Zhao, L. (2020). High level production of a *Bacillus amlyoliquefaciens* chitosanase in *Pichia pastoris* suitable for chitooligosaccharides preparation. *International Journal of Biological Macromolecules*, *149*, 1034–1041. https://doi.org/10.1016/j.ijbiomac.2020.02.001.

Lynd, L. R., Weimer, P. J., Van Zyl, W. H., & Isak, S. (2002). Microbial cellulose utilization: Fundamentals and biotechnology. *Microbiology and Molecular Biology Reviews*, *66*(3), 506–577. Downloaded from http://mmbr.asm.org/on February 6, 2013 by Indian Institute of Technology Madras https://doi.org/10.1128/MMBR.66.3.506.

Ma, X., Gözaydın, G., Yang, H., Ning, W., Han, X., Poon, N. Y., et al. (2020). Upcycling chitin-containing waste into organonitrogen chemicals via an integrated process. *Proceedings of the National Academy of Sciences of the United States of America*, *117*(14), 7719–7728. https://doi.org/10.1073/pnas.1919862117.

Ma, J., Shi, S., Jia, X., Xia, F., Ma, H., Gao, J., et al. (2019). Advances in catalytic conversion of lignocellulose to chemicals and liquid fuels. *Journal of Energy Chemistry*, *36*(x), 74–86. https://doi.org/10.1016/j.jechem.2019.04.026.

Mahmoodi, P., Karimi, K., & Taherzadeh, M. J. (2018). Efficient conversion of municipal solid waste to biofuel by simultaneous dilute-acid hydrolysis of starch and pretreatment of lignocelluloses. *Energy Conversion and Management*, *166*(April), 569–578. https://doi.org/10.1016/j.enconman.2018.04.067.

Mahmoodi, M., Najafpour, G. D., & Mohammadi, M. (2019). Bioconversion of agroindustrial wastes to pectinases enzyme via solid state fermentation in trays and rotating drum bioreactors. *Biocatalysis and Agricultural Biotechnology*, *21*(May), 101280. https://doi.org/10.1016/j.bcab.2019.101280.

Matei, A., & Doehlemann, G. (2016). Cell biology of corn smut disease—*Ustilago maydis* as a model for biotrophic interactions. *Current Opinion in Microbiology*, *34*, 60–66. https://doi.org/10.1016/j.mib.2016.07.020.

Meents, M. J., Watanabe, Y., & Samuels, A. L. (2018). The cell biology of secondary cell wall biosynthesis. *Annals of Botany*, *121*(6), 1107–1125. https://doi.org/10.1093/aob/mcy005.

Meghana, M., & Shastri, Y. (2020). Sustainable valorization of sugar industry waste: Status, opportunities, and challenges. *Bioresource Technology*, *303*(November 2019), 122929. https://doi.org/10.1016/j.biortech.2020.122929.

Meng, Q. S., Zhang, F., Liu, C. G., Zhao, X. Q., & Bai, F. W. (2020). Identification of a novel repressor encoded by the putative gene ctf 1 for cellulase biosynthesis in *Trichoderma reesei* through artificial zinc finger engineering. *Biotechnology and Bioengineering*. https://doi.org/10.1002/bit.27321.

Michely, S., Gaillardin, C., Nicaud, J. M., & Neuvéglise, C. (2013). Comparative physiology of oleaginous Species from the *Yarrowia Clade*. *PLoS One*, *8*(5), 1–10. https://doi.org/10.1371/journal.pone.0063356.

Mondala, A. H. (2015). Direct fungal fermentation of lignocellulosic biomass into itaconic, fumaric, and malic acids: Current and future prospects. *Journal of Industrial Microbiology and Biotechnology*, *42*(4), 487–506. https://doi.org/10.1007/s10295-014-1575-4.

Niglio, S., Procentese, A., Russo, M. E., Piscitelli, A., & Marzocchella, A. (2019). Integrated enzymatic pretreatment and hydrolysis of apple pomace in a bubble column bioreactor. *Biochemical Engineering Journal*, *150*(July), 107306. https://doi.org/10.1016/j.bej.2019.107306.

Nizami, A. S., Rehan, M., Waqas, M., Naqvi, M., Ouda, O. K. M., Shahzad, K., et al. (2017). Waste biorefineries: Enabling circular economies in developing countries. *Bioresource Technology*, *241*, 1101–1117. https://doi.org/10.1016/j.biortech.2017.05.097.

Oh, Y. H., Eom, I. Y., Joo, J. C., Yu, J. H., Song, B. K., Lee, S. H., et al. (2015). Recent advances in development of biomass pretreatment technologies used in biorefinery for the production of bio-based fuels, chemicals and polymers. *Korean Journal of Chemical Engineering*, *32*(10), 1945–1959. https://doi.org/10.1007/s11814-015-0191-y.

Park, S. Y., Kim, J. Y., Youn, H. J., & Choi, J. W. (2018). Fractionation of lignin macromolecules by sequential organic solvents systems and their characterization for further valuable applications. *International Journal of Biological Macromolecules*, *106*, 793–802. https://doi.org/10.1016/j.ijbiomac.2017.08.069.

Paz, A., Outeiriño, D., Pérez Guerra, N., & Domínguez, J. M. (2019). Enzymatic hydrolysis of brewer's spent grain to obtain fermentable sugars. *Bioresource Technology*, *275*(December 2018), 402–409. https://doi.org/10.1016/j.biortech.2018.12.082.

Perez, C. L., Casciatori, F. P., & Thoméo, J. C. (2019). Strategies for scaling-up packed-bed bioreactors for solid-state fermentation: The case of cellulolytic enzymes production by a thermophilic fungus. *Chemical Engineering Journal*, *361*(December 2018), 1142–1151. https://doi.org/10.1016/j.cej.2018.12.169.

Prajapati, B. P., Jana, U. K., Suryawanshi, R. K., & Kango, N. (2020). Sugarcane bagasse saccharification using *Aspergillus tubingensis* enzymatic cocktail for 2G bio-ethanol production. *Renewable Energy*, *152*, 653–663. https://doi.org/10.1016/j.renene.2020.01.063.

Qian, Y., Zhong, L., Gao, J., Sun, N., Wang, Y., Sun, G., et al. (2017). Production of highly efficient cellulase mixtures by genetically exploiting the potentials of *Trichoderma reesei* endogenous cellulases for hydrolysis of corncob residues. *Microbial Cell Factories*, *16*(1), 1–16. https://doi.org/10.1186/s12934-017-0825-3.

Qian, Y., Zhong, L., Sun, Y., Sun, N., Zhang, L., Liu, W., et al. (2019). Enhancement of cellulase production in *Trichoderma reesei* via disruption of multiple protease genes identified by comparative secretomics. *Frontiers in Microbiology*, *10*(December), 1–12. https://doi.org/10.3389/fmicb.2019.02784.

Qiao, K., Imam Abidi, S. H., Liu, H., Zhang, H., Chakraborty, S., Watson, N., et al. (2015). Engineering lipid overproduction in the oleaginous yeast *Yarrowia lipolytica*. *Metabolic Engineering*, *29*, 56–65. https://doi.org/10.1016/j.ymben.2015.02.005.

Qin, L. N., Cai, F. R., Dong, X. R., Huang, Z. B., Tao, Y., Huang, J. Z., et al. (2012). Improved production of heterologous lipase in *Trichoderma reesei* by RNAi mediated gene silencing of an endogenic highly expressed gene. *Bioresource Technology*, *109*, 116–122. https://doi.org/10.1016/j.biortech.2012.01.013.

Rantasalo, A., Landowski, C. P., Kuivanen, J., Korppoo, A., Reuter, L., Koivistoinen, O., et al. (2018). A universal gene expression system for fungi. *Nucleic Acids Research*, *46*(18). https://doi.org/10.1093/nar/gky558.

Rantasalo, A., Vitikainen, M., Paasikallio, T., Jäntti, J., Landowski, C. P., & Mojzita, D. (2019). Novel genetic tools that enable highly pure protein production in *Trichoderma reesei*. *Scientific Reports*, *9*(1), 1–12. https://doi.org/10.1038/s41598-019-41573-8.

Regestein, L., Klement, T., Grande, P., Kreyenschulte, D., Heyman, B., Maßmann, T., et al. (2018). From beech wood to itaconic acid: Case study on biorefinery process integration. *Biotechnology for Biofuels*, *11*(1), 1–11. https://doi.org/10.1186/s13068-018-1273-y.

Rouches, E., Herpoël-Gimbert, I., Steyer, J. P., & Carrere, H. (2016). Improvement of anaerobic degradation by white-rot fungi pretreatment of lignocellulosic biomass: A review. *Renewable and Sustainable Energy Reviews*, *59*, 179–198. https://doi.org/10.1016/j.rser.2015.12.317.

Ruiz-Herrera, J., Leon, C. G., Guevara-Oivera, L., & Carabez-Trejo, A. (1995). Yeast-mycelial dimorphism of haploid and diploid strains of *Ustiago maydis*. *Microbiology*, *141*(1995), 695–703. www.microbiologyresearch.org.

Ruiz-Leza, H. A., Rodriguez-Jasso, R. M., Rodriguez-Herrera, R., Contreras-Esquivel, J. C., & Aguilar, C. N. (2007). Bio-reactors desing for solid state fermentation. *Revista Mexicana de Ingeniería Química*, *6*(1), 33–40.

Sartori, T., Tibolla, H., Prigol, E., Colla, L. M., Costa, J. A. V., & Bertolin, T. E. (2015). Enzymatic saccharification of lignocellulosic residues by cellulases obtained from solid state fermentation using *Trichoderma viride*. *BioMed Research International*, *2015*, 1–10. https://doi.org/10.1155/2015/342716.

Scheller, H. V., & Ulvskov, P. (2010). Hemicelluloses. *Annual Review of Plant Biology*, *61*(1), 263–289. https://doi.org/10.1146/annurev-arplant-042809-112315.

Schwartz, C., Curtis, N., Löbs, A.-K., & Wheeldon, I. (2018). Multiplexed CRISPR activation of cryptic sugar metabolism enables *Yarrowia lipolytica* growth on cellobiose. *Biotechnology Journal*, *13*(9), 1700584. https://doi.org/10.1002/biot.201700584.

Sharma, P., Gaur, V. K., Kim, S. H., & Pandey, A. (2020). Microbial strategies for bio-transforming food waste into resources. *Bioresource Technology*, *299*(December 2019), 122580. https://doi.org/10.1016/j.biortech.2019.122580.

Shida, Y., Furukawa, T., & Ogasawara, W. (2016). Deciphering the molecular mechanisms behind cellulase production in *Trichoderma reesei*, the hyper-cellulolytic filamentous fungus. *Bioscience, Biotechnology and Biochemistry*, *80*(9), 1712–1729. https://doi.org/10.1080/09168451.2016.1171701.

Silva, D. F., Hergesel, L. M., Campioni, T. S., Carvalho, A. F. A., & Oliva-Neto, P. (2018). Evaluation of different biological and chemical treatments in agroindustrial residues for the production of fungal glucanases and xylanases. *Process Biochemistry*, *67*(December 2017), 29–37. https://doi.org/10.1016/j.procbio.2018.02.008.

Singh, B. (2014). *Myceliophthora thermophila* syn. *Sporotrichum thermophile*: A thermophilic mould of biotechnological potential. *Critical Reviews in Biotechnology*, *36*(1), 59–69. https://doi.org/10.3109/07388551.2014.923985.

Soccol, C. R., da Costa, E. S. F., Letti, L. A. J., Karp, S. G., Woiciechowski, A. L., & de Vandenberghe, L. P. S. (2017). Recent developments and innovations in solid state fermentation. *Biotechnology Research and Innovation*, *1*(1), 52–71. https://doi.org/10.1016/j.biori.2017.01.002.

Sperandio, G. B., & Ferreira Filho, E. X. (2019). Fungal co-cultures in the lignocellulosic biorefinery context: A review. *International Biodeterioration and Biodegradation*, *142*(December 2018), 109–123. https://doi.org/10.1016/j.ibiod.2019.05.014.

Srivastava, U., Singh, Z., & Saini, P. (2020). Solid-state fermentation. In S. Yıkmış (Ed.), *Technological developments in food preservation, processing and storage* (pp. 188–204). IGI Global. Issue December.

Steiger, M. G., Mattanovich, D., & Sauer, M. (2017). Microbial organic acid production as carbon dioxide sink. *FEMS Microbiology Letters*, *364*(21), 1–4. https://doi.org/10.1093/femsle/fnx212.

Steiger, M. G., Rassinger, A., Mattanovich, D., & Sauer, M. (2019). Engineering of the citrate exporter protein enables high citric acid production in *Aspergillus niger*. *Metabolic Engineering*, *52*, 224–231. https://doi.org/10.1016/j.ymben.2018.12.004.

Stoffels, P., Müller, M. J., Stachurski, S., Terfrüchte, M., Schröder, S., Ihling, N., et al. (2020). Complementing the intrinsic repertoire of *Ustilago maydis* for degradation of the pectin backbone polygalacturonic acid. *Journal of Biotechnology*, *307*(November 2019), 148–163. https://doi.org/10.1016/j.jbiotec.2019.10.022.

Takkellapati, S., Li, T., & Gonzalez, M. A. (2018). An overview of biorefinery-derived platform chemicals from a cellulose and hemicellulose biorefinery. *Clean Technologies and Environmental Policy*, *20*(7), 1615–1630. https://doi.org/10.1007/s10098-018-1568-5.

Tolbert, A., Akinosho, H., Khunsupat, R., Naskar, A. K., & Ragauskas, A. J. (2014). Characterization and analysis of the molecular weight of lignin for biorefining studies. *Biofuels, Bioproducts and Biorefining*. https://doi.org/10.1002/bbb.1500.

Tong, Z., Zheng, X., Tong, Y., Shi, Y. C., & Sun, J. (2019). Systems metabolic engineering for citric acid production by *Aspergillus niger* in the post-genomic era. *Microbial Cell Factories*, *18*(1), 1–15. https://doi.org/10.1186/s12934-019-1064-6.

Ubando, A. T., Felix, C. B., & Chen, W. H. (2020). Biorefineries in circular bioeconomy: A comprehensive review. *Bioresource Technology*, *299*(November 2019). https://doi.org/10.1016/j.biortech.2019.122585.

Vea, E. B., Romeo, D., & Thomsen, M. (2018). Biowaste valorisation in a future circular bioeconomy. *Procedia CIRP*, *69*(May), 591–596. https://doi.org/10.1016/j.procir.2017.11.062.

Wang, B. T., Hu, S., Yu, X. Y., Jin, L., Zhu, Y. J., & Jin, F. J. (2020). Studies of cellulose and starch utilization and the regulatory mechanisms of related enzymes in Fungi. *Polymers*, *12*(3), 1–17. https://doi.org/10.3390/polym12030530.

Wang, G., Huang, D., Li, Y., Wen, J., & Jia, X. (2015). A metabolic-based approach to improve xylose utilization for fumaric acid production from acid pretreated wheat bran by *Rhizopus oryzae*. *Bioresource Technology, 180*, 119–127. https://doi.org/10.1016/j.biortech.2014.12.091.

Wierckx, N., Agrimi, G., Lübeck, P. S., Steiger, M. G., Mira, N. P., & Punt, P. J. (2020). Metabolic specialization in itaconic acid production: a tale of two fungi. *Current Opinion in Biotechnology, 62*, 153–159. https://doi.org/10.1016/j.copbio.2019.09.014.

Wu, C., Chen, Y., Qiu, Y., Niu, X., Zhu, N., Chen, J., et al. (2020). A simple approach to mediate genome editing in the filamentous fungus *Trichoderma reesei* by CRISPR/Cas9-coupled in vivo gRNA transcription. *Biotechnology Letters, 42*. https://doi.org/10.1007/s10529-020-02887-0.

Wu, X., Liu, Q., Deng, Y., Li, J., Chen, X., Gu, Y., et al. (2017). Production of itaconic acid by biotransformation of wheat bran hydrolysate with *Aspergillus terreus* CICC40205 mutant. *Bioresource Technology, 241*, 25–34. https://doi.org/10.1016/j.biortech.2017.05.080.

Xu, Y., Shan, L., Zhou, Y., Xie, Z., Ball, A. S., Cao, W., et al. (2019). Development of a Cre-loxP-based genetic system in *Aspergillus niger* ATCC1015 and its application to construction of efficient organic acid-producing cell factories. *Applied Microbiology and Biotechnology, 103*(19), 8105–8114. https://doi.org/10.1007/s00253-019-10054-3.

Zambanini, T., Hosseinpour Tehrani, H., Geiser, E., Merker, D., Schleese, S., Krabbe, J., et al. (2017). Efficient itaconic acid production from glycerol with *Ustilago vetiveriae* TZ1. *Biotechnology for Biofuels, 10*(1), 1–15. https://doi.org/10.1186/s13068-017-0809-x.

Zhang, N., Li, S., Xiong, L., Hong, Y., & Chen, Y. (2015). Cellulose-hemicellulose interaction in wood secondary cell-wall. *Modelling and Simulation in Materials Science and Engineering, 23* (8), 85010. https://doi.org/10.1088/0965-0393/23/8/085010.

Zhang, J., Wu, C., Wang, W., Wang, W., & Wei, D. (2018). A versatile *Trichoderma reesei* expression system for the production of heterologous proteins. *Biotechnology Letters, 40*(6), 965–972. https://doi.org/10.1007/s10529-018-2548-x.

Zhang, H., Zhang, J., & Bao, J. (2016). High titer gluconic acid fermentation by *Aspergillus niger* from dry dilute acid pretreated corn stover without detoxification. *Bioresource Technology, 203*, 211–219. https://doi.org/10.1016/j.biortech.2015.12.042.

Zhou, D., Shen, D., Lu, W., Song, T., Wang, M., Feng, H., et al. (2020). Production of 5-hydroxymethylfurfural from chitin biomass: A review. *Molecules, 25*(3), 1–15. https://doi.org/10.3390/molecules25030541.

Zhou, X., & Zhu, H. (2020). Current understanding of substrate specificity and regioselectivity of LPMOs. *Bioresources and Bioprocessing, 7*(1), 11. https://doi.org/10.1186/s40643-020-0300-6.

Zugenmaier, P. (2001). Conformation and packing of various crystalline cellulose fibers. *Progress in Polymer Science (Oxford), 26*(9), 1341–1417. https://doi.org/10.1016/S0079-6700(01)00019-3.

Chapter 8

Bioelectrochemical technologies: Current and potential applications in agriculture resource recovery

Hai The Pham

Research group for Physiology and Applications of Microorganisms (PHAM group), GREENLAB, Center for Life Science Research (CELIFE) and Department of Microbiology, Faculty of Biology, VNU University of Science, Vietnam National University, Hanoi, Vietnam

Chapter outline

List of acronyms

BES bioelectrochemical system
MDC microbial desalination cell
MEC microbial electrolysis cell
MES microbial electrosynthesis system
MFC microbial fuel cell
MSC microbial solar cell
OsMFC osmotic MFC
AEM anion exchange membrane

Recent Advancement in Microbial Biotechnology. https://doi.org/10.1016/B978-0-12-822098-6.00002-1

CEM	cation exchange membrane
PEM	proton exchange membrane
PCR-DGGE	polymerase chain reaction—denaturing gradient gel electrophoresis
COD	chemical oxygen demand
TAN	total ammonia nitrogen
VFA	volatile fatty acid
MCFA	medium chain fatty acid
NADH	nicotinamide adenine dinucleotide
PTFE	polytetrafluoroethylene
RVC	reticulated vitreous carbon
SHE	standard hydrogen electrode
CMS	cattle manure slurry
CSP	corn stover powder
AGS	artificial garbage slurry
WAS	waste activated sludge
CFT	carbon fiber textiles
GDE	gas diffusion electrode
MEA	membrane-electrode assembly
SSM	stainless steel mesh
SSF	stainless steel foil
RED	reverse electrodialysis
ANNAMOX	anaerobic ammonium oxidation
AD	anaerobic digestion
CSTR	continuous stirred-tank reactor
SRB	sulfate-reducing bacteria
DET	direct electron transfer
TMCS	transmembrane chemisorption
FO	forward osmosis
ES	electrochemical system
CAE	circular agricultural economy

1 Introduction

The ever-growing human population of the world will ultimately face severe challenges in terms of energy supply, food supply, resource exhaustion, and ecological sustainability. Therefore mankind is under increasing pressure to use nonrenewable resources in more efficient manners (producing more useful products and less waste) and to use more and more renewable resources (Chu, Liang, Jiang, & Zeng, 2020). In the end, the ideal target is a circular economy where waste can be recycled to recover resources and to produce products with high energy efficiencies (Bian, Bajracharya, Xu, Pant, & Saikaly, 2020). For agriculture, in particular, this is a must as our soil and water budgets are limited, but the demand for food keeps increasing exponentially (Brink, Densmore, &

Hill, 1977; Zwart & Bastiaanssen, 2004). In other words, resource recovery in agriculture is vital for a sustainable future society.

For resource recovery in general and particularly in agriculture, biocatalytic approaches are preferred as they have many advantages, such as low price, good stability under long-term operation, no pollution, high product selectivity, and high energy efficiency (Li et al., 2020). Conventional bioconversions of agricultural waste solely involve microbial fermentation that can offer the refinery of a number of useful products, such as sugars, organic acids, amino acids, solvents, etc. However, they have certain limitations, especially in the spectrum of products they can produce or yield/concentration thresholds (Cai et al., 2016; Cheng & Logan, 2007; Lalaurette, Thammannagowda, Mohagheghi, Maness, & Logan, 2009; Pham et al., 2006; Van Eerten-Jansen, Heijne, Buisman, & Hamelers, 2012; Zuo, Maness, & Logan, 2006). Bioelectrochemical systems (BESs), a recent emerging technology platform based on the catalytic activities of electroactive microorganisms, can offer new resource recovery potentials with more controllable options at hand (Bajracharya et al., 2016;Bian et al., 2020 ; Rabaey & Rozendal, 2010). An attractive feature of BESs is that they can even be used to capture CO_2, a major waste commonly emitting into the atmosphere, causing a significant loss of carbon and the global warming effect (Bian et al., 2020; Rabaey & Rozendal, 2010). BESs, therefore, open up many new technological options for agriculture resource recovery. This chapter will provide an overview of up-to-date advances in applying BESs for upgrading agricultural waste to energy or valuable products, under the theme of resource recovery. As studies on BESs have been extensively reviewed (Bajracharya, Sharma, et al., 2016; Bian et al., 2020; Clauwaert et al., 2008; Kelly & He, 2014; Logan et al., 2006; Pham et al., 2006; Prévoteau, Carvajal-Arroyo, Ganigué, & Rabaey, 2020; Rabaey et al., 2007; Rabaey, Angenent, Schroder, & Keller, 2009; Rabaey & Rozendal, 2010; Rabaey & Verstraete, 2005; Rozendal, Hamelers, Rabaey, Keller, & Buisman, 2008; Saratale et al., 2017; Wang & Ren, 2014), the author's approach will be from a practical point of view, providing the readers with the most practical information (highlighted in *italics* among the text), which is about key technical approaches ("how to do") to achieve specific resource recovery goals ("for what") and their respective technical limits and limitations. With that approach, the author wishes to provide a "manual-like" material for those who want to start using BESs for a certain application purpose in agriculture resource recovery, rather than to provide insightful calculations or analyses of the BESs developed to date. Furthermore, the outlook of the future application potentials of BESs in agriculture resource recovery will also be discussed.

2 BESs

BESs are, by nature of their names, electrochemical systems in which at least one electrochemical reaction is catalyzed by biological agent(s) (Bajracharya,

Sharma, et al., 2016; Rabaey et al., 2009). Each BES generally consists of two electrodes, allowing oxidation-reduction reactions to occur, and the biocatalysts may be involved in the reaction(s) either at one electrode or at both. The biocatalysts here can be microorganisms or enzymes, but as microbial electrochemical systems have been studied more intensively than enzymatic systems, especially in recent years, BESs are generally regarded as microbial electrochemical systems. For an in-depth exploration of what BESs are, the readers are recommended to read a book by Rabaey et al. (2009), with additional reference to a recent review by Bajracharya, Vanbroekhoven, Buisman, Pant, and Strik (2016). For a summary of that, please read Supplemental Material.

Depending on the objectives of using it and how it is operated, a BES can be a microbial fuel cell (MFC) or an enzymatic fuel cell, a microbial electrolysis cell (MEC), a microbial electrosynthesis cell (MES), a microbial solar cell (MSC), or a microbial desalination cell (MDC). With many such designations, a generalized term of MXC was even proposed to denote all these different kinds of microbial electrochemical systems (Harnisch & Schröder, 2010). Indeed, before the term "BESs" was generally used, researchers in the field were more familiar with MFCs, as these are the very first type of BESs, from which other types are inspiringly developed. A microbial fuel cell converts chemical energy, usually in energy-rich substrates as electron donors at the anode, to electrical energy, by taking advantages of the electrochemically active (electroactive) microorganism as the catalysts (Logan et al., 2006; Rabaey et al., 2009) (Fig. 1A). These microbial catalysts oxidize the substrates and transfer the extracted electrons to the anode, enabling the generation of an electrical current when the electrons flow through an external circuit to the cathode and eventually accepted by a strong oxidant (usually oxygen). Likewise, an MFC takes advantage of natural electron flow to harvest energy, and thus do not require energy input. When one wants to use the system to drive an upflow energy reaction by supplying some external power to the system, an MFC becomes an MEC if the additional voltage is sufficient to half-electrolyze water to produce hydrogen, or becomes an MES if the additional power is to control electrode potential(s) to allow the formation of other products such as methane, fatty acids, alcohols, etc. from organics or CO_2 (Harnisch & Schröder, 2010; Rabaey et al., 2009) (Fig. 1D and E). If the electron donors of an MFC come from photosynthesis, i.e., the energy source is actually from sunlight, then the MFC becomes an MSC, which can be considered as a bioelectrochemical platform to convert light to electricity (Strik et al., 2011; Strik, Hamelers, Snel, & Buisman, 2008) (Fig. 1B). MFCs can also be used for desalination and are so-called MDCs, by exploiting the electrochemical processes in MFCs to separate cations and anions and using multiple membranes (Kim & Logan, 2013; Saeed et al., 2015) (Fig. 1C). As can be seen, many such varieties of BESs or MXCs are actually based on the MFC framework with two operational principles: (i) without energy input and electrode potential control (MFC, MSC, and

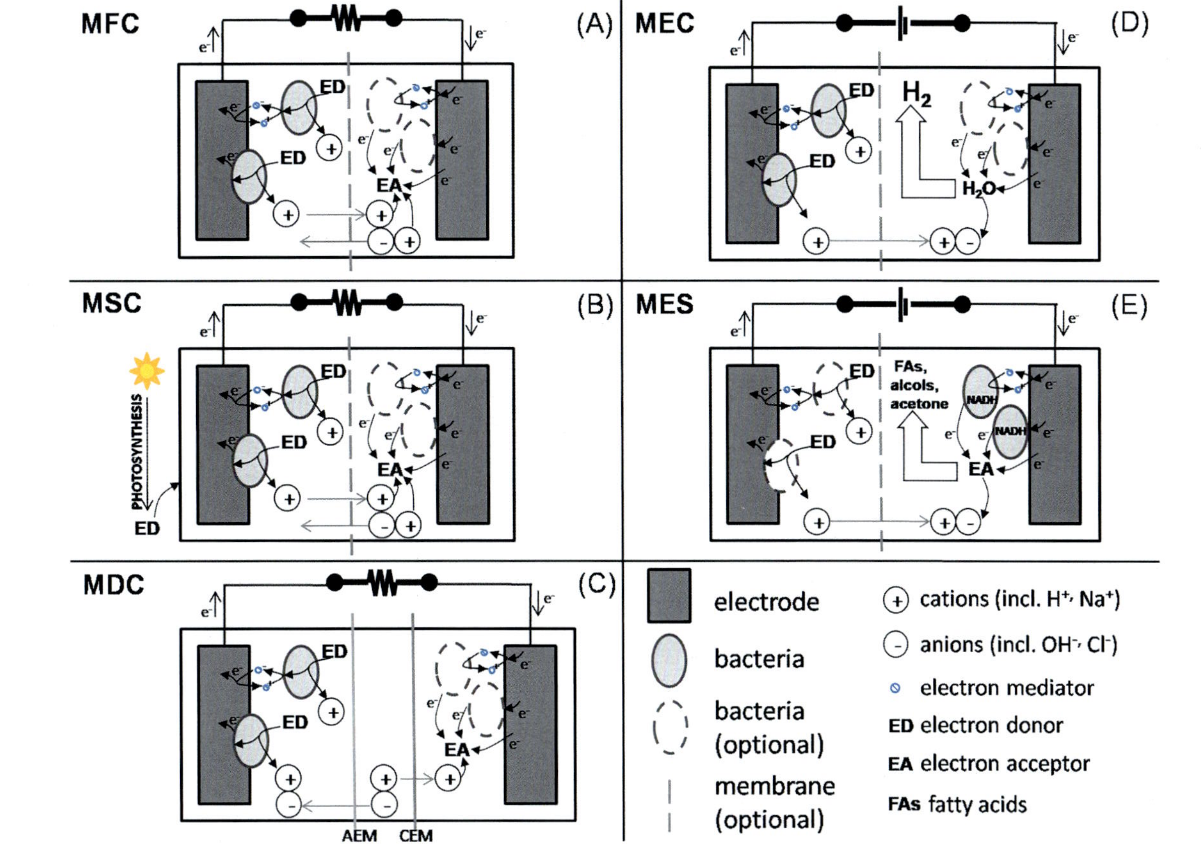

FIG. 1 Schematic descriptions of different types of BESs. Notes: *MFC*, microbial fuel cell; *MSC*, microbial solar cell; *MDC*, microbial desalination cell; *MEC*, microbial electrolysis cell; *MES*, microbial electrosynthesis cell; *AEM*, anion exchange membrane; *CEM*, cation exchange membrane (including PEM (proton exchange membrane)). Components (bacteria, membrane) depicted with dashed line are optional, meaning their presence is not required. When a membrane is lacking, the BES becomes a membrane-less system, which can be a sediment BES as reported in several studies. For MFC-typed systems, the electron donor (ED) can be glucose, acetate, sulfide, even cellulose or any reduced compounds associated with agricultural wastes and metabolizable by electroactive bacteria, while the electron acceptor (EA) can be oxygen or ferricyanide or permanganate, etc. For MECs, the ED can be similar as in MFC-typed systems, and the EA is water that is split to generate hydrogen. For MESs, the ED can be water, ferrocyanide or can be similar as in MFC-typed systems if a bioanode is used, while the EA can be a fermentation product/intermediate such as acetate, butyrate, etc. or can be carbon dioxide.

MDC), and (ii) with energy input and electrode potential control (MEC and MES). Depending on the technical objectives, one can decide to use a suitable system and apply a suitable operational principle. A comprehensive illustration of the different BES types is in Fig. 1.

3 BESs in recovering energy from agricultural wastes

Agricultural wastes, rich in organic contents, are an excellent source of renewable energy. The energy in agricultural wastes can be recovered directly as electricity by using MFCs or as fuel gases (i.e., hydrogen and methane) by using MECs/MESs.

3.1 Direct generation of electricity

3.1.1 Electricity generation from animal wastes

Animal wastes contain high nitrogen, phosphorus, and organic contents in the forms of residual feed and animal excreta that can cause severe pollutions, i.e., eutrophication, if discarded to the environment without treatment (Min, Kim, Oh, Regan, & Logan, 2005). On the other hand, the substances in animal wastes still store a great amount of energy in their chemical bonds that can be harvested as electricity in BESs, more specifically in microbial fuel cells (MFCs). Thus MFCs can be used not only to treat these wastes but also to recover the energy from them. Indeed, for the aim of producing electricity from waste, the MFC is the only BES type to be used (Bajracharya, Sharma, et al., 2016; Rabaey et al., 2009).

Treating animal wastewaters

Min et al. (2005) were the first to test the use of an MFC to treat animal wastewater, which is swine wastewater in their study. The strength of the tested wastewater was up to $8320 \pm 190\,\mathrm{mg\,L^{-1}}$ soluble chemical oxygen demand (sCOD). Two technical options were tested: option 1: a two-chamber bottle-typed reactor with a proton exchange membrane (PEM), a carbon paper anode, a Pt-coated carbon paper cathode and a buffered catholyte; option 2: a single-chamber reactor (having no membrane) with a similar anode, an air-breathing Pt-PTFE-coated cathode and fed directly the wastewater. When tested with full-strength wastewater, option 1 resulted in only a maximum power density of $45\,\mathrm{mW\,m^{-2}}$, which was well comparable to those of MFC systems operated with other substrates reported to that date (Min et al., 2005). However, this value were six times lower than that obtained with option 2 when tested also with full-strength wastewater: $261\,\mathrm{mW\,m^{-2}}$ (Min et al., 2005) (Table 1). The reason for this performance difference is believed to be the lower internal resistance of the single-chamber system compared to the two-chamber. This implies that the single-chamber MFC already performed well with full-strength and raw swine

TABLE 1 Different BES technical achievements in direct electricity generation from agriculture-associated wastes.

Waste type	Waste pretreament	Applied BES technical features	Aver. I density ($mA m^{-2}$ anode surface area)	Aver. P density ($mW m^{-2}$ anode surface area)	Max. P density ($mW m^{-2}$ anode surface area)	Coulombic efficiency (%)	Reference
Swine WW	No	Two-chamber MFC; PEM; carbon paper anode, Pt-coated carbon paper cathode, buffered catholyte, $R = 1000\ \Omega$	56.3 ± 0.4	7.1 ± 0.2	45		(Min et al., 2005)
Swine WW	No	Single-chamber MFC; no membrane; carbon paper anode, Pt-coated and PTFE-coated carbon paper cathode, $R = 1000\ \Omega$	159 ± 0.4	57 ± 0.4	261	8	(Min et al., 2005)
High-strength swine WW	Sieved (1 mm)	Single-chamber MFC; no membrane; carbon cloth anode, Pt-coated and PTFE-coated carbon cloth cathode, $R = 2200\ \Omega$	n.a.	33.4	n.a.	1.5	(Lin, Wu, Nelson, Miller, & Zhu, 2016)
High-strength swine WW	Sieved (1 mm); diluted 10-fold	Single-chamber MFC; no membrane; carbon cloth anode, Pt-coated and PTFE-coated carbon cloth cathode, $R = 2200\ \Omega$	n.a.	*c.* 234	n.a.		(Lin et al., 2016)

Continued

TABLE 1 Different BES technical achievements in direct electricity generation from agriculture-associated wastes.—cont'd

Waste type	Waste pretreament	Applied BES technical features	Aver. I density ($mA\,m^{-2}$ anode surface area)	Aver. P density ($mW\,m^{-2}$ anode surface area)	Max. P density ($mW\,m^{-2}$ anode surface area)	Coulombic efficiency (%)	Reference
Dairy WW	No pretreated but diluted with water to requisite desired organic loads in the anode influents	Single-chamber MFC; with an MEA; $R = 100\ \Omega$	n.a.	n.a.	n.a. ($1.1\,W\,m^{-3}$)	4.3	(Venkata Mohan, Mohanakrishna, Velvizhi, Babu, & Sarma, 2010)
Cow waste slurry	Diluted 10-fold, sieved and centrifuged	Single-chamber MFC; PEM; graphite anode, Pt-coated graphite cathode, $R = 460\ \Omega$	n.a.	n.a.	0.34	0.22	(Yokoyama, Ohmori, Ishida, Waki, & Tanaka, 2006)
Farm manure slurry	Grounded and mixed with water	Membrane-less; carbon cloth anode; Pt-coated and PTFE-coated carbon paper cathode; varied R	n.a.	n.a.	11	n.a.	(Scott & Murano, 2007)
Cattle manure in water (cattle manure slurry)	Screened and mixed with water	Three-chamber; central chamber containing slurry and perforated screens allowing leachate to cross to the other two chambers; each of the other two chambers containing a graphite brush anode and an air-breathing Pt-coated carbon cloth cathode; $R = 470\ \Omega$	n.a.	n.a.	67	n.a.	(Zheng & Nirmalakhandan, 2010)

Cattle manure in water (cattle manure slurry)	Screened and mixed with water	Twin-chamber; no membrane; two anode-cathode sets at both sides of the chamber; graphite brush anode and Pt-coated carbon cloth cathode; R = 470 Ω	n.a.	n.a.	67	n.a.	(Lee & Nirmalakhandan, 2011)
Cattle dung slurry	No	Two-chamber; an anaerobic digester and the anode chamber combined; graphite-granules bound anode and cathode; $KMnO_4$ as the catholyte; R = 1000 Ω	n.a.	n.a.	220	n.a.	(Zhao et al., 2012)
Dairy manure slurry	No	Three-chamber: a cylinder anode chamber with stirrer and contacting two cubic cathode chambers; two graphite brush anodes; graphite brush embedded in graphite granules serving as the matrix for biocathode; $R = 100\ \Omega$	n.a.	n.a.	n.a. (15.1 $W m^{-3}$ anode liquid volume)	30–40	(Zhang et al., 2012)

Continued

TABLE 1 Different BES technical achievements in direct electricity generation from agriculture-associated wastes.—cont'd

Waste type	Waste pretreament	Applied BES technical features	Aver. I density ($mA\,m^{-2}$ anode surface area)	Aver. P density ($mW\,m^{-2}$ anode surface area)	Max. P density ($mW\,m^{-2}$ anode surface area)	Coulombic efficiency (%)	Reference
Cattle manure slurry (CMS)	Screened and mixed with water	Using electrode cassettes as modules to be "plugged" in and out any CMS container; an electrode cassette consists of two air-breathing Pt-coated carbon paper cathodes facing each other and sharing a same narrow air chambers (purged with air) and two respective carbon felt anodes contacting the cathodes through unwoven cloths (replacing the PEM); varied R	n.a.	n.a.	600–700	26–29	(Inoue et al., 2013)
Cattle manure slurry (CMS)	Sieved	Using allochthonous inocula from activated sludge and domestic sewage; single-chamber air-cathode MFC: graphite fiber anode, PTFE-coated activated carbon cathode; $R = 1000\ \Omega$	n.a.	n.a.	1259	n.a.	(Xie et al., 2017)

Cattle manure slurry (CMS) (mixed with cornstalk)	Air dried, smashed, sieved, and mixed to achieve a C/N ratio of 30; and then treated with ultrasonication	Two-chamber (rectangular shape); with KCl salt bridge, graphite rod anode and cathode; $R = 1000\ \Omega$	102	n.a.	102	n.a.	(Shen et al., 2018)
Raw animal wastes	Mixed with sea sand (1:1, w/w); then watersaturated	Stainless steel anodes and copper cathodes inserted directly into (50) pots containing the raw manure mixture; connected in series; $R = 150\ \Omega$	n.a. (0.25–0.5 mA)	n.a.	n.a. ($8\ W\ m^{-3}$)	n.a.	(El-Nahhal et al., 2020)
Corn stover waste	Soaked with water (neutral); then steam exploded	Single-chamber MFC; no membrane; carbon paper anode, Pt-coated carbon cloth cathode, $R = 250\ \Omega$	n.a.	n.a.	371 ± 13	20–30	(Zuo et al., 2006)
Corn stover waste	Soaked with water (neutral); then steam exploded	Single-chamber MFC; no membrane; carbon paper anode, Pt-coated and gas-diffusion-layer-coated carbon cloth cathode, conductivity adjusted to $20\ mS\ cm^{-1}$, $R = 250\ \Omega$	n.a.	n.a.	933	20–30	(Zuo et al., 2006)
Corn stover waste	Soaked with H_2SO_4 (acid); then steam exploded	Single-chamber MFC; no membrane; carbon paper anode, Pt-coated carbon cloth cathode, $R = 250\ \Omega$	n.a.	n.a.	367 ± 13	20–30	(Zuo et al., 2006)

Continued

TABLE 1 Different BES technical achievements in direct electricity generation from agriculture-associated wastes.—cont'd

Waste type	Waste pretreament	Applied BES technical features	Aver. I density ($mA\,m^{-2}$ anode surface area)	Aver. P density ($mW\,m^{-2}$ anode surface area)	Max. P density ($mW\,m^{-2}$ anode surface area)	Coulombic efficiency (%)	Reference
Corn stover waste	Soaked with H_2SO_4 (acid); then steam exploded	Single-chamber MFC; no membrane; carbon paper anode, Pt-coated and gas-diffusion-layer-coated carbon cloth cathode, conductivity adjusted to 20 $mS\,cm^{-1}$, $R=250\,\Omega$	n.a.	n.a.	971	20–30	(Zuo et al., 2006)
Corn stover waste	Grounded to produce raw corn stover powder (CSP)	Single-chamber bottle-type MFC; carbon paper anode, Pt-coated and gas-diffusion-layer-coated carbon cloth cathode, $R=1000\,\Omega$	n.a.	n.a.	331	n.a. ($323 \pm 20\,C\,g^{-1}$ substrate)	Wang, Feng, et al. (2009)
Corn stover waste	Grounded to produce raw corn stover powder (CSP), then pretreated by steam explosion, producing corn stover residual solids (CSRS)	Single-chamber bottle-type MFC; carbon paper anode, Pt-coated and gas-diffusion-layer-coated carbon cloth cathode, $R=1000\,\Omega$	n.a.	n.a.	406	n.a. ($264 \pm 10\,C\,g^{-1}$ substrate)	Wang, Feng, et al. (2009)

Corn stover waste	Pretreated in continuous-horizontal-screw reactor: 158°C, 24mg sulfuric acid/g dry biomass, 5min residence time; subject to cellulase treatment at 50°C for 4days	Biorefinery (fermentation with *Zymomonas mobiliz* for 3days) prior to MFC treatment; MFC type: air-cathode single chamber containing a membrane-electrode assembly (carbon felt anode, PEM, Pt-deposited carbon cathode)	n.a. (maximum current: 3.4 A m^{-2} at 8% loading)	n.a.	1200	<10 (at 8% loading and higher) (maximum 40% at 1% loading)	(Borole, Hamilton, & Schell, 2013)
Wheat straw waste	Pretreated a three-step reactor: first soaked at 80°C->hemicelluloses extracted at 170–180°C->celluloses hydrolyzed by enzymes at 195°C to produce hydrolysate	H-typed two-chamber; with PEM; carbon paper anode, Pt-coated carbon paper cathode; buffered ferricyanide catholyte; $R = 1000\ \Omega$	n.a.	79.6	124	25	(Zhang, Min, Huang, & Angelidaki, 2009)
Rice mill wastewater	No	Two-chamber; earthen pot anode chamber placed in a plastic bucket as cathode chamber (the pot wall functioned as PEM!); stainless steel anode; anolyte pH = 8; graphite plate cathode; R = 100 Ω	n.a.	n.a.	49 mW m^{-2} (2310 mW m^{-3} unit volume)	21.2	(Behera, Jana, More, & Ghangrekar, 2010)

Notes: *Aver.*, average; *Max.*, maximum; *I*, current; *P*, power; *WW*, wastewater; *MFC*, microbial fuel cell; *PEM*, proton exchange membrane; *R*, external resistance; *n.a.*, data not available.

wastewater. Indeed, MFCs with this configuration were also reported to have similar performances in several other studies (Kiely et al., 2011; Velasquez-Orta, Head, Curtis, & Scott, 2011). Thus this *single-chamber reactor design* appeared to be a technical improvement of MFC technology for power generation from animal waste treatment. Furthermore, it was found that such a system could remove up to 83% of ammonium in the wastewater (Min et al., 2005). Later, another study of the same group revealed that ammonium was removed by transforming to ammonia due to pH elevation near the cathode and then by the evaporation of ammonia (Kim, Zuo, Regan, & Logan, 2008). That is, ammonium was not removed in the system by being consumed as a substrate for electricity generation. On the other hand, in the case when high-strength swine wastewater was used, ammonia evaporation might not be fast enough, and the accumulation of ammonium might inhibit the activity of electroactive bacteria, causing reduced power generation although a single-chamber configuration was employed (Lin et al., 2016). It was reported that the MFCs operated with 10-fold diluted swine wastewater generated power densities sevenfold higher than those of the MFCs operated with the undiluted wastewater (Table 1). The toxicity of such (undiluted) high strength wastewater was also attributed to high concentrations of volatile fatty acids (VFAs) (Lin et al., 2016). Therefore a BES should be operated at *an optimal organic load,* e.g., 4.44 kg COD m^{-3} for a single-chamber MFC system tested by Venkata Mohan et al. (2010), to achieve maximum protein removal, turbidity reduction, and power density. Besides, other technical measures to improve MFC power generation with high-strength swine wastewater were proposed: (i) *zeolite pretreatment* to remove NH_4^+ and (ii) *activated carbon pretreatment* to remove VFAs. By applying these pretreatments, the power output of the MFCs could increase by more than 80% (Lin et al., 2016).

Treating animal waste slurries

Solid excreta of animals are difficult to be treated directly, and usually have to be mixed with water to form slurries that are more accessible to microorganisms. However, treating these slurries is still different from treating animal wastewaters due to higher solid contents still in the slurries. Yokoyama et al. (2006) attempted to treat cow waste slurry by also using a *single-chamber MFC* system. The system is in general similar to that used by Min et al. (2005) except that it has a PEM attached to the cathode and a cubic-shaped chamber. Surprisingly the system had a poor performance in terms of electricity generation, with a maximum power density of only 0.34 mW m^{-2} and a Coulombic efficiency of only 0.22% (Yokoyama et al., 2006) (Table 1). The reason for this poor performance was not discussed, but probably the accessibility of the substrates and the effect of competing reactions (such as sulfate reduction) might be involved. However, it is encouraging that the system could remove up to 70% COD and 84% BOD of the slurry. Unlike in the study by Min et al. (2005), 84% total nitrogen, 70% total phosphorus, and 91% total potassium still retained in the treated slurry. Moreover, the concentration of ammonium-nitrogen in the slurry even increased by 1.9-fold after treatment, and the authors proposed that the treated slurry was usable as a fertilizer.

In another study, Scott and Murano (2007) tested different *"geometric designs"* of MFCs for the treatment of farm manure slurry having the solid content of up to 20% (*w*/w). These membraneless systems have a carbon cloth anode placed horizontally at the bottom sludge layer and a carbon paper cathode placed horizontally in the water layer on top. Different electrode positions were tried, as well as another cathode material such as nickel and a multiple anode design or the use of a catalyst layer applied onto the sludge substrate. However, the systems still had a relatively poor performance, with a maximum power density of about $4\,mW\,m^{-2}$ and a maximum current density of $40\,mA\,m^{-2}$ (Scott & Murano, 2007) (Table 1). Using a Pt-coated and PTFE-coated carbon paper cathode could boost the maximum power density up to $11\,mW\,m^{-2}$, but this level is still low compared to values reported for other systems. By calculation, the group determined that oxygen mass transport, oxygen solubility, and cathode reaction are the key factors limiting the performance of the MFCs.

The above-mentioned cathode-reaction-associated problems indeed can be partially overcome by using a *three-chamber system* that is actually a combination of two air-cathode single-chamber reactors (Zheng & Nirmalakhandan, 2010). In such a system, the central chamber contains manure in water (the slurry), and has two perforated screens that allow the slurry leachate to cross into the two other chambers on two sides. Each of the side chambers has a graphite brush anode and a Pt-coated carbon cloth cathode contacting air. Such a design can enhance oxygen transport and also the cathode reaction, and a maximum power density of $67\,mW\,m^{-2}$ could be achieved (Zheng & Nirmalakhandan, 2010) (Table 1). The latter can be increased by 40% if a mediator such as humic acid was added to the electrolyte. These power outputs are comparable with those of some MFCs operated with readily biodegradable substrates, and thus suggest the feasibility of producing electricity with manure slurry. Indeed, the power density of $67\,mW\,m^{-2}$ was also achieved by *a twin-compartment MFC* system with a reactor volume of up to 1.85 L (Lee & Nirmalakhandan, 2011) (Table 1). This system actually resembles a single-chamber system but it has two 1-cm-distanced anode-cathode sets at both sides and no membrane between the electrodes. The use of graphite brush anodes (enabling higher surface area) and air-breathing Pt-coated carbon cloth cathodes was also believed to contribute to enhancing the performance of the system. It was calculated that this system could produce up to $24.3\,mW\,kg^{-1}$ dry manure upon being tested with different manure-to-liquid ratios (Lee & Nirmalakhandan, 2011).

To further improve the energy recovery from solid animal waste, other modified setups of MFCs were tested. Zhao et al. (2012) used an innovated setup that *combined an anaerobic digester and the anode chamber of an MFC* in one compartment. Thus the compartment was in contact with a PEM in turn contacting the cathode chamber. Such a system can, at the same time, carry out anaerobic acidogenesis and electricity generation from the produced fatty acids. The system was even operated with fresh cattle dung (mixed with water to create 8% total solid slurry). Its performance was quite impressive with a generated voltage up to 900–1000 mV (1180 open circuit voltage) (after 30 days of operation) and a maximum power density of up to 220 mW m^{-2} (Zhao et al., 2012) (Table 1). Total COD removal of the system reached up to *c.* 74%, showing an attractive bioconversion, although the Coulombic efficiency was only *c.* 2.8%, which was believed to be the result of the competition by anaerobic digestion (Zhao et al., 2012). To increase the Coulombic efficiency when treating animal waste slurry, Zhang et al. (2012) proposed the use of an innovated MFC configuration with a *cylindrical anode chamber contacting two cubic biocathode chambers*. These authors' strategy is to boost the maximum power output by increasing the linear extrapolated open-circuit voltage (E_b) of the cell while reducing its internal resistance (R_{int}). This might be achieved by establishing the biocathodes with a high contact surface (using a graphite brush embedded in graphite granules). A stirred anode chamber was also believed to enhance the anode reaction through reducing mass transfer loss. Interestingly, the power output of the system increased with the increasing percentage of total solids (TS) in the anolyte, reaching a stable density of *c.* 8 W m^{-3} (anode liquid volume) and a maximum density of 15.1 W m^{-3} (anode liquid volume) at 6% TS (Table 1). Its maximum Coulombic efficiency even reached the highest reported level of 30%–40%. These are properly due to the fact that this design could lower the internal resistance of the system (down to *c.* 30 Ω—the lowest reported to that date), and increase its E_b (up to 1.04 V, which was significantly higher than reported). Astoundingly, in the anode of this MFC, known electrogens such as *Ochrobactrum*, *Pseudomonas*, *Comamonas*, and *Desulfobulbus* were rare and instead, Clostridia dominated (Zhang et al., 2012).

Indeed, Clostridia were also found on the electrode surface in another innovated BES having *electrode cassettes* (Inoue et al., 2013). Such an electrode cassette consists of two air-breathing Pt-coated carbon paper cathodes facing each other and sharing the same narrow air chambers (purged with air), and two respective carbon felt anodes contacting the cathodes through unwoven cloths (replacing the PEM). Each such cassette can function like a flexible module that can be immersed in or withdrawn off any chamber containing animal waste slurry, and the chamber becomes an anode chamber. The key concept here is that multiple modules can be combined in one reactor to produce more power (Shimoyama et al., 2008). Such a system could produce power densities of *c.* 10–16 W m^{-3}, which are comparable to those reported by Zhang et al. (2012). However, its power densities normalized to anode surface could reach 600–700 mW m^{-2}, which were the highest at that moment for cattle manure MFCs (Table 1). A Coulombic efficiency of up to 26%–29% could be achieved with the MFC. It is even more interesting that the system degraded 70% of holocellulose and lignin contents in the cattle manure slurry (Inoue et al., 2013).

However, it seemed that without the lignocellulose content in the slurry, the power generation is very poor, indicating a high adaptation of the anode microbial community to these substrates. In the biofilm formed on the anode surface of the system, bacteria belonging to Geobacteraceae were enriched and believed to interact with other bacteria in the anode chamber, including the above-mentioned Clostridia, through direct interspecies electron transfer (Lovley, 2017). This unique bacterial interaction might be the result of the unique configuration of the system. Thus the configuration of an MFC treating animal waste appears to be a key factor in determining its power generation, based on all the findings of the works discussed earlier.

Apart from efforts to improve MFC configuration, several other studies also proposed other technical measures to increase the power generation of MFCs while treating animal waste slurries, such as the use of proper inocula or additional treatments of the wastes before loaded into MFCs. Xie et al. (2017) proved that using *allochthonous inocula* could increase the power output of their single-chamber air-cathode MFCs while treating cattle manure slurry (CMS). Accordingly, the MFCs inoculated with activated sludge and domestic sewage could produce power densities of up to *c.* 1.26 W m^{-2} and 1.1 W m^{-2}, respectively, while the uninoculated produced only *c.* 0.9 W m^{-2} (Xie et al., 2017) (Table 1). The MFC inoculated with activated sludge even achieved a COD removal efficiency of *c.* 85%, which was the highest report to that date (Xie et al., 2017). The presence of more fermentative bacteria (e.g., *Bacteroides*) and nitrogen fixation bacteria (e.g., *Azoarcus* and *Sterolibacterium*) in the allochthonous inocula were believed to explain the more efficient degradation of the slurry. Indeed, the slurry, if *pretreated properly,* e.g., with ultrasonication, can become more degradable, which can lead to higher power production (up to 102 mW m^{-2}) by MFCs, although this pretreatment seemed effective only in the initial period of MFC operation (Shen et al., 2018).

Treating raw solid animal wastes

The only study reported thus far on treating raw manure with MFCs is by El-Nahhal et al. (2020). Interestingly, the authors simply "plugged" stainless steel anodes and copper cathodes into (50) pots containing a mixture of raw (chicken, cattle or horse) manure and sea sand (1:1, *w*/w), saturated with water. Thus each pot actually functioned like a single-chamber MFC with the electrolyte being the matrix inside the pot. By connecting the pots in series, the authors could harvest a voltage of up to 20–25 V and a power density of up to 8 W m^{-3}, although the current was low (0.25–0.5 mA) (El-Nahhal et al., 2020) (Table 1). Moreover, the system removed 98%–99% COD, up to 90% total nitrogen, and 80%–90% organic carbon of the raw manure, although after a long time of 60 days. However, based on the results of only culture-based analyses of bacteria performed in this study, it is difficult to understand the mechanism of electricity generation and biodegradation in the system.

3.1.2 Electricity generation from lignocellulosic wastes

Lignocellulosic wastes actually represent another type of agricultural wastes that are very much abundant while they themselves are actually huge energy resource reservoirs to be harvested. Thus under the pressure of finding new energy sources to replace fossil fuels, the interest in extracting energy from lignocellulosic biomass, especially in wastes, is increasing. Indeed, quite a number of chemical and biological approaches for energy recovery from lignocellulosic wastes are available but many of them encounter technical and economical limitations (Zhang et al., 2009). The use of MFCs is promising in overcoming some of those limitations.

Treating corn-derived lignocellulosic wastes

Corn stover waste can become more accessible for microbial treatments in general, and for MFC treatments in particular, through some *pretreatments*. Zuo et al. (2006) were the first to investigate the possibility of using corn stover hydrolysate as a substrate for an MFC. In their study, corn stover waste was pretreated by being soaked with water or H_2SO_4, before subjected to a high-pressure steam and high temperature (190–220°C) treatment (a process called steam explosion). The resulted hydrolysate was diluted, and its pH and conductivity adjusted before it was treated by a *single-chamber MFC* (the typical configuration developed by Liu, Ramnarayanan, and Logan (2004). Such an MFC, enriched with glucose as the substrate, could perform relatively well with the corn stover hydrolysate in its anode medium, producing a maximum power density of *c.* $371 \pm 13\,mW\,m^{-2}$ when operated with the neutral hydrolysate and $367 \pm 13\,mW\,m^{-2}$ with the acid hydrolysates (Table 1). The Coulombic efficiencies of the MFCs reached 20%–30%, while their BOD removal efficiencies could reach up to 93%–94% and COD removal efficiencies 60%–70% (Zuo et al., 2006). Coating the cathode with a gas diffusion layer could boost the power output by twofold, up to $810 \pm 3\,mW\,m^{-2}$ (for the neutral hydrolysate) or $861 \pm 37\,mW\,m^{-2}$ (for the acid hydrolysate) (Table 1). Proper adjustments of the conductivity of the electrolyte could further increase the power densities up to $933\,mW\,m^{-2}$ and $971\,mW\,m^{-2}$, respectively. These results demonstrate the prospect of recovering energy from corn stover (up to $4.6 \times 10^{10}\,kWh\,year^{-1}$), although the yield (10%) was still low in comparison with ethanol production by fermentation (47%) and direct combustion (20%–30%) (Zuo et al., 2006). This prospect encouraged more innovations to further improve the power generation from corn stover waste treatment by MFCs and their electron efficiency. For instance, Wang et al. (2009) proposed a *bioaugmentation* approach to improving the performance of also a single-chamber-type MFC that could even generate electricity when fed with raw corn stover powder (CSP). Accordingly, the MFC with its microbial community enriched using domestic wastewater could be bioaugmented with a "H-C mixed culture" preacclimated to the cellulose substrate and thus had a high saccharification rate. It was reported that the MFC that was not bioaugmented performed poorly when fed with CSP, producing only a power density of $2 \pm 1\,mW\,m^{-2}$, but the bioaugmented one had a distinctively better performance, producing up to $300\,mW\,m^{-2}$ (Wang et al., 2009) (Table 1). The latter could remove up to 42% cellulose, 17% hemicellulose, and 4% lignin in the CSP. It performed even better with corn stover residual solids resulted from steam explosion pretreatment of CSP, removing up to 60% cellulose and 11% lignin, but recovering 20% less charge from the substrate. The higher performance was believed to be attributed to the activity of the "H-C culture," which might contain the microorganisms that can degrade the cellulosic substrates very efficiently, producing abundant soluble fermentation products that can be easily utilized by electroactive bacteria. Thus the bioaugmentation of MFCs can be a technical measure to be considered when dealing with lignocellulosic biomass, which is in general insoluble and recalcitrant. However, the system still required further improvements in terms of energy recovery to become more comparable to available technologies (Wang et al., 2009). Nevertheless, the prospect of using MFCs to treat corn stover waste, even its solid form, is evident.

To achieve a high energy recovery from corn stover waste, *combining biorefinery and bioelectrochemical conversion* is a good option. Borole et al. (2013) combined a biorefinery system with an MFC system to treat corn stover waste, with a goal of using the MFC to remove byproducts and residual sugars in the biorefinery stream. Accordingly, the corn stover waste was pretreated with strong acids and at high temperatures in a continuous-horizontal-screw reactor. The obtained product (in a slurry-like form) was treated with cellulase for 4 days before its pH was adjusted, and nutrients were added to allow fermentation with the glucose-xylose fermenting bacterium *Zymomonas mobiliz* to occur. The products of that biorefinery process, including residual sugars, acetate, sugar-degradation products, and many phenolic compounds, were subsequently fed to a single-chamber MFC inoculated with mixed cultures preenriched in existing MFCs. The power generation of the MFC could reach high levels when the corn refinery substrate ratio in the anode influent was low (from 1% to 8%) (Borole et al., 2013). The peak power density of the system could reach $1200\,mW\,m^{-2}$ at the substrate ratio of 8% (Table 1). However, the power density of the MFC decreased rapidly when the substrate loading rate increased, and became very low if the substrate contained 100% corn refinery stream. Interestingly, the authors found that potential fermentation inhibitors such as furfural and other phenolic compounds were removed simultaneously with the sugars. Therefore the poor performance of the MFC at high loading rates were believed to be due to massive biomass-mass transfer limitation and particularly the competition from excessive nonelectroactive microorganisms present in the stream. Nonetheless, the study suggests that such a combinatorial approach can improve the energy efficiencies of both systems: (i) the energy recovered by the MFC can somehow compensate for the energy lost for extensive cooling in biorefinery; (ii) the biorefinery process can improve the degradability of the substrates for use in the MFC, which is generally low and leads to low energy efficiency of the MFC itself.

Treating wheat straw lignocellulosic wastes

Wheat straw is one of the most abundant renewable resources and can be easily hydrolyzed. Thus energy recovery from wheat straw, especially by using MFCs to overcome the hurdles of other available technologies, has been of particular interest. Zhang et al. (2009) investigated the possibility of producing electricity by the *combination of heat liquefaction and MFC treatment* of wheat straw. Accordingly, wheat straw waste was pretreated with heat in a three-step reactor, in which the waste was first soaked at 80°C, then its hemicelluloses extracted at 170–180°C, and finally, its celluloses hydrolyzed by enzymes at 195°C with a water countercurrent (Zhang et al., 2009). Such a pretreatment was very efficient in producing a hydrolysate that was rich in sugars, which is a perfect substrate for MFCs. A two-chamber MFC system operated with this hydrolysate, a carbon paper anode, and a Pt-coated carbon paper cathode in a buffered ferricyanide catholyte could produce a stable power density of $79.6\,mW\,m^{-2}$ and a maximum power density of $124\,mW\,m^{-2}$ (with $1000\,mg\,L^{-1}$ COD loaded to the anode) (Zhang et al., 2009) (Table 1). Altogether, this study proved the feasibility of producing electricity from wheat straw waste by using MFCs, although much work remained to be done to achieve high power outputs.

Treating rice mill wastewater

Rice-associated wastes are abundant in rice-producing countries. They also represent a significant fraction of lignocellulosic biomass wastes that can be recycled, especially for energy recovery. Although there have been a number of studies on energy recovery from rice-associated wastes, few of them investigating the use of BESs, or specifically MFCs. A noteworthy report by Behera et al. (2010) presented the use of a cost-effective MFC system for the treatment and energy recovery of rice mill wastewater (WW). This rich energy source could be harvested by feeding the wastewater to the anode chamber of the MFC, which in this study is simply an earthen pot placed in a bigger plastic bucket functioning as the cathode chamber. In this case, the authors *use the earthen wall of the pot to replace the PEM* usually used in other MFC systems, thereby creating a very low-cost MFC. Such an MFC system, operated with a stainless steel anode, a graphite plate cathode, and with either oxygen or permanganate as the oxidant, could function even slightly better than the one having a PEM when treating rice mill WW. For instance, the authors reported that at the optimal pH of 8, the former could remove *c*. 96% COD of the undiluted WW and 93% of COD of the diluted WW, while the latter could only do 93% and 85% (Behera et al., 2010) (Table 1). Similarly, the former could remove lignin and phenol more efficiently than the latter (*c*. 84% and 81% vs. 79% and 77%, respectively), which is an impressive performance not observed in other systems, e.g., the one treating corn stover (Zuo et al., 2006). In terms of electricity generation, the earthen pot MFC produced power densities that were 11-fold, 7.6-fold, and 4.4-fold higher than those produced by the PEM MFC when they were operated at pHs 6, 7, and 8, respectively. At the optimal pH of 8, the former generated *c*. 49 mW m^{-2} or 2310 mW m^{-3} (unit volume) vs. 15.6 mW m^{-2} or 528 mW m^{-3} (unit volume) produced by the latter. Thus the earthen pot wall appeared to be comparable to or better than the PEM in cation-exchanging performance while the cost of the former is much lower. Therefore the Behera et al.'s (2010) study demonstrated a great potential of using an inexpensive MFC platform for energy recovery from rice mill WW treatment, which is very suitable to be applied in rice-producing developing countries. Nevertheless, the power output level of the system is still low, and thus, further improvements are required. A potential improvement is to address the cathode reaction by replacing the oxidant, such as oxygen with permanganate, which increased the power generation by twofold, as reported by the authors.

3.2 Production of fuel gases

As direct electricity generation from agricultural wastes using BESs (MFCs) still has low output levels, another approach of using BESs to recover energy from wastes is to produce energy-storing chemicals. By supplying small amounts of additional energy to control the cathode potential in the desired range, one can produce fuel gases such as hydrogen and methane. Due to the catalytic activities of electroactive bacteria in the anode or in the cathode or in both, the additional energy input is actually much lower than the activation energy that is usually required for such energy upflow reactions to occur. In these cases, BESs are actually operated as microbial electrolysis cells (MECs) or microbial electrosynthesis cells (MESs). Agricultural wastes can be degraded in the anode reaction to provide energy for the formation of the fuel gases at the cathode, or these gases can be produced by the reduction of VFAs indirectly generated from agricultural wastes (Cerrillo, Viñas, & Bonmatí, 2017). CO_2, being actually a by-product released in various agricultural processes, can also be captured to produce the high-energy CH_4 instead of being wasted into the atmosphere causing the "greenhouse" effect. Thus to produce fuel gases by BESs is considered to be a prospective option for energy recovery.

3.2.1 Production of hydrogen

Hydrogen is a clean and renewable energy carrier, especially for transportation, when produced using sustainable energy sources such as biomass. Hydrogen has the highest combustion heat in the weight of all fuels (120 MJ/kg compared with 44 MJ/kg of gasoline) (Lalaurette et al., 2009). To date, hydrogen has been mainly produced via thermochemical processes that deplete significant amounts of fossil fuels. Thus bioproduction of hydrogen from biomass is gaining more and more interest. Multiple technological platforms for biohydrogen production are available, such as biophotolysis, photo-fermentation, and dark-fermentation (Lalaurette et al., 2009). Among them, dark-fermentation has higher hydrogen yields that can reach up to 3 mol H_2/mol glucose in practice, but this is only about 25% of the theoretical yield (Cheng & Logan, 2007). Fermentation end products can be further converted to hydrogen, but such a conversion requires a low hydrogen partial pressure to be maintained by hydrogen consumption by methanogens or sulfate-reducing bacteria (Schink, 1997; Wu, Jain, & Zeikus, 1994). In order to further recover hydrogen from fermentation when the concentrations of hydrogen are high, external energy must be supplied to make the reactions thermodynamically favorable. This can be achieved through the use of BESs.

In theory, hydrogen can be produced at a redox potential lower than $E° = -410$ mV vs. SHE at pH 7.0. Considering the fact that the open circuit anode potential of MFCs operated with acetate as the electron donor was around −300 mV vs. SHE, Liu, Grot, and Logan (2005) hypothesized that it might be feasible to produce hydrogen by applying a circuit voltage greater than 110 mV to the MFCs. This voltage is substantially lower than that needed for hydrogen derived from the electrolysis of water, which is theoretically 1210 mV vs. SHE at neutral pHs. Thus Liu et al. (2005) were the first to investigate the possibility of using a BES to produce hydrogen by exploiting voltage generated from the oxidation of acetate. They used two-chamber reactors with plain carbon cloth anodes receiving electrons from electroactive bacteria-oxidizing acetate and Pt-coated carbon paper cathodes on which hydrogen was produced. Indeed, the input voltage was >0.25 V, which is higher than the theoretical value of 0.11 V but still far lower than the voltage needed for water hydrolysis (1.21 V) (Liu et al., 2005). Thus the so-called *bioelectrochemically assisted microbial reactor (BEAMR)* is considered to be energy-efficient for hydrogen production. This is particularly convincing when taking into account that the system recovered more than 90% current as H_2, and yielded 2.9 mol H_2 mol^{-1} acetate while the maximum theoretical yield was 4 mol H_2 mol^{-1} acetate (Table 2). Similar findings were also reported by Rozendal, Hamelers, Euverink, Metz, and Buisman (2006) almost at the same time, but the cathodic hydrogen efficiency in their study was only 57%, probably due to cathode reaction overvoltage (Table 2). The group also proposed to designate the system by the term "microbial electrolysis cell" or MEC, which has been widely used since then (Rozendal et al., 2006).

Hydrogen production by MECs has its prominent characteristics compared with other technologies, such as the efficient conversion of substrates without the limit of thermodynamics and easy production control (Lu, Xing, & Ren, 2012). This implies a great potential of using MECs to efficiently produce hydrogen from diverse substrates, including various agricultural wastes. Indeed, hydrogen can be produced from cellulosic biomass or starch-processing wastewater directly by MECs or indirectly by the integration of MECs with dark fermentation.

TABLE 2 Different BES technical achievements in production of fuel gases from agriculture-associated wastes.

Fuel gas produced	Applied BES technical features	Waste type	Cathodic gas recovery (%)	Energy recovery (%)	Production rate (m^3 gas day^{-1} m^{-3} reactor volume)	Reference
Hydrogen	MEC: two-chamber; with AEM; NH_3-pretreated graphite granule anode with enriched bacteria preacclimated in a MFC (fed cellulose); Pt-coated carbon cloth cathode; applied $V = 0.2–0.8\,V$	Cellulose	68	63	0.11	(Cheng & Logan, 2007)
Hydrogen	Two-stage process integrating: (i) fermentation of acid-hydrolyzed corn stalk; and (ii) MEC: single-chamber: NH_3-pretreated graphite brush anode with a biofilm preenriched in a MFC, Pt-coated carbon cloth cathode, applied $V = 0.5\,V$	Corn stover waste	n.a.	230	1.00 ± 0.19	(Lalaurette et al., 2009)
Hydrogen	Two-stage process integrating: (i) fermentation of corn stover waste pretreated by dilute-acid hydrolysis; and (ii) single-chamber MEC with a special anode arrangement: single-chamber: two graphite felt anodes on both sides of a Pt-coated carbon cloth cathode in the middle, applied $V = 0.8\,V$	Corn stalk (pretreated by air drying, milling, sieving, diluting with H_2SO_4 (0.5 wt%) at 1:10 (*w/v*) and hydrolyzing at 121°C for 60 min before cooling and adjusting pH to 7)	88	n.a.	3.43 ± 0.12	(Li, Liang, Bai, Fan, & Hou, 2014)

Continued

TABLE 2 Different BES technical achievements in production of fuel gases from agriculture-associated wastes.—cont'd

Fuel gas produced	Applied BES technical features	Waste type	Cathodic gas recovery (%)	Energy recovery (%)	Production rate (m^3 gas day^{-1} m^{-3} reactor volume)	Reference
Hydrogen	Integrated system comprising: (i) a 4-L dark fermentor having a gas-impermeable rubber septum; (ii) 2–3 single-chamber MFCs connected in series with (iii) one 75-mL single-chamber MEC, all of them having graphite brush anodes and Pt-coated carbon cloth cathodes (applied $V > 0.435$ V generated by the MFCs)	Wood crumbs	n.a.	23	0.28	(Wang et al., 2011)
Methane	Electrofermentation: two-chamber with PEM; carbon plate cathode (working electrode) coated with *carbon fiber textiles* (CFT); abiotic carbon bar anode; applied $V = 0.8$ V	Artificial garbage slurry supplemented with rice straw	~90	>100	8.18	(Sasaki et al., 2013)
Methane	Electrofermentation: two-chamber with PEM; bottle-type cathode chamber containing a CFT-coated carbon plate cathode; a PEM bag (anode chamber) with the carbon plate anode inside; cathode potential poised at −0.8 V vs. Ag/AgCl	Tomato plant residues (pretreated by air-drying, shredding, and mixing with medium to 10%)	n.a.	n.a.	0.197	(Hirano & Matsumoto, 2018)

Methane	Electrofermentation: single-chamber; graphite rod electrodes; supplied $V = 40$ mV	Wastewater containing 5% wheat straw (air dried and pulverized)	n.a.	n.a.	0.201	(Prajapati & Singh, 2020b)
Methane	CO_2 reduction: single-chamber and two-chamber MECs; graphite fiber brush anode, multiple carbon cloth cathodes with biofilm; cathode potential poised at −0.2 to −1.2 V vs. Ag/AgCl	CO_2	96	n.a.	n.a. (656 mmolCH_4 day^{-1} m^{-2})	(Cheng, Xing, Call, & Logan, 2009)
Methane	CO_2 reduction: two-chamber bottle-type MECs; with PEM; carbon rod anode, carbon paper cathode; cathode potential poised at −0.65 to −0.9 V vs. Ag/AgCl	CO_2	>80	n.a.	n.a. (400 mmolCH_4 day^{-1} m^{-2} at −0.9 V vs. Ag/AgCl)	(Villano et al., 2010)
Methane	CO_2 reduction: two-chamber bottle-type MECs; with PEM; graphite granule anode with *Geobacter sulfurreducens* oxidizing acetate, graphite granule cathode; cathode potential poised at −0.85 vs. Ag/AgCl	CO_2	n.a.	30	0.018	(Villano, Monaco, Aulenta, & Majone, 2011)
Methane	CO_2 reduction: single-chamber bottle-type MEC; plain carbon felt electrodes; operated with *Methanobacterium* medium; cathode potential poised at −1.0 V vs. Ag/AgCl	CO_2	95	n.a.	n.a. (182 mmol CH_4 day^{-1} m^{-2})	(Kobayashi et al., 2013)

Continued

TABLE 2 Different BES technical achievements in production of fuel gases from agriculture-associated wastes.—cont'd

Fuel gas produced	Applied BES technical features	Waste type	Cathodic gas recovery (%)	Energy recovery (%)	Production rate (m^3 gas day^{-1} m^{-3} reactor volume)	Reference
Methane	CO_2 reduction: two-chamber MRMCs (microbial reverse electrodialysis methanogenic cells); with reverse electrodialysis (RED) containing 6 CEMs and 6 AEMs providing external voltage; bioanode based on heat-treated graphite brush; Pt-coated stainless steel cathode	CO_2	75 ± 2	n.a.	n.a. ($\sim 2\,mmol\,L^{-1}\,day^{-1}$)	(Luo et al., 2014)
Methane	CO_2 reduction: two-chamber MRMCs (microbial reverse electrodialysis methanogenic cells); with reverse electrodialysis (RED) containing 6 CEMs and 6 AEMs providing external voltage; bioanode based on heat-treated graphite brush; carbon black layer-coated carbon cloth cathode	CO_2	69 ± 3	n.a.	n.a. ($\sim 1.25\,mmol\,L^{-1}\,day^{-1}$)	(Luo et al., 2014)

Methane	CO_2 reduction: three-chamber MECs with CEM; two cathode chambers sharing one anode chamber; Pt foil anode; heat-treated stainless steel felt; cathode potential poised at −1.3 V vs. Ag/AgCl	CO_2	60.8	21.9	n.a. (7.2 L CH_4 day^{-1} m^{-2})	(Liu, Zheng, Buisman, & ter Heijne, 2017)
Methane	CO_2 reduction: two-chamber MECs with CEM; bioanode based on carbon felt and oxidizing sulfide, Pt-coated carbon felt cathode; powered with two similar sulfide-oxidizing MFCs	CO_2	51	n.a.	0.0085 (0.354 mL CH_4 h^{-1} L^{-1})	(Jiang, Su, & Li, 2014)
Methane	CO_2 reduction: two-chamber MECs with bipolar membrane; TiO_2-based photoanode, Ti-wired carbon cloth cathode with biofilm preenriched from existing methane-producing BESs; $R = 1\ \Omega$	CO_2	95.2 ± 1.8	0.1	n.a. (1.92 ± 0.04 L day^{-1} m^{-2} (illuminated photoanode area))	(Fu et al., 2018)
Methane	CO_2 reduction: two-chamber MECs with CEM; Pt foil anode (counter), Pt-wired carbon felt cathode (working); cathode potential poised at −0.7 vs. Ag/AgCl	Gaseous CO_2	22	n.a.	n.a.	(Schlager et al., 2017)

Notes: *MEC*, microbial electrolysis cell; *AEM*, anion exchange membrane; *CEM*, cation exchange membrane; *n.a.*, data not available.

Production of hydrogen directly from cellulosic biomass with MECs

Cellulosic biomass is typically a category of abundant agricultural waste. It is also a rich energy source readily accessible to microorganisms for recovery. Cheng and Logan (2007) were the first to try producing hydrogen directly in an MEC while degrading cellulose at the anode. They also used *a two-chamber system* (similar to that of Liu et al., 2005) but *with an anion exchange membrane* (AEM) (instead of a CEM). With such technical innovations, they proved that direct H_2 production from cellulose degradation was feasible. Their MEC could achieve an H_2 recovery of up to 68%, an energy recovery of 63%, and a production rate of $0.11\,m^3\,H^2\,day^{-1}\,m^{-3}$ (Table 2). These numbers are still low in comparison with those when producing hydrogen from acetate, but they will demonstrate the potential of producing H_2 in an MEC with cellulose as the anode substrate. Indeed, the production rate could be increased by increasing the hydrolysis rate of cellulosic particles to produce more VFAs, probably by enriching more cellulolytic microorganisms (Cheng & Logan, 2007). Another way is the intensive pretreatment (such as *hydrothermal liquefaction*) of cellulosic materials, combining with the use of special anode materials such as carbon nanotubes to create a large anode contact surface (Shen et al., 2016). The hydrogen production rate could be even further increased by using a single-chamber MEC, which was already shown to be more efficient in producing hydrogen from acetate (Call & Logan, 2008).

Production of hydrogen by integrating fermentation and MECs

Since the process efficiency and the rates of direct hydrogen production from cellulose were low compared to those achieved with single VFAs, Lalaurette et al. (2009) reasoned that fermentation of cellulose and electrohydrogenesis should be carried out and optimized in separate reactors to achieve a more efficient process. In their study, corn stover waste, which is rich in cellulosic contents, was pretreated by dilute-acid hydrolysis before fermented by *Clostridium thermocellum*, and the fermentation product was fed to the anode of an MEC. The MEC used was a single-chamber type as used by Call and Logan (2008), which had an ammonia-pretreated graphite brush anode with a biofilm pre-enriched in an MFC on one side and a Pt-coated carbon cloth cathode on the opposite side. When supplied with 0.5 V voltage, such a system could generate hydrogen efficiently with corn stover fermentation effluent at a rate ($1.00 \pm 0.19\,m^{-3}\,H_2\,day^{-1}\,m^{-3}$ reactor volume) comparable to that with cellobiose ($0.96 \pm 0.16\,m^{-3}\,H_2\,day^{-1}\,m^{-3}$ reactor volume) (Lalaurette et al., 2009) (Table 2). It was also encouraging that the gas produced from corn stover fermentation effluent contained up to 69% of H_2 and less methane than those produced from cellobiose. Combining fermentation and MEC, the energy efficiency could reach 220%–230%, which showed a very efficient production of H_2 considering the amount of energy invested (Lalaurette et al., 2009).

Furthermore, the hydrogen production rate when using corn stover fermentation effluent increased by a factor 10 compared to when using cellulose particles. Altogether, the results of this study demonstrated that *a two-stage process combining fermentation and MEC* is more efficient for hydrogen production than each single process alone.

Hydrogen production in such a two-stage process could be even at higher rates by further improvements of both the fermentor and the MEC. Li et al. (2014) could produce more than $3\,m^3\,H_2\,m^{-3}\,day^{-1}$ by using a *stirred anaerobic bioreactor* for fermenting acid-hydrolyzed corn stalk and feeding the products to an *MEC with a special anode arrangement.* The latter is a cubic-shaped single-chambered reactor having two graphite felt anodes on both sides of a Pt-coated carbon cloth cathode in the middle (Liang et al., 2011). Such an electrode arrangement and the inoculation of the anode using a hydrogen fermentation effluent could result in an MEC that produced hydrogen at a rate of $1.73\,m^3\,H_2\,m^{-3}\,day^{-1}$, a yield of $129.8\,mL\,H_2\,g^{-1}$ corn stalk, and a purity of 46.3%–54.2% (*v*/v) (Li et al., 2014). Integrating both fermentation and MEC processes, the hydrogen production rate was calculated to be a record value of $3.43 \pm 0.12\,m^3\,m^{-3}\,day^{-1}$, the yield reached $387.1\,mL\,H_2\,g^{-1}$ corn stalk, and the hydrogen recovery could achieve 88% (Li et al., 2014) (Table 2). These impressive figures confirmed the advantages of a two-state strategy combining fermentation and MEC technology in hydrogen production from cellulosic biomass.

The urge for hydrogen production from cellulosic biomass with less energy investment even led to the idea of *integrating dark fermentation with MECs as well as MFCs* (Wang et al., 2011). MFCs can be used to provide additional energy for MEC(s) instead of using an external power supply. In such a way, renewable energy from waste is more thoroughly leveraged. Thus in their study, Wang et al. (2011) used several MFCs connected in series to supply additional voltage to an MEC producing hydrogen, while all the MFCs and the MEC were fed with the effluent from a cellulose-fermenting reactor. The serially connected MFCs-MEC system could produce $0.28\,m^3\,H_2\,m^{-3}\,day^{-1}$, $9.8\,mmol\,H_2\,g^{-1}$ COD, which were 18 times higher than those achieved with a previously reported MFC-MEC system operated with acetate (Sun et al., 2008) (Table 2). The hypothesis here is that the better performance was due to the larger volume of MEC and the optimal number of MFCs, which was 2, used by Wang et al. (2011). The integrated fermentation-MEC-MFC process could yield $14.3\,mmol\,H_2\,g^{-1}$ cellulose, which was 41% more than fermentation alone and reach an energy efficiency of 23% (vs. 16% of fermentation alone) (Table 2). Therefore this integrated approach seems to be an energy-efficient solution for high-rate hydrogen production from cellulosic biomass. However, COD regeneration reducing hydrogen yield and the decreasing voltage of the MFCs over time are the remaining issues to consider about this approach.

3.2.2 Production of methane

Methane is a competitive energy compound and is usually produced biologically by anaerobic digestion (Pham et al., 2006). Compared to hydrogen, methane contains less energy per mass unit, but the production of methane also requires less energy (Clauwaert & Verstraete, 2009). In theory, there are two main biochemical pathways of methanogenesis: (i) hydrogenotrophic methanogenesis where hydrogen produced in fermentation is consumed to form methane according to the following equation: $CO_2 + 4H_2 \rightarrow CH_4 + 2H_2O$ ($\Delta G° = -137$ kJ); and (ii) acetoclastic methanogenesis where acetate, the usually final product in fermentation acidogenesis, is converted to methane according to the following equation: $CH_3COO^- + H^+ \rightarrow CH_4 + CO_2$ ($\Delta G° = -31$ kJ) (Madigan, Martinko, & Parker, 2004). Methanogenesis allows further conversions of fermentation products, especially VFAs, and thus results in complete energy recovery. However, this process in an anaerobic digester usually fails when the substrate concentrations are low (Pham et al., 2006). The use of a BES (MEC or MES) can fill this gap (Clauwaert & Verstraete, 2009). BESs can produce methane in essentially similar pathways but with the aid of an electrical current, i.e., either via electrofermentation consuming acetate or via CO_2 reduction. Using a BES, less voltage investment is required to produce methane than to produce hydrogen, meaning also a lower capital cost (Clauwaert & Verstraete, 2009).

Production of methane via electrofermentation

The term "electrofermentation" was first used by Rabaey and Rozendal (2010). In electrofermentation, redox potentials of the environment are controlled to drive fermentation towards a desired direction and obtain high yields of target products via electrochemical reactions on electrodes (Moscoviz, Toledo-Alarcón, Trably, & Bernet, 2016). One of the first studies that applied the electrofermentation principle to produce methane from agriculture-associated waste is the work of Sasaki et al. (2013) with artificial garbage slurry (AGS) supplemented with rice straw. The unique technical feature used in their BES setting is the use of *carbon fiber textiles* (CFT) to coat the carbon plate cathode (working electrode) to increase the cathode reaction rate through increasing cathode contact surface. A two-chamber BES consisting of such a cathode, an abiotic carbon bar anode, and a PEM to separate the two chambers was operated at 55°C with AGS to produce methane while the cathode was poised at −0.8 V. The system could produce methane at a rate of up to *c.* 8179 mL gas L^{-1} day^{-1} (equivalent to 8.18 m^3 m^{-3} day^{-1}) (with the gas containing *c.* 54%–55% CH_4) even at high organic loading rates (Sasaki et al., 2013) (Table 2). Although such production rate was significantly higher than that when the cathode potential was not regulated, the authors found that most of CH_4 was produced not from direct electrical COD reduction. Another noteworthy point about this system is that its "equivalent" power output (372.6 kJ L^{-1} day^{-1}) was hundreds of times higher than the power input (0.6 kJ L^{-1} day^{-1}). Taking advantage of this system, Hirano and Matsumoto (2018) created an innovated reactor design having a large bottle-type cathode chamber containing a CFT-coated carbon plate cathode and a PEM bag with the counter electrode (anode) inside. Such a system, with the cathode potential poised at −0.8 V, was used to electroferment tomato plant residues after inoculated with an anaerobic digester sludge. It could remove 57.3 ± 3.6% cellulose in the residues, produced 197.2 ± 11.0 mL gas L^{-1} day^{-1}, and accumulated very little acetate, while the control reactor not having its cathode potential regulated only removed 53.1 ± 4.1% cellulose, produced only 180.9 ± 13.4 mL gas L^{-1} day^{-1}, and still accumulated some acetate (Hirano & Matsumoto, 2018) (Table 2). It was also calculated that 11 times larger energy amount and 2.5 times more cost benefit could be acquired with the electrofermentative BES with CFT compared to mere fermentation by a continuous stirred-tank reactor.

The electrofermentation principle can be applied to generate methane even from solid agricultural wastes such as wheat straw without the need of pretreatment (Prajapati & Singh, 2020b). A *membraneless bottle-type single-chamber reactor* could be used for this purpose. The reactor simply containing graphite rod electrodes and operated anaerobically with wastewater loaded with 5% air-dried and pulverized wheat straw could generate methane at various supplied voltages of 20, 40, 80, and 120 mV. Surprisingly it was found that the highest CH_4 yield (175.17 mL g^{-1} COD removed) and the highest CH_4 production rate of 401.98 ± 61.53 mL day^{-1} (equivalent to about 0.201 m^3 day^{-1} m^{-3} reactor volume) could be obtained with the supplied voltage as low as 40 mV (Prajapati & Singh, 2020b) (Table 2). In addition, more soluble COD was removed and more VFAs were produced under that condition. It was hypothesized that the applied voltage of 40 mV might create optimal redox conditions for methanogens to grow and degrade more COD contents. The authors even applied this approach for the co-digestion of a "fully solid" mixture of sewage sludge, food waste (cooked food), and agricultural waste (pulverized wheat straw) (Prajapati & Singh, 2020a). In this case, when varying the supplied voltage, the authors only obtained methane with 120 mV, while with the lower voltages, hydrogen was the main biogas product. Fewer types of VFAs produced and higher cellulase activities when additional voltages were applied suggested that bioelectrolysis aided hydrolysis of substrates and might also reduce inhibitory effects, like in some other reports (Borole et al., 2013). Hence, in general bioelectrocatalysis can enhance the degradation efficiency as well as the biogas production of fermentation but the process thermodynamics seems to be determined not only by the redox conditions but also the nature of the substrates.

Production of methane via only the reduction of carbon dioxide

Capturing carbon dioxide to produce a fuel gas such as methane is of particular interest to reduce the CO_2 emission from agricultural wastes and simultaneously recover resources from them. It was hypothesized that bioelectrochemical methane production from CO_2 can be either through: (i) a direct electrode reaction according to the following equation: $CO_2 + 8H^+ + 8e^- \rightarrow CH_4 + 2H_2O$ ($E_{cathode} = -0.24$ V vs. SHE); or (ii) hydrogenotrophic methanogenesis (as presented earlier) from hydrogen abiotically or biotically produced at an electrode (Van Eerten-Jansen et al., 2012). Hence, all technological attempts to produce methane from CO_2 reduction have been thus far developed and optimized by experimenting around those processes.

Cheng et al. (2009) were the first to produce methane from direct CO_2 reduction on the electrode surface without the need for precious metal catalysts. They used both *single-chamber MECs and two-chamber MECs,* each containing a graphite fiber brush anode and several carbon cloth cathodes having biofilms pre-enriched in existing MECs. The single-chamber system could produce methane at a rate consistent with bioelectricity generation when fed only with CO_2 as the carbon source and electron acceptor. This was evident as no CH_4 was produced when using abiotic cathodes even at the applied cathode potential of -1.0 V. Direct methane production from the cathode CO_2 reduction was evident in the two-chamber system at a rate up to 656 mmol-CH_4 day^{-1} m^{-2} when the cathode potential was set to -1.2 V vs. Ag/AgCl (Cheng et al., 2009) (Table 2). The system even achieved a 96% current capture efficiency. On the other hand, Villano et al. (2010) proved that hydrogen was produced less and methane was produced more when the cathode potential became more negative than -0.7 V vs. Ag/AgCl, suggesting that methane was partially formed from hydrogen. Similarly, Kobayashi et al. (2013) found that the applied potential ranging from -0.75 to -1.25 V (vs. Ag/AgCl) was optimal for the production of methane. That is, methane generation from CO_2 can still be explained by both direct electron transfer and interspecies hydrogenotrophic methanogenesis (Kobayashi et al., 2013; Villano et al., 2010). The latter was even believed to be the main mechanism in some systems (Batlle-Vilanova et al., 2015). Thus to have a system with a stable performance, Villano et al. (2011) tried enriching *a methanogenic biofilm* on the cathode surface of a two-chamber MEC having also an *acetate-oxidizing bioanode* governed by *Geobacter sulfulreducens*. This approach seemed to be efficient in creating a well-performing cathode biofilm (producing 0.018 m^3 CH_4 day^{-1} m^{-3}) (Table 2) and a BES requiring less energy input (Villano et al., 2011). However, it seemed that in this case the methane production rate could be limited by acetate oxidation kinetics, which is another factor that needs optimizing. *Controlling cathode pH* was proposed as an improvement that could actually increase methane production and reduce hydrogen production (Samarakoon, Dinamarca, Nelabhotla, Winkler, & Bakke, 2020). Indeed, applying this measure could lead to a current-to-methane recovery of $79 \pm 2\%$ and an energy efficiency of 75% (Table 2), but the system performance was still limited by acetate oxidation at the anode (Villano, Scardala, Aulenta, & Majone, 2013).

As the anode reaction may limit the overall performance of a methane-producing MEC, other technical solutions were sought to save the energy supplied to the system. Some of those addressed the membrane used in MECs. A promising solution is the use of *reverse electrodialysis (RED)* that generates a redox potential difference between two solutions of different salinities created by using waste heat (Luo et al., 2014). With this measure, heat generated from many industrial or agricultural processes can be recycled to power electromethanogenesis, instead of being wasted. Instead of having a membrane, an MEC could be operated with a RED stack containing several pairs of CEM and AEM that created chambers fed alternately with high concentration (HC) and low concentration (LC) thermolytic solutions. In the study by Luo et al. (2014), the salt solutions were a thermolytic 1.7 M NH_4HCO_3 solution (the HC), and an LC solution obtained from diluting the HC to a salinity of 75. When a modified two-chamber type reactor with such a RED stack was operated with a Pt-coated stainless steel cathode, a $75 \pm 2\%$ Coulombic recovery could be obtained (Table 2). Thus a RED stack appeared to work well to supply energy for electromethanogenesis. However, it is not always feasible to apply RED in a BES due to its complexity and the requirement of heat supply to create the salinity solutions. Therefore *careful selection of the membrane type* to be used in methane-producing BESs can be another technical measure to enhance methane production while investing less energy. Babanova et al. (2016) tested the replacement of the Nafion-117 CEM with the Ultrex CMI-7000 CEM and observed significant improvements in the performance of a two-chamber glass reactor BES producing methane from CO_2 reduction. It was interesting that the resistance of the Ultrex CMI-7000 was only half of that of the Nafion-117, leading to a respectively twofold lower internal resistance of the BES operated with the former compared to that of the BES operated with the latter. This also led to five- to sixfold more current consumption (with respect to Coulombic and energy efficiencies), larger membrane-cross pH gradient and less methane production of the BES with Nafion-117 compared to those of the BES with Ultrex CMI-7000 (Babanova et al., 2016). Those findings suggest that the type of the membrane can affect the performance of methane-producing BES and should be a technical feature to consider in technology development.

Next to the membrane, cathode material might be another factor to address for improving the performance of methane-producing BESs. Materials with improved biocompatibility (to allow better growths of methanogenic biofilms) and low cost are desired. Indeed, while testing the RED stack, Luo et al. (2014) also tested different types of cathode materials and found that the highest methane production rate was achieved with the Pt-coated stainless steel cathode, followed by the carbon cloth coated with a carbon black layer and the heat-treated graphite brush. The Coulombic recoveries obtained with these cathodes were $75 \pm 2\%$, $69 \pm 3\%$, and $36 \pm 2\%$, respectively (Table 2). Thus their system performed the best with Pt-coated stainless steel cathode but when considering the cost/performance balance, the *carbon cloth cathode coated with a carbon black layer* would be preferred, especially for practical applications (Luo et al., 2014). In another study, *heat-treated stainless steel felt (HSSF)* was shown to be an innovative cathode material with excellent performance and reasonable cost (Liu et al., 2017). A three-chamber MEC (with two cathode chambers sharing one anode chamber), when operated with HSSF cathode, could produce methane at the rates comparable to those when operated with graphite felt cathode (e.g., at 1.0 L methane day^{-1} m^{-2} (cathode projection area) when the cathodes poised at −1.1 V vs. Ag/AgCl). Considering that HSSF has a higher durability, a better electrocatalytic activity for hydrogen evolution (leading to a faster start-up of the biocathode), and also a reasonable cost in comparison with graphite felt, HSSF is recommended as a good cathode material candidate used in methane-producing BESs (Liu et al., 2017).

In order to save more energy investment and also make bioelectromethanogenesis from CO_2 more sustainable, another approach is to combine it with other bioanode processes for waste treatment and to use MFCs or solar cells to power the methane-producing MEC. A striking example is the work by Jiang et al. (2014) in which a *sulfide-oxidizing bioanode* was integrated into a methane-producing MEC and two sulfide-oxidizing MFCs were used to supply an additional voltage for the MEC. The *MFCs-MEC coupled system* could produce up to 0.354 mL CH_4 h^{-1} L^{-1} (or 0.0085 m^3 day^{-1} m^{-3} reactor volume) with a 51% Coulombic efficiency (Jiang et al., 2014) (Table 2), which is considerable and demonstrates the potential of this approach. To further increase the Coulombic efficiency of bioelectromethanogenesis and save nonrenewable energy, Fu et al. (2018) even proposed an innovative *hybrid system having a solar cell anode and a biocathode* producing methane from CO_2. The system had two chambers separated by a bipolar membrane, including the cathode chamber harboring a biocathode preacclimated in existing methane-producing BESs and the anode chamber harboring a TiO_2-based photoanode working in a 0.2-M KOH anolyte. It could generate sufficient power to maintain a constant cathode potential at −0.24 V vs. SHE, allowing the production of methane at a rate of 1.92 ± 0.04 L day^{-1} m^{-2} (illuminated photoanode area) and achieving a Coulombic efficiency of $95.2 \pm 1.8\%$, which was the highest to that date (Fu et al., 2018) (Table 2). Notably, the solarenergy-to-fuel conversion efficiency of the system was around 0.1%, which was approximately half of that of natural photosynthesis. Thus this so-called artificial photosynthesis system actually allows a direct use of solar energy to reduce CO_2 and thereby enabling a sustainable method of bioelectromethanogenesis.

As discussed earlier, an optimal range of applied cathode potentials for methane production can be determined. However, BESs operated with cathode potentials within such an optimal range still encounter the issue of low product selectivity. Jiang et al. (2013) tested a CO_2-reducing two-chamber BES operated with cathode potentials scanned from −0.85 V to −1.15 V (vs. Ag/AgCl) and discovered that the *set potential determined the products*. More specifically, the cathode potentials set from −0.85 to −0.95 V (vs. Ag/AgCl) led to the production of only methane and hydrogen, but those less negative than −0.95 V (vs. Ag/AgCl) led to the production of methane, hydrogen, and also acetate (Jiang et al., 2013). Moreover, the value of the potential could even determine the production rate. It was found that the cathode potentials at −1.0 V and − 1.1 V (vs. Ag/AgCl) could respectively result in the methane production rates of 63.6 mL day^{-1} and 101.9 mL day^{-1}, which were 6.6 and 13.9 times, respectively, as high as the previous results reported for similar systems (Jiang et al., 2013) (Table 2). Thus the value of the cathode potential to be applied should be a key technical feature to be considered when one wants to enhance the rate of bioelectrochemical methane production from CO_2. Indeed, it was found that the start-up potential also determined the electron transfer behavior and therefore the establishment of the microbial community of the biocathode (Jiang et al., 2013; Li et al., 2020; Xu et al., 2017). All these findings strongly supported the hypothesis that start-up potential is a key factor in determining the composition of a biocathode as well as its function.

An approach one may consider enhancing the performance of bioelectro-methanogenesis is to *improve the procedure of enrichment* of the working microbial community on the biocathode. Baek, Kim, Lee, and Lee (2017) tried enriching a methanogenesis biocathode "from scratch" instead of using a pre-acclimated one as in other studies. Their study pointed out that the subcultured biocathodes performed better than the initially enriched one (Baek et al., 2017), which actually supports the use of preacclimated biocathodes to start up BESs. Thus innovating the procedure of enrichment of biocathodic microorganisms is one way to improve bioelectrochemical methane production.

4 BESs in upgrading agricultural wastes to valuable products

Similar to methane, various organic chemicals can be produced in microbial electrosynthesis cells (MESs), depending on the technical features of the systems and the applied cathode potentials. There are also two main routes to bioelectrochemically synthesize valuable products from agricultural wastes: (i) reduction of VFAs indirectly generated from agricultural wastes; or (ii) reduction of CO_2 to sequester CO_2 emitted from agricultural activities (e.g., waste treatment/degradation, ruminous activities, etc.). It is hypothesized that the latter was executed by chemoautotrophic microorganisms through Wood-Ljungdahl (WL), reverse tricarboxylic acid, and variants of hydroxypropionate/hydroxybutyrate pathways (Bian et al., 2020). The products of microbial electrosynthesis (or "bioelectrosynthesis" as generalized by many researchers) can be commodity chemicals such as acetate or ethanol or other compounds with longer chain molecules such as butyrate, butanol, caproate, caprylate, etc. They can serve as precursory compounds that can be further upgraded (biochemically or chemically) to longer chain carboxylates, biofuels, bioplastics, polysaccharides, and proteins (Chu et al., 2020). Bioelectrosynthesis is actually attractive in industry because it can be operated under mild reaction conditions (e.g., ambient temperature and pressure) (Schlager et al., 2017). Furthermore, biocatalysts can be self-regenerated and are therefore highly suitable for long-term operations of BESs (Schlager et al., 2017).

4.1 Production of acetate

Production of acetate by BESs has been probably studied the most in comparison with the productions of other products. It should be noted that acetate is a commodity chemical that is widely used in various industries. In theory, electrochemical synthesis of acetate from hydrogen and carbon dioxide is thermodynamically not practical without substantial external energy supply and/or the use of expensive catalysts (Aulenta, Reale, Catervi, Panero, & Majone, 2008). The use of an electrode feeding electrons to acetogens to produce acetate by CO_2 reduction is, on the other hand, practically feasible and thus appears to be promising (Nevin, Woodard, Franks, Summers, & Lovley, 2010). Such production of acetate can be carried out by using a single culture or a mixed culture at the cathode, with various kinds of cathode materials, which may actually affect the production rate, the product titer as well as electron recovery (LaBelle, Marshall, Gilbert, & May, 2014; Marshall, Ross, Fichot, Norman, & May, 2012; Marshall, Ross, Fichot, Norman, & May, 2013; Nevin et al., 2010; Nevin et al., 2011).

4.1.1 Enhancing acetate production in BESs

Several approaches have been proposed to enhance acetate production in BESs by focusing on improving the biocathodes in the systems. LaBelle et al. (2014) tested *extended incubation and enrichment* of acetogens in a BES with a continuous supply of CO_2 and achieved a production rate ranging from *c.* 1 to $3\,g\,L_{catholyte}^{-1}\,day^{-1}$, with applied cathode potentials ranging from -0.6 to $-0.8\,V$ vs. SHE and proper pH controls. Sealing the cathode chamber could further increase the rate to $10.8\,g\,m^{-2}\,day^{-1}$ (LaBelle et al., 2014) (Table 3). In some other studies, Jourdin et al. (2018) proposed a *forced flow-through mode* of operation to create *a thick biofilm* on the carbon felt cathode poised at $-0.85\,V$ vs. SHE in a MES. The system could produce acetate at a rate of up to $9.85 \pm 0.65\,g\,L^{-1}\,day^{-1}$ ($371 \pm 5\,g\,m^{-2}\,day^{-1}$) and with a 46.2% electron-to-acetate recovery, although it also produced compounds other than acetate (Jourdin et al., 2018) (Table 3).

The bioelectrochemical production of acetate can be enhanced by *lowering the applied cathode potential*. For instance, the production rate and titer by *C. ljungdahlii* increased 10-fold when the cathode potential was lowered to $-0.695\,V$ (vs. SHE) (Bajracharya et al., 2015) (Table 3). At a similar potential of $-0.69\,V$ (vs. SHE), the acetogen *Acetobacterium woodii* could produce acetate at a high rate of $12.9 \pm 5.4\,g\,m^{-2}\,day^{-1}$ ($536 \pm 226\,mg\,m^{-2}\,h^{-1}$), resulting in a maximum titer of *c.* $1.4\,g\,L^{-1}$, although it was previously reported

TABLE 3 Different BES technical achievements in production of value-added products from agriculture-associated wastes.

Valuable product	Applied BES technical features	Waste type	Cathodic electron recovery (%)	Projected cathode area-based production ($g m^{-2} day^{-1}$)	Production rate ($mM day^{-1}$)	Maximum product titer ($g L^{-1}$)	Reference
Acetate	MES: two-chamber; with *PEM, graphite stick electrodes; cathode poised at − 0.4 V* vs. *SHE, inoculated with H2-pregrown Sporomusa ovata*	CO_2 (in the form of HCO_3^-)	86 ± 21	7.3	n.a.	0.063	(Nevin et al., 2010)
Acetate	MES: two-chamber; with *PEM, Pt-welded Ti plate anode, stainless steel wired graphite felt cathode; cathode poised at − 0.695 V* vs. *SHE, inoculated with Clostridum ljungdahlii*	CO_2 (in the form of HCO_3^-)	40	7.51	2.4	0.6	(Bajracharya et al., 2015)
Acetate	MES: two-chamber; with *CEM*, TaO_2/IrO_2 (35/65%)-coated Ti mesh anode, *stainless steel felt cathode, cathode poised at −0.69 V* vs. *SHE, inoculated with Acetobacterium woodii*	CO_2 (in the form of HCO_3^-)	113 ± 48	12.9 ± 5.4	n.a.	~1.4	(Arends, 2013)

Continued

TABLE 3 Different BES technical achievements in production of value-added products from agriculture-associated wastes.—cont'd

Valuable product	Applied BES technical features	Waste type	Cathodic electron recovery (%)	Projected cathode area-based production ($g\,m^{-2}\,day^{-1}$)	Production rate ($mM\,day^{-1}$)	Maximum product titer ($g\,L^{-1}$)	Reference
Acetate	MES: two-chamber; with *CEM*, TaO_2/IrO_2 (35/65%)-coated Ti mesh anode, *carbon felt cathode, cathode poised at* -0.69 *V* vs. *SHE, inoculated with Acetobacterium woodii*	CO_2 (in the form of HCO_3^-)	69 ± 27	8.16 ± 3.14	n.a.	~0.8	(Arends, 2013)
Acetate	MES: two-chamber; with *PEM*, graphite stick anode, *carbon cloth cathode modified with chitosan, cathode poised at* -0.6 *V* vs. *Ag/AgCl, inoculated with Sporomusa ovata*	CO_2	86 ± 12	2.7 ± 0.7	n.a. ($229 \pm 56\,mM\,m^{-2}\,day^{-1}$)	n.a.	(Zhang et al., 2013)

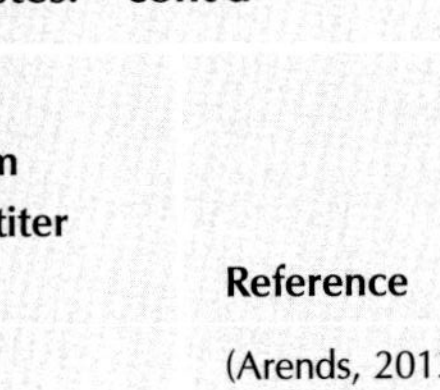
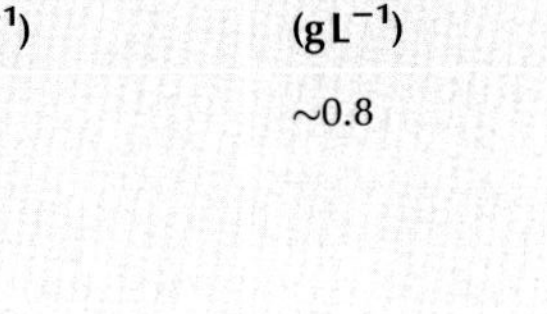
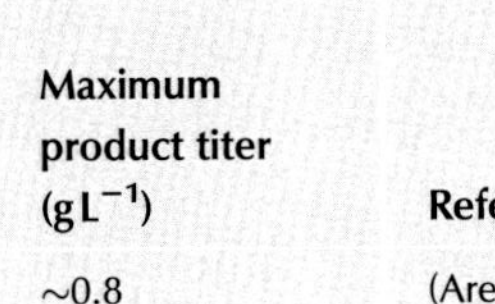

Acetate	MES: two-chamber; with *PEM,* graphite stick anode, *carbon cloth cathode modified with* cyanuric chloride, *cathode poised at −0.6 V* vs. *Ag/AgCl, inoculated with Sporomusa ovata*	CO_2	81 ± 16	2.4 ± 0.6	n.a. (205 ± 50 mM m^{-2} day^{-1})	n.a.	(Zhang et al., 2013)
Acetate	MES: two-chamber; with *PEM,* graphite stick anode, *carbon cloth cathode modified with Au-nanoparticles, cathode poised at −0.6 V* vs. *Ag/AgCl, inoculated with Sporomusa ovata*	CO_2	83 ± 14	2.14 ± 0.52	n.a. (181 ± 44 mM m^{-2} day^{-1})	n.a.	(Zhang et al., 2013)
Acetate	MES: two-chamber; with *PEM,* graphite stick anode, *carbon cloth cathode modified with carbon nanotube textile composite, cathode poised at −0.6 V* vs. *Ag/AgCl, inoculated with Sporomusa ovata*	CO_2	~83	~1.18	n.a. (~100 mM m^{-2} day^{-1})	n.a.	(Zhang et al., 2013)

Continued

TABLE 3 Different BES technical achievements in production of value-added products from agriculture-associated wastes.—cont'd

Valuable product	Applied BES technical features	Waste type	Cathodic electron recovery (%)	Projected cathode area-based production ($g m^{-2} day^{-1}$)	Production rate ($mM day^{-1}$)	Maximum product titer ($g L^{-1}$)	Reference
Acetate	MES: two-chamber; with *PEM*, graphite stick anode, *graphite felt cathode; cathode poised at* −*0.6 V* vs. *Ag/AgCl, inoculated with Sporomusa ovata*	CO_2	82 ± 14	16.6 ($0.068 g L^{-1} day^{-1}$)	n.a.	n.a.	(Nie et al., 2013)
Acetate	MES: two-chamber; with *PEM,* photoanode, *Si nanowired* TiO_2-*based cathode; with light, cathode poised at* −*0.595 V* vs. *Ag/AgCl; Sporomusa ovata enriched on cathode nanowires*	CO_2	90	n.a. ($0.24 g L^{-1} day^{-1}$)	n.a.	1.2	(Liu et al., 2015)
Acetate	MES: two-chamber; with *CEM*, graphite-rod-connected graphite granule electrodes; *cathode poised at* −*0.6 to* −*0.8 V* vs. *SHE; mixed culture enriched on cathode*	CO_2	n.a.	10.8 ($3.1 g L_{catholyte}^{-1} day^{-1}$)	15	n.a.	(LaBelle et al., 2014)

Acetate	MES: two-chamber; with CEM, Pt wire anode (counter), reticulated vitreous carbon cathode modified with multiwalled carbon nanotubes; *cathode poised at* −0.85 V vs. SHE; *mixed culture selectively enriched on cathode*	CO_2	100 ± 4	685 ± 30	n.a.	11	(Jourdin et al., 2015)
Acetate	MES: three-chamber; with the middle chamber between a CEM contacting the anolyte and an AEM contacting the catholyte; Ir oxide-coated titanium mesh anode, anolyte pH 2; stainless-steel-framed carbon felt cathode; *applied fixed current at* −50 mA; *mixed culture enriched on cathode*	CO_2	~60–70	20.4 (0.58 g L^{-1} day^{-1})	n.a.	13.5	(Gildemyn et al., 2015)

Continued

TABLE 3 Different BES technical achievements in production of value-added products from agriculture-associated wastes.—cont'd

Valuable product	Applied BES technical features	Waste type	Cathodic electron recovery (%)	Projected cathode area-based production ($g\,m^{-2}\,day^{-1}$)	Production rate ($mM\,day^{-1}$)	Maximum product titer ($g\,L^{-1}$)	Reference
Acetate	MES operated in forced flow-through mode: two-chamber; with CEM; Pt/IrO_2-coated Ti mesh anode, carbon felt cathode; *cathode poised at* −0.85 V vs. SHE; *mixed culture enriched on cathode*	CO_2	46.2	371 ± 5 ($9.85 \pm 0.65\,g\,L^{-1}\,day^{-1}$)	n.a.	8.2 ± 0.4	(Jourdin, Raes, Buisman, & Strik, 2018)
Acetate	MES operated in recirculation mode: three-chamber; with CEM; Pt-Ti anode; a VITO-Core *gas diffusion cathode* with a PTFE-based gas diffusion layer facing an additional chamber supplying gas and a catalyst layer facing the catholyte; *cathode poised at* −1.1 V vs. Ag/AgCl; *mixed culture enriched on cathode*	CO_2	35	36.6 ($61\,mg\,L^{-1}\,day^{-1}$)	n.a.	2.89	(Bajracharya, Vanbroekhoven, et al., 2016)

Ethanol	MES: flat plate two-chamber reactor: with AEM; graphite electrodes; 0.2 M hexacyanoferrate as anolyte, growth medium containing acetate as catholyte; catholyte supplemented with methyl viologen as mediator; cathode poised at −0.55 V vs. SHE; operated with mixed culture	Acetate	74.6 ± 6	0.017 (0.058 $g\,m^{-3}\,day^{-1}$)	n.a.	0.62 ± 0.03 (13.5 ± 0.7 mM)	(Steinbusch, Hamelers, Schaap, Kampman, & Buisman, 2010)
Ethanol	MES: H-type two-chamber; with PEM; graphite stick electrodes; growth medium DSMZ 311 as electrolytes; cathode poised at −0.69 V vs. SHE; tungstate added to catholyte 10-fold more than the default amount in the medium; operated with *S. ovata*	Acetate produced from CO_2	87.6 ± 6.5	0.18 ± 0.05 (4.0 ± 1.2 $mmol\,m^{-2}\,day^{-1}$)	0.048 ± 0.015	1.5 ± 0.5	(Ammam, Tremblay, Lizak, & Zhang, 2016)

Continued

TABLE 3 Different BES technical achievements in production of value-added products from agriculture-associated wastes.—cont'd

Valuable product	Applied BES technical features	Waste type	Cathodic electron recovery (%)	Projected cathode area-based production ($g m^{-2} day^{-1}$)	Production rate ($mM day^{-1}$)	Maximum product titer ($g L^{-1}$)	Reference
Isopropanol	MES: single-chamber; CoP_i stainless steel anode, Ni-Mo-Zn stainless steel cathode; applied voltage $\geq 2.7 V$ supplied by photovoltaic cell; operated with *Ralstonia eutropha* mutant *Re*2133-pEG12 engineered to produce isopropanol	CO_2	3.9	n.a.	n.a.	0.216 ± 0.017	(Torella et al., 2015)
Isopropanol (together with acetate and butyrate)	MES: two-chamber; with CEM; Ir-MMO-coated Ti-mesh anode, biotic carbon felt cathode operated in a modified homoacetogenic medium; continuous mode, HRT > 5d, pH < 5.5; applied a fixed current of $-5 A m^{-2}$	CO_2	n.a.	3.3	n.a.	0.82	(Arends, Patil, Roume, & Rabaey, 2017)

Isopropanol (together with acetate and butyrate)	MES: two-chamber; with CEM; Ir-MMO-coated Ti-mesh anode, biotic carbon felt cathode operated in a modified homoacetogenic medium; continuous mode, HRT=5d, pH=5.0; applied a fixed current of $-5\,A\,m^{-2}$	CO_2	3	1.17±0.34	n.a.	0.29±0.08	(Arends et al., 2017)
Butanol	MES electrofermentor: H-type, two-chamber; with PEM; graphite felt electrodes; fermentation medium (118 g glucose L^{-1}) as catholyte; cathode poised at −0.54 V vs. Ag/AgCl; operated with *C. pasteurianum* ATCC 6013	Sugars (from sugar-rich agricultural wastes or their fermented effluents)	182	n.a.	n.a.	13.3	(Khosravanipour Mostafazadeh et al., 2016)

Continued

TABLE 3 Different BES technical achievements in production of value-added products from agriculture-associated wastes.—cont'd

Valuable product	Applied BES technical features	Waste type	Cathodic electron recovery (%)	Projected cathode area-based production ($g\,m^{-2}\,day^{-1}$)	Production rate ($mM\,day^{-1}$)	Maximum product titer ($g\,L^{-1}$)	Reference
Butyrate (together with acetate, ethanol, and butanol)	MES: two-chamber; with CEM; Ti rod anode, biotic carbon cloth cathode operated in mineral medium supplemented with methanogen inhibitor; *cathode poised at* −0.8 V vs. SHE; inoculated with a mixed culture from a syngas fermentor	CO_2	n.a.	n.a.	4.1 mM C day^{-1} (for all products)	~20 mM C	(Ganigué, Puig, Batlle-Vilanova, Balaguer, & Colprim, 2015)
Butyrate	MES: tubular two-chamber reactor; with CEM; outer Ti-MMO anode, inner biotic carbon cloth cathode operated in mineral medium supplemented with methanogen inhibitor; *cathode poised at* −0.8 V vs. SHE; inoculated with a mixed culture from a syngas fermentor	CO_2	n.a.	n.a.	7.2 mM C day^{-1} (only butyrate)	87.5 mM C (concentrated to 252.4 ± 65.8 mM C with liquid membrane pretreated with dodecanol/dodecane)	(Batlle-Vilanova et al., 2017)

Butyrate (together with other products)	MES operated in *forced flow-through* mode: two-chamber; with CEM; Pt/ IrO_2-coated Ti mesh anode, carbon felt cathode; *cathode poised at* -0.85 V vs. SHE; *mixed culture enriched on cathode*	CO_2	28.6	120.6 ± 0.8 (3.2 ± 0.1 g L^{-1} day^{-1})	n.a.	2.2	(Jourdin et al., 2018)
Succinate/ succinic acid	MES: two-chamber; with CEM; carbon felt electrodes; *cathode poised at* -0.8 V vs. Ag/AgCl; operated with *Actinobacillus succinogenes* NJ113 as a biocathode catalyst	Corncob hydrolysate (not pretreated)	n.a.	n.a.	n.a.	3.84 ± 0.03 (25% more than the nonelectric MES)	(Zhao et al., 2016)
Succinate/ succinic acid	MES: two-chamber; with CEM; carbon felt electrodes; *cathode poised at* -1.8 V vs. Ag/AgCl; operated with *Actinobacillus succinogenes* NJ113 as a biocathode catalyst	Corncob hydrolysate (pretreated with $Ca(OH)_2$ and NaOH)	n.a.	n.a.	n.a.	~7	(Zhao et al., 2016)

Continued

TABLE 3 Different BES technical achievements in production of value-added products from agriculture-associated wastes.—cont'd

Valuable product	Applied BES technical features	Waste type	Cathodic electron recovery (%)	Projected cathode area-based production ($g m^{-2} day^{-1}$)	Production rate ($mM day^{-1}$)	Maximum product titer ($g L^{-1}$)	Reference
Succinate/ succinic acid	MES: two-chamber; with CEM; carbon felt electrodes; *cathode poised at* −1.8 V vs. Ag/AgCl; operated with *Actinobacillus succinogenes* NJ113 as a biocathode catalyst; pH regulated with NaOH	Corncob hydrolysate (pretreated with $Ca(OH)_2$ and NaOH)	n.a.	n.a.	n.a.	18.09	(Zhao et al., 2016)
Succinate	MES: single-chamber electrofermentor; operated with minimal medium containing glucose and HCO_3^- as main and additional C sources; cathode poised at −0.65 V vs. Ag/AgCl; using Neutral Red as mediator; using strain *E. coli* T110	Glucose (probably as a product from agricultural waste conversion) and CO_2 (in the form of HCO_3^-)	100–300	n.a.	n.a.	3.12 (26.39 mM)	(Wu et al., 2019)

	expressing mtrABC, fccA, and cymA gene clusters conferring electroactivity and fumarate-reducing activity of *S. oneidensis* MR-1						
Succinate	MES: single-chamber electrofermentor; operated with minimal medium containing glucose and HCO_3^- as main and additional C sources; cathode poised at −0.65 V vs. Ag/AgCl; using Neutral Red as mediator; using a T110 mutant overexpressing bicarbonate transporter and carbonic anhydrase genes	Glucose (probably as a product from agricultural waste conversion) and CO_2 (in the form of HCO_3^-)	>300	n.a.	n.a.	3.6 (30.56 mM)	(Wu et al., 2019)

Continued

TABLE 3 Different BES technical achievements in production of value-added products from agriculture-associated wastes.—cont'd

Valuable product	Applied BES technical features	Waste type	Cathodic electron recovery (%)	Projected cathode area-based production ($g\,m^{-2}\,day^{-1}$)	Production rate ($mM\,day^{-1}$)	Maximum product titer ($g\,L^{-1}$)	Reference
Caproate; caprylate	MES: flat plate two-chamber reactor: with CEM; Pt-coated titanium mesh anode, graphite felt cathode; recirculated anolyte (0.1 M hexacyanoferrate), minimal medium containing acetate as catholyte; cathode poised at −0.9 V vs. SHE; operated with mixed culture dominated with *Clostridum kluyveri*	Acetate (can be as a product of fermentation of agricultural wastes)	26; <1	n.a.	n.a.	0.739; 0.036	(Van Eerten-Jansen et al., 2013)
Caproate (together with other products)	MES operated in *forced flow-through* mode: two-chamber; with CEM; Pt/ IrO_2-coated Ti mesh anode, carbon felt cathode; *cathode poised at* − 0.85 V vs. SHE; *mixed culture enriched on cathode*	CO_2	12.8	35.8 ± 0.4 ($0.95 \pm 0.05\,g\,L^{-1}\,day^{-1}$)	n.a.	1	(Jourdin et al., 2018)

Solvent mixture (ethanol, methanol, propanol, butanol, and acetone)	MES: single-chamber; Pt/Ir wire anode, Pt/Ir-wired carbon-PTFE cathode; cathode (working electrode) poised at −0.85 V; operated with mixed culture	Acetate and butyrate (can be as products of fermentation of agricultural wastes)	n.a.	n.a.	n.a.	~0.003 (for butanol), ~0.023 (for propanol), 0.004 (for ethanol), ~0.024 (for methanol), ~0.04 (for acetone)	(Sharma et al., 2013)
Alcoholic mixture (ethanol and butanol)	MES: two-chamber; with PEM; Pt-coated carbon paper anode, carbon brush cathode; applied voltage = 1.5 V (cathode potential of −0.68 ± 0.018 V vs. Ag/AgCl); operated with mixed culture	Anaerobic digestor effluent (can be from fermentation of agricultural wastes)	n.a.	n.a.	n.a.	0.031 for ethanol; 0.057 for butanol	(Kondaveeti & Min, 2015)
Alcoholic mixture (butanol, isopropanol, and other alcohols)	MES: two-chamber; with PEM; Pt-coated carbon paper anode, carbon brush cathode; applied voltage = 2 V (cathode potential of −0.84 ± 0.032 V vs. Ag/AgCl); operated with mixed culture	Anaerobic digestor effluent (can be from fermentation of agricultural wastes)	n.a.	n.a.	n.a.	0.091 for butanol; 0.06 for isopropanol; ~0.05 for other alcohols	(Kondaveeti & Min, 2015)

Continued

TABLE 3 Different BES technical achievements in production of value-added products from agriculture-associated wastes.—cont'd

Valuable product	Applied BES technical features	Waste type	Cathodic electron recovery (%)	Projected cathode area-based production ($g\,m^{-2}\,day^{-1}$)	Production rate ($mM\,day^{-1}$)	Maximum product titer ($g\,L^{-1}$)	Reference
Alcoholic mixture (ethanol and methanol)	MES: H-type two-chamber; carbon fiber brush electrodes; applied voltage $-1.5\,V$; operated with a mixed culture enriched from anaerobic digestion sludge: operated at initial VFAs [COD] = $4\,g\,L^{-1}$	Acetate and butyrate (can be as products of fermentation of agricultural wastes)	n.a.	n.a.	n.a.	0.052 ± 0.006 ($1.13 \pm 0.13\,mM$) for ethanol; 0.164 ± 0.005 ($5.13 \pm 0.16\,mM$) for methanol	(Gavilanes, Reddy, & Min, 2019)
Alcoholic mixture (ethanol, methanol, and propanol)	MES: single-chamber; carbon fiber brush electrodes; applied voltage = $0.8\,V$; operated with a mixed culture enriched from anaerobic digestion sludge; operated with pH controlled at 5 and at initial VFAs [COD] = $8\,g\,L^{-1}$	Acetate and butyrate (can be as products of fermentation of agricultural wastes)	n.a.	n.a.	n.a.	0.318 ± 0.005 ($6.9 \pm 0.11\,mM$) for ethanol; 0.236 ± 0.001 ($7.39 \pm 0.038\,mM$) for methanol; 0.053 ± 0.001 ($0.88 \pm 0.013\,mM$) for propanol	(Gavilanes, Noori, & Min, 2019)

Mixture of fatty acids (acetate, butyrate, isobutyrate, and caproate) and alcohols (ethanol, butanol, isobutanol, and hexanol)	MES: two-chamber glass reactor; with tubular CEM; Pt wire anode, graphite granule cathode; cathode poised at −0.8 V vs. SHE; pH regulated at 5.2	CO_2	n.a.	n.a.	4.8 mM C day^{-1} for acetate; 3.3 mM C day^{-1} for butyrate; 1.9 mM C day^{-1} for isobutyrate; 2.0 mM C day^{-1} for caproate; n.a. for alcohols	4.9 (82.5 mM or 165.0 mM C) for acetate; 3.1 (35.7 mM or 142.8 mM C) for butyrate; 1.6 (18.5 mM or 74.1 mM C) for isobutyrate; 1.2 (10.7 mM or 64.2 mM C) for caproate; 1.3 (28.1 mM or 56.1 mMC) for ethanol; 0.8 (11 mM or 44 mM C) for butanol; 0.2 (2.8 mM or 11.3 mM C) for isobutanol; 0.2 (2.1 mM or 12.8 mM C) for hexanol	(Vassilev et al., 2018)

Notes: *MES*, microbial electrosynthesis system (or cell); *PEM*, proton exchange membrane; *AEM*, anion exchange membrane; *CEM*, cation exchange membrane; *SHE*, standard hydrogen electrode; *MMO*, mixed metal oxide; *VFA*, volatile fatty acid; *n.a.*, data not available.

unable to consume current (Arends, 2013). However, it should be noted that lowering the cathode potential to more negative values can result in increased productions of other products (Kondaveeti & Min, 2015).

The most promising and must-do approaches to improve acetate bioelectrosynthesis are those that *innovate the cathode material and the cathode configuration*. The aim of these approaches is to render more biocompatibility of the cathode (or, in other words, to make it more "friendly" to microorganisms) as well as to better assist the attachment of microorganisms, biofilm formation, and the microbe-electrode interactions. Zhang et al. (2013) were the first to carry out an intensive study testing various *cathode modifications* in an acetate-producing MES operated with *Sporomusa ovata*. They coated the carbon cloth with various materials including chitosan, some amino acid derivatives, metal nanoparticles (of Au, Pd, and Ni), and some carbon nanotube textile (CNT) composites, etc. Among these materials, *chitosan, cyanuric chloride, 3-aminopropyltriethoxysilane, polyaniline, Au-nanoparticles, Pd-nanoparticles, Ni-nanoparticles, and CNT composites* significantly increased the acetate production rate approximately 7.6-, 6.8-, 3.2-, 3-, 6-, 4.7-, 4.5-, and 3.3-fold, respectively, compared to the untreated carbon cloth (Zhang et al., 2013). The maximum production rate achieved with the chitosan-modified cathode could reach $229 \pm 56\,mM\,m^{-2}\,day^{-1}$ (equal to *c.* $2.7\,g\,m^{-2}\,day^{-1}$) (Table 3). However, this is not the highest production rate to date as BES researchers keep looking for more efficient cathode-engineering solutions. For example, using a *Ni-nanowire-network-coated graphite cathode* instead of the plain graphite one, Nie et al. (2013) could increase the acetate production rate of a BES operated with *S. ovata* 10-fold with respect to cathode area (*c.* $16.6\,g\,m^{-2}\,day^{-1}$) or sevenfold with respect to volume ($0.068\,g\,L^{-1}\,day^{-1}$) (Table 3). The nanowire network is believed to significantly increase the microbe-electrode contact area and thereby, the density of electrogens enriched on the electrode (Nie et al., 2013). Furthermore, an even more interesting innovation was the use of a *TiO_2-based photocathode coated with Si nanowires* and also a *photoanode* in a so-called unassisted solar BES (Liu et al., 2015), to reduce the energy investment (by exploiting solar energy) and thus the cost of acetate production, following the concept previously proposed by Nevin et al. (2010). When operated with *S. ovata* at the cathode poised at $-0.595\,V$ vs. Ag/AgCl, the system could produce up to $0.24\,g$ acetate $L^{-1}\,day^{-1}$, or $1.2\,g\,L^{-1}$ in terms of titer, which were the highest values reported to that date for acetate production from CO_2 reduction by pure cultures (Liu et al., 2015) (Table 3).

For acetate-producing BESs operated with mixed cultures, a number of performance-improving approaches also focused on engineering the cathode configuration or the reactor configuration. An exceptionally high production rate of $685 \pm 30\,g\,m^{-2}\,day^{-1}$ was achieved with a system operated with *reticulated vitreous carbon modified with multiwalled carbon nanotubes (NanoWeb-RVC) as the cathode material* (Jourdin et al., 2015). This unique improvement in terms of electrode material results in a superiorly high surfacearea-to-volume ratio and the abundance of nanostructures on the electrode surface, which greatly enhances microbe-electrode interactions, bacterial attachment, biofilm development, and microbial extracellular electron transfer. Operated with such a biocathode, the system, a closed two-chamber reactor with its cathode poised at $-0.85\,V$ vs. SHE, could have a long-lasting and stable performance with the mentioned high acetate production rate, an acetate titer reaching $11\,g\,L^{-1}$, a CO_2 recovery reaching $94 \pm 2\%$, and an electron recovery reaching $100 \pm 4\%$ (Jourdin et al., 2015) (Table 3). Another reason for the enhanced performance of the system was believed to be *a well-acclimated electrogenic community*, enriched by using a methanogen-inhibitor (2-bromoethanesulfonic acid) (Patil et al., 2015). Furthermore, a striking advantage of this approach is that the cathode material can be produced by electrophoretically depositing multiwalled carbon nanotubes on reticulated vitreous carbon, which is a method feasible to be deployed for large-scale production. Thus far, this system has still been the one with the highest acetate production rate. However, in terms of titer, it has been beaten by a *three-chamber system* developed by Gildemyn, Luther, Andersen, Desloover, and Rabaey (2015) with the third chamber enabling the extraction and concentration of acetate produced from the cathode. Such a three-chamber system, operated with only a carbon felt cathode and a fixed current of $-50\,mA$, could achieve an acetate titer of up to $13.5\,g\,L^{-1}$ (the highest to date) and a relatively high production rate ($0.58\,g\,L^{-1}\,day^{-1}$ on average and $0.7\,g\,L^{-1}\,day^{-1}$ at maximum) (Gildemyn, Verbeeck, et al., 2015) (Table 3). It must be interesting to create a similar three-chamber system but with a Nanoweb-RVC-based biocathode discussed earlier to investigate whether a BES with both a high acetate production rate and a high acetate recovery efficiency can be obtained.

In addition to innovating the cathode material and the reactor design, configuration improvements of acetate-producing BESs should also address the limited mass transfer of CO_2 at the gas-liquid-solid interface of the cathode. This is because the direct use of CO_2 in the BESs is preferred for real practical applications, while the direct CO_2 capture efficiency at the cathode interface is low ($<10\%$) (Bajracharya, Vanbroekhoven, et al., 2016). That poor CO_2 capture could be improved by the *use of a gas diffusion electrode (GDE) having a PTFE-based gas diffusion layer* (similar to an air-cathode in an MFC). The latter enabled a twofold increased CO_2 mass transfer rate compared to conventional CO_2 sparging, resulting in a twofold increase of maximum acetate titer (to $2.89\,g\,L^{-1}$) and an average acetate production rate ($61\,mg\,L^{-1}\,day^{-1}$) that is comparable to those of BESs operated with HCO_3^- (Bajracharya, Vanbroekhoven, et al., 2016) (Table 3). It was reported that a maximum production rate of $650\,mg\,L^{-1}\,day^{-1}$ could be achieved in certain periods during the operation of the system with a gas mixture containing 80% CO_2. Although such a production rate was impressive, it is believed that CO_2 mass transfer and utilization in a BES can be even further improved through attention to the *concentration and the bubble size of CO_2* sparged into the cathode (Bian et al., 2020).

4.2 Production of products other than acetate

Although the bioelectrosynthesis of acetate has been studied the most, it is not very attractive in the economic aspect. Even when solar energy is used to power the process, the price to produce acetate by BESs is still at 406 EUR/ton, which is still considerably higher than those of available acetate-producing technologies (Bian et al., 2020). Therefore the bioelectrochemical productions of other value-added products from agriculture-associated wastes are preferred. These products include alcoholic compounds (ethanol, methanol, isopropanol, and butanol), fatty acids (butyrate, succinate, caproate, caprylate, etc.), acetone, and other fermentation products. These compounds can be produced by "electrofermentation" in an anode chamber or a cathode chamber at appropriate redox potentials, i.e., fermentation can be driven by the redox potentials towards producing them (probably from VFAs). They can also be produced simply by reducing acetate under the impact of the biocathode in a BES. As such, they can be produced as additional products when acetate is bioelectrochemically produced from CO_2 reduction. The production of these products is gaining more interest as they are also useful precursors for organic synthesis and attractive commodity goods. Ethanol and other alcoholic compounds are also excellent biofuels to replace fossil fuels in the future. The key here is still how to achieve a good selectivity of bioelectrosynthesis to produce the desired products at high productivities and yields.

4.2.1 Production of ethanol in a BES anode

Ethanol could be produced from glucose (in winery wastes) in a BES anode using the ethanol-producing yeast *Saccharomyces cerevisiae*, together with electricity generation if a mediator, methylene blue, was added (Sugnaux et al., 2016). It was believed that ethanol production was due to the *Crabtree effect*, i.e., ethanol fermentation can still occur while electron acceptors for respiration are available. Accordingly, when operated as an MFC, a two-chamber *BES with reticulated vitreous carbon foam-based bioanode* could produce ethanol with a concentration of up to *c.* 6% and an efficiency reaching ~50%, regardless the value of the external resistance used (Table 3). When operated with an applied voltage of 1.3 V, a single-chamber BES could produce 96.3% the theoretical amount of ethanol, compared to only 80.6% achieved without additional voltage. However, higher applied voltages resulted in declines in ethanol production. Therefore the bioanode reaction seems to enhance ethanol production by yeast. Hence, combining it with fermentation can be a promising technical approach for enhanced bioproduction of ethanol.

4.2.2 Production of ethanol by reducing acetate

Acetate, the product of fermentation or bioelectrosynthesis, can be reduced to a product with added values such as ethanol. Steinbusch et al. (2010) were the first to try producing ethanol by bioelectrochemically reducing acetate by a mixed culture. Their BES was a flat plate two-chamber reactor with graphite electrodes, an AEM separating an anode chamber containing hexacyanoferrate as the anolyte and a cathode chamber harboring a medium containing acetate as the catholyte. While controlling the *cathode potential at* $-0.55\,V$ *vs. SHE*, they tried to steer the cathode reactions towards producing more ethanol by *using mediators* to enhance the low rate of acetate reduction and to inhibit the formation of other products. Interestingly *methyl viologen* was a good candidate for that, accelerating ethanol production sixfold and well inhibited the formation of *n*-butyrate and methane (Steinbusch et al., 2010). The ethanol production efficiency of the system operated with methyl viologen could reach $74.6 \pm 6\%$ (Table 3). Indeed, this mediator was also reported to enhance the production of ethanol and butanol in fermentation 28- and 12-fold, respectively, probably by diverting the electron flow towards the production of these more reduced metabolites (Yarlagadda, Gupta, Dodge, & Francis, 2012).

Indeed, how to drive bioelectrosynthesis towards the production of desired products is always the key to this technology because the biocathode reactions usually produce multiple products. Ethanol cannot normally be produced in an MES reducing CO_2, as acetate, hydrogen, and/or methane are the thermodynamically favored products. However, by *the addition of tungstate* into the catholyte of a conventional two-chamber MES operated with *Sporomusa ovata* at the cathode poised at $-0.69\,V$ vs. SHE, Ammam et al. (2016) could generate ethanol from the acetate produced. According to the authors, a 10-fold increase in the concentration of added tungstate enabled the production of ethanol at a rate of $0.18 \pm 0.05\,g\,m^{-2}\,day^{-1}$ (or $4.0 \pm 1.2\,mmol\,m^{-2}\,day^{-1}$ or $48.0 \pm 14.6\,\mu M\,day^{-1}$) (Table 3), while no ethanol production was feasible with the original concentration (Ammam et al., 2016). Although ethanol did not accumulate in the cathode chamber, the finding of this study suggests a way to *divert metabolic pathways* of the electroactive biocatalysts towards the production of ethanol. The authors further revealed that tungstate was essential for the key enzymes of *the acetate re-asssimilation pathway* that leads to the formation of ethanol. Further genetic engineering is perhaps promising to promote even more bioelectrochemical production of ethanol.

4.2.3 Production of isopropanol from CO_2

An excellent example of how genetic engineering can be applied to divert bioelectrochemical conversion towards the desired product is the production of isopropanol from CO_2 demonstrated by Torella et al. (2015). Like ethanol, isopropanol is also an attractive liquid fuel to store energy converted from renewable energies (from sunlight, wind, etc.). Indeed, Torella et al. (2015) used photovoltaic cells to supply power to a fermentor-like single-chamber BES having a CoP_i stainless steel anode and a Ni-Mo-Zn stainless steel cathode to produce isopropanol. Such production was feasible by the biocatalytic activity of the *Ralstonia eutropha* mutant *Re*2133-pEG12, which is *metabolically engineered to redirect acetyl-CoA towards the synthesis of isopropanol*, instead of towards PHB synthesis as in the wild-type. With the applied cell potential of $\geq$2.7 V, the bacterium could grow consistently and produced 216 $\pm$ 17 mg L^{-1} isopropanol, which was the highest yield of fuel reported for a BES to that date (Torella et al., 2015) (Table 3). It is also encouraging that the product selectivity, in this case, was strikingly high ($\sim$90% yield).

In order to improve the product selectivity to obtain more isopropanol from bioelectrosynthesis, another simple strategy is to apply proper catholyte pH and hydraulic retention time (HRT) (Arends et al., 2017). The production of isopropanol, together with butyrate, through further reductions of acetate, the main product of bioelectrosynthesis, could be enhanced by *lowering the catholyte pH* (to <5.5) and *operating the BES in continuous mode* with a high HRT (>5d). Under these conditions, a two-chamber MES could produce up to 0.82 g isopropanol L^{-1}, in addition to acetate and butyrate, when operated with an Ir-MMO-coated Ti-mesh anode and a biotic carbon felt cathode in a modified homoacetogenic medium. It was also reported that a stable production of isopropanol (at a rate of 1.17 $\pm$ 0.34 g m^{-2} day^{-1} or 0.06 $\pm$ 0.02 g L^{-1} day^{-1}) could be achieved at pH 5 and HRT of 5d (Table 3). The enhanced production of isopropanol might be explained by a shift of the microbial community at the cathode towards the dominance of *Rummeliibacillus*, due to the change in operation to continuous mode. In addition, low pHs might trigger the switch from acidogenesis to solventogenesis, which produces solvents, including alcohols (Arends et al., 2017). However, the reason why only isopropanol but not ethanol or butanol was produced remains unclear.

4.2.4 Production of butanol by electrofermentation

As mentioned, ethanol, isopropanol, and also butanol are alcoholic compounds that are high-quality biofuels. Compared to ethanol or isopropanol, butanol is a more advantageous biofuel as it has higher energy density, lower volatility, and other physicochemical properties that are more appropriate for use in compression ignition engines (http://biobutanol.com). However, the conventional fermentation-based bioproduction of butanol has encountered several challenges, the most serious of which is low productivity, which can be expectedly enhanced by bioelectrosynthesis. As the formation of butanol involves complex metabolic pathways, the use of selective single cultures capable of such specific metabolism in bioelectrosynthesis seems required. For that reason, *Clostridia appear to be the most suitable candidates* as biocatalysts for bioelectrosynthesis of butanol. It was reported that among a number of *Clostridium* species, *C. pasteurianum* was the most electroactive for cathodic reduction in BES (Khosravanipour Mostafazadeh et al., 2016). Accordingly, *C. pasteurianum* ATCC 6013 displayed an excellent butanol production performance when tested in a H-type two-chamber BES operated with graphite electrodes and fermentation medium containing glucose as the sole carbon source. Butanol production by the system notably increased by 260% vs. that by only fermentation while butyrate and acetate decreased by 25% and 37%, respectively. Under optimal operational conditions of 118 g glucose L^{-1} medium, cathode potential at −0.54 V vs. Ag/AgCl and temperature at *c.* 33.5°C, the BES operated with *C. pasteurianum* ATCC 6013 could produce 13.3 g butanol L^{-1}, which approximates the theoretical value (Table 3). This performance is very competitive, suggesting a great potential of applying bioelectrosynthesis for efficient production of butanol from sugar-rich agricultural wastes (e.g., apple or grape pomaces). However, separation of butanol from the product mixture remains a major challenge (Khosravanipour Mostafazadeh et al., 2016).

4.2.5 Production of butyrate from CO_2

Butyrate can be used to produce fuel through esterification or for organic synthesis and pharmaceutical industry. Thus it is more attractive if this value-added product can be co-produced with acetate, the main but economically unattractive product of bioelectrosynthesis. Ganigué et al. (2015) were the first to report the production of butyrate from CO_2 reduction in BESs. Their system was a two-chamber H-type reactor having a titanium rod anode and a carbon cloth cathode poised at −0.8 V vs. SHE and purged occasionally with pure CO_2. Butyrate production in the system was enabled probably by using methanogenesis inhibitors and *inoculating the cathode with a mixed culture from a syngas fermentor*. It was observed that butyrate production began when *pH decreased to 6.53*, together with the detection of ethanol as well as butanol and the decrease of acetate production. The rate of CO_2 fixation to products could reach 4.1 mM C day^{-1} (equivalent to that of syngas fermentation), while the concentrations of butyrate, ethanol, and butanol could reach *c.* 20, 30.8, and 7.3 mM C, respectively (Ganigué et al., 2015) (Table 3). The authors believed that the capability of the system to produce C4 compounds was due to (i) the more reducing power supplied, and (ii) the distinct syngas-fermentor-originated community that contained *Clostridium kluyveri*, which can carry out chain elongation reactions combining acetate and ethanol to form butyrate. Such a production mechanism is ultimately driven by cathodic hydrogen formation and, therefore, can be seriously affected by the partial pressure of hydrogen (pH_2) at the cathode. To overcome this issue, Batlle-Vilanova et al. (2017) *tested a modified tubular reactor design* and *a strictly regulated CO_2 feeding strategy*, which apparently improved butyrate production. The tubular-design reactor, operated with components similar to the H-type (mentioned earlier) but without CO_2 feeding regulation, could already produce up to 59.7 mM C of butyrate because it was able to maintain a high pH_2. However, the pH decrease in that system was still slow, limiting the formation of ethanol and thus butyrate. The pH decrease could be accelerated by carefully controlling the amount of CO_2 supplied to the system to maintain just a sufficient pH_2 but still promote more production of acids. This CO_2 feeding regulation strategy could result in more ethanol production and also an enhanced butyrate production, with the concentration and the production rate of the latter reaching up to 87.5 mM C and 7.2 mM C day^{-1}, respectively (Batlle-Vilanova et al., 2017) (Table 3). These values are over fourfold those achieved with the H-type BES operated without CO_2 feeding regulation. Such a butyrate bioelectrosynthesis performance was impressive until it was later outcompeted by the performance of the BES operated with *a thick cathodic biofilm*, created by Jourdin et al. (2018) as mentioned earlier. These authors could enrich a thick and even biofilm on both sides of the cathode by operating the BES in a *forced flow through mode* (continuous mode). It was found that electron uptake was significantly enhanced due to the thick cathode biofilm (from $-1\,A\,m^{-3}$ before the biofilm formation to -8 to $-10\,A\,m^{-3}$ after that). In terms of VFA production, the rate increased three- to fourfold with the thick biofilm. Particularly, the production of butyrate (together with caproate) was detected only after the biofilm was well established, with the rate reaching $3.2 \pm 0.1\,g\,L^{-1}\,day^{-1}$, the electron recovery reaching 28.6%, and the concentration reaching $1.5–2.2\,g\,L^{-1}$, respectively (Table 3). Such a butyrate production rate was reported to be 14 times higher than that reported by Batlle-Vilanova et al. (2017), and is still the most impressive to date.

4.2.6 Production of succinate/succinic acid

Like butyrate, succinate or succinic acid is one important platform chemical used as a precursor for the production of many other industrial chemicals. The production of succinic acid requires a high metabolic specificity and thus far has been only carried out by single cultures having that metabolism, obtained either in nature or by genetic engineering. The first report about bioelectrochemical production of succinate is by Zhao et al. (2016), who used a naturally occurring single culture (*Actinobacillus succinogenes* NJ113), having a great potential of producing succinic acid, as a cathode biocatalyst to electroferment corncob hydrolysate in the catholyte of a two-chamber reactor having carbon felt electrodes. When the cathode was poised at -0.8 V vs. Ag/AgCl, electrofermentation could enhance the production of succinic acid by *c.* 25% vs. nonelectric fermentation, with the titers reaching $3.84 \pm 0.03\,g\,L^{-1}$ vs. $2.94 \pm 0.17\,g\,L^{-1}$, respectively (Zhao et al., 2016) (Table 3). However, such production was still limited because of metabolic inhibitors (phenol, furfural, etc.) present in corncob hydrolysate. Therefore *proper pretreatments* of the hydrolysate, e.g., with a mixture of $Ca(OH)_2$ and NaOH, could significantly further improve the production of succinic acid, achieving a titer of up to $6.85\,g\,L^{-1}$. Moreover, *optimizations of the applied cathode potential and the use of pH regulator* were also efficient improvements (Zhao et al., 2016). Accordingly, the production of succinic acid could reach its peak (with a titer of *c.* $7\,g\,L^{-1}$) at the cathode potential of -1.8 V vs. Ag/AgCl (Table 3).

With respect to pH regulators, NaOH appeared to be the optimal one, resulting in a maximum succinic acid concentration of 18.09 g L^{-1} and the highest succinic acid yield of 60%.

Natural strains capable of high yield bioelectrochemical production of succinic acid or succinate, like *A. succinogenes* NJ113, are rare. Therefore another approach to efficiently produce succinate in a BES is to genetically engineer existing strains to create novel strains capable of producing this compound at enhanced rates. Following this strategy, Wu et al. (2019) attempted to create *Escherichia coli* strains that expressed the key gene clusters associated with the capability of utilizing electricity to reduce fumarate (producing succinate) from *Shewanella oneidensis* MR-1. Successfully-created recombinant strains could generate much higher current and reduce more fumarate to produce more succinate than the wildtype when tested in a three-electrode bioelectrochemical reactor with glucose and HCO_3^- as main and additional carbon sources, respectively, and Neutral Red as an electron mediator. For instance, one of the recombinant strains, T110, could produce up to 26.39 mM succinate (or 3.12 g succinate L^{-1}), which was 39.7% more than the amount produced by the wild type (Wu et al., 2019) (Table 3). Overexpressing the bicarbonate transporter and carbonic anhydrase genes in T110 could further increase the efficiency of succinate production, with the titer of succinate reaching 30.56 mM or 3.6 g L^{-1}. Thus genetic engineering is clearly one way to go for direct microbial electrosynthesis of targeted products from glucose, as also recommended for the productions of ethanol and isopropanol above (Ammam et al., 2016; Torella et al., 2015).

4.2.7 Production of medium chain fatty acids (caproate and/or caprylate)

Unlike C2 (acetic) and C4 (butyric and succinic) acids, medium chain fatty acids (MCFAs) are fatty acids having 6–12 carbon atoms in their molecules. They are also important platform chemicals, and can be used for other application purposes such as feed additives, plastics, and lubricants (Agler, Wrenn, Zinder, & Angenent, 2011). In addition, they can be converted to liquid biofuels, which are superior to VFAs or ethanol because of the higher energy density of their molecules (Petrus & Noordermeer, 2006). However, like VFAs, the production of MCFAs by fermentation is currently also limited in yield. It is expected that bioelectrosynthesis could be a strategy to improve that situation but to ensure a good selectivity towards MCFA products, microbial catalysts having MCFA-specific synthesis pathways must be used. For instance, Van Eerten-Jansen et al. (2013) reported that the use of *a mixed culture dominated with C. kluyveri for cathode inoculation* was required to produce caproate as the predominant bioelectrosynthesis product. They used a BES that had a typical flat-plate two-chamber design developed by the Wageningen University group (Steinbusch et al., 2010), with a Pt-coated titanium mesh anode and a graphite felt cathode inoculated with the mixed culture and poised at −0.9 V vs. SHE. Operated with a minimal medium catholyte containing acetate and not containing any mediator, the system could produce 0.739 g caproate L^{-1} (69% of the recovered carbon amount) and 0.036 g caprylate L^{-1}, while simultaneously producing butyrate (0.263 g L^{-1}) and ethanol (0.027 g L^{-1}) (Van Eerten-Jansen et al., 2013) (Table 3). It was hypothesized that caproate and caprylate were the products of chain elongation reactions of butyrate and ethanol, which were, in turn the products of acetate reduction. Therefore in order to gain more MCFAs by bioelectrosynthesis, one approach was to promote those chain elongation reactions through *creating and leveraging a thick cathodic biofilm*, as mentioned earlier (Jourdin et al., 2018). Particularly, the production of caproate was detected only after the biofilm was well established (by a forced flow through operation), with the rate reaching 0.95 ± 0.05 g L^{-1} day^{-1}, the electron recovery reaching 12.8%, and the concentration reaching 0.7–1 g L^{-1}, respectively (Table 3). Furthermore, no other products (ethanol, lactate, methanol, or other organics) were detected, indicating a high bioelectrosynthetic selectivity of the biofilm. Interestingly, the essential role of the biofilm, together with the observations that no ethanol and no hydrogen buildups were detected in continuous mode, suggest a possible new bioelectrochemical pathway to add CO_2 directly to acetate and further to elongate the molecular chain (Jourdin et al., 2018). Nevertheless, separation to recover the targeted products remains a challenge (Van Eerten-Jansen et al., 2013).

4.2.8 Other BESs producing mixed products other than acetate

As mentioned, the selection of biocatalysts to be used at the BES cathode can help steer bioelectrosynthesis towards more products other than acetate. For example, Sharma et al. (2013) *used sulfate-reducing bacteria (SRB)* in an MES and could produce an assortment of solvents including ethanol, methanol, propanol, butanol, and acetone from the reduction of acetate and butyrate at a cathode potential of −0.85 V vs. Ag/AgCl. Their system was a single-chamber reactor operated with a Pt/Ir wire anode and a carbon-PTFE cathode, which was asserted to be low in cost. Although no data were available about the production rates, the SRB-originated consortium was considered highly robust as it could tolerate the concentrations of VFAs up to 0.1 M each and also high salt concentrations (up to 10 g L^{-1}) (Sharma et al., 2013) (Table 3).

Bioelectrosynthesis can be driven towards the productions of alcohols also by *regulating the cathode potential*. This was demonstrated by Kondaveeti and Min (2015), who investigated the possibility of using a mixed culture MES to produce alcohol by reducing the VFAs in an anaerobic digestor (AD) effluent. The system had a classical two-chamber BES configuration with a Pt-coated carbon paper anode, a PEM, and a carbon brush cathode (Kondaveeti & Min, 2015). No alcohols were produced when a voltage of 1 V was supplied (leading to a cathode potential of -0.45 ± 0.02 V vs. Ag/AgCl), probably because the cathode potential was not negative enough. When the applied voltage was 1.5 V (leading to a cathode potential of -0.68 ± 0.018 V vs. Ag/AgCl), ethanol was well produced together with butanol but at low titers (0.031 and $0.057\,g\,L^{-1}$, respectively) (Table 3), probably due to low VFA contents in the AD effluent. A higher voltage supply of 2 V (leading to a cathode potential of -0.84 ± 0.032 V vs. Ag/AgCl) resulted in the production of more diverse products including butanol ($0.091\,g\,L^{-1}$), isopropanol ($0.06\,g\,L^{-1}$), and some other alcohols ($\sim 0.05\,g\,L^{-1}$) (Table 3), which was similar to Sharma et al. (2013). However, the system was still considered to have a poor performance in terms of producing alcohol, as hydrogen and methane were still produced at significant amounts. It was reported that further increases in the supplied voltage only led to more production of lactate but no more productions of alcohols (Kondaveeti & Min, 2015).

The bioelectrochemical productions of alcoholic compounds from VFAs are advantageous because VFAs are considered end products of anaerobic digestion (AD), as they may inhibit methane production, and further VFAs-to-alcohols fermentative conversions currently have limited yields. Therefore more research attempts have been made to improve the bioelectrochemical reduction of VFAs to produce alcohols. Gavilanes, Noori, and Min (2019) tested the effect of different high COD (VFAs) concentrations on the bioalcohol production performance of their BES, which was a typical H-type two-chamber system having carbon fiber brush electrodes and operated with a mixed culture enriched from an AD sludge. By applying an external voltage of 1.5 V, the authors found that *the initial COD concentration* could affect the microbial activity (as presented by the changes of the electric current) and, more importantly, the cathode potential. A maximum current density, as well as a maximum alcohol production, was achieved at the initial COD concentration of $4\,g\,L^{-1}$. At this COD concentration, the titers of methanol (the predominant product) and ethanol reached $0.164 \pm 0.005\,g\,L^{-1}$ (5.13 ± 0.16 mM) and $0.052 \pm 0.006\,g\,L^{-1}$ (1.13 ± 0.13 mM), respectively (Gavilanes, Reddy, & Min, 2019) (Table 3). However, methane production was still detected and even increased with time and pH. Thus the authors proposed that methane production should be suppressed, probably by *controlling the catholyte pH at low levels*, to enhance the production of alcohols and to avoid the use of toxic chemicals. Indeed, the concentrations of produced alcohols significantly increased by continuously maintaining the pH at 5, while no methane was produced (Gavilanes, Noori, & Min, 2019). Accordingly, under such a condition, the concentrations of ethanol, methanol, and propanol could reach $0.318 \pm 0.005\,g\,L^{-1}$ (6.9 ± 0.11 mM), $0.236 \pm 0.001\,g\,L^{-1}$ (7.39 ± 0.038 mM), and $0.053 \pm 0.001\,g\,L^{-1}$ (0.88 ± 0.013 mM), respectively, at the initial COD concentration of $8\,g\,L^{-1}$ (Table 3). Nonetheless, decreases in the productions of the alcohols with time were observed.

To improve the bioelectrochemical production of alcohols and/or longer chain fatty acids, *combining different technical advances* can be an approach. For instance, Vassilev et al. (2018) used a two-chamber glass reactor with a tubular CEM harboring a Pt wire anode inside and a large cathodic chamber (outside the membrane) containing graphite granules to maximize the biocathode reaction surface area. These authors also regulated the pH of catholyte at 5.2 by CO_2 supply and used 2-bromoethanesulfonate to inhibit methanogenesis while employing an open reactor microbiome (mixed culture) for more versatility. Operated with the cathode potential of $-0.8\,V$ vs. SHE, the system could produce a variety of alcohols (including ethanol, butanol, isobutanol, and even hexanol) and also a variety of fatty acids (including acetate, butyrate, isobutyrate, and even caproate) from CO_2 reduction. Acetate was still the key product of the system, with its titer reaching $4.9\,g\,L^{-1}$ (82.5 mM or 165.0 mM C). The productions of other fatty acids were also considerable: 3.1 g butyrate L^{-1} (35.7 mM or 142.8 mM C), 1.6 g isobutyrate L^{-1} (18.5 mM or 74.1 mM C), and 1.2 g caproate L^{-1} (10.7 mM or 64.2 mM C), which was the highest caproate titer reported to that date (Vassilev et al., 2018) (Table 3). The corresponding alcohols of these acids were produced at various degrees, with ethanol being the main product, with its titer reaching $1.3\,g\,L^{-1}$ (28.1 mM or 56.1 mMC) in the first batch, but gradually decreasing afterward. Butanol, isobutanol, and hexanol were produced more in later batches of operation, reaching the respective titers of $0.8\,g\,L^{-1}$ (11 mM or 44 mM C), $0.2\,g\,L^{-1}$ (2.8 mM or 11.3 mM C), and $0.2\,g\,L^{-1}$ (2.1 mM or 12.8 mM C) (Vassilev et al., 2018). It was proven that the enriched microbiome in the BES was responsible for such productions of various products and, that Clostridia dominated and might be key players in this microbiome, as also hypothesized in other studies. This is interesting for practical applications as the reactor was operated in an open environment but it could naturally enrich Clostridia, unlike in other studies where single *Clostridium* cultures or preenriched Clostridia-dominated mixed cultures were forcefully used in BESs. Indeed, the bioelectrochemical production of a mixture with so various products all the way from CO_2 was reported for the first time and encouraging for future studies. However, how to separate each product from the mixture in a cost-efficient manner is still a question.

5 BES for the recovery of nutrients from agricultural wastes

5.1 Recovery of nitrogen

Agricultural activities can produce nitrogen-rich wastes, such as unused fertilizer components or animal excreta, which contain nitrogen mainly in the forms of ammonium and nitrate. These wastes can be released into the environment through runoff and/or wastewater discharge. They are not only serious environmental pollution causing eutrophication to receiving water bodies but also a substantial loss of resource as the nitrogen in them can be otherwise recovered and reused for agriculture (as fertilizers, feed additives, etc.). Therefore technologies for nitrogen removal and recovery from those wastes are required. The conventional nitrification/denitrification process has been long used to remove nitrogen from waste, but it is energy-intensive, costly in operation, and low in removal efficiency (Kuntke et al., 2018). The more recently-developed ANNAMOX-based technology requires significantly less energy and offers lower cost, but it is still low in efficiency, time-consuming (due to the slow growth of ANNAMOX bacteria), and not sustainable (Kuntke et al., 2018). Moreover, the significant disadvantage of these two processes is that nitrogen is lost in the form of dinitrogen (N_2) instead of being recovered in reactive forms (Kuntke et al., 2018). Under this circumstance, the use of BESs as a novel technology for nitrogen removal appears to be promising because: (i) the use of BESs is less energy-intensive, (ii) the electric current in BESs allows for concentrating ammonium at the cathode, and (iii) it is easy to couple BESs with recovery methods, such as stripping (Arredondo et al., 2015).

The working principle of the removal and recovery of nitrogen (ammonium) by a BES, whether being an MFC or an MEC, is solely based on the drive of cations (including NH_4^+) from anode to cathode by the electric current, the transformation of ammonium to ammonia due to pH elevation near the cathode, and the evaporation of ammonia (Kim et al., 2008; Kuntke et al., 2011; Min et al., 2005). In the cathode compartment, ammonium can be removed and recovered either by stripping, membrane processes, ion exchange, or struvite precipitation (Arredondo et al., 2015). Therefore in most of the cases, BESs for nitrogen recovery are two-chambered reactors each containing a CEM to separate the chambers, unless otherwise modified for special purposes. Hence, thus far technical attempts have been less focused on BES aspects (mainly on achieving proper electric currents), and more focused on improving ammonium recovery.

A limitation of the common two-chambered BES is its low ammonium transport rate. To increase it, increasing current density could be a strategy (Kuntke, Sleutels, Saakes, & Buisman, 2014). This strategy could be realized by *applying an external voltage*, modifying the system with *frames containing flow fields* to improve the electrode contact area, and using phosphate buffer as the catholyte (Kuntke et al., 2014). Such an operation could enable the system to remove $27.8 \pm 10.8\%$ total ammonia nitrogen (TAN) from synthetic wastewater and 33%–34% TAN from diluted urine with an applied current density of $\sim$23 A m^{-2} (Table 4). However, the major problem of the system is the back transport of NH_4^+ to the anode, especially when the cathode was operated in batch mode. This makes the recovery of TAN, e.g., by stripping, in such system not feasible. Therefore more efficient ways of recovering TAN at the cathode were required.

5.1.1 Nitrogen recovery by BESs and innovative stripping methods

Innovative stripping methods can be integrated with BESs to improve the recovery of TAN from BES cathode. Kuntke et al. (2012) proposed to use a *single-chamber MFC* with a membrane-electrode assembly (MEA), of which the air-cathode operated in a gas chamber so that the *ammonia produced was*

TABLE 4 Different BES technical achievements in nutrient recovery from agriculture-associated wastes.

Recovered nutrient	Applied BES and recovery technical features	Waste type	Removal/ recovery rate ($g\,m^{-2}\,day^{-1}$)	Removal/ recovery efficiency (%)	Net recovery efficiency (as per final usable products) (%)	Maximum product conc. ($g\,L^{-1}$)	Reference
Total ammonia nitrogen	MFC: two-chamber with CEM, graphite plate electrodes; ferricyanide as catholyte	Synthetic wastewater containing high NH_4^+	8.893 mmol day^{-1}	5–7	n.a.	0.005 (0.27 mM)	(Kuntke et al., 2011)
Total ammonia nitrogen	MEC: two-chamber with PEM, graphite felt anode; platinized flow field frames as current collectors and cathode; phosphate buffer as catholyte; applied current density up to 23 A m^{-2}	Synthetic wastewater containing high NH_4^+	16.8	27.8 ± 10.8	n.a.	n.a.	(Kuntke et al., 2014)
Total ammonia nitrogen	MEC: two-chamber with PEM, graphite felt anode; platinized flow field frames as current collectors and cathode; phosphate buffer as catholyte; applied current density up to 23 A m^{-2}	5 × diluted urine	186.1 (can be stable at 167.8 if catholyte was recirculated)	34.3 ± 2.3	n.a.	n.a.	(Kuntke et al., 2014)

Total ammonia nitrogen	MFC: air-cathode single-chamber with MEA including graphite felt anode and Pt-coated titanium felt cathode; platinized flow field frames for anode; NH_3 stripped by cathode aeration and recovered by acid absorption	Synthetic urine	3.29	>50	1.6	n.a.	(Kuntke et al., 2012)
Total ammonia nitrogen	MEC: two-chamber: with CEM; carbon felt anode, steel wire cathode; bioanode fed with synthetic wastewater containing acetate; cathode fed with reject water; operated in galvanostatic mode: 5, 10, 15, or 20 mA was tested; NH_3 stripped by cathode aeration and recovered by 2 M HCl; remaining water-soluble NH_3 captured by air sparging and recovered by 2 M HCl	Synthetic reject water	n.a.	94	94	n.a.	(Wu & Modin, 2013)
Total ammonia nitrogen	MEC: two-chamber: with CEM; carbon felt anode, steel wire cathode; bioanode fed with synthetic wastewater containing acetate; cathode fed with reject water; operated in galvanostatic mode: 5, 10, 15, or 20 mA were tested; NH_3 stripped by cathode aeration and recovered by 2 M HCl; remaining water-soluble NH_3 captured by air sparging and recovered by 2 M HCl	Real reject water	n.a.	79	79	n.a.	(Wu & Modin, 2013)

Continued

TABLE 4 Different BES technical achievements in nutrient recovery from agriculture-associated wastes.—cont'd

Recovered nutrient	Applied BES and recovery technical features	Waste type	Removal/recovery rate ($g\,m^{-2}\,day^{-1}$)	Removal/recovery efficiency (%)	Net recovery efficiency (as per final usable products) (%)	Maximum product conc. ($g\,L^{-1}$)	Reference
Total ammonia nitrogen	MFC and MEC: two-chamber; with CEM, carbon felt mesh anode, stainless steel mesh cathode; cathode compartment coupled to stripping/absorption (H_2SO_4) unit; phosphate buffer as catholyte; applied $V=0.1$–$0.8\,V$; batch mode	Pig slurry	n.a.	49.9 (at applied voltage = 0.6 V)	49.9 (at applied voltage = 0.6 V)	n.a.	(Sotres, Cerrillo, Viñas, & Bonmatí, 2015)
Total ammonia nitrogen	MFC and MEC: two-chamber; with CEM, carbon felt mesh anode, stainless steel mesh cathode; cathode compartment coupled to stripping/absorption (H_2SO_4) unit; NaCl ($0.1\,g\,L^{-1}$) as catholyte; applied $V=0.1$–$0.8\,V$; continuous mode	Pig slurry	25.5	94.3 (at applied voltage = 0.8 V)	94.3 (at applied voltage = 0.8 V)	n.a.	(Sotres et al., 2015)

Total ammonia nitrogen	Submersible MDC (SMDC): two chambers turning their backs on each other: carbon paper-based bioanode in anode chamber having AEM, cathode chamber having MEA (carbon cloth cathode with CEM); AEM and CEM facing out and contacting the wastewater in a CSTR; NH_3 stripped by cathode aeration and recovered by acid absorption	Anaerobic digestion wastewater (in a continuous stirred-tank reactor (CSTR)) (can be associated with agriculture waste)	80 (max. 97.8 at 8.8 A m^{-2})	91.7	n.a.	n.a.	(Zhang & Angelidaki, 2015c)
Total ammonia nitrogen	Same as above	Wastewater (in a biogas plant)	86	n.a.	40.8	n.a.	(Zhang & Angelidaki, 2015a)
Total ammonia nitrogen	Bipolar bioelectrodialysis system: four-chamber: anode chamber with carbon fiber brush-based bioanode contacting a bipolar membrane, an acid production chamber between the bipolar membrane and an AEM, a desalination chamber between the AEM and an CEM, cathode chamber with Pt-coated stainless steel woven mesh cathode; NH_3 stripped by hydrogen flow and recovered by acid absorption; applied $V=0.8$–1.4 V	Cattle manure	n.a.	94	94	n.a.	(Zhang & Angelidaki, 2015b)

Continued

TABLE 4 Different BES technical achievements in nutrient recovery from agriculture-associated wastes.—cont'd

Recovered nutrient	Applied BES and recovery technical features	Waste type	Removal/ recovery rate ($g\,m^{-2}\,day^{-1}$)	Removal/ recovery efficiency (%)	Net recovery efficiency (as per final usable products) (%)	Maximum product conc. ($g\,L^{-1}$)	Reference
Total ammonia nitrogen	Stack MEC: two-chamber each; with CEM; titanium frames having platinized flow fields; graphite felt anode, cathode frame itself as cathode; applied $V=0.7$–0.9 V vs. Ag/AgCl; 10 mM NaCl as catholyte, recirculated between the cathode chamber and a TMCS unit; NH_3 was recovered as NH_4SO_4 at the TMCS unit	5 × diluted urine	n.a.	70	49	25.5	(Kuntke, Zamora, Saakes, Buisman, & Hamelers, 2016)
Total ammonia nitrogen	Up-scaled MEC: two-chamber; with CEM; mixed metal oxide-coated Ti plate electrodes; 2.5 L chamber volume; applied $V=0.7$–0.9 V vs. Ag/AgCl; NH_3 from catholyte recovered in a TMCS module	Diluted urine	n.a.	31 ± 59	31 ± 59	n.a.	(Zamora et al., 2017)

Total ammonia nitrogen	Tubular MEC: two-chamber; with CEM; Pt/C-coated carbon cloth cathode in the cathode chamber inside the CEM tube, carbon brush anode outside the CEM; applied $V = 0.8\,V$; NH_3 stripped by cathode aeration; NH_3 (cathode) $+CO_2$ (anode)->absorption bottle to make NH_4HCO_3, remaining NH_3 going to recovery bottle and recovered with acid; anode effluent fed to FO: NH_4HCO_3 solution used to draw water, concentrate anode effluent recirculated to anode	Synthetic piggery wastewater	7.6	79.7 ± 2.0	79.7 ± 2.0	14.6	(Qin & He, 2014)
Total ammonia nitrogen	Same as above	Landfill leachate (can be from agriculture wastes)	n.a.	63.7 ± 6.6	63.7 ± 6.6	n.a.	(Qin, Molitor, Brazil, Novak, & He, 2016)
Total ammonia nitrogen	Osmotic MFC: two-chamber; with FO membrane instead of CEM; carbon brush anode, Pt-coated carbon cloth cathode; catholyte contained NaCl as the draw solute; varied R tested; TAN in catholyte can be recovered by various methods	Mimicked livestock waste digestate	n.a.	85.3 ± 3.5 (with $35\,g\,L^{-1}$ NaCl in the catholyte)	n.a.	n.a.	(Qin, Hynes, Abu-Reesh, & He, 2017)

Continued

TABLE 4 Different BES technical achievements in nutrient recovery from agriculture-associated wastes.—cont'd

Recovered nutrient	Applied BES and recovery technical features	Waste type	Removal/recovery rate ($g\,m^{-2}\,day^{-1}$)	Removal/recovery efficiency (%)	Net recovery efficiency (as per final usable products) (%)	Maximum product conc. ($g\,L^{-1}$)	Reference
Phosphorus	MFC: two-chamber; with PEM; carbon felts anode, carbon felts (or RVC) cathode; anode cultivated with *E. coli* + methylene blue as mediator; catholyte contained $FePO_4 \cdot H_2O$	Sewage sludge	n.a.	82	73.8 (= 82% x 90%)	n.a.	(Fischer, Bastian, Happe, Mabillard, & Schmidt, 2011)
Phosphorus	MFC; air-cathode single-chamber; carbon felt based bioanode, Pt/C-coated carbon paper cathode; polyester nonwoven cloth separating cathode from the anolyte; $R = 10\,\Omega$	Swine wastewater	n.a.	82	27	n.a.	(Ichihashi & Hirooka, 2012)
Phosphorus	MEC: single-chamber: heat-treated graphite fiber brush bioanode; stainless steel flat plate (SSF) cathode; applied $V = 0.75$, 0.90, and 1.05 V	Mimicked anaerobic digestion supernatant (containing ammonium phosphate)	12.7 ± 2.16 (at applied $V = 1.05$ V)	26 ± 13 (at applied $V = 1.05$ V)	26 ± 13 (at applied $V = 1.05$ V)	n.a.	(Cusick & Logan, 2012)

Phosphorus	MEC: single-chamber: heat-treated graphite fiber brush bioanode; stainless steel mesh (SSM) cathode; applied $V = 0.75$, 0.90, and 1.05 V	Mimicked anaerobic digestion supernatant (containing ammonium phosphate)	20.4 ± 2.16 (at applied $V = 1.05$ V)	40.3 ± 8 (at applied $V = 1.05$ V)	40.3 ± 8 (at applied $V = 1.05$ V)	n.a.	(Cusick & Logan, 2012)
Phosphorus	MEC: two-chamber (H-type); with PEM, plain carbon cloth-based bioanode, Pt-coated carbon cloth cathode; applied $V = 1.1$ V; COD concentration at anode = 1500 mg L^{-1}	Simulated reject wastewater (can be from agriculture)	n.a.	95	95	n.a.	(Almatouq & Babatunde, 2017)
Phosphorus	Tubular MEC with fluidized bed cathode: two-chamber; with tubular CEM; outer anode chamber harboring a carbon mesh based bioanode, cathode chamber inside CEM tube and harboring a fluidized bed cathode; applied $V = 0.8$, 1.0, and 1.4	Raw digestate	n.a. (210 ± 14 g-P m^{-3} cathode day^{-1} at applied $V = 1.4$ V)	70–85	66–71	n.a.	(Cusick, Ullery, Dempsey, & Logan, 2014)
Phosphorus	Single-chamber MEC: graphite fiber brush-based bioanode, stainless steel mesh (SSM) cathode (multiple pieces); inclined to assist struvite precipitation and hydrogen recovery; applied $V = 1.2$ V	Dewatering centrate	n.a.	70–82	68.2	n.a.	(Yuan & Kim, 2017)

Continued

TABLE 4 Different BES technical achievements in nutrient recovery from agriculture-associated wastes.—cont'd

Recovered nutrient	Applied BES and recovery technical features	Waste type	Removal/ recovery rate ($g m^{-2} day^{-1}$)	Removal/ recovery efficiency (%)	Net recovery efficiency (as per final usable products) (%)	Maximum product conc. ($g L^{-1}$)	Reference
Phosphorus	Single-chamber MEC: graphite fiber brush-based bioanode, stainless steel foil (SSF) cathode (one piece); inclined to assist struvite precipitation and hydrogen recovery; applied $V = 1.2 V$	Dewatering centrate	n.a.	92	96 (after 7 days)	n.a.	(Yuan & Kim, 2017)
Phosphorus	Fenton pretreatment + anaerobic digestion + MEC Single-chamber MEC: no membrane, carbon brush anode, stainless steel mesh cathode; applied $V = 1.0 V$	Waste activated sludge (WAS)	n.a.	n.a.	18	n.a. (1.72 g/g total solids)	(Hou et al., 2020)
Phosphorus	Single-chamber MEC: anode consisting of Pt-coated Ti mesh disk and graphite felt disk; Ti plate cathode; anode poised at −0.35 V vs. Ag/AgCl; recovering P as $Ca_3(PO_4)_2$	Artificial wastewater containing acetate, phosphate, and calcium (applicable for similar agricultural wastewaters)	n.a.	74	66 (by increasing phosphate concentration)	n.a.	(Lei et al., 2019)

Total ammonia nitrogen and phosphorus	BES coupled with FO: two-chamber MFC/MEC: carbon brush anode, Pt/C-coated carbon cloth cathode; catholyte = 10 mM NaCl; applied $V = 0.8$ V when operated as MEC; NH_3 transported to catholyte was recovered by H_2 stripping and acid absorption; anode effluent was concentrated in FO unit and recirculated to anode chamber; P in anode effluent recovered by a struvite precipitation unit	Digestion centrate	n.a.	n.a.	99.7 ± 13.0 for TAN; $79.5 \pm 0.5\%$ for P	n.a.	(Zou, Qin, Moreau, & He, 2017)
Total ammonia nitrogen and phosphorus	MFC/MEC: single-chamber; graphite fiber brush-based bioanode; Pt-coated PTFE-coated wet-proof carbon cloth cathode (MFC) or stainless steel mesh (SSM) or Pt/C cathode (MEC); effluent precipitation in a separate unit with [$MgCl_2$] adjusted to achieve a Mg:P ratio of 1.6:1	Digestate	n.a.	n.a.	83.1 ± 3.7 for PO_4^{3-}; 14.7 ± 0.6 for NH_4^+	n.a.	(Pepè Sciarria, Vacca, Tambone, Trombino, & Adani, 2019)

Continued

TABLE 4 Different BES technical achievements in nutrient recovery from agriculture-associated wastes.—cont'd

Recovered nutrient	Applied BES and recovery technical features	Waste type	Removal/ recovery rate ($g\,m^{-2}\,day^{-1}$)	Removal/ recovery efficiency (%)	Net recovery efficiency (as per final usable products) (%)	Maximum product conc. ($g\,L^{-1}$)	Reference
Total ammonia nitrogen and phosphorus	MFC/MEC: single-chamber; graphite fiber brush-based bioanode; Pt-coated PTFE-coated wet-proof carbon cloth cathode (MFC) or stainless steel mesh (SSM) or Pt/C cathode (MEC); effluent precipitation in a separate unit added with sea water bitterns to achieve a Mg:P ratio of 1.6:1	Digestate	n.a.	n.a.	87.7 ± 2.85 for PO_4^{3-}; 14.7 ± 0.5 for NH_4^+	n.a.	(Pepè Sciarria et al., 2019)
Total ammonia nitrogen, phosphorus, and potassium	Three-compartment MDC-like reactor: graphite granule-based bioanode, graphite granule cathode; third chamber (between the CEM and the AEM) to concentrate ions; anode potential poised at 0.0V vs. SHE	Urine	n.a. (7.18 kg of NH_4-N $m^{-3}\,day^{-1}$, 0.52 kg of PO_4-P $m^{-3}\,day^{-1}$, and 1.62 kg of K^+ $m^{-3}\,day^{-1}$)	49.5 ± 1.8% of N, 42.8 ± 1.0% of P, and 54.7 ± 1.3% of K	14.8	26.2 ± 0.3 (for N), 27.55 ± 0.95 (for P), and 7.0 ± 0.4 (for K)	(Ledezma, Jermakka, Keller, & Freguia, 2017)

Notes: *MFC*, microbial fuel cell; *MDC*, microbial desalination cell; *MEC*, microbial electrolysis cell; *MEA*, membrane electrode assembly; *PEM*, proton exchange membrane; *AEM*, anion exchange membrane; *CEM*, cation exchange membrane; *SHE*, standard hydrogen electrode; *CSTR*, continuous stirred tank reactor; *TMCS*, transmembrane chemisorption; *FO*, forward osmosis; *n.a.*, data not available.

stripped away by aeration. The escaping gas was channeled through boric acid and sulfuric acid to recover nitrogen as ammonium (NH_4^+). Also, with a flow field anode frame and a Pt-coated titanium felt cathode, the MFC could produce a current density of up to $0.5\,A\,m^{-2}$, resulting in a removal rate of $3.29\,g\,N\,day^{-1}\,m^{-2}$ (Kuntke et al., 2012) (Table 4). The rate seemed proportional to the current density of the MFC. However, the ammonium recovery efficiency only reached 1.6% (Kuntke et al., 2018). Nevertheless, the ammonium recovery by MFCs and air stripping requires no chemical addition (for stripping) and (10 times) less energy for aeration than conventional technologies. Indeed, an alternative stripping method is to take advantage of the *flow of hydrogen produced in the cathode of an MEC* to "pull" away NH_3 in the catholyte with it before the latter was captured by an acid solution (2 M HCl) (Wu & Modin, 2013). The system also achieved higher recoveries with higher current levels, with the maximum recovery reaching 94% with 15 mA when treating synthetic reject water and 79% with 20 mA when treating real reject water (Table 4). Technically, ammonium can be recovered electrochemically, but the bioelectrochemical recovery of ammonium appears to be more efficient and cost-effective (Gildemyn, Luther, et al., 2015). Thus the approach of integrating BESs and stripping for ammonium recovery is promising and has been applied to recover nitrogen from pig slurry (Sotres et al., 2015). A typical two-chamber BES coupled to an ammonia stripping/absorption unit (Desloover, Abate Woldeyohannis, Verstraete, Boon, & Rabaey, 2012) was able to recover up to 49.9% ammonium in pig slurry (when the BES was operated in batch mode with an applied voltage of 0.6 V and phosphate buffer as the catholyte). It could even recover 94.3% ammonium in pig slurry at a very high rate ($25.5\,g\,N\,m^{-2}\,day^{-1}$) when operated in continuous mode with sodium chloride as the catholyte (Table 4), probably due to a larger magnitude of pH increase of that catholyte compared to phosphate buffer. These figures demonstrate the feasibility of using BES-gas stripping integrative systems for recovering nitrogen from real (agricultural) wastes.

The BES-stripping integration can be applied flexibly to recover nitrogen from various kinds of wastes. For example, Zhang and Angelidaki (2015a) used a modified BES called *"submersible microbial desalination cell" or SMDC* to remove and recover TAN from the treated wastewater in a continuous stirred-tank reactor (CSTR). Unlike the conventional MDC, their SMDC had its AEM (of the anode chamber) and CEM (of the cathode chamber) facing out and contacting the wastewater when the reactor was submersed into the CSTR (Zhang & Angelidaki, 2015c). In such a way, the bulk solution of the CSTR is actually a "third chamber" of the SMDC, from which ammonium can be transported into the cathode chamber and removed as ammonia by stripping before recovered by acid absorption. Such an SMDC integrated with stripping could recover TAN at an average rate of $80\,g\,N\,m^{-2}\,day^{-1}$ and achieved higher recoveries with higher current densities: with a peak current density at 8.8 A m^{-2}, it could recover 91.7% TAN at a rate of $97.8\,g\,N\,m^{-2}\,day^{-1}$ (Table 4). The recovery was also affected by the initial ammonium concentration in the wastewater and the presence of other cations, as reported previously (Kuntke et al., 2012). Nonetheless, the SMDC-stripping integrative system offers no requirement of chemical addition for ammonium recovery, no ammonium inhibition to microorganisms, lower energy investment, and scalability. It was therefore applied successfully to recover 86 g N m^{-2} day^{-1} (at a current density reaching 4.33 A m^{-2}) and thereby reducing ammonium toxicity in agriculture-associated biogas plants, leading to 112% extra biogas production (Zhang & Angelidaki, 2015a) (Table 4). To remove ammonium together with sulfate, another anaerobic digestion inhibitor, Zhang and Angelidaki (2015b) even modified the MDC to create a *bipolar bioelectrodialysis system.* In addition to an AEM and a CEM, such a system had a bipolar membrane that allows the creation of an acid production chamber between the anode chamber and the desalination chamber, for the removal and recovery of sulfate, while gas stripping was applied for ammonium recovery. It performed very well on cattle manure, recovering more than 94% of NH_4^+ and SO_4^{2-} in the manure (Zhang & Angelidaki, 2015b) (Table 4).

5.1.2 Nitrogen recovery by BESs and transmembrane chemisorption (TMCS)

An alternative for the integration of BESs and gas stripping for ammonium recovery is the coupling of BESs with transmembrane chemisorption (TMCS), which is believed to be technically less complex and therefore more robust than the former. TMCS is a method for NH_3 recovery by transporting NH_3 gas through a gas-permeable hydrophobic membrane before it is absorbed as ammonium (NH_4^+) in a suitable acid solution (Kuntke et al., 2016). This method does not require high gas flow rates (as in the stripping method) and thus demands less energy for ammonia recovery. Moreover, the nanometer-sized gas-filled pores of the membrane enable a more efficient capture of ammonia into the acid. Therefore TMCS is considered more advantageous than stripping and has been tested in BESs, together with other technical improvements, for more efficient TAN recoveries.

The first integration of TMCS within a BES for enhanced TAN recovery from diluted urine or animal excreta was reported by Kuntke et al. (2016). These authors used *stack two-chamber MECs each having titanium frames with platinized flow fields* (an innovation of the group mentioned earlier) and *recirculated the catholyte* between the cathode compartment and a *TMCS unit*, inside which H_2SO_4 was used for absorption of ammonia. The TMCS unit could recover 95% of the ammonium-nitrogen in the catholyte and thus the system could achieve the highest reported NH_4^+-N concentration recovered in the acid, which was 22.5 g N L^{-1} (45-fold the concentration in the influent, being diluted urine) (Kuntke et al., 2016) (Table 4). Overall, with four stacks, the recovery of ammonium-nitrogen from the influent was 49%, only slightly lower than that of an equivalent electrochemical system (ES), which was 57%, but the BES only consumed half of the energy consumed by the ES. This impressive performance of the BES-TMCS integrative system encouraged further attempts to *scale up the system* for nitrogen recovery from (agriculture) wastes. For instance, Zamora et al. (2017) built an MEC with 2.5 L chambers coupled with a TMCS module, which could be operated with up to 30 L of wastewater at a time, for TAN recovery from urine. Such a two-chamber apparatus, with titanium plate electrodes coated with mixed metal oxide catalysts, could recover up to $31 \pm 59\%$ of ammonium nitrogen from the influent (Zamora et al., 2017) (Table 4). This performance is relatively impressive for an up-scaled system as the laboratory-scale recovery was only about 20% higher. Interestingly, this up-scaled system consumed less energy than the electrochemical counterpart. However, scale-up-associated issues did exist, including the limited COD removal of the system, the presence of oxygen in the cathode and the unstable bioanode.

5.1.3 Nitrogen recovery by BESs and forward osmosis (FO)

After the treatment with BESs and stripping or TMCS, a significant portion of ammonium still remains in the anode effluent, i.e., the ammonium recovery is still limited. Therefore a new approach involving the integration of forward osmosis (FO) has been proposed to concentrate the TAN of the anode effluent and hence allowing for additional nitrogen recovery to further enhance the recovery efficiency (Qin et al., 2016; Qin & He, 2014; Zou et al., 2017). In forward osmosis coupled with a BES, a high-solute-concentration solution is used to draw water off the anode effluent, which subsequently returns to the anode for additional nitrogen removal. In addition to the effect of concentrating ammonium for improved recovery, an added value of FO is water recovery, which is important as water will be becoming a limited resource in the future. Qin and He (2014) were the first *to integrate FO with an MEC* to recover nitrogen from synthetic piggery wastewater. To maximize the ammonium transport and recovery, *a tubular two-chamber configuration* was used, consisting of Pt/C-coated carbon cloth cathode placed in the cathode chamber inside a CEM tube and a carbon brush anode placed in the anode chamber between the CEM and the glass cover of the reactor. The system was operated with an applied voltage of 0.8 V and its cathode aerated to strip NH_3 before this gas was combined with CO_2 from the anode in an absorption bottle to produce NH_4HCO_3. The remaining NH_3 was recovered by acid absorption as described, and the produced *NH_4HCO_3 was used as the draw solute* in an FO unit fed with the anode effluent. Once the NH_4HCO_3 solution drew water off the anode effluent, the concentrated anolyte was recirculated to the anode chamber for additional treatment. Such a draw solute self-supplying system could recover up to $79.7 \pm 2.0\%$ of the TAN in the anode, which is about 30% more than the recovery by other methods (Table 4). Its recovery rate could reach

$7.6\,g\,m^{-2}\,day^{-1}$ (Kuntke et al., 2018), and the recovered ammonium concentration could reach a striking value of $14.6\,g\,L^{-1}$ (0.86 M) (Table 4). With such an excellent performance, the system was also successfully applied to treat landfill leachate, achieving an ammonium recovery of $63.7 \pm 6.6\%$ while additionally recovering 51% of the water in the anode effluent (Qin et al., 2016) (Table 4). The lower ammonium recovery obtained with landfill leachate (compared to that with synthetic piggery wastewater above) was probably due to the complex composition and the other cations of that waste.

The BES-FO integrative systems described earlier are relatively complex in terms of technical operation. Hence, Qin et al. (2017) even proposed a *direct integration of FO in a BES* to simplify the setup while still achieving efficient ammonium recovery. In such a BES, designated as the osmotic MFC (OsMFC), an FO membrane is used instead of the CEM, and NaCl is used in the catholyte as the draw solute (Qin et al., 2017). Ammonium is also transported from the anode to the cathode through the FO membrane due to the electric current as the driving force. This current increases with the water flux in the OsMFC, and so does the ammonium transport. The OsMFC developed by Qin et al. (2017), having two chambers with a carbon brush-based bioanode and a Pt-coated carbon cloth cathode, could recover more ammonium when the NaCl concentration in the catholyte increased, i.e., the water flux increased due to increased osmotic pressure. According to the authors, by increasing the NaCl concentration from 2 to $35\,g\,L^{-1}$, the current density of the OsMFC increased from $1.1 \pm 0.1\,A\,m^{-2}$ to $2.6 \pm 0.1\,A\,m^{-2}$, and its ammonium removal efficiency increased from $23.5 \pm 3.5\%$ to $52.5 \pm 4.7\%$ (Qin et al., 2017) (Table 4). When the OsMFC was operated in continuous mode, the ammonium removal efficiency even reached $85.3 \pm 3.5\%$. The ammonium removal mechanism was found to mainly involve the following processes: diffusion, ion exchange, water flux co-transport, and current-driven migration, of which the latter two are tightly associated with the electricity generation.

5.1.4 The attention to the load ratio when using BESs for nitrogen recovery

The load ratio concept was proposed by Wageningen University BES research group who defined it as the ratio between the current density and the TAN loading rate (Kuntke et al., 2018; Rodríguez Arredondo, Kuntke, ter Heijne, & Buisman, 2019). The load ratio is a crucial parameter to consider when one wants to optimize the recovery of TAN from waste using BESs. When this ratio is <1, the current is not sufficient to transport all ammonium across the membrane; when it is >1, the current is higher than required for the transport, and thus, the process is not energyefficient. However, in practice, due to the effect of many factors, the optimum load ratio may not be 1. For example, (Rodríguez Arredondo et al., 2019) found that the maximum TAN removal efficiency of their urine-treating BES ($\sim$61%) was achieved at a load ratio of 0.7. Therefore it is essential that this parameter should be experimentally tuned when using any BES for TAN recovery. For detailed information about the load ratio concept, the readers are referred to the mentioned materials.

5.2 Recovery of phosphorus

Together with nitrogen, phosphorus is an important nutrient in agriculture. Considerable amount of phosphorus is discharged to the environment in waste streams as the results of the extensive use of fertilizers in plantation and the livestock activities. The presence of excessive phosphorus in the environment, together with nitrogen, can cause adverse environmental incidents, such as eutrophication, as mentioned earlier. Moreover, the phosphorus source (phosphate rock) of the world is very limited but being depleted quickly due to increasing exploitation, especially for agriculture to meet the increasing demand for food (Pepè Sciarria et al., 2019). Therefore the recovery of phosphorus is considered crucial for sustainable development. Phosphorus can be recovered by various chemical methods as phosphate precipitates, among which the most common method is struvite precipitation, based on the reactions of free NH_4^+, Mg^{2+}, and PO_4^{3-} ions in waste, according to the following equation:

$$NH_4^+ + Mg^{2+} + PO_4^{3-} + 6H_2O \rightarrow MgNH_4PO_4 \cdot 6H_2O + 2H^+ \quad (*)$$

With its composition, struvite is ideal to be reused as a fertilizer in agriculture, and thus, its production is preferred. Conventional chemical methods for struvite precipitation are costly due to their high energy consumption and the requirement of chemical addition (such as Mg). BESs are believed to reduce the cost of phosphorus recovery, as the electric current generated in BESs can assist the precipitation process, reducing the energy required. Moreover, when MECs are used, energy in hydrogen produced can offset some energy investment (Cusick et al., 2014). Hence, the use of BESs for phosphorus recovery has been constantly optimized in recent years and becoming more and more promising.

Phosphorus recovery as struvite by BESs was demonstrated by a number of studies. Fischer et al. (2011) reported that ferric phosphate in digested sewage sludge could be reduced in a *mediator-assisted MFC* operated with *E. coli*. This system had a typical conventional two-chamber configuration with carbon felt electrodes and could remove up to 82% phosphorus in the catholyte, 90% of which was recovered as struvite (Table 4), while producing a voltage of 0.3 V (Fischer et al., 2011). The system was even up-scaled to a 3-L MFC/MEC operated by the same principle but with a biofilm enriched on a reticulated vitreous carbon anode (Happe et al., 2016). With proper applied voltage and cathode pH, the up-scaled system could recover up to 67% of P in sewage sludge as precipitate (Table 4). The great advantage of the produced precipitate is that it contained little amount of toxic metals such as As, Cd, Pb, Cr, etc. These advantages were also displayed by the phosphorus recovery performance of an air-cathode single-chamber MFC (Ichihashi & Hirooka, 2012). This system also removed up to 82% phosphorus from swine wastewater, but only 27% was precipitated as struvite (Table 4), probably due to the redissolution of phosphate caused by reactions in the wastewater.

The struvite formation on BES cathode surface was believed to be assisted by the local high pH at the cathode (Fischer et al., 2011; Happe et al., 2016). Such cathode pH increases were also evident in a *membraneless single-chamber MEC* aimed at recovering phosphorus from anaerobic digestion effluent (Cusick & Logan, 2012). This system, with a graphite fiber brush-based bioanode, could remove more phosphate (40.3 ± 8%) when operated with a *stainless steel mesh (SSM) cathode* than when operated with a stainless steel flat plate cathode (26 ± 13%) (Table 4). Indeed, similar performances were also observed with other single-chamber MECs operated with SSM cathodes while treating digestate (Pepè Sciarria et al., 2019). The performance with the SSM cathode was even comparable with that with a Pt/C cathode, and thus, it was recommended that the former could replace the latter. The rate of struvite precipitation on these cathodes increased with the applied voltage, which clearly demonstrates the effect of the electric current in a BES to improve struvite formation.

5.2.1 Enhanced phosphorus recovery by optimizing BES operational parameters

Phosphorus recovery in BESs can be further improved as affecting parameters such as COD load, chamber volume, aeration, and the concentrations of NH_4^+, Mg^{2+}, etc. can be optimized. Hirooka and Ichihashi (2013) proved that both *NH_4^+ and Mg^{2+}* were required for P recovery as struvite, which in a BES might be optimum at *higher molar ratios of N:P* compared to the theoretical value of 1:1. They also found that precipitation could decrease cathode performance, and thus, attention should be paid to this for long-term uses of BESs for phosphorus recovery (Hirooka & Ichihashi, 2013). Another group claimed that *COD load, chamber volume, and aeration* could also affect the performance of a conventional two-chamber-typed MFC in P recovery (Almatouq & Babatunde, 2016). According to that, increasing COD load (from $0.7\,g\,L^{-1}$ to $1.5\,g\,L^{-1}$) led to increased current densities, and thus, to increased cathode pHs (from 7.4 to 8.3), which finally resulted in increased struvite precipitation (from 8% to 38%) (Table 4). Also, increasing anode and cathode volumes could offer more COD to electroactive bacteria, leading to increased current densities and also increased phosphorus recoveries. Furthermore, aeration could increase the current density of the BES by 40%, causing the cathode pH to increase from 7.5 to 8.5 and the P recovery to increase by 25%. Almatouq and Babatunde (2017) also operated their BES as an MEC and found that the *applied voltage* could also affect phosphorus recovery performance of the system. Specifically, increasing the voltage (from 0.4 to 0.8 V) led to the increases in cathode pH from 8 to 9.1 and P recovery from 45 ± 5% to 90 ± 7% (Almatouq & Babatunde, 2017).

However, further increases in the voltage led to reduced pHs and P recoveries, probably due to the inhibition of high currents on the bacterial activity. Noticeably, while applying a voltage of 1.1 V, increasing COD load (to 1500 mg L^{-1}) in such an MEC could enable a very high cathode pH and hence achieve a 95% phosphorus recovery (Table 4).

5.2.2 *Enhanced phosphorus recovery by other technical improvements*

A significant challenge in phosphorus recovery using BESs is cathode scaling and membrane scaling when precipitation is saturated (Almatouq & Babatunde, 2017). Scale formation causes technical troubles for treatment systems, and thus, the additional cost is required to remove scales. In addition, conventional designs may limit the current density of BESs and thus their P recoveries. Therefore Cusick et al. (2014) proposed the use of a *tubular two-chamber MEC with a fluidized bed cathode* for efficient P recovery from raw digestate. This system contained an outer anode chamber harboring a carbon mesh-based bioanode, a tubular CEM, and an inner cathode chamber harboring a cathode in a fluidized bed form, which was previously reported to be very efficient for crystallized phosphate recovery with minimized scale formation. This MEC could remove 70%–85% and recover (precipitate) 66%–71% of phosphorus from the raw digestate (Table 4). At the applied voltage of 1.4 V (corresponding to a current density of 2.4 ± 0.4 A m^{-2} membrane) and the flow rate of 1.0 mL min^{-1}, the average precipitation rate could reach 210 ± 14 g-P m^{-3} cathode day^{-1} (Cusick et al., 2014). Such performance of the system is impressive, considering that the energy required (10 ± 1.0 Wh g^{-1} P) can be offset by hydrogen production.

A decisive factor to struvite precipitation is the cathode material and structure, as struvite crystals deposit on the cathode surface. In addition to the use of fluidized bed cathode discussed earlier, several other cathode alternations were proposed. For instance, Yuan and Kim (2017) discovered that the stainless steel mesh (SSM) cathode, as proposed by Cusick and Logan (2012), still had a limited capacity to hold produced struvite crystals (due to its open mesh spaces) and should be replaced by the *stainless steel foil (SSF) cathode*. In their study, a single-chamber MEC operated with the SSF cathode and with high current densities (resulted from high COD (acetate) loads) removed 53% of phosphorus in dewatering centrate in the first operation day and achieved the maximum recovery of 96% in 7 days, while another one operated with the SSM cathode could remove only up to 70%–80% and recover only 68.2% of P in the feed (Yuan & Kim, 2017) (Table 4). This performance of the BES with dewatering centrate is remarkable as dewatering centrate (or filtrate), particularly those from agricultural wastewater often containing concentrated phosphorus and other components, can be untreatable in wastewater treatment processes. Moreover, as calculated by the authors, the net energy requirement of the system is substantially smaller than the previously reported ones.

Like dewatering centrate, another type of wastes that can also contain a high content of phosphorus is waste activated sludge (WAS), which is a common product of wastewater treatment plants, including those treating agricultural wastes. The high solid content of WAS is a big obstacle for recovery, even for BESs, as solid particles strongly interfere with the struvite formation process (Hou et al., 2020). Therefore Hou et al. (2020) proposed the *combination of Fenton pretreatment, anaerobic digestion, and microbial electrolysis* as an efficient strategy to recover phosphorus from WAS. Accordingly, WAS was pretreated with Fe(III)/PCA/H_2O_2 to lyse biomass by the Fenton reaction, and subsequently anaerobically digested to generate soluble COD that can be used for the current-generation when the resulted digestate was finally fed to an MEC. The MEC used by the group was a single-chamber reactor with the most advanced improvements based on previous findings, including a carbon brush anode, a stainless steel mesh cathode, and an applied voltage of 1 V. It was clear that without pretreatment (especially without anaerobic digestion), the P recovery was very poor, only reaching $<1\%$ of total P in WAS. With only anaerobic digestion before MEC treatment, the recovery could already increase to about 11.4% (1.09 g/g TS), and when Fenton treatment was included, the recovery further increased to 18% (1.72 g/g TS) (Hou et al., 2020) (Table 4). Another noteworthy advantage of the integrated approach is that its struvite product had a significantly higher purity than the product obtained without pretreatment (74.4 vs. 44.64, wt%). Thus Fenton pretreatment and anaerobic digestion significantly improves phosphorus recovery by BES and should be applied when treating solid wastes such as WAS.

5.2.3 Phosphorus recovery by MEC-induced calcium phosphate precipitation

Phosphorus recovery as calcium phosphate precipitation can be an alternative to struvite precipitation as the former has been demonstrated feasible by using electrochemical systems (Lei, Remmers, Saakes, van der Weijden, & Buisman, 2018). Indeed, the same mechanism could also work in bioelectrochemical systems (Lei et al., 2019). This was demonstrated by the performance of a single-chamber MEC operated with a multi-Pt/Ti/C-disk based bioanode (poised at -0.35 V vs. Ag/AgCl) and a titanium plate cathode that is fed with artificial wastewater containing acetate, phosphate, and calcium. The system could remove more P with increasing COD (acetate) concentrations in feed, achieving a maximum P removal of 74% when 10 mM acetate is fed (Table 4). This behavior is actually similar to that of struvite-forming BESs discussed earlier. However, it was found that the precipitation of calcium phosphate could be strongly interfered by the formation of calcite ($CaCO_3$) due to the abundance of carbonate resulted from the pH increase associated with acetate oxidation. That is, this calcium phosphate precipitation-based process is better used for the recovery of phosphorus from wastewater(s) that have high P concentrations (Lei et al., 2019). Clearly, further works should be done to improve technological aspects of this phosphorus recovery approach to render it more competitive.

5.3 Simultaneous recovery of different nutrients

Agriculture-associated waste streams usually contain different nutrients (N, P, K, etc.) all together. Thus different BES approaches can be combined to concomitantly remove the nutrients. For example, a *BES can be coupled with an FO system* not only to improve nitrogen recovery through concentrating ammonium in anode effluent but also to enable efficient phosphorus recovery through more struvite precipitation in the concentrated anolyte (Zou et al., 2017). In their study, Zou et al. (2017) used a two-chamber MFC/MEC consisting of a carbon brush-based bioanode, a CEM, and a Pt/C-coated carbon cloth cathode to treat digestion centrate. The anode effluent of the reactor was fed to an FO unit that could remove $54.2 \pm 1.9\%$ of the anode effluent water before the concentrated anode effluent was fed back to the anode chamber. Ammonia was recovered in the system by ammonium transport from the anode to the cathode and subsequent stripping with hydrogen produced at the cathode (when the system was applied with a 0.8-V voltage). At the same time, phosphorus in the anode effluent was recovered by struvite precipitation in a separate chemical precipitation unit as a part of the anode effluent recirculation loop. Such a recirculation BES-FO hybrid system could recover $99.7 \pm 13.0\%$ of net ammonium nitrogen and $79.5 \pm 0.5\%$ of phosphorus as struvite (Table 4), which indicate a promising potential of using the system for simultaneous recovery of both nitrogen and phosphorus from nutrient-rich wastes. A challenge of such a simultaneous recovery is the difference in treatment capacities for N and P, which causes difficulties to coordinate the whole process.

The simultaneous recovery of nitrogen and phosphorus can be achieved by struvite formation (as explained for many BESs earlier). However, little attention seemed to be paid to the crystallization process, while this can be critical in determining the recovery efficiency. Thus *optimization of crystallization in BES-associated struvite precipitation* can be an approach to achieve efficient recoveries of both N and P from waste. This was demonstrated by Pepè Sciarria et al. (2019) when using BESs for treating digestate and recovering nutrients. Digestate, the product of anaerobic digestion, can be used directly as a renewable fertilizer as it contains high levels of N and P. However, when these levels exceed the crop requirements, the digestate needs to be treated to reduce them to matching levels (Pepè Sciarria et al., 2019). Fortunately, digestate is an ideal substrate for BESs, and thus, its nutrients can be recovered with BESs (MFCs and MECs). The MFCs used by Pepè Sciarria et al. (2019) were single-chamber reactors, each containing a graphite fiber brush-based bioanode and a Pt-coated PTFE-coated wet-proof carbon cloth cathode. Such an MFC, when operated with a $1000\,\Omega$ resistance, could recover $35.8 \pm 1.2\%$ of

PO_4^{3-} and 10.1 ± 0.5% of NH_4^+ of the fed digestate (Pepè Sciarria et al., 2019). These recoveries were significantly enhanced to 83.1 ± 3.7% and 14.7 ± 0.6%, respectively, when the precipitation of the anode effluent was carried out in a separate flask with the addition of Mg in the form of $MgCl_2$ to achieve a *Mg:P molar ratio* of 1.6:1 (Table 4). Similar recovery trends were observed with the MEC with similar configuration but operated with a stainless steel mesh cathode or a Pt/C cathode. The separate precipitation with the addition of Mg to adjust the Mg:P molar ratio to 1.6:1 also increased the recoveries of PO_4^{3-} and NH_4^+ by the MEC from 20%–30% and < 5%, respectively, to 68%–74% and 7%–9%, respectively. The addition of sea water bitterns instead of $MgCl_2$ could lead to similar or even better recovery performances of both the MFC and the MEC. Therefore optimization of the struvite crystallization process by supplying sufficient magnesium could improve the recoveries of both nitrogen and phosphorus by BESs while reducing the cost (as sea water bitterns are naturally available magnesium salts).

Nitrogen and phosphorus can be recovered even together with potassium in MDC-like BESs (Ledezma et al., 2017; Sleutels, Hamelers, & Buisman, 2010). These BESs also contained a third chamber in addition to the anode and cathode chambers but it can extract ions from the electrode chambers and thereby concentrating them, unlike the third chamber in an MDC losing ions to the electrode chambers. The concentrated solution in that third chamber (working chamber) can be further treated with carbonate and subjected to "flash-cooling" to promote crystallization of the nutrients for recovery (Ledezma et al., 2017). The movement of ions (NH_4^+ and K^+ from the anode chamber and PO_4^{3-} from the cathode chamber) to the working chamber is driven by the electric current, and thus, this nutrient-recovering process is called *"bioelectroconcentration"*. Such a bioelectroconcentration system, when operated with a fixed anode potential at 0.0 V vs. SHE, could recover 49.5 ± 1.8% of N, 42.8 ± 1.0% of P, and 54.7 ± 1.3% of K from the influent, at the respective rates of 7.18 kg of NH_4-N m^{-3} day^{-1}, 0.52 kg of PO_4-P m^{-3} day^{-1}, and 1.62 kg of K^+ m^{-3} day^{-1} (Ledezma et al., 2017) (Table 4). The highest concentrations of the nutrients that this technology could achieve were 26.2 ± 0.3 g NH_4-N L^{-1} (or 1.87 ± 0.02 M), 27.55 ± 0.95 g PO_4-P L^{-1} (or 0.29 ± 0.01 M), and 7.0 ± 0.4 g K^+ L^{-1}, which were *c.* 4.5-fold, 12.2-fold, and 3.8-fold the influent concentrations, respectively. However, the recoveries as crystals by flash-cooling were still low, e.g., only 14.8% for nitrogen. Further improvements are therefore needed to achieve efficient bioelectroconcentrations, although such simultaneous nutrient removal processes are very attractive.

6 General remarks

As detailed in the earlier sections, various agriculture wastes can be used as substrates for BESs, from which different resources can be recovered. In BESs, substrates can be either oxidized to harvest electrons for electricity production

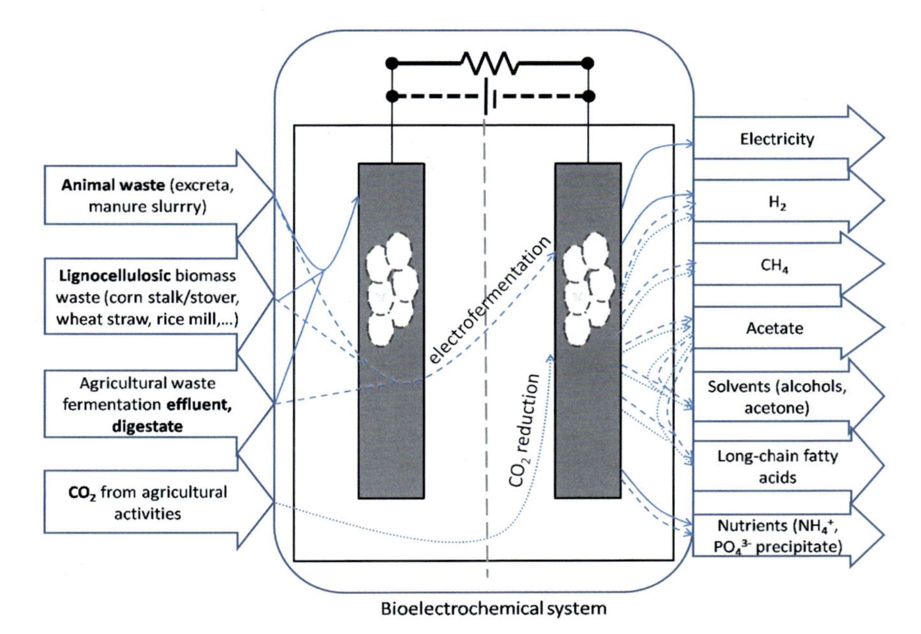

FIG. 2 Schematic of the bioconversions of various agriculture wastes as substrates in BESs to the resources that can be recovered. Notes: *arrow symbols* on the left illustrate the substrates (wastes) entering the BES; *arrow symbols* on the right illustrate outputs of the BES; *solid line arrows* indicate bioelectrogenesis- and bioelectrolysis-associated pathways; *dashed line arrows* indicate electrofermentation-associated pathways (usually occurring in single-chamber BESs having no membrane); *dotted line arrows* indicate CO_2-reduction-associated pathways. The *central dashed line* indicates the membrane, the presence of which is optional. *Gray bars* are symbols of electrodes while *dashed-line white oval shapes* indicate microbial cells, the presence of which depends on the BES type.

or hydrogen production, or "electrofermented", or even reduced (as for CO_2 or acetate, etc.) to synthesize various products, including hydrogen, methane, acetate, organic acids, solvents (alcohols and acetones), nutrients (N, P, and K), etc., depending on the chemical natures of substrates and the BES operational conditions (Fig. 2). Almost all kinds of agriculture wastes can be feedstock to BESs as they are generally rich in organic contents and nutrients. It is important to note that *BESs push biodegradation beyond the limits of existing technologies* that are solely based on anaerobic digestion because the end products of the latter can be further degraded in BESs to recover energy as electricity, fuel gases, or valuable products. With such a powerful waste-to-resource conversion capacity, BESs are therefore a very promising and important technology for resource recovery from agriculture wastes (please read also Supplemental Material for new promising BES applications).

Despite its promising potentials, the practical applications of the BES technology are still not realized, e.g., in agriculture, because of its limitations, as discussed here and there earlier. In terms of electricity generation, the low power output of BESs makes them less viable compared to other technologies (Arends & Verstraete, 2012). Although the target BES power output of $1\,kW\,m^{-3}$ has been achieved at laboratory-scale (Fan, Hu, & Liu, 2007; Nevin et al., 2008), up-scaled systems have thus far only displayed modest performances, far away from that level. Therefore BES studies have been directed from bioelectrogenesis to bioelectrosynthesis and/or bioelectrochemical nutrient recovery, especially for the last 10 years. However, bioelectrosynthesis and bioelectrochemical nutrient recovery face their own issues such as low selectivity, high cost, and lack of efficient methods for product recovery (Kelly & He, 2014; Kuntke et al., 2018; Nancharaiah, Mohan, & Lens, 2016; Prévoteau et al., 2020). As a result, the key tasks of ongoing BESs studies are still to overcome these limitations.

7 BESs and the prospect of a circular agricultural economy

A circular economy enables self-sustainability, i.e., waste is constantly recycled and reused, and thus, zero-waste production or zero environmental impact is expected. This principle should also apply for a circular agricultural economy (CAE), i.e., a circular economy solely based on agriculture, especially for developing countries (for further reading, please refer to Supplemental Material). With all the available BES technologies, it is possible to achieve a CAE by using only BES technologies as long as they are fully developed (Fig. 3), or by combining BES technologies with other sustainable technologies. This is because agriculture wastes are solely organics, which are highly compatible substrates for BESs. Accordingly, any wastes produced in agriculture can be treated to recover energy in them as electricity or chemical energy in biofuels by using MFCs or MECs. The nutrients in the waste can also be recovered in the same ways, with additional technical modules for harvesting. Waste, with proper pretreatments, can even be bioelectro-refined (or electro-fermented) to various valued chemicals or biomass by using MFCs or MESs. By CO_2 capture and fixation in MESs or MSCs, valued chemicals or biomass can also be produced from carbon dioxide released from bioelectrochemical degradation of waste or during any other agricultural activities. In such ways, all the materials involved in agricultural activities can be recycled or, in other words, recirculated in a closed-loop, theoretically resulting in zero-waste production. Indeed, any other sustainable technologies can take part in any step of this loop, to complement BES technologies, especially when considering that different BES technologies are, at different degrees, not yet optimized and/or economically viable. Nonetheless, the possibility of using BES technologies to accomplish CAE indicates the great potentials of these technologies to be exploited for our future society.

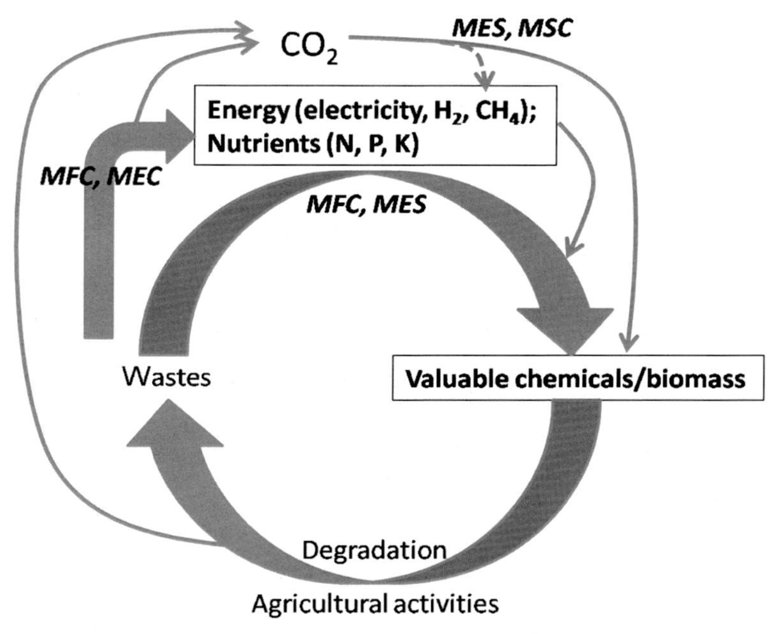

FIG. 3 The participations of BES technologies in a circular agricultural economy model. Notes: *MFC*, microbial fuel cell; *MEC*, microbial electrolysis cell; *MES*, microbial electrosynthesis system; *MSC*, microbial solar cell.

8 Conclusions

BESs have recently emerged as novel sustainable technologies for agriculture resource recovery to complement existing ones. BESs offer unique resource recovery applications such as extracting energy from agriculture wastes as electricity and biofuels, upgrading the wastes to valuable chemicals or biomass, or recovering nutrients from the wastes. These unique applications are feasible due to BESs' unique working mechanisms determined by their electroactive biocatalysts, their abiotic catalysts as well as their configurational and operational characteristics. By varying these factors, diverse technical options are available for BES applications in resource recovery from various agricultural wastes. Although significant technical advances have been achieved with BESs, in general, their performances are still up to certain limits, and significant improvement breakthroughs are required for their large-scale applications. When scaling up BESs, how to minimize losses, and how to recover products in cost-efficient manners are major challenges. Particularly, the cost vs. performance balance is the key issue for engineering considerations. Additional measures to enhance BES performances in agriculture resource recovery should also include efficient pretreatments of agriculture wastes and/or proper integrations of BESs with other technologies (anaerobic digestion, biorefinery, etc.). Nevertheless, BESs still show great potentials for use to achieve a circular agricultural economy, to cope with resource exhaustions, and to ensure sustainable development for the future, especially for agriculture-based developing countries.

References

Agler, M. T., Wrenn, B. A., Zinder, S. H., & Angenent, L. T. (2011). Waste to bioproduct conversion with undefined mixed cultures: The carboxylate platform. *Trends in Biotechnology, 29*(2), 70–78.

Almatouq, A., & Babatunde, A. O. (2016). Concurrent phosphorus recovery and energy generation in mediator-less dual chamber microbial fuel cells: Mechanisms and influencing factors. *International Journal of Environmental Research and Public Health, 13*(4), 375.

Almatouq, A., & Babatunde, A. O. (2017). Concurrent hydrogen production and phosphorus recovery in dual chamber microbial electrolysis cell. *Bioresource Technology, 237*, 193–203. https://doi.org/10.1016/j.biortech.2017.02.043.

Ammam, F., Tremblay, P.-L., Lizak, D. M., & Zhang, T. (2016). Effect of tungstate on acetate and ethanol production by the electrosynthetic bacterium Sporomusa ovata. *Biotechnology for Biofuels, 9*(1), 163. https://doi.org/10.1186/s13068-016-0576-0.

Arends, J. (2013). *Optimizing the plant microbial fuel cell: Diversifying applications and product outputs*. Ghent University.

Arends, J. B. A., Patil, S. A., Roume, H., & Rabaey, K. (2017). Continuous long-term electricity-driven bioproduction of carboxylates and isopropanol from CO2 with a mixed microbial community. *Journal of CO_2 Utilization, 20*, 141–149. https://doi.org/10.1016/j.jcou.2017.04.014.

Arends, J. B., & Verstraete, W. (2012). 100 years of microbial electricity production: Three concepts for the future. *Microbial Biotechnology, 5*(3), 333–346. https://doi.org/10.1111/j.1751-7915.2011.00302.x.

Arredondo, M. R., Kuntke, P., Jeremiasse, A., Sleutels, T., Buisman, C., & Ter Heijne, A. (2015). Bioelectrochemical systems for nitrogen removal and recovery from wastewater. *Environmental Science: Water Research & Technology, 1*(1), 22–33.

Aulenta, F., Reale, P., Catervi, A., Panero, S., & Majone, M. (2008). Kinetics of trichloroethene dechlorination and methane formation by a mixed anaerobic culture in a bio-electrochemical system. *Electrochimica Acta, 53*(16), 5300–5305. https://doi.org/10.1016/j.electacta.2008.02.084.

Babanova, S., Carpenter, K., Phadke, S., Suzuki, S., Ishii, S.i., Phan, T., et al. (2016). The effect of membrane type on the performance of microbial electrosynthesis cells for methane production. *Journal of the Electrochemical Society, 164*(3), H3015.

Baek, G., Kim, J., Lee, S., & Lee, C. (2017). Development of biocathode during repeated cycles of bioelectrochemical conversion of carbon dioxide to methane. *Bioresource Technology, 241*, 1201–1207. https://doi.org/10.1016/j.biortech.2017.06.125.

Bajracharya, S., Sharma, M., Mohanakrishna, G., Dominguez Benneton, X., Strik, D. P. B. T. B., Sarma, P. M., et al. (2016). An overview on emerging bioelectrochemical systems (BESs): Technology for sustainable electricity, waste remediation, resource recovery, chemical production and beyond. *Renewable Energy, 98*, 153–170. https://doi.org/10.1016/j.renene.2016.03.002.

Bajracharya, S., ter Heijne, A., Dominguez Benetton, X., Vanbroekhoven, K., Buisman, C. J. N., Strik, D. P. B. T. B., et al. (2015). Carbon dioxide reduction by mixed and pure cultures in microbial electrosynthesis using an assembly of graphite felt and stainless steel as a cathode. *Bioresource Technology*, *195*, 14–24. https://doi.org/10.1016/j.biortech.2015.05.081.

Bajracharya, S., Vanbroekhoven, K., Buisman, C. J., Pant, D., & Strik, D. P. (2016). Application of gas diffusion biocathode in microbial electrosynthesis from carbon dioxide. *Environmental Science and Pollution Research*, *23*(22), 22292–22308.

Batlle-Vilanova, P., Ganigué, R., Ramió-Pujol, S., Bañeras, L., Jiménez, G., Hidalgo, M., et al. (2017). Microbial electrosynthesis of butyrate from carbon dioxide: Production and extraction. *Bioelectrochemistry*, *117*, 57–64. https://doi.org/10.1016/j.bioelechem.2017.06.004.

Batlle-Vilanova, P., Puig, S., Gonzalez-Olmos, R., Vilajeliu-Pons, A., Balaguer, M. D., & Colprim, J. (2015). Deciphering the electron transfer mechanisms for biogas upgrading to biomethane within a mixed culture biocathode. *RSC Advances*, *5*(64), 52243–52251. https://doi.org/10.1039/c5ra09039c.

Behera, M., Jana, P. S., More, T. T., & Ghangrekar, M. M. (2010). Rice mill wastewater treatment in microbial fuel cells fabricated using proton exchange membrane and earthen pot at different pH. *Bioelectrochemistry*, *79*(2), 228–233. https://doi.org/10.1016/j.bioelechem.2010.06.002.

Bian, B., Bajracharya, S., Xu, J., Pant, D., & Saikaly, P. E. (2020). Microbial electrosynthesis from CO2: Challenges, opportunities and perspectives in the context of circular bioeconomy. *Bioresource Technology*, *302*, 122863. https://doi.org/10.1016/j.biortech.2020.122863.

Borole, A. P., Hamilton, C. Y., & Schell, D. J. (2013). Conversion of residual organics in corn stover-derived biorefinery stream to bioenergy via a microbial fuel cell. *Environmental Science & Technology*, *47*(1), 642–648. https://doi.org/10.1021/es3023495.

Brink, R., Densmore, J., & Hill, G. (1977). Soil deterioration and the growing world demand for food. *Science*, *197*(4304), 625–630.

Cai, W., Liu, W., Yang, C., Wang, L., Liang, B., Thangavel, S., et al. (2016). Biocathodic methanogenic community in an integrated anaerobic digestion and microbial electrolysis system for enhancement of methane production from waste sludge. *ACS Sustainable Chemistry & Engineering*, *4*(9), 4913–4921. https://doi.org/10.1021/acssuschemeng.6b01221.

Call, D., & Logan, B. E. (2008). Hydrogen production in a single chamber microbial electrolysis cell lacking a membrane. *Environmental Science & Technology*, *42*(9), 3401–3406. https://doi.org/10.1021/es8001822.

Cerrillo, M., Viñas, M., & Bonmatí, A. (2017). Startup of electromethanogenic microbial electrolysis cells with two different biomass inocula for biogas upgrading. *ACS Sustainable Chemistry & Engineering*, *5*(10), 8852–8859. https://doi.org/10.1021/acssuschemeng.7b01636.

Cheng, S., & Logan, B. E. (2007). Sustainable and efficient biohydrogen production via electrohydrogenesis. *Proceedings of the National Academy of Sciences*, *104*(47), 18871–18873. https://doi.org/10.1073/pnas.0706379104.

Cheng, S., Xing, D., Call, D. F., & Logan, B. E. (2009). Direct biological conversion of electrical current into methane by electromethanogenesis. *Environmental Science & Technology*, *43*(10), 3953–3958. https://doi.org/10.1021/es803531g.

Chu, N., Liang, Q., Jiang, Y., & Zeng, R. J. (2020). Microbial electrochemical platform for the production of renewable fuels and chemicals. *Biosensors and Bioelectronics*, *150*, 111922.

Clauwaert, P., Aelterman, P., Pham, T. H., De Schamphelaire, L., Carballa, M., Rabaey, K., et al. (2008). Minimizing losses in bio-electrochemical systems: The road to applications [review]. *Applied Microbiology and Biotechnology*, *79*(6), 901–913.

Clauwaert, P., & Verstraete, W. (2009). Methanogenesis in membraneless microbial electrolysis cells. *Applied Microbiology and Biotechnology*, *82*(5), 829–836. https://doi.org/10.1007/s00253-008-1796-4.

Cusick, R. D., & Logan, B. E. (2012). Phosphate recovery as struvite within a single chamber microbial electrolysis cell. *Bioresource Technology*, *107*, 110–115. https://doi.org/10.1016/j.biortech.2011.12.038.

Cusick, R. D., Ullery, M. L., Dempsey, B. A., & Logan, B. E. (2014). Electrochemical struvite precipitation from digestate with a fluidized bed cathode microbial electrolysis cell. *Water Research*, *54*, 297–306. https://doi.org/10.1016/j.watres.2014.01.051.

Desloover, J., Abate Woldeyohannis, A., Verstraete, W., Boon, N., & Rabaey, K. (2012). Electrochemical resource recovery from digestate to prevent ammonia toxicity during anaerobic digestion. *Environmental Science & Technology*, *46*(21), 12209–12216.

El-Nahhal, Y. Z., Al-Agha, M. R., El-Nahhal, I. Y., El Aila, N. A., El-Nahal, F. I., & Alhalabi, R. A. (2020). Electricity generation from animal manure. *Biomass and Bioenergy*, *136*, 105531. https://doi.org/10.1016/j.biombioe.2020.105531.

Fan, Y. Z., Hu, H. Q., & Liu, H. (2007). Sustainable power generation in microbial fuel cells using bicarbonate buffer and proton transfer mechanisms. *Environmental Science & Technology*, *41* (23), 8154–8158.

Fischer, F., Bastian, C., Happe, M., Mabillard, E., & Schmidt, N. (2011). Microbial fuel cell enables phosphate recovery from digested sewage sludge as struvite. *Bioresource Technology*, *102*(10), 5824–5830. https://doi.org/10.1016/j.biortech.2011.02.089.

Fu, Q., Xiao, S., Li, Z., Li, Y., Kobayashi, H., Li, J., et al. (2018). Hybrid solar-to-methane conversion system with a faradaic efficiency of up to 96%. *Nano Energy*, *53*, 232–239. https://doi.org/10.1016/j.nanoen.2018.08.051.

Ganigué, R., Puig, S., Batlle-Vilanova, P., Balaguer, M. D., & Colprim, J. (2015). Microbial electrosynthesis of butyrate from carbon dioxide. *Chemical Communications*, *51*(15), 3235–3238. https://doi.org/10.1039/c4cc10121a.

Gavilanes, J., Noori, M. T., & Min, B. (2019). Enhancing bio-alcohol production from volatile fatty acids by suppressing methanogenic activity in single chamber microbial electrosynthesis cells (SCMECs). *Bioresource Technology Reports*, *7*, 100292. https://doi.org/10.1016/j.biteb.2019.100292.

Gavilanes, J., Reddy, C. N., & Min, B. (2019). Microbial electrosynthesis of bioalcohols through reduction of high concentrations of volatile fatty acids. *Energy & Fuels*, *33*(5), 4264–4271. https://doi.org/10.1021/acs.energyfuels.8b04215.

Gildemyn, S., Luther, A. K., Andersen, S. J., Desloover, J., & Rabaey, K. (2015). Electrochemically and bioelectrochemically induced ammonium recovery. *Journal of Visualized Experiments*, (95), e52405.

Gildemyn, S., Verbeeck, K., Slabbinck, R., Andersen, S. J., Prévoteau, A., & Rabaey, K. (2015). Integrated production, extraction, and concentration of acetic acid from CO2 through microbial electrosynthesis. *Environmental Science & Technology Letters*, *2*(11), 325–328. https://doi.org/10.1021/acs.estlett.5b00212.

Happe, M., Sugnaux, M., Cachelin, C. P., Stauffer, M., Zufferey, G., Kahoun, T., et al. (2016). Scale-up of phosphate remobilization from sewage sludge in a microbial fuel cell. *Bioresource Technology*, *200*, 435–443. https://doi.org/10.1016/j.biortech.2015.10.057.

Harnisch, F., & Schröder, U. (2010). From MFC to MXC: Chemical and biological cathodes and their potential for microbial bioelectrochemical systems. *Chemical Society Reviews*, *39*(11), 4433–4448.

Hirano, S.-I., & Matsumoto, N. (2018). Analysis of a bio-electrochemical reactor containing carbon fiber textiles for the anaerobic digestion of tomato plant residues. *Bioresource Technology*, *249*, 809–817. https://doi.org/10.1016/j.biortech.2017.09.206.

Hirooka, K., & Ichihashi, O. (2013). Phosphorus recovery from artificial wastewater by microbial fuel cell and its effect on power generation. *Bioresource Technology*, *137*, 368–375. https://doi.org/10.1016/j.biortech.2013.03.067.

Hou, H., Li, Z., Liu, B., Liang, S., Xiao, K., Zhu, Q., et al. (2020). Biogas and phosphorus recovery from waste activated sludge with protocatechuic acid enhanced Fenton pretreatment, anaerobic digestion and microbial electrolysis cell. *Science of the Total Environment*, *704*, 135274. https://doi.org/10.1016/j.scitotenv.2019.135274.

Ichihashi, O., & Hirooka, K. (2012). Removal and recovery of phosphorus as struvite from swine wastewater using microbial fuel cell. *Bioresource Technology*, *114*, 303–307. https://doi.org/10.1016/j.biortech.2012.02.124.

Inoue, K., Ito, T., Kawano, Y., Iguchi, A., Miyahara, M., Suzuki, Y., et al. (2013). Electricity generation from cattle manure slurry by cassette-electrode microbial fuel cells. *Journal of Bioscience and Bioengineering*, *116*(5), 610–615. https://doi.org/10.1016/j.jbiosc.2013.05.011.

Jiang, Y., Su, M., & Li, D. (2014). Removal of sulfide and production of methane from carbon dioxide in microbial fuel cells–microbial electrolysis cell (MFCs–MEC) coupled system. *Applied Biochemistry and Biotechnology*, *172*(5), 2720–2731. https://doi.org/10.1007/s12010-013-0718-9.

Jiang, Y., Su, M., Zhang, Y., Zhan, G., Tao, Y., & Li, D. (2013). Bioelectrochemical systems for simultaneously production of methane and acetate from carbon dioxide at relatively high rate. *International Journal of Hydrogen Energy*, *38*(8), 3497–3502. https://doi.org/10.1016/j.ijhydene.2012.12.107.

Jourdin, L., Grieger, T., Monetti, J., Flexer, V., Freguia, S., Lu, Y., et al. (2015). High acetic acid production rate obtained by microbial electrosynthesis from carbon dioxide. *Environmental Science & Technology*, *49*(22), 13566–13574. https://doi.org/10.1021/acs.est.5b03821.

Jourdin, L., Raes, S. M. T., Buisman, C. J. N., & Strik, D. P. B. T. B. (2018). Critical biofilm growth throughout unmodified carbon felts allows continuous bioelectrochemical chain elongation from CO2 up to caproate at high current density [original research]. *Frontiers in Energy Research*, *6*(7). https://doi.org/10.3389/fenrg.2018.00007.

Kelly, P. T., & He, Z. (2014). Nutrients removal and recovery in bioelectrochemical systems: A review. *Bioresource Technology*, *153*, 351–360.

Khosravanipour Mostafazadeh, A., Drogui, P., Brar, S. K., Tyagi, R. D., Le Bihan, Y., Buelna, G., et al. (2016). Enhancement of biobutanol production by electromicrobial glucose conversion in a dual chamber fermentation cell using C. pasteurianum. *Energy Conversion and Management*, *130*, 165–175. https://doi.org/10.1016/j.enconman.2016.10.050.

Kiely, P. D., Cusick, R., Call, D. F., Selembo, P. A., Regan, J. M., & Logan, B. E. (2011). Anode microbial communities produced by changing from microbial fuel cell to microbial electrolysis cell operation using two different wastewaters. *Bioresource Technology*, *102*(1), 388–394. https://doi.org/10.1016/j.biortech.2010.05.019.

Kim, Y., & Logan, B. E. (2013). Microbial desalination cells for energy production and desalination. *Desalination*, *308*, 122–130. https://doi.org/10.1016/j.desal.2012.07.022.

Kim, J. R., Zuo, Y., Regan, J. M., & Logan, B. E. (2008). Analysis of ammonia loss mechanisms in microbial fuel cells treating animal wastewater. *Biotechnology and Bioengineering*, *99*(5), 1120–1127. https://doi.org/10.1002/bit.21687.

Kobayashi, H., Saito, N., Fu, Q., Kawaguchi, H., Vilcaez, J., Wakayama, T., et al. (2013). Bioelectrochemical property and phylogenetic diversity of microbial communities associated with bioelectrodes of an electromethanogenic reactor. *Journal of Bioscience and Bioengineering*, *116*(1), 114–117. https://doi.org/10.1016/j.jbiosc.2013.01.001.

Kondaveeti, S., & Min, B. (2015). Bioelectrochemical reduction of volatile fatty acids in anaerobic digestion effluent for the production of biofuels. *Water Research*, *87*, 137–144. https://doi.org/10.1016/j.watres.2015.09.011.

Kuntke, P., Geleji, M., Bruning, H., Zeeman, G., Hamelers, H. V. M., & Buisman, C. J. N. (2011). Effects of ammonium concentration and charge exchange on ammonium recovery from high strength wastewater using a microbial fuel cell. *Bioresource Technology*, *102*(6), 4376–4382. https://doi.org/10.1016/j.biortech.2010.12.085.

Kuntke, P., Sleutels, T. H. J. A., Rodríguez Arredondo, M., Georg, S., Barbosa, S. G., ter Heijne, A., et al. (2018). (Bio)electrochemical ammonia recovery: Progress and perspectives. *Applied Microbiology and Biotechnology*, *102*(9), 3865–3878. https://doi.org/10.1007/s00253-018-8888-6.

Kuntke, P., Sleutels, T. H. J. A., Saakes, M., & Buisman, C. J. N. (2014). Hydrogen production and ammonium recovery from urine by a microbial electrolysis cell. *International Journal of Hydrogen Energy*, *39*(10), 4771–4778. https://doi.org/10.1016/j.ijhydene.2013.10.089.

Kuntke, P., Śmiech, K. M., Bruning, H., Zeeman, G., Saakes, M., Sleutels, T. H. J. A., et al. (2012). Ammonium recovery and energy production from urine by a microbial fuel cell. *Water Research*, *46*(8), 2627–2636. https://doi.org/10.1016/j.watres.2012.02.025.

Kuntke, P., Zamora, P., Saakes, M., Buisman, C. J. N., & Hamelers, H. V. M. (2016). Gas-permeable hydrophobic tubular membranes for ammonia recovery in bio-electrochemical systems. *Environmental Science: Water Research & Technology*, *2*(2), 261–265. https://doi.org/10.1039/c5ew00299k.

LaBelle, E. V., Marshall, C. W., Gilbert, J. A., & May, H. D. (2014). Influence of acidic pH on hydrogen and acetate production by an electrosynthetic microbiome. *PLoS One*, *9*(10). https://doi.org/10.1371/journal.pone.0109935, e109935.

Lalaurette, E., Thammannagowda, S., Mohagheghi, A., Maness, P.-C., & Logan, B. E. (2009). Hydrogen production from cellulose in a two-stage process combining fermentation and electrohydrogenesis. *International Journal of Hydrogen Energy*, *34*(15), 6201–6210. https://doi.org/10.1016/j.ijhydene.2009.05.112.

Ledezma, P., Jermakka, J., Keller, J., & Freguia, S. (2017). Recovering nitrogen as a solid without chemical dosing: Bio-electroconcentration for recovery of nutrients from urine. *Environmental Science & Technology Letters*, *4*(3), 119–124. https://doi.org/10.1021/acs.estlett.7b00024.

Lee, Y., & Nirmalakhandan, N. (2011). Electricity production in membrane-less microbial fuel cell fed with livestock organic solid waste. *Bioresource Technology*, *102*(10), 5831–5835. https://doi.org/10.1016/j.biortech.2011.02.090.

Lei, Y., Du, M., Kuntke, P., Saakes, M., van der Weijden, R., & Buisman, C. J. N. (2019). Energy efficient phosphorus recovery by microbial electrolysis cell induced calcium phosphate precipitation. *ACS Sustainable Chemistry & Engineering*, *7*(9), 8860–8867. https://doi.org/10.1021/acssuschemeng.9b00867.

Lei, Y., Remmers, J. C., Saakes, M., van der Weijden, R. D., & Buisman, C. J. N. (2018). Is there a precipitation sequence in municipal wastewater induced by electrolysis? *Environmental Science & Technology*, *52*(15), 8399–8407. https://doi.org/10.1021/acs.est.8b02869.

Li, J., Li, Z., Xiao, S., Fu, Q., Kobayashi, H., Zhang, L., et al. (2020). Startup cathode potentials determine electron transfer behaviours of biocathodes catalysing CO2 reduction to CH4 in microbial electrosynthesis. *Journal of CO_2 Utilization*, *35*, 169–175. https://doi.org/10.1016/j.jcou.2019.09.013.

Li, X.-H., Liang, D.-W., Bai, Y.-X., Fan, Y.-T., & Hou, H.-W. (2014). Enhanced H2 production from corn stalk by integrating dark fermentation and single chamber microbial electrolysis cells with double anode arrangement. *International Journal of Hydrogen Energy*, *39*(17), 8977–8982. https://doi.org/10.1016/j.ijhydene.2014.03.065.

Liang, D.-W., Peng, S.-K., Lu, S.-F., Liu, Y.-Y., Lan, F., & Xiang, Y. (2011). Enhancement of hydrogen production in a single chamber microbial electrolysis cell through anode arrangement optimization. *Bioresource Technology*, *102*(23), 10881–10885. https://doi.org/10.1016/j.biortech.2011.09.028.

Lin, H., Wu, X., Nelson, C., Miller, C., & Zhu, J. (2016). Electricity generation and nutrients removal from high-strength liquid manure by air-cathode microbial fuel cells. *Journal of Environmental Science and Health, Part A*, *51*(3), 240–250. https://doi.org/10.1080/10934529.2015.1094342.

Liu, C., Gallagher, J. J., Sakimoto, K. K., Nichols, E. M., Chang, C. J., Chang, M. C., et al. (2015). Nanowire–bacteria hybrids for unassisted solar carbon dioxide fixation to value-added chemicals. *Nano Letters*, *15*(5), 3634–3639.

Liu, H., Grot, S., & Logan, B. E. (2005). Electrochemically assisted microbial production of hydrogen from acetate. *Environmental Science & Technology*, *39*(11), 4317–4320.

Liu, D., Zheng, T., Buisman, C., & ter Heijne, A. (2017). Heat-treated stainless steel felt as a new cathode material in a methane-producing bioelectrochemical system. *ACS Sustainable Chemistry & Engineering*, *5*(12), 11346–11353. https://doi.org/10.1021/acssuschemeng.7b02367.

Logan, B. E., Hamelers, B., Rozendal, R., Schrorder, U., Keller, J., Freguia, S., et al. (2006). Microbial fuel cells: Methodology and technology. *Environmental Science & Technology*, *40*(17), 5181–5192.

Lovley, D. R. (2017). Happy together: Microbial communities that hook up to swap electrons. *The ISME Journal*, *11*(2), 327–336. https://doi.org/10.1038/ismej.2016.136.

Lu, L., Xing, D., & Ren, N. (2012). Pyrosequencing reveals highly diverse microbial communities in microbial electrolysis cells involved in enhanced H2 production from waste activated sludge. *Water Research*, *46*(7), 2425–2434.

Luo, X., Zhang, F., Liu, J., Zhang, X., Huang, X., & Logan, B. E. (2014). Methane production in microbial reverse-electrodialysis methanogenesis cells (MRMCs) using thermolytic solutions. *Environmental Science & Technology*, *48*(15), 8911–8918. https://doi.org/10.1021/es501979z.

Madigan, M. T., Martinko, J., & Parker, J. (2004). *Brock biology of microorganisms* (10 ed.). Pearson Education Inc.

Marshall, C. W., Ross, D. E., Fichot, E. B., Norman, R. S., & May, H. D. (2012). Electrosynthesis of commodity chemicals by an autotrophic microbial community. *Applied and Environmental Microbiology*, *78*(23), 8412–8420.

Marshall, C. W., Ross, D. E., Fichot, E. B., Norman, R. S., & May, H. D. (2013). Long-term operation of microbial electrosynthesis systems improves acetate production by autotrophic microbiomes. *Environmental Science & Technology*, *47*(11), 6023–6029. https://doi.org/10.1021/es400341b.

Min, B., Kim, J., Oh, S., Regan, J. M., & Logan, B. E. (2005). Electricity generation from swine wastewater using microbial fuel cells. *Water Research*, *39*(20), 4961–4968. https://doi.org/10.1016/j.watres.2005.09.039.

Moscoviz, R., Toledo-Alarcón, J., Trably, E., & Bernet, N. (2016). Electro-fermentation: How to drive fermentation using electrochemical systems. *Trends in Biotechnology*, *34*(11), 856–865.

Nancharaiah, Y., Mohan, S. V., & Lens, P. (2016). Recent advances in nutrient removal and recovery in biological and bioelectrochemical systems. *Bioresource Technology*, *215*, 173–185.

Nevin, K. P., Hensley, S. A., Franks, A. E., Summers, Z. M., Ou, J., Woodard, T. L., et al. (2011). Electrosynthesis of organic compounds from carbon dioxide is catalyzed by a diversity of acetogenic microorganisms. *Applied and Environmental Microbiology*, *77*(9), 2882. https://doi.org/10.1128/aem.02642-10.

Nevin, K. P., Richter, H., Covalla, S., Johnson, J., Woodard, T., Orloff, A., et al. (2008). Power output and columbic efficiencies from biofilms of Geobacter sulfurreducens comparable to mixed community microbial fuel cells. *Environmental Microbiology*, *10*(10), 2505–2514.

Nevin, K. P., Woodard, T. L., Franks, A. E., Summers, Z. M., & Lovley, D. R. (2010). Microbial electrosynthesis: Feeding microbes electricity to convert carbon dioxide and water to multicarbon extracellular organic compounds. *mBio*, *1*(2). e00103-00110.

Nie, H., Zhang, T., Cui, M., Lu, H., Lovley, D. R., & Russell, T. P. (2013). Improved cathode for high efficient microbial-catalyzed reduction in microbial electrosynthesis cells. *Physical Chemistry Chemical Physics*, *15*(34), 14290–14294. https://doi.org/10.1039/c3cp52697f.

Patil, S., Arends, J., Vanwonterghem, I., Meerbergen, J., Guo, K., Tyson, G., et al. (2015). Selective enrichment establishes a stable performing community for microbial electrosynthesis of acetate from CO2. *Environmental Science & Technology*, *49*, 8833–8843. https://doi.org/10.1021/es506149d.

Pepè Sciarria, T., Vacca, G., Tambone, F., Trombino, L., & Adani, F. (2019). Nutrient recovery and energy production from digestate using microbial electrochemical technologies (METs). *Journal of Cleaner Production*, *208*, 1022–1029. https://doi.org/10.1016/j.jclepro.2018.10.152.

Petrus, L., & Noordermeer, M. A. (2006). Biomass to biofuels, a chemical perspective. *Green Chemistry*, *8*(10), 861–867.

Pham, T. H., Rabaey, K., Aelterman, P., Clauwaert, P., De Schamphelaire, L., Boon, N., et al. (2006). Microbial fuel cells in relation to conventional anaerobic digestion technology. *Engineering in Life Sciences*, *6*(3), 285–292.

Prajapati, K. B., & Singh, R. (2020a). Bio-electrochemically hydrogen and methane production from co-digestion of wastes. *Energy*, *198*, 117259. https://doi.org/10.1016/j.energy.2020.117259.

Prajapati, K. B., & Singh, R. (2020b). Enhancement of biogas production in bio-electrochemical digester from agricultural waste mixed with wastewater. *Renewable Energy*, *146*, 460–468. https://doi.org/10.1016/j.renene.2019.06.154.

Prévoteau, A., Carvajal-Arroyo, J. M., Ganigué, R., & Rabaey, K. (2020). Microbial electrosynthesis from CO2: Forever a promise? *Current Opinion in Biotechnology*, *62*, 48–57. https://doi.org/10.1016/j.copbio.2019.08.014.

Qin, M., & He, Z. (2014). Self-supplied ammonium bicarbonate draw solute for achieving wastewater treatment and recovery in a microbial electrolysis cell-forward osmosis-coupled system. *Environmental Science & Technology Letters*, *1*(10), 437–441. https://doi.org/10.1021/ez500280c.

Qin, M., Hynes, E. A., Abu-Reesh, I. M., & He, Z. (2017). Ammonium removal from synthetic wastewater promoted by current generation and water flux in an osmotic microbial fuel cell. *Journal of Cleaner Production*, *149*, 856–862. https://doi.org/10.1016/j.jclepro.2017.02.169.

Qin, M., Molitor, H., Brazil, B., Novak, J. T., & He, Z. (2016). Recovery of nitrogen and water from landfill leachate by a microbial electrolysis cell–forward osmosis system. *Bioresource Technology*, *200*, 485–492. https://doi.org/10.1016/j.biortech.2015.10.066.

Rabaey, K., Angenent, L., Schroder, U., & Keller, J. (2009). *Bioelectrochemical systems*. IWA publishing.

Rabaey, K., Rodriguez, J., Blackall, L. L., Keller, J., Gross, P., Batstone, D., et al. (2007). Microbial ecology meets electrochemistry: Electricity-driven and driving communities. *ISME Journal*, *1* (1), 9–18.

Rabaey, K., & Rozendal, R. A. (2010). Microbial electrosynthesis—Revisiting the electrical route for microbial production. *Nature Reviews. Microbiology*, *8*(10), 706–716 (doi:nrmicro2422 [pii] https://doi.org/10.1038/nrmicro2422.

Rabaey, K., & Verstraete, W. (2005). Microbial fuel cells: Novel biotechnology for energy generation. *Trends in Biotechnology*, *23*(6), 291–298.

Rodríguez Arredondo, M., Kuntke, P., ter Heijne, A., & Buisman, C. J. N. (2019). The concept of load ratio applied to bioelectrochemical systems for ammonia recovery. *Journal of Chemical Technology & Biotechnology*, *94*(6), 2055–2061. https://doi.org/10.1002/jctb.5992.

Rozendal, R. A., Hamelers, H. V. M., Euverink, G. J. W., Metz, S. J., & Buisman, C. J. N. (2006). Principle and perspectives of hydrogen production through biocatalyzed electrolysis. *International Journal of Hydrogen Energy*, *31*(12), 1632–1640.

Rozendal, R. A., Hamelers, H. V. M., Rabaey, K., Keller, J., & Buisman, C. J. N. (2008). Towards practical implementation of bioelectrochemical wastewater treatment [review]. *Trends in Biotechnology*, *26*(8), 450–459.

Saeed, H. M., Husseini, G. A., Yousef, S., Saif, J., Al-Asheh, S., Fara, A. A., et al. (2015). Microbial desalination cell technology: A review and a case study. *Desalination*, *359*, 1–13.

Samarakoon, G., Dinamarca, C., Nelabhotla, A., Winkler, D., & Bakke, R. (2020). *Modelling bioelectrochemical CO2 reduction to methane*.

Saratale, R. G., Saratale, G. D., Pugazhendhi, A., Zhen, G., Kumar, G., Kadier, A., et al. (2017). Microbiome involved in microbial electrochemical systems (MESs): A review. *Chemosphere*, *177*, 176–188. https://doi.org/10.1016/j.chemosphere.2017.02.143.

Sasaki, D., Sasaki, K., Watanabe, A., Morita, M., Igarashi, Y., & Ohmura, N. (2013). Efficient production of methane from artificial garbage waste by a cylindrical bioelectrochemical reactor containing carbon fiber textiles. *AMB Express*, *3*(1), 17. https://doi.org/10.1186/2191-0855-3-17.

Schink, B. (1997). Energetics of syntrophic cooperation in methanogenic degradation. *Microbiology and Molecular Biology Reviews*, *61*(2), 262–280.

Schlager, S., Haberbauer, M., Fuchsbauer, A., Hemmelmair, C., Dumitru, L. M., Hinterberger, G., et al. (2017). Bio-electrocatalytic application of microorganisms for carbon dioxide reduction to methane. *ChemSusChem*, *10*(1), 226–233. https://doi.org/10.1002/cssc.201600963.

Scott, K., & Murano, C. (2007). A study of a microbial fuel cell battery using manure sludge waste. *Journal of Chemical Technology & Biotechnology*, *82*(9), 809–817. https://doi.org/10.1002/jctb.1745.

Sharma, M., Aryal, N., Sarma, P. M., Vanbroekhoven, K., Lal, B., Benetton, X. D., et al. (2013). Bioelectrocatalyzed reduction of acetic and butyric acids via direct electron transfer using a mixed culture of sulfate-reducers drives electrosynthesis of alcohols and acetone. *Chemical Communications*, *49*(58), 6495–6497. https://doi.org/10.1039/c3cc42570c.

Shen, R., Liu, Z., He, Y., Zhang, Y., Lu, J., Zhu, Z., et al. (2016). Microbial electrolysis cell to treat hydrothermal liquefied wastewater from cornstalk and recover hydrogen: Degradation of organic compounds and characterization of microbial community. *International Journal of Hydrogen Energy*, *41*(7), 4132–4142. https://doi.org/10.1016/j.ijhydene.2016.01.032.

Shen, J., Wang, C., Liu, Y., Hu, C., Xin, Y., Ding, N., et al. (2018). Effect of ultrasonic pretreatment of the dairy manure on the electricity generation of microbial fuel cell. *Biochemical Engineering Journal*, *129*, 44–49. https://doi.org/10.1016/j.bej.2017.10.013.

Shimoyama, T., Komukai, S., Yamazawa, A., Ueno, Y., Logan, B., & Watanabe, K. (2008). Electricity generation from model organic wastewater in a cassette-electrode microbial fuel cell. *Applied Microbiology and Biotechnology*, *80*, 325–330. https://doi.org/10.1007/s00253-008-1516-0.

Sleutels, T. H. J. A., Hamelers, H. V. M., & Buisman, C. J. N. (2010). Reduction of pH buffer requirement in bioelectrochemical systems. *Environmental Science & Technology*, *44*(21), 8259–8263. https://doi.org/10.1021/es101858f.

Sotres, A., Cerrillo, M., Viñas, M., & Bonmatí, A. (2015). Nitrogen recovery from pig slurry in a two-chambered bioelectrochemical system. *Bioresource Technology*, *194*, 373–382. https://doi.org/10.1016/j.biortech.2015.07.036.

Steinbusch, K. J. J., Hamelers, H. V. M., Schaap, J. D., Kampman, C., & Buisman, C. J. N. (2010). Bioelectrochemical ethanol production through mediated acetate reduction by mixed cultures. *Environmental Science & Technology*, *44*(1), 513–517. https://doi.org/10.1021/es902371e.

Strik, D. P. B. T. B., Hamelers, H. V. M., Snel, J. F. H., & Buisman, C. J. N. (2008). Green electricity production with living plants and bacteria in a fuel cell. *International Journal of Energy Research*, *32*(9), 870–876. https://doi.org/10.1002/er.1397.

Strik, D. P. B. T. B., Timmers, R. A., Helder, M., Steinbusch, K. J. J., Hamelers, H. V. M., & Buisman, C. J. N. (2011). Microbial solar cells: Applying photosynthetic and electrochemically active organisms. *Trends in Biotechnology*, *29*(1), 41–49. https://doi.org/10.1016/j.tibtech.2010.10.001.

Sugnaux, M., Happe, M., Cachelin, C. P., Gloriod, O., Huguenin, G., Blatter, M., et al. (2016). Two stage bioethanol refining with multi litre stacked microbial fuel cell and microbial electrolysis cell. *Bioresource Technology*, *221*, 61–69. https://doi.org/10.1016/j.biortech.2016.09.020.

Sun, M., Sheng, G.-P., Zhang, L., Xia, C.-R., Mu, Z.-X., Liu, X.-W., et al. (2008). An MEC-MFC-coupled system for biohydrogen production from acetate. *Environmental Science & Technology*, *42*(21), 8095–8100.

Torella, J. P., Gagliardi, C. J., Chen, J. S., Bediako, D. K., Colón, B., Way, J. C., et al. (2015). Efficient solar-to-fuels production from a hybrid microbial–water-splitting catalyst system. *Proceedings of the National Academy of Sciences*, *112*(8), 2337. https://doi.org/10.1073/pnas.1424872112.

Van Eerten-Jansen, M. C. A. A., Heijne, A. T., Buisman, C. J. N., & Hamelers, H. V. M. (2012). Microbial electrolysis cells for production of methane from CO2: Long-term performance and perspectives. *International Journal of Energy Research*, *36*(6), 809–819. https://doi.org/10.1002/er.1954.

Van Eerten-Jansen, M. C. A. A., Ter Heijne, A., Grootscholten, T. I. M., Steinbusch, K. J. J., Sleutels, T. H. J. A., Hamelers, H. V. M., et al. (2013). Bioelectrochemical production of caproate and caprylate from acetate by mixed cultures. *ACS Sustainable Chemistry & Engineering*, *1*(5), 513–518. https://doi.org/10.1021/sc300168z.

Vassilev, I., Hernandez, P. A., Batlle-Vilanova, P., Freguia, S., Krömer, J. O., Keller, J., et al. (2018). Microbial electrosynthesis of isobutyric, butyric, caproic acids, and corresponding alcohols from carbon dioxide. *ACS Sustainable Chemistry & Engineering*, *6*(7), 8485–8493. https://doi.org/10.1021/acssuschemeng.8b00739.

Velasquez-Orta, S. B., Head, I. M., Curtis, T. P., & Scott, K. (2011). Factors affecting current production in microbial fuel cells using different industrial wastewaters. *Bioresource Technology*, *102*(8), 5105–5112. https://doi.org/10.1016/j.biortech.2011.01.059.

Venkata Mohan, S., Mohanakrishna, G., Velvizhi, G., Babu, V. L., & Sarma, P. N. (2010). Biocatalyzed electrochemical treatment of real field dairy wastewater with simultaneous power generation. *Biochemical Engineering Journal*, *51*(1), 32–39. https://doi.org/10.1016/j.bej.2010.04.012.

Villano, M., Aulenta, F., Ciucci, C., Ferri, T., Giuliano, A., & Majone, M. (2010). Bioelectrochemical reduction of CO2 to CH4 via direct and indirect extracellular electron transfer by a hydrogenophilic methanogenic culture. *Bioresource Technology*, *101*(9), 3085–3090. https://doi.org/10.1016/j.biortech.2009.12.077.

Villano, M., Monaco, G., Aulenta, F., & Majone, M. (2011). Electrochemically assisted methane production in a biofilm reactor. *Journal of Power Sources*, *196*(22), 9467–9472. https://doi.org/10.1016/j.jpowsour.2011.07.016.

Villano, M., Scardala, S., Aulenta, F., & Majone, M. (2013). Carbon and nitrogen removal and enhanced methane production in a microbial electrolysis cell. *Bioresource Technology*, *130*, 366–371. https://doi.org/10.1016/j.biortech.2012.11.080.

Wang, H., & Ren, Z. J. (2014). Bioelectrochemical metal recovery from wastewater: A review. *Water Research*, *66*, 219–232.

Wang, A., Sun, D., Cao, G., Wang, H., Ren, N., Wu, W.-M., et al. (2011). Integrated hydrogen production process from cellulose by combining dark fermentation, microbial fuel cells, and a microbial electrolysis cell. *Bioresource Technology*, *102*(5), 4137–4143. https://doi.org/10.1016/j.biortech.2010.10.137.

Wu, W.-M., Jain, M. K., & Zeikus, J. G. (1994). Anaerobic degradation of normal-and branched-chain fatty acids with four or more carbons to methane by a syntrophic methanogenic triculture. *Applied and Environmental Microbiology*, *60*(7), 2220–2226.

Wu, X., & Modin, O. (2013). Ammonium recovery from reject water combined with hydrogen production in a bioelectrochemical reactor. *Bioresource Technology*, *146*, 530–536. https://doi.org/10.1016/j.biortech.2013.07.130.

Wu, Z., Wang, J., Liu, J., Wang, Y., Bi, C., & Zhang, X. (2019). Engineering an electroactive Escherichia coli for the microbial electrosynthesis of succinate from glucose and CO2. *Microbial Cell Factories*, *18*(1), 15. https://doi.org/10.1186/s12934-019-1067-3.

Xie, B., Gong, W., Ding, A., Yu, H., Qu, F., Tang, X., et al. (2017). Microbial community composition and electricity generation in cattle manure slurry treatment using microbial fuel cells: Effects of inoculum addition. *Environmental Science and Pollution Research*, *24*(29), 23226–23235. https://doi.org/10.1007/s11356-017-9959-4.

Xu, H., Giwa, A. S., Wang, C., Chang, F., Yuan, Q., Wang, K., et al. (2017). Impact of antibiotics pretreatment on bioelectrochemical CH4 production. *ACS Sustainable Chemistry & Engineering*, *5*(10), 8579–8586. https://doi.org/10.1021/acssuschemeng.7b00923.

Yarlagadda, V. N., Gupta, A., Dodge, C. J., & Francis, A. J. (2012). Effect of exogenous electron shuttles on growth and fermentative metabolism in Clostridium sp. BC1. *Bioresource Technology*, *108*, 295–299. https://doi.org/10.1016/j.biortech.2011.12.040.

Yokoyama, H., Ohmori, H., Ishida, M., Waki, M., & Tanaka, Y. (2006). Treatment of cow-waste slurry by a microbial fuel cell and the properties of the treated slurry as a liquid manure. *Animal Science Journal*, *77*(6), 634–638. https://doi.org/10.1111/j.1740-0929.2006.00395.x.

Yuan, P., & Kim, Y. (2017). Increasing phosphorus recovery from dewatering centrate in microbial electrolysis cells. *Biotechnology for Biofuels*, *10*(1), 70. https://doi.org/10.1186/s13068-017-0754-8.

Zamora, P., Georgieva, T., Ter Heijne, A., Sleutels, T. H. J. A., Jeremiasse, A. W., Saakes, M., et al. (2017). Ammonia recovery from urine in a scaled-up microbial electrolysis cell. *Journal of Power Sources*, *356*, 491–499. https://doi.org/10.1016/j.jpowsour.2017.02.089.

Zhang, Y., & Angelidaki, I. (2015a). Counteracting ammonia inhibition during anaerobic digestion by recovery using submersible microbial desalination cell. *Biotechnology and Bioengineering*, *112*(7), 1478–1482. https://doi.org/10.1002/bit.25549.

Zhang, Y., & Angelidaki, I. (2015b). Recovery of ammonia and sulfate from waste streams and bioenergy production via bipolar bioelectrodialysis. *Water Research*, *85*, 177–184. https://doi.org/10.1016/j.watres.2015.08.032.

Zhang, Y., & Angelidaki, I. (2015c). Submersible microbial desalination cell for simultaneous ammonia recovery and electricity production from anaerobic reactors containing high levels of ammonia. *Bioresource Technology*, *177*, 233–239. https://doi.org/10.1016/j.biortech.2014.11.079.

Zhang, Y., Min, B., Huang, L., & Angelidaki, I. (2009). Generation of electricity and analysis of microbial communities in wheat straw biomass-powered microbial fuel cells. *Applied and Environmental Microbiology*, *75*(11), 3389. https://doi.org/10.1128/aem.02240-08.

Zhang, T., Nie, H., Bain, T. S., Lu, H., Cui, M., Snoeyenbos-West, O. L., et al. (2013). Improved cathode materials for microbial electrosynthesis. *Energy & Environmental Science*, *6*(1), 217–224.

Zhang, G., Zhao, Q., Jiao, Y., Wang, K., Lee, D.-J., & Ren, N. (2012). Biocathode microbial fuel cell for efficient electricity recovery from dairy manure. *Biosensors and Bioelectronics*, *31*(1), 537–543. https://doi.org/10.1016/j.bios.2011.11.036.

Zhao, Y., Cao, W., Wang, Z., Zhang, B., Chen, K., & Ouyang, P. (2016). Enhanced succinic acid production from corncob hydrolysate by microbial electrolysis cells. *Bioresource Technology*, *202*, 152–157. https://doi.org/10.1016/j.biortech.2015.12.002.

Zhao, G., Ma, F., Wei, L., Chua, H., Chang, C.-C., & Zhang, X.-J. (2012). Electricity generation from cattle dung using microbial fuel cell technology during anaerobic acidogenesis and the development of microbial populations. *Waste Management*, *32*(9), 1651–1658. https://doi.org/10.1016/j.wasman.2012.04.013.

Zheng, X., & Nirmalakhandan, N. (2010). Cattle wastes as substrates for bioelectricity production via microbial fuel cells. *Biotechnology Letters*, *32*(12), 1809–1814. https://doi.org/10.1007/s10529-010-0360-3.

Zou, S., Qin, M., Moreau, Y., & He, Z. (2017). Nutrient-energy-water recovery from synthetic sidestream centrate using a microbial electrolysis cell—Forward osmosis hybrid system. *Journal of Cleaner Production*, *154*, 16–25. https://doi.org/10.1016/j.jclepro.2017.03.199.

Zuo, Y., Maness, P.-C., & Logan, B. E. (2006). Electricity production from steam-exploded corn stover biomass. *Energy & Fuels*, *20*(4), 1716–1721. https://doi.org/10.1021/ef060033l.

Zwart, S. J., & Bastiaanssen, W. G. (2004). Review of measured crop water productivity values for irrigated wheat, rice, cotton and maize. *Agricultural Water Management*, *69*(2), 115–133.

Liu, H., Ramnarayanan, R., & Logan, B. E. (2004). Production of electricity during wastewater treatment using a single chamber microbial fuel cell. *Environmental Science & Technology*, *38*(7), 2281–2285.

Wang, X., Feng, Y., Wang, H., Qu, Y., Yu, Y., Ren, N., et al. (2009). Bioaugmentation for electricity generation from corn stover biomass using microbial fuel cells. *Environmental Science & Technology*, *43*(15), 6088–6093.

Chapter 9

Purple nonsulfur bacteria: An important versatile tool in biotechnology

Azka Asif[a], Hareem Mohsin[b], and Yasir Rehman[c]
[a]*School of Biological Sciences, University of the Punjab, Lahore, Pakistan,* [b]*Department of Allied Health Sciences, The Superior College Lahore, Lahore, Pakistan,* [c]*Department of Life Sciences, School of Science, University of Management and Technology, Lahore, Pakistan*

Chapter outline

1 Introduction

Microorganisms are highly diverse in terms of their habitat, genetics, physiological, biochemical, molecular, and metabolic properties. One major group of microorganisms comprises photosynthetic bacteria that can absorb light

 https://doi.org/10.1016/B978-0-12-822098-6.00003-3

and convert light energy into chemical energy for their propagation. This group is further divided into oxygenic photosynthetic bacteria (taking in carbon dioxide and releasing oxygen) and anoxygenic photosynthetic bacteria (taking in carbon dioxide but not releasing oxygen). The anoxygenic photosynthetic bacteria covers a diverse group of purple bacteria, which further comprises purple sulfur (PSB) and purple nonsulfur bacteria (PNSB).

Approximately 50 genera are known of these purple bacteria. Both groups, PSB and PNSB, coexist in aquatic environments. PSB and PNSB differ from each other regarding metabolism and phylogeny. PSB can carry out photoautotrophy more effectively as compared to photoheterotrophy, but it cannot grow in the dark. On the contrary, PNSB are fully equipped for metabolism and growth in the dark as they are capable of carrying out both kinds of metabolic pathways, i.e., photoautotrophy and photoheterotrophy (Madigan & Jung, 2009).

1.1 Systematics of Anoxygenic phototrophic purple Bacteria

Originally, PNSB were distinguished from PSB on the basis of physiology regarding the growth and utilization of sulfide. PNSB were reported unable to tolerate and oxidize micro quantities of sulfide. However, it became the distinguishing feature between PNSB and PSB because PNSB did not internalize the sulfide; rather, it was deposited on the outside of the cell (Brune, 1995; Hansen & van Gemerden, 1972). Therefore, when grown on a medium containing sulfide, PNSB can be easily identified by microscopic examination of sulfide globules in their cells. Literature also reports the presence of PNSB in sulfide-rich habitats where PNSB exhibit tolerance against sulfur. For example, both the species *Rhodobacter sulfidophilus* and *Rhodoferax antarcticus* have been reported to tolerate sulfide concentration above 4.0 mM (Jung, Achenbach, Karr, Takaichi, & Madigan, 2004).

This sulfide-based classification of purple bacteria is also supported by molecular criteria. Furthermore, the comparative 16S rRNA sequencing followed by phylogenetic analysis also revealed PSB belonging to the group of gamma-proteobacteria and PNSB to the group of alpha-proteobacteria or beta-proteobacteria (Imhoff, Hiraishi, & Süling, 2005).

1.1.1 Purple nonsulfur Bacteria

Madigan and Jung (2009) have reported the identification of 20 genera of PNSB. The species of *Rhodopseudomonas* and *Rhodobacter* are used as model organisms for the research on anoxygenic photosynthesis in the laboratory. Among these genera, unusual microbial species related to the habitat were also identified. Such unusual habitats include hot, highly saline, acidic, and alkaline environments. Further, the sequence analysis of photocomplex proteins proves that lateral gene transfer is involved due to which there is a similarity in photocomplex proteins in all the genera (Jung & Jung, 2009; Nagashima, Hiraishi, Shimada, & Matsuura, 1997).

1.2 Habitats of PNSB

PNSB inhabit at least 11% of the marine habitats (Idris, 2014). They are widely distributed in soil and activated sludge (Overmann & Garcia-Pichel, 2013). Literature also reports a wide distribution of PNSB in habitats like swamps, coastal regions, and waste lagoons (Madigan & Jung, 2009; Soon, Al-Azad, & Ransangan, 2014). PNSB are present in the form of macroscopic blooms but are difficult to identify as compared to PSB, owing to their physiological sensitivity towards sulfur compounds (Madigan & Jung, 2009). The presence and growth of PNSB in microenvironments depends on the amount of light, low oxygen tension, nutrients, and available organic compounds (Okubo, Futamata, & Hiraishi, 2006; Urmeneta, Navarrete, Huete, & Guerrero, 2003).

1.3 Pigmentation of PNSB

PNSB are pigmented organisms primarily due to the synthesis of various carotenoids and bacteriochlorophylls. This pigmented apparatus serves as an important tool to macroscopically recognize and initially characterize these microorganisms. The absorption ranges of these pigments lie between 450 and 550 nm. Their colors range from brown to reddish-brown to red and pink. The pink or red color is due to the presence of the carotenoid lycopene, rhodopsin (463, 490, 524 nm), and spirolloxanthin (486, 515, 552 nm). Reddish-brown color is imparted due to rhodopsin. Okenone (521 nm) is responsible for purple-red coloration, whereas compounds like spheroidene (450, 482, 514 nm) give brownish-red coloration under reducing microenvironment conditions (Mehrabi, Ekanemesang, Aikhionbare, Kimbro, & Bender, 2001).

2 Isolation and characterization methods of PNSB

Laboratory conditions for enrichment and growth of PNSB usually include a temperature range of 25–30°C in the presence of continuous illumination with an anoxic/anaerobic microenvironment (Luongo et al., 2017). The approximate wavelength required for enrichment lies in the infrared region from 800 nm to 900 nm. It has been reported by Kim, Ito, and Takahashi (1982) that an increase in light intensity increases the growth and proliferation of PNSB during enrichment up to a certain value. However, a decrease in pigment production is reported by Getha, Chong, and Vikineswary (1998) with increasing intensities of light. This decrease is due to the degradation caused by the heat dissipated from the light sources (Jalal, Zaima Azira, Rahman, Kamaruzzaman, & Faizul, 2014). The excessive heat also affects normal growth conditions of cells, thus reducing the cell count, as previously mentioned.

3 Metabolic variety in PNSB

Inhabiting a wide range of areas makes PNSB adaptable to several metabolic pathways. Hence, they can switch between different modes of metabolisms (Imhoff, 2005). Their metabolic versatility ranges from photoautotrophic to photoheterotrophic to chemoautotrophic. Predominantly, they are photoheterotrophic under anaerobic conditions. In this mode, they use organic compounds as electron donors in the presence of light (Poretsky, 2003).

3.1 Photoheterotrophy

Most of the PNSB are typically photoheterotrophs and use various carbon sources. Different strains are known to grow best at different carbon sources (Sojka, 1978). Among the more commonly used carbon sources are the readily consumable malate or pyruvate and other organic acids. Preference of nitrogen sources may also vary, ammonium chloride being the most preferred, followed by urea, L-tyrosine, and sodium glutamate. The latter can also serve as a carbon source in the media (Jung & Jung, 2009).

The most common addition to the media for PNSB is yeast extract (Biebl & Pfennig, 1981), which abundantly supplements the medium with a variety of organic compounds suitable for the heterotrophic growth of the bacteria. Most importantly, it is a source of B-complex vitamins, which is a requirement for good photoheterotrophic growth by many PNSB species. The most important of these are B1, B3, B7, and *p*-aminobenzoic acid. The lesser needed is the growth factor, vitamin B12, which is important for only a few PNSB (Kompantseva, Komova, & Kostrikina, 2010; Siefert & Koppenhagen, 1982).

Carbon sources might vary a great deal for the photoheterotrophic growth of PNSB, ranging from organic acids, fatty acids, amino acids, alcohols, carbohydrates, and various carbon-numbered compounds (Milford, Achenbach, Jung, & Madigan, 2000). Malate, fumarate, acetate, succinate, and pyruvate can be easily used, with a few exceptions, as they can enter the citric acid cycle at different steps. Ethanol, propionate, and lactate are also known to be metabolized by PNSB. Further, some aromatic compounds like benzoate, derivatives of benzoate, cyclohexane carboxylate, etc., are assimilated phototrophically by the PNSB (Ghosh, Dairkee, Chowdhury, & Bhattacharya, 2017). While *Pheospirillum,* now known as *Rhodospirillum,* is isolated using benzoate as the sole source of carbon in enrichment culture (Gibson & Harwood, 1995; Liu et al., 2019). Toluene is one of the aromatic compounds that allow the growth of strains like *Blastochloris sulfoviridis* photoheterotrophically (Madigan et al., 2019; Manisha, Kudle, Singh, Merugu, & Rudra, 2017).

3.2 Photoautotrophy

If an inorganic source of electrons and CO_2 is provided, PNSB can grow photoautotrophically. The PNSB are known to possess the Calvin Benson cycle, which is used in their autotrophic mode. They fix carbon dioxide using the enzyme RuBisCo, ribulose 1, 5-Biphosphate Carboxylase/oxygenase (Larimer et al., 2004). This mode switching is actually due to the presence of the Calvin cycle in these bacteria, which is important for autotrophic as well as photoheterotrophic growth (Falcone & Tabita, 1993; Mcewan, 1994; McKinlay & Harwood, 2010a; Ormerod & Sirevag, 1983; Tabita, 1995).

3.3 Chemoheterotrophy

The organic compounds that are used by the PNSB phototrophically, those same compounds, can also be used in the process of chemoheterotrophy. The oxygen tolerance varies among the species of PNSB, but the *Rhodobacter* species can thrive in higher concentrations of oxygen (Pérez, Dorador, Molina, Yáñez, & Hengst, 2018). Other modes, like fermentation or anaerobic respiration, are also used by the PNSB for growth in the absence of oxygen. In the case of fermentative growth, compounds like pyruvate (Kim, Kim, & Lee, 2008) and other sugars are utilized primarily by *Rhodospirillum rubrum* and *Rhodobacter capsulatus* (Madigan & Imhoff, 2001). Accessory chemicals can be added for the enrichment of selective strains like dimethyl sulfoxide, or trimethylamine-N oxide (Shaw et al., 1999) supports the abundant growth of *R. capsulatus*. This specie has the ability to reduce the nitrate to nitrogen gas with the carbon sources acting as electron donors that cannot be fermented (Demtröder, Pfänder, & Masepohl, 2020).

Electron donors like hydrogen or $S_2O_3^{2-}$ are used in chemoheterotrophic growth without light. Media supplemented with the gases hydrogen, oxygen, and carbon dioxide will support the growth as these gases will be acting as electron donors, electron acceptors, and carbon sources, respectively (Madigan & Gest, 1979). Chemoheterotrophy being an important mode of metabolism is still an ambiguity, but this ability to extract energy from metabolizing inorganic sources gives purple bacteria a physiological edge over nonphototrophic bacteria.

3.4 Nitrogen fixation

If there are anoxic, dark conditions, they can grow well by fermentation. They can utilize nitrogen (N_2) as a nitrogen source by way of a nitrogenase enzyme (McKinlay & Harwood, 2010b; Schultz, Gotto, Weaver, & Yoch, 1985; Wang & Norén, 2006). PNSB have the ability to fix nitrogen except some of the species (Masepohl, 2017). *Rhodobacter capsulatus* and *Rhodobacter sphaeroides* are able to grow at a rapid rate in the presence of nitrogen gas as a sole source of nitrogen. PNSB, especially the species *Rhodobacter* have an advantage of thriving in the anoxic conditions with a limited amount of fixed nitrogen (Erkal et al., 2019).

Adding to the versatile nature of PNSB, they are also able to produce hydrogen in the presence of organic carbon sources, i.e., photoheterotrophy.

However, they can also consume hydrogen in the presence of inorganic carbon sources, i.e., photoautotrophy (Koku, Eroğlu, Gündüz, Yücel, & Türker, 2002). The nitrogenase enzyme of PNSB takes part in the hydrogen production by the photofermentation. Production can be increased by maintaining a proper carbon to nitrogen ratio, the intensity of light, and the volume of the inoculum, as well as minimum oxygen exposure as nitrogenases, are oxygen-sensitive (Harwood, 2008; Show, Lee, Tay, Lin, & Chang, 2012). Species like *Rhodopseudomonas palustris, Rhodospirillum ruburum, Rhodobacter spheroids* (Wang et al., 2018), and *Rhodobacter capsulatus* are reported to be mostly used for bio-hydrogen production (Basak, Jana, Das, & Saikia, 2014).

3.5 Oxygen requirements for PNSB

PNSB are primarily facultative anaerobes, which means that they can survive under oxic conditions. This leads to a reduction in pigment production. Bacteriochlorophyll-dependent photosynthesis is performed in specialized membranous structures. In this context, *Rhodobacter sphaeroids* and *Rhodobacter capsulatus* are extensively studied (Naylor, Addlesee, Gibson, & Hunter, 1999). The presence of low oxygen concentrations leads to morphological changes, including the formation of an invaginated intracytoplasmic membrane known as chromatophores (Fiedor, Ostachowicz, Baster, Lankosz, & Burda, 2016). Due to the formation of these invaginations, the surface area for light absorption and energy production is increased as these photosynthetic sites contain complexes for the transfer of electrons. Three major complexes comprise the photosynthetic apparatus of these organisms: light-gathering complexes I (B875) and II (B800–850) and reaction centers along with other necessary components required in the electron transport chain (Naylor et al., 1999). Therefore, it is evident that the PNSB photosynthesize in the presence of illumination and synthesize bacteriochlorophyll and carotenoids under anoxic/anaerobic conditions. However, in dark and low oxygen concentrations, there is lesser pigmentation (Hunter, Daldal, Thurnauer, & Beatty, 2008; Willows & Kriegel, 2009).

3.6 Model PNSB species and their metabolism

Rhodobacter capsulatus has been reported to be the most versatile, easily switching between metabolic pathways according to their environments. This specie grows phototrophically by utilizing organic compounds in the presence of light. It shifts its metabolism to chemolithotrophy and fermentation in dark conditions (Jung & Jung, 2009). Another model microorganism, *Rhodopseudomonas palustris*, is reported to grow with all possible modes of metabolism both in the oxygenic or anoxygenic environment (Larimer et al., 2004). It uses chemohetrotrophy in the presence of oxygen, thus undergoing aerobic respiration. However, the phototrophic mode is switched on in low oxygen concentrations or anoxygenic environment (Baars, Morel, & Zhang, 2018).

Rhodospirillaceae is a model organism for studying anoxygenic photosynthesis, the evolution of photosynthesis, and their photosynthetic structure (Karrasch, Bullough, & Ghosh, 1995). They are of interest due to their versatile metabolic behaviors and their consequent adaptability to a range of environmental conditions (Larimer et al., 2004; Mcewan, 1994; Tabita, 1995). *R. capsulatus* is known to grow photoautotrophically, photoheterotrophically, chemoautotrophically, and, using various electron acceptors and sugars, also, chemoheterotrophically (Madigan & Gest, 1979).

3.7 Interaction of PNSB with heavy metals

The versatility of PNSB allows them to interact with heavy metals in the environment by using different processes. They can oxidize heavy metals in the environment and thereby reduce their toxicity (Mukhopadhyay, Rosen, Phung, & Silver, 2002; Silver & Phung, 2005). They can also uptake metallic ions from the environment and use them as electron receptors in the electronic transport chain during anaerobic photoheterotrophic conditions in the presence of uninterrupted illumination (Kessi, 2006). This property makes PNSB a promising candidate for bioremediation.

4 Biotechnological application of PNSB

4.1 PNSB as biological control agent

Rhodospirillum centenum can be used as a biological control agent. A study was conducted to explore the potential of *R. centenum* as a growth stimulant and an antagonist against the disease of leaf spot of rice caused by *Curvularia lunata*. The rice varieties were inoculated with *R. centenum* culture, where it showed its capability of producing amylase, protease, and lipase. Additionally, this antagonized *C. lunata* rice varieties showed better germination and enhanced root length. Thus, *R. centenum* played a chief role in the elimination of disease and rejuvenation of plant health. Such PNSB acting as a biological control agent hold great potential in the agricultural field (Vareeket & Soytong, 2020). Salicylic acid is also an important extracellular metabolite that can be used for the control of phytopathogen. *Rubrivivax gelatinosus* strain RASN4 was isolated from rice rhizospheres and screened for the production of salicylic acid. Quantitative analysis revealed the production as 27.3 mg/l (Gupta & Sinha, 2020).

4.2 PNSB in plant production

Other roles that PNSB takes on in the agriculture industry involve biofertilizers, conferring resistance against stress conditions, and reducing greenhouse gases (Sakarika et al., 2019). These are discussed below:

PNSB holds the potential to increase plant growth and product yield by acting as fertilizers, both directly and indirectly. In the case of a direct process, dead PNSB biomass is used (Spanoghe et al., 2020). The major objective is to provide a suitable concentration of nitrogen, phosphorous, and potassium (NPK) content. When applied to the soil, the dead biomass undergoes decomposition by the autochthonous soil microbes and releases nutrients that are taken up by the plant (Coppens et al., 2016). This process of direct fertilization also helps in the improvement of soil structure, ensuring sustainable usage of land. When viable PNSB cells are applied, the same effect is reported. The NPK content delivered to the soil measured in the dry weight of *Rhodobacter* sp. was 8.5%, 2.4%, and 0.5% respectively. It was noticeable that the phosphorous content was much higher as compared to microalgae (8.1%–1.3%–1.4%). In case of an indirect process, living cells ensure nutrient availability for the plant because PNSB, like *Rhodopseudomonas* and *Rhodobacter* sp., play a role in biogeochemical cycles like nitrogen fixation and phosphorous chelation (Lai, Liang, Hsu, Hsieh, & Hung, 2017). It has been reported by (Hsu, Lo, Fang, Lur, & Liu, 2015) that *R. palustris* PS3 had led to a 17% increase in nitrogen uptake efficiency of lettuce (Sakarika et al., 2019).

PNSB also plays a vital role in conferring resistance against biotic and abiotic stresses. They succor in roots, stem, flower, leaves and pigment formation, growth and development, and fruit ripening processes. Plant growth-promoting substances (PGPS) produced by PNSB lead to the expression or suppression of genes and the production of enzymes and metabolites (Khuong et al., 2020). Indole-3-acetic acid (IAA) is a common PNSB-derived PGPS produced by *Rhodopseudomonas* sp. via indole-3-pyruvate and tryptamine pathways (Khuong et al., 2020; Nookongbut, Kantachote, Khuong, & Tantirungkij, 2020). *R. rubrum* is reported to produce melatonin (Tan et al., 2010), which plays a defensive role against oxidative stress (Tan et al., 2013). *R. palustris* is reported to produce 5-aminolevulinic acid (ALA), which serves as a plant growth regulator (Nunkaew, Kantachote, Kanzaki, Nitoda, & Ritchie, 2014). Nunkaew et al. (2014) reported that increased ALA was produced when PNSB was inoculated into saline rice fields, resulting in increased plant growth and productivity (Kantha, Kantachote, & Klongdee, 2015).

Methane is one of the commonly mentioned greenhouse gas. Methanogenic archaea can sequester methane, but PNSB proves to be more beneficial. PNSB are capable of alleviating methane levels in rice paddy fields (Powlson, Whitmore, & Goulding, 2011). It has also been reported that PNSB can additionally consume carbon dioxide as a source of carbon and, its growth and efficiency are increased in the presence of continuous light. In the laboratory condition, they can reduce carbon dioxide levels by up to 47% (Kantha et al., 2015; Sakpirom, Kantachote, Nunkaew, & Khan, 2017).

4.3 Protein source

PNSB are rich in proteins along with essential amino acids, co-factors, vitamins, and lesser nucleic acids (Prasad, Vasavi, Girisham, & Reddy, 2008; Sasikala &

Ramana, 1995). The cellular protein content of PNSB is comparable to egg proteins or soybean (Ponsano, Lacava, & Pinto, 2002) and is also higher than other single-cell proteins due to the abundance of essential amino acids (Azad, Vikineswary, Ramachandran, & Chong, 2001). The uses of PNSB are diverse. For instance, in the pisciculture and poultry industry (Salma, Miah, Tareq, Maki, & Tsujii, 2007), PNSB are recommended as single-cell protein (SCP) feeds.

4.4 Medically and industrially important compounds

PNSB have clinical and medical applications as well, as these are a source of Vitamin B12, hopanoids (Nagumo, Takanashi, Hojo, & Suzuki, 1991), and ubiquinone Q10 (Sasaki, Watanabe, & Tanaka, 2002). Mitsui, Matsunaga, Ikemoto, and Renuka (1985) has discussed the use of photosynthetic bacteria like PNSB as an economical source of carbohydrate for methane-producing bacteria. *Rhodobacter sphaeriodes*, due to its vibrant pigments and carotenoids, are used in food dyes and natural dyes (Menxu, 1991). *Rhodopseudomonas palustris* is also used as a feed for aquatic farms and is mass-produced for this purpose (Kim & Lee, 2000). *Rhodobacter sphaeriodes* O.U.001 is also known to leach ammonia in its resting stage (Sasikala, Ramana, Rao, & Subrahmanyam, 1990), which is a property of various photosynthetic bacteria and can be used in metallurgy (Takabatake, Suzuki, Ko, & Noike, 2004). Photosynthetic bacteria are a good source of porphyrins (Ishii, Hiraishi, Arai, & Kitamura, 1990), and an example is *Rhodobacter sphaeroides* CR386, which produces porphyrins in aerobic conditions (Utsunomiya, Yamane, Watanabe, & Sasaki, 2003). Aminolevulinic acid, used as a bioherbicide, is another product reported from photosynthetic bacteria like *Rhodobacter shpaeroides* (Sasaki, Tanaka, Nishizawa, & Hayashi, 1991) and other salt-tolerant PNSB (Nunkaew, Kantachote, Nitoda, & Kanzaki, 2015). Production of extracellular polymeric substance (EPS) is also seen in several PNSB. For instance, *Rhodovulvum* spp. S88 (Watanabe, Shiba, Sasaki, Nakashimada, & Nishio, 1998) and *Rhodopseudomonas acidophila* produces ESP on their cell surfaces (Sheng, Yu, & Yue, 2006). There is also sufficient evidence of indole acetic acid production by PNSB (Merugu, Girisham, & Reddy, 2010; Ramchander, Rudra, Atthapu, Girisham, & Reddy, 2011; Sasikala & Ramana, 1995). Phytohormones, produced by the photobiotransformation of indole acetic acid is also carried out by some PNSB (Sasikala & Ramana, 1995). An example is the phytohormone rhodestrin, isolated from *Rhodobacter sphaeroides* OU5, which was found to be a metabolite of anthranilate (Sunayana, Sasikala, & Ramana, 2005).

4.5 Recycling of nitrogen in soil

Anoxygenic photosynthetic bacteria are also found in paddy soil microbiomes, where they take part in soil fertility primarily by fixing nitrogen

(Habte & Alexander, 1980). Diazotrophic growth is an important contribution of PNSB to the ecosystem (Chalam, Sasikala, Ramana, Uma, & Rao, 1997). The genus *Rhodopseudomonas* quite promptly assimilates nitrogen (Tarabas, Hnatush, Moroz, & Kovalchuk, 2019). It is reported that the nitrogenase activity of PNSB is curbed by the use of pesticides (Chalam et al., 1997). However, some species like *Rhodopseudomonas palustris* is reported to assimilate a crop fungicide carbendazim as their source of nitrogen and carbon (Rajkumar & Lalithakumari, 1992). Several other instances of resistance to pesticides have been reported. For instance, species of *Rhodoferax* degrade phenoxy herbicides (Ehrig, Müller, & Babel, 1997).

4.6 Polyhydroxyalkonates production

Poly β-hydroxybutyrate (PHB) is a complex polymer accumulated intracellularly under stressed environments and incubations of many bacteria and is an important resource in the biodegradable materials industry. *Rhodobacter sphaeroides* strain RV under alkaline conditions accumulates PHB in nitrogen deprivation, using lactate as a sole carbon source (Khatipov, Miyake, Miyake, & Asada, 1998). Similarly, another study reported phosphate, sulphate and nitrogen limitations for harvesting PHB from PNSB (R. Merugu, Girisham, & Reddy, 2010a, 2010b).

Rhodobacter sphaeroides and *Rhodobacter sphaeroides* IL 106 also produce Polyhydroxyalkanoate (PHA) by using acetic acid (Noparatnaraporn, Takeno, & Sasaki, 2001). Ideally, for the production of PHA, a photosynthetic bacterium with an optimum temperature of 37–40°C and anaerobic dark culture conditions makes a good candidate (Lorrungruang, Martthong, Sasaki, & Noparatnaraporn, 2006). *Rhodobacter Sphaeroides* produces 97% PHB and only 3% polyhydroxyvalerate (PHV) in anaerobic phototrophic growth (Brandl, Gross, Lenz, Lloyd, & Fuller, 1991). *Rhodovulum sulfidophilum* is a marine PNSB that produces PHA as well (Higuchi-Takeuchi & Numata, 2019).

4.7 Bio-fertilizers

PNSB are also mass-produced as biofertilizers. For example, diazotrophs like *Rhodopseudomonas capsulatus* are important for rice development up to the ear stage, especially in the in limiting nitrogen conditions (Bali, Blanco, Hill, & Kennedy, 1992; Elbadry & Elbanna, 1999). The benefits of *Rhodopseudomonas capsulatus* for four rice varieties has been demonstrated using hydroponic cultures (Elbadry & Elbanna, 1999). The quality of tomatoes is known to increase using lyophilized *Rhodobacter sphaeroides* as bacterial soil inoculums (Kondo, Nishihara, & Nakata, 2006). *Rhodopseudomonas palustris* and *Rubrivivax gelatinosus* are known to produce plant growth-promoting phytohormones and also reduce greenhouse gases in rice paddies (Sakpirom et al., 2017).

4.8 Industrially important enzymes

PNSB are also known to produce industrially important hydrolytic enzymes. *Rhodocyclus gelatinosus* is known to produce amylases, the product of which are maltooligosaccharides, which is used in chemical as well as food industries (Buranakarl, Cheng-Ying, Ito, Izaki, & Takahashi, 1985; Munjam, Vasavi, Girisham, & Reddy, 2003). Amylases from anoxygenic phototrophic bacteria are versatile in their substrates. Substrates of amylases from PNSB include oyster glycogen, pullulan, and amylose, including the usually seen substrate; soluble starch (Buranakarl et al., 1985). PNSB are also known to produce proteolytic enzymes (Seangtumnor, Kantachote, Nookongbut, & Sukhoom, 2018). *Rhodopseudomonas palustris* has been reported to produce carboxykinases (Inui, Nakata, Roh, Zahn, & Yukawa, 1999). PNSB also holds importance in the shrimp industry as they produce 'anti-*vibrio*' compounds (*Rhodovulum sulfidophilum* PS342), which reduce industry losses by preventing pathogen invasions (Seangtumnor et al., 2018).

4.9 Bio-hydrogen production

Hydrogen production can be achieved by mechanisms such as photoelectrolysis, electrolysis of water, biophotolysis, and photocatalysis; however, biological hydrogen production by phototrophy is the most employable. PNSB and their light-driven hydrogen production have been well studied (Ghosh et al., 2017; Harwood, 2008). Several upsides of using biological hydrogen production over fossil fuels are that there are no pollutants, the risk of depletion of nonrenewable reserves runs low, and the process produces water (Kapdan & Kargi, 2006; Melis, 2002). Due to the versatility of PNSB in regards to their carbon sources, wastewater is a well-studied medium for the production of biohydrogen as it bioremediates the wastewater by reduction of organic compounds (Lee, Chen, Wang, & Tung, 2002; Wu, Hay, Kong, Juan, & Jahim, 2012). Wastewater from the food industry is an important example of a medium for photobiological hydrogen production (Lin, Juan, & Hsien, 2011; Lin, Zheng, & Juan, 2012). Integrated systems for photobiological production using green algae and photosynthetic bacteria producing biomass for anaerobic bacteria, which then generate small organic acids for the former has been discussed by Melis and Melnicki (2006). *Rsp. rubrum* can produce up to 87% stoichiometric yield of hydrogen from CO while using a variety of carbon sources: acetate, malate, formate, glucose, fructose, and sucrose via fermentation (Najafpour, Younesi, & Mohamed, 2006). Similarly, *Rubrivivax gelatinosus* CS can also oxidize CO for hydrogen production with and without illumination (Maness & Weaver, 2002). Another important member of PNSB in hydrogen production is *Rhodopseudomonas palustris*, which is studied to have produced a generous amount of hydrogen when immobilized in nonporous latex in an argon atmosphere (Gosse et al., 2007). Immobilizing PNSB cells with PVA cryogels is a method reported to produce high biohydrogen yields. The cryogel matrix contains glycerol and allows near-infrared light to pass through, which is a basic requirement for PNSB growth. It has been reported that this procedure yielded hydrogen for continuous 67 days. Thus, the addition of glycerol with matrix provides additional advantage (du Toit & Pott, 2020). *Rhodobacter* genus has been mostly reported to generate generous amount of hydrogen growing in a variety of media like short-chain volatile acids (acetate and butyrate), ground wheat, glucose, and sucrose with maximum yields of about 2614 ± 121.76 ml H_2/L achieved by sugar plant wastewater (Assawamongkholsiri, Reungsang, Plangkang, & Sittijunda, 2018; Tiang et al., 2020).

4.10 Metal resistance

Members of the *Rhodospirillaceae* family are known for their resistance against oxides and oxyanions of toxic heavy metals (Moore & Kaplan, 1992). For example, in a batch culture, *Rhodovulvum* sp. and *Rhodobacter sphaeroides* are seen to remove cadmium from the medium (Watanabe, Kawahara, Sasaki, & Noparatnaraporn, 2003). The process of biosorption of lead and cadmium ions by *Alcaligenes eutrophus* H16 and *Rhodobacter sphaeroides* has also been reported (Seki, Suzuki, & Mitsueda, 1998). Similarly, phosphorus removal from sediments of the oyster farm was observed for *Rhodobacter sphaeroides* IL 106 (Takeno, Sasaki, Watanabe, Kaneyasu, & Nishio, 1999), and removal and detoxification of elemental selenium is reported for *Rhodospirillum rubrum* (Kessi, Ramuz, Wehrli, Spycher, & Bachofen, 1999).

PNSB, while resisting these metals, also utilize them to their advantage (Merugu, Prasad, Girisham, & Reddy, 2008a, 2008b). Trace elements are known to act as cofactors, and so reactions like hydrogen production increases in the presence of molybdenum for *Rhodobacter sphaeroides* (Kars, Gündüz, Yücel, Türker, & Eroglu, 2006). Other metals' tolerance by *Rhodobacter sphaeroides* includes iron and cobalt (Giotta, Agostiano, Italiano, Milano, & Trotta, 2006). However, in some cases of resisting metal toxicity, a price has to be paid. For example, nickel and cobalt are known to reduce the amount of light-harvesting complexes in *Rhodobacter sphaeroides* (Balsalobre et al., 1993; Giotta et al., 2006; Italiano et al., 2011).

4.11 Bioremediation

For their metal resistance and pollutant degradative abilities, PNSB are coveted organisms for bioremediation. Studies on immobilized *Rhodobacter capsulatus* revealed their ability to remove organic carbon, phosphate, and ammonium ions from the medium within 19–22 days (Sawayama, Rao, & Hall, 1998). Similarly, a consortium of *Rhodobacter sphaeroides* and *Rhodopseudomonas palustris* immobilized in ceramic and cultured in synthetic wastewater was involved in the removal of nitrates, phosphates, and H_2S (Nagadomi, Takahasi,

Sasaki, & Yang, 2000). Phosphorus removal for *Rhodocyclus* spp. is also reported (Zilles, Peccia, Kim, Hung, & Noguera, 2002). Therefore, PNSB are also useful in organic waste degradation such as agricultural or food industry wastes (molasses) (Sagir, Ozgur, Gunduz, Eroglu, & Yucel, 2017; Sasikala, Ramana, & Rao, 1992), citric acid fermentation wastes (Zhi, Yang, Berthold, Doetsch, & Shen, 2010), lactic acid wastes (Sasikala & Ramana, 1995; Sasikala, Ramana, & Subrahmanyam, 1991), refinery wastes like oil containing sewage water (Takeno, Yamaoka, & Sasaki, 2005), and high organic content wastewater (Ogbonna, Yoshizawa, & Tanaka, 2000).

4.12 Wastewater treatment

Oil spills and its ascribed pollution and danger to wildlife is a known reality, and photosynthetic bacteria have been known to help treat wastewater contaminated by oil. PNSB, like *Rhodobacter sphaeroides*, have been studied for their treatment of palm oil (Hassan et al., 1997). Tributyl phosphate is used in the chemical industry and nuclear fuel processing and can be degraded by *Rhodopseudomonas palustris* (Berne, Allainmat, & Garcia, 2005). A consortium of microorganisms like *Chlorella sorokiniana*, *Spirulina platensis*, and *Rhodobacter sphaeroides* are often used for better digestion of wastewater content (Ogbonna et al., 2000). *Rhodopseudomonas palustris* strain B1 is also involved in the processing of sago effluents (Ibrahim, Vikineswary, Al-Azad, & Chong, 2006). Yeast plant waste has been reported to be successfully treated by using *Rhodopseudomonas yavorovii* IMV B-7620 (Tarabas, Hnatush, Moroz, & Kovalchuk, 2019). Complex organic substrates like aromatic compounds, i.e., benzoate and its derivatives (Wright & Madigan, 1991), thiols (Visscher & Taylor, 1993), amines (Cabello et al., 2004), and phenols (Blasco & Castillo, 1992) can also be utilized by PNSB. Heavy metals like chromium [usually Cr (VI)] are often a component of pesticides and are considered as an environmental pollutant, and it can be reduced by *Rhodobacter sphaeroides* (Nepple, Kessi, & Bachofen, 2000) and *Rhodobacter capsulatus* (Ramchander Merugu, Pratap Rudra, Girisham, & Reddy, 2011) from Cr (VI) to Cr (III), the less toxic state of chromium. Similarly, mercury resistant *Afifella marina* is used to biosorb mercury from effluents (Mukkata, Kantachote, Wittayaweerasak, Megharaj, & Naidu, 2019). Arsenic resistance in *Rhodopseudomonas palustris* C1 (Nookongbut, Kantachote, Krishnan, & Megharaj, 2017) and lead resistance by *Rhodobacter sphaeroides* have also been reported (Li, Peng, Jia, Lu, & Fan, 2016). Further, sulfide, which is an odorous pollutant of water, can be removed using in situ treatment by PNSB (Sadi & Firmansyah, 2020; Sun, Pang, Xi, & Hu, 2019). Cadmium-polluted sludge has recently yielded a novel species of the *Rhodobacteraceae* family known as *Pseudogemmobacter bohemicus* having high resistance to cadmium (Suman et al., 2019).

4.13 Microbial fuel cells for energy generation

Waste matter can be used as a fuel for the generation of electricity in microbial fuel cells (MFCs). This waste biomass consists of a number of various types of organic compounds like acids, etc. Apart from this, a mixture of waste substances from different sources can be used for electricity production. Due to their unique metabolism, PNSB are also employed in microbial fuel cells. One such example is of *Rhodopseudomonas palustris* G11 from activated sludge, which accumulates polyphosphates and poly-β-hydroxybutyrate in their stationary growth using light, and upon transferring to fresh medium, with the help of a photomicrobial fuel cell (PMFC), released 0.03 Vs (Lai et al., 2017). *Rhodopseudomonas palustris* DX-1, originally isolated from an MFC, is reported as an exoelectrogen (capable of transferring electrons across its membrane to an anode) with power generations as high as $2720 \pm 60\,mW/m^2$ (Xing, Zuo, Cheng, Regan, & Logan, 2008). It is versatile in its usage of carbon sources: acetate, lactate, fumarate, ethanol, glycerol, etc. along with consuming entire cells of cyanobacteria (Inglesby, Beatty, & Fisher, 2012). The oxidation of organic compounds results in hydrogen production. Therefore, a mutation aimed at the suppression of hydrogen production will result in greater power production (Morishima et al., 2007). *Rhodobacter* is also often found in association with *Rhodopseudomonas*. Both of them produce compounds soluble in the medium, which can facilitate the process of electron transfer to the anode, called mediators or, more appropriately, electron mediators (Xing, Cheng, Regan, & Logan, 2009). Light is also found to aid in electricity generation (Cao, Huang, Boon, Liang, & Fan, 2008).

Rhodobacter capsulatus can be utilized in a proton exchange membrane (PEM) fuel cell, which is actually the coupling of photoheterotrophic hydrogen production of this microorganism in a photobioreactor, and a MFC deriving that hydrogen as a substrate (He, Bultel, Magnin, & Willison, 2006). Or, alternatively, by placing phototrophic bacteria in the anodic chamber of the fuel cell (Cho et al., 2008; Rosenbaum, Schröder, & Scholz, 2005a; Rosenbaum, Schröder, & Scholz, 2005b). Other ideas of a coupled MFC can be seen in the example of a two-step biohydrogen production for further electricity generation using organic compounds. This was accomplished by linking the dark fermentation of *Escherichia coli* with the photofermentation of *Rhodobacter sphaeroides* (Rosenbaum et al., 2005a). MFCs hold great value for long-term expeditions, for instance, space travel. A proposal suggesting the use of a recombinant *Rhodobacter sphaeroides* (hydrogenases of *Rhodospirillium rubrum*) using human waste to generate electricity has been given (Hull, 2009).

A quick overview of all the biotechnological applications of PNSB is shown in Fig. 1, whereas biotechnologically important microorganisms are summarized in Table 1.

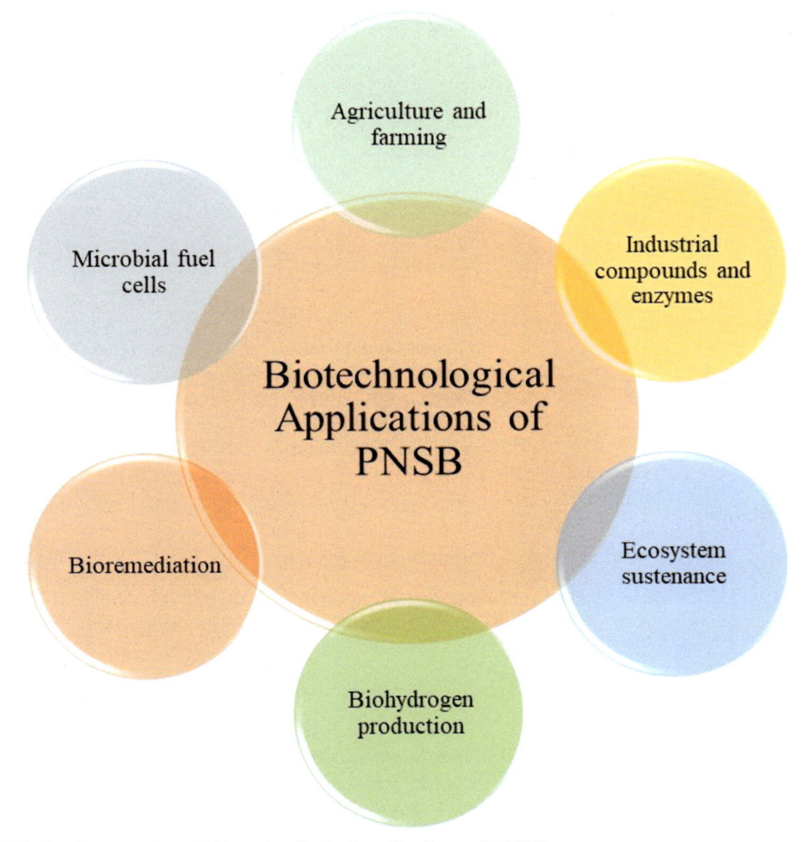

FIG. 1 An overview of biotechnological applications of PNSB.

5 Conclusion

PNSB are a group of bacteria with multiple modes of metabolism, i.e., photoheterotrophy, chemoheterotrophy, and photoautotrophy. They are useful in metal detoxification and bioremediation due to their diverse range of nutrition requisites, making them able to grow on municipal wastewater, activated sludge as well as chemical and industrial wastes. Many PNSB are hydrogen producers as well as electricity generators, when used in microbial fuel cells. PNSB can be used as single-cell protein feeds and biofertilizers and are a rich source of carotenoids, ubiquinone, polyhydroxyalkanoates, enzymes, and therapeutic compounds like hopanoids and B12. The biotechnological applications of PNSB provide an endless horizon for research.

TABLE 1 Some of the biotechnologically important PNSB and their applications.

Sr. no.	Application	Bacterial strains	Reference
Ecosystem			
1	Nitrogen gas/ nitrogen fixation	*Rhodobacter capsulatus, Rhodobacter sphaeroides*	(Erkal et al., 2019)
Biohydrogen			
2	Biohydrogen production	*Rhodopseudomonas palustris, Rhodospirillum ruburum, Rhodobacter spheroids, Rhodobacter capsulatus, Rubrivivax gelatinosus*	(Assawamongkholsiri et al., 2018; Basak et al., 2014; Maness & Weaver, 2002; Tiang et al., 2020; Wang et al., 2018)
Metabolites			
3	Salicylic acid	*Rubrivivax gelatinosus*	(Gupta & Sinha, 2020)
4	Indole acetic acid (IAA)	*Rhodopseudomonas*	(Khuong et al., 2020; Nookongbut et al., 2020)
5	Melatonin; oxidative stress	*Rhodospirillum rubrum*	(Tan et al., 2010)
6	Aminoluvlenic acid (ALA)	*Rhodopseudomonas palustris, Rhodobacter shpaeroides*	(Kantha et al., 2015; Nunkaew et al., 2014; Sasaki et al., 1991)
7	Pigments and carotenoids; dyes	*Rhodobacter sphaeriodes*	(Menxu, 1991)
8	Aquatic farm feed	*Rhodopseudomonas palustris*	(Kim & Lee, 2000)
9	Porphyrins	*Rhodobacter sphaeroides*	(Utsunomiya et al., 2003)
10	Extracellular polymeric substances (EPS)	*Rhodovulvum, Rhodopseudomonas acidophila*	(Sheng et al., 2006; Watanabe et al., 1998)
11	Phytohormone rhodestrin	*Rhodobacter sphaeroides*	(Sunayana et al., 2005)
12	Polyhydroxybutyrate (PHB)	*Rhodobacter sphaeroides*	(Khatipov et al., 1998)
13	Polyhydroxy alkanoates (PHA)	*Rhodobacter sphaeroides, Rhodobacter sphaeroides, Rhodovulum sulfidophilum*	(Higuchi-Takeuchi & Numata, 2019; Noparatnaraporn et al., 2001)

TABLE 1 Some of the biotechnologically important PNSB and their applications—cont'd

Sr. no.	Application	Bacterial strains	Reference
Agriculture			
14	Biofertilizer	*Rhodopseudomonas capsulatus, Rhodobacter sphaeroides, Rhodopseudomonas palustris, Rubrivivax gelatinosus*	(Bali et al., 1992; Elbadry & Elbanna, 1999; Hsu et al., 2015; Kondo et al., 2006; Lai et al., 2017; Sakpirom et al., 2017)
15	Biological control agent	*Rhodospirillum centenum*	(Vareeket & Soytong, 2020)
Enzymes			
16	Amylases	*Rhodocyclus gelatinosus*	(Buranakarl et al., 1985; Munjam et al., 2003)
17	Carboxykinases	*Rhodopseudomonas palustris*	(Inui et al., 1999)
Antibacterials			
18	"Anti-*vibrio*" compounds	*Rhodovulum sulfidophilum*	(Seangtumnor et al., 2018)
Metal resiatance and bioremediation			
19	Carbendazim assimilation	*Rhodopseudomonas palustris*	(Rajkumar & Lalithakumari, 1992)
20	Phenoxy herbicides decay	*Rhodoferax*	(Ehrig et al., 1997)
21	Sulfide resistance	*Rhodobacter sulfidophilus, Rhodoferax antarcticus*	(Jung et al., 2004)
22	Ammonia leaching	*Rhodobacter sphaeriodes, Rhodobacter capsulatus*	(Sasikala et al., 1990; Sawayama et al., 1998; Takabatake et al., 2004)
23	Benzene utilization	*Rhodospirillum*	(Gibson & Harwood, 1995)
24	Toluene	*Blastochloris sulfoviridis*	(Madigan et al., 2019; Manisha et al., 2017)

Continued

TABLE 1 Some of the biotechnologically important PNSB and their applications—cont'd

Sr. no.	Application	Bacterial strains	Reference
25	Pyruvate utilization	*Rhodospirillum rubrum, Rhodobacter capsulatus*	(Kim et al., 2008; Madigan & Imhoff, 2001)
26	Cadmium removal	*Rhodovulvum, Rhodobacter sphaeroides, Pseudogemmobacter bohemicus, Alcaligenes eutrophus, Rhodobacter sphaeroides*	(Seki et al., 1998; Suman et al., 2019; Watanabe et al., 2003)
27	Lead biosorption	*Alcaligenes eutrophus, Rhodobacter sphaeroides*	(Li et al., 2016; Seki et al., 1998)
28	Phosphorus removal	*Rhodobacter sphaeroides, Rhodobacter capsulatus, Rhodocyclus*	(Sawayama et al., 1998; Takeno et al., 1999; Zilles et al., 2002)
29	Selenium	*Rhodospirillum rubrum*	(Kessi et al., 1999)
30	Iron and cobalt	*Rhodobacter sphaeroides*	(Giotta et al., 2006)
31	Oil degradation	*Rhodobacter sphaeroides*	(Hassan et al., 1997)
32	Tributyl phosphate degradation	*Rhodopseudomonas Palustris*	(Berne et al., 2005)
33	Sago effluent treatment	*Rhodopseudomonas palustris*	(Ibrahim et al., 2006)
34	Yeast plant waste	*Rhodopseudomonas yavorovii*	(Tarabas, Hnatush, & Moroz, 2019)
35	Chromium reduction	*Rhodobacter sphaeroides, Rhodobacter capsulatus*	(Nepple et al., 2000)
36	Mercury resistance	*Afifella marina*	(Mukkata et al., 2019)
37	Arsenic resistance	*Rhodopseudomonas palustris*	(Nookongbut et al., 2017)

TABLE 1 Some of the biotechnologically important PNSB and their applications—cont'd

Sr. no.	Application	Bacterial strains	Reference
Microbial fuel cells (MFCs)			
38	Photo-microbial fuel cell (PMFC)	*Rhodopseudomonas palustris*	(Lai et al., 2017)
39	Exoelectrogen	*Rhodopseudomonas palustris*	(Xing et al., 2008)
40	Proton exchange membrane (PEM) fuel cell	*Rhodobacter capsulatus*	(He et al., 2006)
41	Simultaneous biohydrogen/ electricity generation	*Rhodobacter sphaeroides*	(Hull, 2009; Rosenbaum et al., 2005b)

References

Assawamongkholsiri, T., Reungsang, A., Plangkang, P., & Sittijunda, S. (2018). Repeated batch fermentation for photo-hydrogen and lipid production from wastewater of a sugar manufacturing plant. *International Journal of Hydrogen Energy*, *43*(7), 3605–3617. https://doi.org/10.1016/j.ijhydene.2017.12.119.

Azad, S., Vikineswary, S., Ramachandran, K., & Chong, V. (2001). Growth and production of biomass of Rhodovulum sulfidophilum in sardine processing wastewater. *Letters in Applied Microbiology*, *33*(4), 264–268.

Baars, O., Morel, F. M., & Zhang, X. (2018). The purple non-sulfur bacterium Rhodopseudomonas palustris produces novel petrobactin-related siderophores under aerobic and anaerobic conditions. *Environmental Microbiology*, *20*(5), 1667–1676.

Bali, A., Blanco, G., Hill, S., & Kennedy, C. (1992). Excretion of ammonium by a nifL mutant of Azotobacter vinelandii fixing nitrogen. *Applied and Environmental Microbiology*, *58*(5), 1711–1718.

Balsalobre, C., Calonge, J., Jiménez, E., Lafuente, R., Mouriño, M., Muñoz, M., et al. (1993). Using the metabolic capacity of Rhodobacter sphaeroides to assess heavy metal toxicity. *Environmental Toxicology and Water Quality*, *8*(4), 437–450.

Basak, N., Jana, A. K., Das, D., & Saikia, D. (2014). Photofermentative molecular biohydrogen production by purple-non-sulfur (PNS) bacteria in various modes: The present progress and future perspective. *International Journal of Hydrogen Energy*, *39*(13), 6853–6871.

Berne, C., Allainmat, B., & Garcia, D. (2005). Tributyl phosphate degradation by Rhodopseudomonas palustris and other photosynthetic bacteria [comparative study]. *Biotechnology Letters*, *27*(8), 561–566. https://doi.org/10.1007/s10529-005-2882-7.

Biebl, H., & Pfennig, N. (1981). Isolation of members of the family *Rhodospirillaceae*. In *The prokaryotes* (pp. 267–273). Springer.

Blasco, R., & Castillo, F. (1992). Light-dependent degradation of nitrophenols by the phototrophic bacterium Rhodobacter capsulatus E1F1. *Applied and Environmental Microbiology*, *58*(2), 690–695.

Brandl, H., Gross, R. A., Lenz, R. W., Lloyd, R., & Fuller, R. C. (1991). The accumulation of poly (3-hydroxyalkanoates) in Rhodobacter sphaeroides. *Archives of Microbiology*, *155*(4), 337–340.

Brune, D. C. (1995). Sulfur compounds as photosynthetic electron donors. In *Anoxygenic photosynthetic bacteria* (pp. 847–870). Springer.

Buranakarl, L., Cheng-Ying, F., Ito, K., Izaki, K., & Takahashi, H. (1985). Production of molecular hydrogen by photosynthetic bacteria with raw starch. *Agricultural and Biological Chemistry*, *49* (11), 3339–3341.

Cabello, P., Pino, C., Olmo-Mira, M. F., Castillo, F., Roldán, M. D., & Moreno-Vivián, C. (2004). Hydroxylamine assimilation by Rhodobacter capsulatus E1F1 requirement of the hcp gene (hybrid cluster protein) located in the nitrate assimilation nas gene region for hydroxylamine reduction. *Journal of Biological Chemistry*, *279*(44), 45485–45494.

Cao, X., Huang, X., Boon, N., Liang, P., & Fan, M. (2008). Electricity generation by an enriched phototrophic consortium in a microbial fuel cell. *Electrochemistry Communications*, *10*(9), 1392–1395.

Chalam, A., Sasikala, C., Ramana, C. V., Uma, N., & Rao, P. R. (1997). Effect of pesticides on the diazotrophic growth and nitrogenase activity of purple nonsulfur bacteria. *Bulletin of Environmental Contamination and Toxicology*, *58*(3), 463–468.

Cho, Y., Donohue, T., Tejedor, I., Anderson, M., McMahon, K., & Noguera, D. (2008). Development of a solar-powered microbial fuel cell. *Journal of Applied Microbiology*, *104*(3), 640–650.

Coppens, J., Grunert, O., Van Den Hende, S., Vanhoutte, I., Boon, N., Haesaert, G., et al. (2016). The use of microalgae as a high-value organic slow-release fertilizer results in tomatoes with increased carotenoid and sugar levels. *Journal of Applied Phycology*, *28*(4), 2367–2377.

Demtröder, L., Pfänder, Y., & Masepohl, B. (2020). Rhodobacter capsulatus AnfA is essential for production of Fe-nitrogenase proteins but dispensable for cofactor biosynthesis and electron supply. *MicrobiologyOpen*, e1033.

du Toit, J.-P., & Pott, R. W. M. (2020). Transparent polyvinyl-alcohol cryogel as immobilisation matrix for continuous biohydrogen production by phototrophic bacteria. *Biotechnology for Biofuels*, *13*(1), 105. https://doi.org/10.1186/s13068-020-01743-7.

Ehrig, A., Müller, R., & Babel, W. (1997). Isolation of phenoxy herbicide-degrading Rhodoferax species from contaminated building material. *Acta Biotechnologica*, *17*(4), 351–356.

Elbadry, M., & Elbanna, K. (1999). Response of four rice varieties to Rhodobacter capsulatus at seedling stage. *World Journal of Microbiology and Biotechnology*, *15*(3), 363–367.

Erkal, N. A., Eser, M. G., Özgür, E., Gündüz, U., Eroglu, I., & Yücel, M. (2019). Transcriptome analysis of Rhodobacter capsulatus grown on different nitrogen sources. *Archives of Microbiology*, *201*(5), 661–671.

Falcone, D. L., & Tabita, F. R. (1993). Complementation analysis and regulation of CO2 fixation gene expression in a ribulose 1,5-bisphosphate carboxylase-oxygenase deletion strain of Rhodospirillum rubrum. *Journal of Bacteriology*, *175*(16), 5066–5077.

Fiedor, J., Ostachowicz, B., Baster, M., Lankosz, M., & Burda, K. (2016). Quantification of purple non-sulphur phototrophic bacteria and their photosynthetic structures by means of total reflection X-ray fluorescence spectrometry (TXRF). *Journal of Analytical Atomic Spectrometry*, *31*(10), 2078–2088.

Getha, K., Chong, V., & Vikineswary, S. (1998). Potential use of the phototrophic bacterium, Rhodopseudomonas palustris, as an aquaculture feed. *Asian Fisheries Science*, *10*, 223–232.

Ghosh, S., Dairkee, U. K., Chowdhury, R., & Bhattacharya, P. (2017). Hydrogen from food processing wastes via photofermentation using purple non-sulfur Bacteria (PNSB)–a review. *Energy Conversion and Management*, *141*, 299–314.

Gibson, J., & Harwood, C. S. (1995). Degradation of aromatic compounds by nonsulfur purple bacteria. In *Anoxygenic photosynthetic bacteria* (pp. 991–1003). Springer.

Giotta, L., Agostiano, A., Italiano, F., Milano, F., & Trotta, M. (2006). Heavy metal ion influence on the photosynthetic growth of Rhodobacter sphaeroides. *Chemosphere*, *62*(9), 1490–1499.

Gosse, J. L., Engel, B. J., Rey, F. E., Harwood, C. S., Scriven, L., & Flickinger, M. C. (2007). Hydrogen production by photoreactive nanoporous latex coatings of nongrowing Rhodopseudomonas palustris CGA009. *Biotechnology Progress*, *23*(1), 124–130.

Gupta, D., & Sinha, S. N. (2020). Production of salicylic acid by a purple non sulfur bacterium Rubrivivax gelatinosus strain RASN4 from rhizosperic soil of paddy fields. *Journal of Global Biosciences*, *9*, 6718–6736.

Habte, M., & Alexander, M. (1980). Nitrogen fixation by photosynthetic bacteria in lowland rice culture. *Applied and Environmental Microbiology*, *39*(2), 342–347.

Hansen, T. A., & van Gemerden, H. (1972). Sulfide utilization by purple nonsulfur bacteria. *Archiv für Mikrobiologie*, *86*(1), 49–56.

Harwood, C. S. (2008). Nitrogenase-catalyzed hydrogen production by purple nonsulfur photosynthetic bacteria. *Bioenergy*, 259.

Hassan, M. A., Shirai, Y., Kusubayashi, N., Karim, M. I. A., Nakanishi, K., & Hasimoto, K. (1997). The production of polyhydroxyalkanoate from anaerobically treated palm oil mill effluent by Rhodobacter sphaeroides. *Journal of Fermentation and Bioengineering*, *83*(5), 485–488.

He, D., Bultel, Y., Magnin, J.-P., & Willison, J. C. (2006). Kinetic analysis of photosynthetic growth and photohydrogen production of two strains of Rhodobacter capsulatus. *Enzyme and Microbial Technology*, *38*(1–2), 253–259.

Higuchi-Takeuchi, M., & Numata, K. (2019). Marine purple photosynthetic bacteria as sustainable microbial production hosts. *Frontiers in Bioengineering and Biotechnology*, *7*, 258. https://doi.org/10.3389/fbioe.2019.00258.

Hsu, S., Lo, K., Fang, W., Lur, H., & Liu, C. (2015). Application of phototrophic bacterial inoculant to reduce nitrate content in hydroponic leafy vegetables. *Crop, Environment & Bioinformatics*, *12*(11).

Hull, N. C. (2009). *Photosynthetic bacterium for long-term space expeditions*.

Hunter, C. N., Daldal, F., Thurnauer, M. C., & Beatty, J. T. (2008). *The purple phototrophic bacteria. Vol. 28*. Springer Science & Business Media.

Ibrahim, S., Vikineswary, S., Al-Azad, S., & Chong, L. (2006). The effects of light intensity, inoculum size, and cell immobilisation on the treatment of sago effluent withRhodopseudomonas palustris strain B1. *Biotechnology and Bioprocess Engineering*, *11*(5), 377–381.

Idris, A. (2014). *Isolation and characterization of hydrogen producing marine photosynthetic bacteria*. Universiti Teknologi Malaysia.

Imhoff, J. F. (2005). Chromatiales ord. nov. In *Bergey's manual of systematic bacteriology* (pp. 1–59). Springer.

Imhoff, J. F., Hiraishi, A., & Süling, J. (2005). Anoxygenic phototrophic purple bacteria. In *Bergey's manual of systematic bacteriology* (pp. 119–132). Springer.

Inglesby, A. E., Beatty, D. A., & Fisher, A. C. (2012). Rhodopseudomonas palustris purple bacteria fed Arthrospira maxima cyanobacteria: Demonstration of application in microbial fuel cells. *RSC Advances*, *2*(11), 4829–4838.

Inui, M., Nakata, K., Roh, J. H., Zahn, K., & Yukawa, H. (1999). Molecular and functional characterization of the *rhodopseudomonas palustris* no. 7 phosphoenolpyruvate carboxykinase gene. *Journal of Bacteriology*, *181*(9), 2689. https://doi.org/10.1128/JB.181.9.2689-2696.1999.

Ishii, K., Hiraishi, A., Arai, T., & Kitamura, H. (1990). Light-dependent porphyrin production by suspended and immobilized cells of Rhodobacter sphaeroides. *Journal of Fermentation and Bioengineering*, *69*(1), 26–32.

Italiano, F., D'Amici, G. M., Rinalducci, S., De Leo, F., Zolla, L., Gallerani, R., et al. (2011). The photosynthetic membrane proteome of Rhodobacter sphaeroides R-26.1 exposed to cobalt. *Research in Microbiology*, *162*(5), 520–527.

Jalal, K., Zaima Azira, Z. A., Rahman, M., Kamaruzzaman, B., & Faizul, N. (2014). Carotenoid contents in Anoxygenic phototrophic purple Bacteria, Marichromatium Sp. and Rhodopseudomonas Sp. of tropical environment, Malaysia. *Oriental Journal of Chemistry*, *30*(2).

Jung, D. O., Achenbach, L. A., Karr, E. A., Takaichi, S., & Madigan, M. T. (2004). A gas vesiculate planktonic strain of the purple non-sulfur bacterium Rhodoferax antarcticus isolated from Lake fryxell, dry valleys, Antarctica. *Archives of Microbiology*, *182*(2–3), 236–243.

Jung, M. T. M., & Jung, D. O. (2009). An overview of purple bacteria: Systematics, physiology, and habitats. In C. N. Hunter, F. Daldal, M. C. Thurnauer, & J. T. Beatty (Eds.), *Vol. 28. The purple phototrophic bacteria*. Dordrecht, The Netherlands: Springer.

Kantha, T., Kantachote, D., & Klongdee, N. (2015). Potential of biofertilizers from selected Rhodopseudomonas palustris strains to assist rice (Oryza sativa L. subsp. indica) growth under salt stress and to reduce greenhouse gas emissions. *Annals of Microbiology*, *65*(4), 2109–2118.

Kapdan, I. K., & Kargi, F. (2006). Bio-hydrogen production from waste materials. *Enzyme and Microbial Technology*, *38*(5), 569–582.

Karrasch, S., Bullough, P. A., & Ghosh, R. (1995). The 8.5 A projection map of the light-harvesting complex I from Rhodospirillum rubrum reveals a ring composed of 16 subunits. *The EMBO Journal*, *14*(4), 631–638.

Kars, G., Gündüz, U., Yücel, M., Türker, L., & Eroglu, İ. (2006). Hydrogen production and transcriptional analysis of nifD, nifK and hupS genes in Rhodobacter sphaeroides OU 001 grown in media with different concentrations of molybdenum and iron. *International Journal of Hydrogen Energy*, *31*(11), 1536–1544.

Kessi, J. (2006). Enzymic systems proposed to be involved in the dissimilatory reduction of selenite in the purple non-sulfur bacteria Rhodospirillum rubrum and Rhodobacter capsulatus. *Microbiology*, *152*(3), 731–743.

Kessi, J., Ramuz, M., Wehrli, E., Spycher, M., & Bachofen, R. (1999). Reduction of selenite and detoxification of elemental selenium by the phototrophic BacteriumRhodospirillum rubrum. *Applied and Environmental Microbiology*, *65*(11), 4734–4740.

Khatipov, E., Miyake, M., Miyake, J., & Asada, Y. (1998). Accumulation of poly-β-hydroxybutyrate by Rhodobacter sphaeroides on various carbon and nitrogen substrates. *FEMS Microbiology Letters*, *162*(1), 39–45.

Khuong, N. Q., Kantachote, D., Nookongbut, P., Onthong, J., Xuan, L. N. T., & Sukhoom, A. (2020). Mechanisms of acid-resistant Rhodopseudomonas palustris strains to ameliorate acidic stress and promote plant growth. *Biocatalysis and Agricultural Biotechnology*, *24*, 101520.

Kim, J. S., Ito, K., & Takahashi, H. (1982). Production of molecular hydrogen in outdoor batch cultures of Rhodopseudomonas sphaeroides. *Agricultural and Biological Chemistry*, *46*(4), 937–941.

Kim, E.-J., Kim, M.-S., & Lee, J. K. (2008). Hydrogen evolution under photoheterotrophic and dark fermentative conditions by recombinant Rhodobacter sphaeroides containing the genes for fermentative pyruvate metabolism of Rhodospirillum rubrum. *International Journal of Hydrogen Energy*, *33*(19), 5131–5136.

Kim, J. K., & Lee, B.-K. (2000). Mass production of Rhodopseudomonas palustris as diet for aquaculture. *Aquacultural Engineering*, *23*(4), 281–293.

Koku, H., Eroğlu, I.n., Gündüz, U., Yücel, M., & Türker, L. (2002). Aspects of the metabolism of hydrogen production by Rhodobacter sphaeroides. *International Journal of Hydrogen Energy*, *27*(11), 1315–1329.

Kompantseva, E. I., Komova, A. V., & Kostrikina, N. A. (2010). Rhodovulum steppense sp. nov., an obligately haloalkaliphilic purple nonsulfur bacterium widespread in saline soda lakes of Central Asia. *International Journal of Systematic and Evolutionary Microbiology*, *60*(5), 1210–1214.

Kondo, K., Nishihara, E., & Nakata, N. (2006). Effect of the purple non-sulfur bacterium (Rhodobacter sphaeroides) on the fruit quality of tomato. In *XXVII International Horticultural Congress-IHC2006: International Symposium on Advances in Environmental Control, Automation 761*.

Lai, Y.-C., Liang, C.-M., Hsu, S.-C., Hsieh, P.-H., & Hung, C.-H. (2017). Polyphosphate metabolism by purple non-sulfur bacteria and its possible application on photo-microbial fuel cell. *Journal of Bioscience and Bioengineering*, *123*(6), 722–730.

Larimer, F. W., Chain, P., Hauser, L., Lamerdin, J., Malfatti, S., Do, L., et al. (2004). Complete genome sequence of the metabolically versatile photosynthetic bacterium Rhodopseudomonas palustris. *Nature Biotechnology*, *22*(1), 55–61. https://doi.org/10.1038/nbt923. http://www.nature.com/nbt/journal/v22/n1/suppinfo/nbt923_S1.html.

Lee, C.-M., Chen, P.-C., Wang, C.-C., & Tung, Y.-C. (2002). Photohydrogen production using purple nonsulfur bacteria with hydrogen fermentation reactor effluent. *International Journal of Hydrogen Energy*, *27*(11 – 12), 1309–1313.

Li, X., Peng, W., Jia, Y., Lu, L., & Fan, W. (2016). Bioremediation of lead contaminated soil with Rhodobacter sphaeroides. *Chemosphere*, *156*, 228–235.

Lin, Y.-H., Juan, M.-L., & Hsien, H.-J. (2011). Effects of temperature and initial pH on biohydrogen production from food-processing wastewater using anaerobic mixed cultures. *Biodegradation*, *22*(3), 551–563.

Lin, Y.-H., Zheng, H.-X., & Juan, M.-L. (2012). Biohydrogen production using waste activated sludge as a substrate from fructose-processing wastewater treatment. *Process Safety and Environmental Protection*, *90*(3), 221–230.

Liu, J., Zhao, Y., Diao, M., Wang, W., Hua, W., Wu, S., et al. (2019). Poly (3-hydroxybutyrate-co-3-hydroxyvalerate) production by Rhodospirillum rubrum using a two-step culture strategy. *Journal of Chemistry, 2019*.

Lorrungruang, C., Martthong, J., Sasaki, K., & Noparatnaraporn, N. (2006). Selection of photosynthetic bacterium Rhodobacter sphaeroides 14F for polyhydroxyalkanoate production with two-stage aerobic dark cultivation. *Journal of Bioscience and Bioengineering*, *102*(2), 128–131.

Luongo, V., Ghimire, A., Frunzo, L., Fabbricino, M., d'Antonio, G., Pirozzi, F., et al. (2017). Photofermentative production of hydrogen and poly-β-hydroxybutyrate from dark fermentation products. *Bioresource Technology*, *228*, 171–175.

Madigan, M. T., & Gest, H. (1979). Growth of the photosynthetic. *Journal of Bacteriology*, *137*, 524–530.

Madigan, M. T., & Imhoff, J. F. (2001). Phylum BXIII. Firmicutes. In *Bergey's manual of systematic bacteriology* (pp. 625–637). Springer.

Madigan, M. T., & Jung, D. O. (2009). An overview of purple bacteria: Systematics, physiology, and habitats. In *The purple phototrophic bacteria* (pp. 1–15). Springer.

Madigan, M. T., Resnick, S. M., Kempher, M. L., Dohnalkova, A. C., Takaichi, S., Wang-Otomo, Z.-Y., et al. (2019). Blastochloris tepida, sp. nov., a thermophilic species of the bacteriochlorophyll b-containing genus Blastochloris. *Archives of Microbiology*, *201*(10), 1351–1359.

Maness, P.-C., & Weaver, P. F. (2002). Hydrogen production from a carbon-monoxide oxidation pathway in Rubrivivax gelatinosus. *International Journal of Hydrogen Energy*, *27*(11–12), 1407–1411.

Manisha, D., Kudle, K. R., Singh, S., Merugu, R., & Rudra, M. P. (2017). Production of biogas from cellulose and benzoate using anaerobic bacterial consortia. *International Journal of Engineering Technologies and Management Research*, *4*(12), 65–70.

Masepohl, B. (2017). Regulation of nitrogen fixation in photosynthetic purple nonsulfur bacteria. In *Modern topics in the phototrophic prokaryotes* (pp. 1–25). Springer.

Mcewan, A. (1994). Photosynthetic Electron-transport and anaerobic metabolism in purple nonsulfur phototrophic Bacteria. *Antonie Van Leeuwenhoek International Journal of General and Molecular*, *66*(1–3), 151–164.

McKinlay, J. B., & Harwood, C. S. (2010a). Carbon dioxide fixation as a central redox cofactor recycling mechanism in bacteria. *Proceedings National Academy of Sciences. United States of America*, *107*(26), 11669–11675.

McKinlay, J. B., & Harwood, C. S. (2010b). Photobiological production of hydrogen gas as a biofuel. *Current Opinion in Biotechnology*, *21*, 244–251.

Mehrabi, S., Ekanemesang, U. M., Aikhionbare, F. O., Kimbro, K. S., & Bender, J. (2001). Identification and characterization of Rhodopseudomonas spp., a purple, non-sulfur bacterium from microbial mats. *Biomolecular Engineering*, *18*, 49–56.

Melis, A. (2002). Green alga hydrogen production: Progress, challenges and prospects. *International Journal of Hydrogen Energy*, *27*(11), 1217–1228.

Melis, A., & Melnicki, M. R. (2006). Integrated biological hydrogen production. *International Journal of Hydrogen Energy*, *31*(11), 1563–1573. https://doi.org/10.1016/j.ijhydene.2006.06.038.

Menxu, Q. X. R. Q. Y. (1991). Study on the factors influencing carotenoid formation in photosynthetic bacterium [J]. *Journal of Microbiology*, *2*.

Merugu, R. C., Girisham, S., & Reddy, S. (2010). Extracellular enzymes of two anoxygenic phototrophic bacteria isolated from leather industry effluents. *BioChemistry: An Indian Journal*, *4*(2), 86–88.

Merugu, R., Girisham, S., & Reddy, S. (2010a). Production of PHB (Polyhydroxybutyrate) by Rhodopseudomonas palustris KU003 and Rhodobacter capsulatus KU002 under phosphate limitation. *International Journal of Applied Biology and Pharmaceutical Technology*, *3*, 746–748.

Merugu, R., Girisham, S., & Reddy, S. (2010b). Production of PHB (Polyhydroxybutyrate) by Rhodopseudomonas palustris KU003 under nitrogen limitation. *International Journal of Applied Biology and Pharmaceutical Technology*, *2*, 686–688.

Merugu, R., Prasad, M., Girisham, S., & Reddy, S. (2008a). Influence of some metals on growth of two anoxygenic phototrophic bacteria. *Nature Environment and Pollution Technology*, *7*(2), 225.

Merugu, R., Prasad, M., Girisham, S., & Reddy, S. (2008b). Phosphate solubilisation by four anoxygenic phototrophic bacteria isolation from leather industry. *Nature, Environment and Pollution Technology*, *7*, 597–599.

Milford, A. D., Achenbach, L. A., Jung, D. O., & Madigan, M. T. (2000). Rhodobaca bogoriensis gen. Nov. and sp. nov., an alkaliphilic purple nonsulfur bacterium from African Rift Valley soda lakes. *Archives of Microbiology*, *174*(1–2), 18–27.

Mitsui, A., Matsunaga, T., Ikemoto, H., & Renuka, B. (1985). Organic and inorganic waste treatment and simultaneous photoproduction of hydrogen by immobilized photosynthetic bacteria. *Developments in Industrial Microbiology; (United States)*, *26*.

Moore, M. D., & Kaplan, S. (1992). Identification of intrinsic high-level resistance to rare-earth oxides and oxyanions in members of the class Proteobacteria: Characterization of tellurite, selenite, and rhodium sesquioxide reduction in Rhodobacter sphaeroides. *Journal of Bacteriology*, *174*(5), 1505–1514.

Morishima, K., Yoshida, M., Furuya, A., Moriuchi, T., Ota, M., & Furukawa, Y. (2007). Improving the performance of a direct photosynthetic. *Journal of Micromechanics and Microengineering*, *17*(9).

Mukhopadhyay, R., Rosen, B. P., Phung, L. T., & Silver, S. (2002). Microbial arsenic: From geocycles to genes and enzymes. *FEMS Microbiology Reviews*, *26*(3), 311–325.

Mukkata, K., Kantachote, D., Wittayaweerasak, B., Megharaj, M., & Naidu, R. (2019). The potential of mercury resistant purple nonsulfur bacteria as effective biosorbents to remove mercury from contaminated areas. *Biocatalysis and Agricultural Biotechnology*, *17*, 93–103.

Munjam, S., Vasavi, D., Girisham, S., & Reddy, S. (2003). Production of amylases (α and β) by anoxygenic phototrophic bacteria. *Journal of Food Science and Technology (Mysore)*, *40*(5), 505–508.

Nagadomi, H., Takahasi, T., Sasaki, K., & Yang, H. (2000). Simultaneous removal of chemical oxygen demand and nitrate in aerobic treatment of sewage wastewater using an immobilized photosynthetic bacterium of porous ceramic plates. *World Journal of Microbiology and Biotechnology*, *16*(1), 57–62.

Nagashima, K. V., Hiraishi, A., Shimada, K., & Matsuura, K. (1997). Horizontal transfer of genes coding for the photosynthetic reaction centers of purple bacteria. *Journal of Molecular Evolution*, *45*(2), 131–136.

Nagumo, A., Takanashi, K., Hojo, H., & Suzuki, Y. (1991). Cytotoxicity of bacteriohopane-32-ol against mouse leukemia L1210 and P388 cells in vitro. *Toxicology Letters*, *58*(3), 309–313.

Najafpour, G., Younesi, H., & Mohamed, A. R. (2006). A survey on various carbon sources for biological hydrogen production via the water-gas reaction using a photosynthetic bacterium (Rhodospirillum rubrum). *Energy Sources, Part A: Recovery, Utilization, and Environmental Effects*, *28*(11), 1013–1026. https://doi.org/10.1080/009083190910541.

Naylor, G. W., Addlesee, H. A., Gibson, L. C. D., & Hunter, C. N. (1999). The photosynthesis gene cluster of Rhodobacter sphaeroides. *Photosynthesis Research*, *62*(2), 121–139.

Nepple, B., Kessi, J., & Bachofen, R. (2000). Chromate reduction by Rhodobacter sphaeroides. *Journal of Industrial Microbiology and Biotechnology*, *25*(4), 198–203.

Nookongbut, P., Kantachote, D., Khuong, N. Q., & Tantirungkij, M. (2020). The biocontrol potential of acid-resistant Rhodopseudomonas palustris KTSSR54 and its exopolymeric substances against rice fungal pathogens to enhance rice growth and yield. *Biological Control*, *104354*.

Nookongbut, P., Kantachote, D., Krishnan, K., & Megharaj, M. (2017). Arsenic resistance genes of as-resistant purple nonsulfur bacteria isolated from as-contaminated sites for bioremediation application. *Journal of Basic Microbiology*, *57*(4), 316–324.

Noparatnaraporn, N., Takeno, K., & Sasaki, K. (2001). Hydrogen and poly-(hydroxy) alkanoate production from organic acids by photosynthetic bacteria. In *Biohydrogen II* (pp. 33–40). Elsevier.

Nunkaew, T., Kantachote, D., Kanzaki, H., Nitoda, T., & Ritchie, R. J. (2014). Effects of 5-aminolevulinic acid (ALA)-containing supernatants from selected Rhodopseudomonas palustris strains on rice growth under NaCl stress, with mediating effects on chlorophyll, photosynthetic electron transport and antioxidative enzymes. *Electronic Journal of Biotechnology*, *17*(1), 4.

Nunkaew, T., Kantachote, D., Nitoda, T., & Kanzaki, H. (2015). Selection of salt tolerant purple nonsulfur bacteria producing 5-aminolevulinic acid (ALA) and reducing methane emissions from microbial rice straw degradation. *Applied Soil Ecology*, *86*, 113–120. https://doi.org/10.1016/j.apsoil.2014.10.005.

Ogbonna, J. C., Yoshizawa, H., & Tanaka, H. (2000). Treatment of high strength organic wastewater by a mixed culture of photosynthetic microorganisms. *Journal of Applied Phycology*, *12*(3–5), 277–284.

Okubo, Y., Futamata, H., & Hiraishi, A. (2006). Characterization of phototrophic purple nonsulfur Bacteria forming colored microbial Mats in a swine wastewater ditch. *Applied and Environmental Microbiology*, 6225–6233.

Ormerod, J. G., & Sirevag, R. (1983). *Essential aspects of carbon metabolism*. Blackwell Scientific Publications.

Overmann, J., & Garcia-Pichel, F. (2013). The phototrophic way of life. In *The prokaryotes* (pp. 203–257). Springer.

Pérez, V., Dorador, C., Molina, V., Yáñez, C., & Hengst, M. (2018). Rhodobacter sp. Rb3, an aerobic anoxygenic phototroph which thrives in the polyextreme ecosystem of the Salar de Huasco, in the Chilean Altiplano. *Antonie Van Leeuwenhoek*, *111*(8), 1449–1465.

Ponsano, E. H. G., Lacava, P. M., & Pinto, M. F. (2002). Isolation of Rhodocyclus gelatinosus from poultry slaughterhouse wastewater. *Brazilian Archives of Biology and Technology*, *45*(4), 445–449.

Poretsky, R. S. (2003). Finding a niche: The habits and habitats ofpurple non-sulfur bacteria. *Microbial Diversity*, 1–19.

Powlson, D. S., Whitmore, A. P., & Goulding, K. W. (2011). Soil carbon sequestration to mitigate climate change: a critical re-examination to identify the true and the false. *European Journal of Soil Science*, *62*(1), 42–55.

Prasad, R. M. M., Vasavi, D., Girisham, S., & Reddy, S. (2008). Production of Asparginases by four anoxygenic phototrophic Bacteria isolated from leather industry effluents. *Ecology, Environment and Conservation*, *14*, 485–487.

Rajkumar, B., & Lalithakumari, D. (1992). Carbendazim as the sole carbon and nitrogen source for a phototrophic bacterium. *Biomedical Letters*, *47*(188), 337–346.

Ramchander Merugu, M., Pratap Rudra, A. T., Girisham, S., & Reddy, S. M. (2011). Chromate reduction by a purple non Sulphur phototrophic bacterium Rhodobacter capsulatus KU002 isolated from tannery effluents. *Journal of Pure and Applied Microbiology*, *5*(2), 66–69.

Ramchander, M., Rudra, M. P., Atthapu, T., Girisham, S., & Reddy, S. (2011). Optimisation of indole acetic acid production by two anoxygenic phototrophic bacteria isolated from tannery effluents. *Journal of Pure and Applied Microbiology*, *5*(2), 929–932.

Rosenbaum, M., Schröder, U., & Scholz, F. (2005a). In situ electrooxidation of photobiological hydrogen in a photobioelectrochemical fuel cell based on Rhodobacter sphaeroides. *Environmental Science & Technology*, *39*(16), 6328–6333.

Rosenbaum, M., Schröder, U., & Scholz, F. (2005b). Utilizing the green alga Chlamydomonas reinhardtii for microbial electricity generation: a living solar cell. *Applied Microbiology and Biotechnology*, *68*(6), 753–756.

Sadi, N., & Firmansyah, F. (2020). The potency of purple sulphur bacteria for in-situ sulphide bioremediation in water. *IOP Conference Series: Earth and Environmental Science*.

Sagir, E., Ozgur, E., Gunduz, U., Eroglu, I., & Yucel, M. (2017). Single-stage photofermentative biohydrogen production from sugar beet molasses by different purple non-sulfur bacteria. *Bioprocess and Biosystems Engineering*, *40*(11), 1589–1601.

Sakarika, M., Spanoghe, J., Sui, Y., Wambacq, E., Grunert, O., Haesaert, G., et al. (2019). Purple non-Sulphur bacteria and plant production: Benefits for fertilization, stress resistance and the environment. *Microbial Biotechnology*.

Sakpirom, J., Kantachote, D., Nunkaew, T., & Khan, E. (2017). Characterizations of purple non-sulfur bacteria isolated from paddy fields, and identification of strains with potential for plant growth-promotion, greenhouse gas mitigation and heavy metal bioremediation. *Research in Microbiology*, *168*(3), 266–275.

Salma, U., Miah, A., Tareq, K., Maki, T., & Tsujii, H. (2007). Effect of dietary Rhodobacter capsulatus on egg-yolk cholesterol and laying hen performance. *Poultry Science*, *86*(4), 714–719.

Sasaki, K., Tanaka, T., Nishizawa, Y., & Hayashi, M. (1991). Enhanced production of 5-aminolevulinic acid by repeated addition of levulinic acid and supplement of precursors in photoheterotrophic culture of Rhodobacter sphaeroides. *Journal of Fermentation and Bioengineering*, *71*(6), 403–406.

Sasaki, K., Watanabe, M., & Tanaka, T. (2002). Biosynthesis, biotechnological production and applications of 5-aminolevulinic acid. *Applied Microbiology and Biotechnology*, *58*(1), 23–29.

Sasikala, C., & Ramana, C. V. (1995). Biotechnological potentials of anoxygenic phototrophic bacteria. I. Production of single-cell protein, vitamins, ubiquinones, hormones, and enzymes and use in waste treatment. In *Vol. 41. Advances in applied microbiology* (pp. 173–226). Elsevier.

Sasikala, K., Ramana, C. V., & Rao, P. R. (1992). Photoproduction of hydrogen from the waste water of a distillery by Rhodobacter sphaeroides OU 001. *International Journal of Hydrogen Energy*, *17*(1), 23–27.

Sasikala, K., Ramana, C. V., Rao, P. R., & Subrahmanyam, M. (1990). Effect of gas phase on the photoproduction of hydrogen and substrate conversion efficiency in the photosynthetic bacterium Rhodobacter sphaeroides OU 001. *International Journal of Hydrogen Energy*, *15*(11), 795–797.

Sasikala, K., Ramana, C. V., & Subrahmanyam, M. (1991). *Photoproduction of hydrogen from waste water of a lactic acid fermentation plant by a purple non-sulfur photosynthetic bacterium, Rhodobacter sphaeroides*.

Sawayama, S., Rao, K. K., & Hall, D. O. (1998). Immobilization of Rhodobacter capsulatus on cellulose beads and water treatment using a photobioreactor. *Journal of Fermentation and Bioengineering*, *86*(5), 517–520.

Schultz, J. E., Gotto, J. W., Weaver, P. F., & Yoch, D. C. (1985). Regulation of nitrogen fixation in Rhodospirillum rubrum grown under dark, fermentative conditions. *Journal of Bacteriology*, *162*, 1322–1324.

Seangtumnor, N., Kantachote, D., Nookongbut, P., & Sukhoom, A. (2018). The potential of selected purple nonsulfur bacteria with ability to produce proteolytic enzymes and antivibrio compounds for using in shrimp cultivation. *Biocatalysis and Agricultural Biotechnology*, *14*, 138–144. https://doi.org/10.1016/j.bcab.2018.02.013.

Seki, H., Suzuki, A., & Mitsueda, S.-I. (1998). Biosorption of heavy metal ions on Rhodobacter sphaeroides and Alcaligenes eutrophus H16. *Journal of Colloid and Interface Science*, *197*(2), 185–190.

Shaw, A. L., Hochkoeppler, A., Bonora, P., Zannoni, D., Hanson, G. R., & McEwan, A. G. (1999). Characterization of DorC from Rhodobacter capsulatus, a c-type cytochrome involved in electron transfer to dimethyl sulfoxide reductase. *Journal of Biological Chemistry*, *274*(15), 9911–9914.

Sheng, G., Yu, H., & Yue, Z. (2006). Factors influencing the production of extracellular polymeric substances by Rhodopseudomonas acidophila. *International Biodeterioration & Biodegradation*, *58*(2), 89–93.

Show, K., Lee, D., Tay, J., Lin, C., & Chang, J. (2012). Biohydrogen production: Current perspectives and the way forward. *International Journal of Hydrogen Energy*, *37*(20), 15616–15631.

Siefert, E., & Koppenhagen, V. (1982). Studies on the vitamin B 12 auxotrophy of Rhodocyclus purpureus and two other vitamin B 12-requiring purple nonsulfur bacteria. *Archives of Microbiology*, *132*(2), 173–178.

Silver, S., & Phung, L. T. (2005). Genes and enzymes involved in bacterial oxidation and reduction of inorganic arsenic. *Applied and Environmental Microbiology*, *71*(2), 599–608.

Sojka, G. (1978). Metabolism of nonaromatic organic compounds. *The Photosynthetic Bacteria*, 707–718.

Soon, T. K., Al-Azad, S., & Ransangan, J. (2014). Isolation and characterization of purple non-sulfur Bacteria, Afifella marina, producing large amount of carotenoids from mangrove microhabitats. *Journal of Microbiology and Biotechnology*, *24*(8), 1034–1043.

Spanoghe, J., Grunert, O., Wambacq, E., Sakarika, M., Papini, G., Alloul, A., et al. (2020). Storage, fertilization and cost properties highlight the potential of dried microbial biomass as organic fertilizer. *Microbial Biotechnology*, *13*.

Suman, J., Zubrova, A., Rojikova, K., Pechar, R., Svec, P., Cajthaml, T., et al. (2019). Pseudogemmobacter bohemicus gen. Nov., sp. nov., a novel taxon from the Rhodobacteraceae family isolated from heavy-metal-contaminated sludge. *International Journal of Systematic and Evolutionary Microbiology*, *69*(8), 2401–2407.

Sun, Z., Pang, B., Xi, J., & Hu, H.-Y. (2019). Screening and characterization of mixotrophic sulfide oxidizing bacteria for odorous surface water bioremediation. *Bioresource Technology*, *290*, 121721.

Sunayana, M., Sasikala, C., & Ramana, C. V. (2005). Production of a novel indole ester from 2-aminobenzoate by Rhodobacter sphaeroides OU5. *Journal of Industrial Microbiology and Biotechnology*, *32*(2), 41–45.

Tabita, F. (1995). *The biochemistry and and metabolic regulation of carbon metabolism and CO2-fixation in purple bacteria*. Kluwer Academic Publishers.

Takabatake, H., Suzuki, K., Ko, I.-B., & Noike, T. (2004). Characteristics of anaerobic ammonia removal by a mixed culture of hydrogen producing photosynthetic bacteria. *Bioresource Technology*, *95*(2), 151–158.

Takeno, K., Sasaki, K., Watanabe, M., Kaneyasu, T., & Nishio, N. (1999). Removal of phosphorus from oyster farm mud sediment using a photosynthetic bacterium, Rhodobacter sphaeroides IL106. *Journal of Bioscience and Bioengineering*, *88*(4), 410–415.

Takeno, K., Yamaoka, Y., & Sasaki, K. (2005). Treatment of oil-containing sewage wastewater using immobilized photosynthetic bacteria. *World Journal of Microbiology and Biotechnology*, *21*(8–9), 1385–1391.

Tan, D. X., Hardeland, R., Manchester, L. C., Paredes, S. D., Korkmaz, A., Sainz, R. M., et al. (2010). The changing biological roles of melatonin during evolution: From an antioxidant to signals of darkness, sexual selection and fitness. *Biological Reviews*, *85*(3), 607–623.

Tan, D. X., Manchester, L. C., Liu, X., Rosales-Corral, S. A., Acuna-Castroviejo, D., & Reiter, R. J. (2013). Mitochondria and chloroplasts as the original sites of melatonin synthesis: a hypothesis related to melatonin's primary function and evolution in eukaryotes. *Journal of Pineal Research*, *54*(2), 127–138.

Tarabas, O., Hnatush, S., Moroz, O., & Kovalchuk, M. (2019). Wastewater bioremediation with using of phototrophic non-sulfur bacteria Rhodopseudomonas yavorovii IMV B-7620. *Ecology and Noospherology*, *30*(2), 63–67.

Tarabas, O., Hnatush, S., & Moroz, O. (2019). The usage of nitrogen compounds by purple non-sulfur bacteria of the Rhodopseudomonas genus. *Regulatory Mechanisms in Biosystems*, *10*(1).

Tiang, M. F., Fitri Hanipa, M. A., Abdul, P. M., Jahim, J. M. D., Mahmod, S. S., Takriff, M. S., et al. (2020). Recent advanced biotechnological strategies to enhance photo-fermentative biohydrogen production by purple non-Sulphur bacteria: An overview. *International Journal of Hydrogen Energy*, *45*(24), 13211–13230. https://doi.org/10.1016/j.ijhydene.2020.03.033.

Urmeneta, J., Navarrete, A., Huete, J., & Guerrero, R. (2003). Isolation and characterization of cyanobacteria from microbial Mats of the Ebro Delta, Spain. *Current Microbiology*, *46*, 199–204.

Utsunomiya, T., Yamane, Y.-i., Watanabe, M., & Sasaki, K. (2003). Stimulation of porphyrin production by application of an external magnetic field to a photosynthetic bacterium, Rhodobacter sphaeroides. *Journal of Bioscience and Bioengineering*, *95*(4), 401–404.

Vareeket, R., & Soytong, K. (2020). Rhodospirillum centenum, A new growth stimulant and antagonistic Bacteria against leaf spot of Rice caused by Curvularia lunata. *AGRIVITA, Journal of Agricultural Science*, *42*(1), 160–167.

Visscher, P. T., & Taylor, B. F. (1993). Organic thiols as organolithotrophic substrates for growth of phototrophic bacteria. *Applied and Environmental Microbiology*, *59*(1), 93–96.

Wang, X., Fang, Y., Wang, Y., Hu, J., Zhang, A., Ma, X., et al. (2018). Single-stage photofermentative hydrogen production from hydrolyzed straw biomass using Rhodobacter sphaeroides. *International Journal of Hydrogen Energy*, *43*.

Wang, H., & Norén, A. (2006). Metabolic regulation of nitrogen fixation in Rhodospirillum rubrum. *Biochemical Society Transactions*, *34*, 160–161.

Watanabe, M., Kawahara, K., Sasaki, K., & Noparatnaraporn, N. (2003). Biosorption of cadmium ions using a photosynthetic bacterium, Rhodobacter sphaeroides S and a marine photosynthetic bacterium, Rhodovulum sp. and their biosorption kinetics. *Journal of Bioscience and Bioengineering*, *95*(4), 374–378.

Watanabe, M., Shiba, H., Sasaki, K., Nakashimada, Y., & Nishio, N. (1998). Promotion of growth and flocculation of a marine photosynthetic bacterium, Rhodovulum sp. by metal cations. *Biotechnology Letters*, *20*(12), 1109–1112.

Willows, R. D., & Kriegel, A. M. (2009). Biosynthesis of bacteriochlorophylls in purple Bacteria. In *Vol. 28. The purple phototrophic bacteria.*

Wright, G. E., & Madigan, M. T. (1991). Photocatabolism of aromatic compounds by the phototrophic purple bacterium Rhodomicrobium vannielii. *Applied and Environmental Microbiology*, *57*(7), 2069–2073.

Wu, T. Y., Hay, J. X. W., Kong, L. B., Juan, J. C., & Jahim, J. M. (2012). Recent advances in reuse of waste material as substrate to produce biohydrogen by purple non-sulfur (PNS) bacteria. *Renewable and Sustainable Energy Reviews*, *16*(5), 3117–3122.

Xing, D., Cheng, S., Regan, J. M., & Logan, B. E. (2009). Change in microbial communities in acetate-and glucose-fed microbial fuel cells in the presence of light. *Biosensors and Bioelectronics*, *25*(1), 105–111.

Xing, D., Zuo, Y., Cheng, S., Regan, J. M., & Logan, B. E. (2008). Electricity generation by Rhodopseudomonas palustris DX-1. *Environmental Science & Technology*, *42*(11), 4146–4151. https://doi.org/10.1021/es800312v.

Zhi, X., Yang, H., Berthold, S., Doetsch, C., & Shen, J. (2010). Potential improvement to a citric wastewater treatment plant using bio-hydrogen and a hybrid energy system. *Journal of Power Sources*, *195*(19), 6945–6953.

Zilles, J. L., Peccia, J., Kim, M.-W., Hung, C.-H., & Noguera, D. R. (2002). Involvement of Rhodocyclus-related organisms in phosphorus removal in full-scale wastewater treatment plants. *Applied and Environmental Microbiology*, *68*(6), 2763–2769.

Chapter 10

Bacterial community response to pesticides polluted soil

Raunak Dhanker[a], Shubham Goyal[b], Krishna Kumar[c], and Touseef Hussain[d]

[a]*Department of Biological Sciences, School of Basic and Applied Sciences GD Goenka University, Gurugram, Haryana, India,* [b]*Amity Institute of Biotechnology, Amity University, Noida, India,* [c]*Department of Biotechnology, School of Chemical and Life Sciences, New Delhi, India,* [d]*Department of Botany, Aligarh Muslim University, Aligarh, Uttar Pradesh, India*

Chapter outline

1 Introduction

Chemicals that are used as pesticides/fertilizers for agriculture are known as agrochemicals. They protect crops from the attack of pests. The present population of India is approximately 1.3 billion, which would increase up to 1.7 billion, by 2050. According to economist's projections, India's economy will increase significantly by 2030, and India will be the third largest economy in the world following China and the United States (http://www.careratings.com/upload/NewsFiles/Studies/Agrochemicals.pdf). More than 50% of India's population stays in urban and semiurban areas which are responsible for the country's changing food requirement. It is estimated that cereal demand will hike 355 million tonnes by 2030. Surplus productivity of crop grains seems challenging to meet the future demand of the population. Increasing the population, the constant development of urban areas, dependence on rain for agriculture, and demising soil fertility of soil are some of the key challenges (Devi, Thomas, & Raju, 2017; Yadav & Dutta, 2019). India is an agricultural country that faces several

Recent Advancement in Microbial Biotechnology. https://doi.org/10.1016/B978-0-12-822098-6.00010-0

challenges in each step (i.e., sowing, growing, and harvesting) of crop production. Key challenges in sowing are soil health analysis, weather forecast info, and seed selection and treatment.

Similarly, the major hindrance to growing crops is an inappropriate information on agrochemicals and incorrect information on crop rotation (due to small farms). The significant challenges in crop harvesting are food storage, distribution, and adherence to classical technology (Poonia, Kumar, & Malik, 2019). Agrochemicals are an essential input in agriculture for crop protection and better yield. According to FY15-18, the increased ingestion of fertilizers per hectare is 11%. Nearly 50% of the crop is damaged due to the attack of pests, insects. and weeds growth around crops. That is why an appropriate use of agrochemicals is highly needed in countries like India, whose significant portion of the population is entirely dependent on agriculture. Agrochemicals prevent the growth of weeds around crops and reduce pest attack effects on the standing crops (Yadav & Dutta, 2019). Microbial metabolic activities sustain soil fertility. Few microbes fix atmospheric nutrients and transfer it further to the food chain, and through the process of decomposition and detrivory, these nutrients are ultimately again released back to the atmosphere. Thus microbes are responsible for maintaining the same amount of nutrients in the atmosphere. Soil is rich with bacteria followed by fungi, algae, and soil protozoan. These microbes play a significant role in the transport of nutrition and elements from soil to plants. The most common example of helpful microbes are rhizospheres (Fernandez et al., 2016; Kyei-Boahen, Slinkard, & Walley, 2001). We must utilize classical techniques innovatively and should focus on the invention of new technologies. More than one discipline such as genomics, system biology, and bioinformatics are involved in metagenomics. The use of agrochemicals helps increase the productivity of crops all over the world; undoubtedly, its excess and continuous use can decrease soil fertility and disturb ecosystems' sustainability. Due to agrochemicals leaching, groundwater, and nearby water bodies are affected that ultimately influences the microflora and fauna and finally reached to human beings causing many diseases (Sekhotha, Monyeki, & Sibuyi, 2016). Therefore, on the one hand, these pest controllers damage the physiochemical parameters of soil; on the other hand, they kill the population of nontarget microorganisms on a large scale.

This is high time to revise existing approaches and introduction of advanced technologies for improving the crop productivity in a sustainable manner that can maintain the soil fertility and can significantly reduce the health risk of human beings (Elkoca, Kantar, & Sahin, 2008) such as the use of microbiota for the degradation of pesticides (Silo-Suh, Lethbridge, & Raffel, 1994; Zhang et al., 2019; Zhang, Wang, Zhang, Teng, & Xu, 2019). This further can improve knowledge of microbial dynamics, symbiotic and nonsymbiotic interaction of microbes, and their interrelationships with their surroundings, to identify their exact niche in their habitat, etc., which can further be utilized for agricultural and industrial applications. This chapter highlights the relevant information on soil microbiota, sustainable use of agrochemicals, and the influence of agrochemicals on soil fertility and new adaptations/technologies for resolving the existing problems of agriculture. Finally, the challenges in sustainable agriculture and appropriate approaches to tackle these challenges are discussed.

2 Currents aspects of agrochemicals in India

To fulfil the food demands of the world's rapidly growing population, the use of agrochemicals in agriculture has gained an enormous boost in the last few decades. These agrochemicals are chemical-based compounds used by the farmers in the management of plants and crops to improve the quality and yields of the crops. India is the seventh largest populated country in the world who is nearly 58% of the population living in rural areas and directly depends on agriculture for their living expenses. India is at the second position in the production of agri-based crops. However, most of the crops are affected by insects, weeds, and various diseases that destroy the major portion of the crops. This is the leading cause of low yield compared to the global crop yield, i.e., 3MT to the global 4MT per hectare (FAOSTAT).

Furthermore, India has maximum diversity in the types of crops production and has a never-ending list of pests and weeds. These scenarios have resulted in establishing of the top agri-companies throughout the country, including DuPont India, Rallies India Limited, Monsanto India, etc. Hence, to compensate for this crop harvest loss, the farmers invest a huge amount of their money in buying the synthetic agrochemicals to increase the yield and protect the crops from pests and weeds. As reported by FICCI, pesticides and herbicides are of utmost required to protect crop during their cultivation and help reduce postharvest loss. The highest consumption of pesticides is found in Jammu and Kashmir, whereas Arunachal Pradesh and Meghalaya are using the least amount of pesticides per annum in India. Only 275 pesticides are registered to be used in India. Out of these, 255 are found to be poisonous for the farmers, and nearly 115 are hazardous in nature (Yadav & Dutta, 2019). So, the use of these commercially available agrochemicals in the crop field and agricultural land has proved to be catastrophic for nature and animal and human health. Due to these limitations, safe and environment-friendly alternatives have to be developed. Thus, the use of biofertilizers and ecofriendly microorganism in agri-fields can be explored as an alternative to synthetic fertilizers (Ramasamy, Geetha, & Yuvaraj, 2020). Hence, microorganisms and biofertilizers can be used as sustainable alternatives to synthetic fertilizers to reduce environmental pollution caused by agrochemicals and improve crop quality, productivity, and soil fertility (Shiri et al., 2020).

3 Role of agrochemicals in agriculture

Agrochemicals are used to increase crop yield and to protect crops from pest attacks. It has been noted that loss of crops due to weeds are much more in India followed by loss due to pests and insects accounting for the total loss of nearly

50% of crop yield (Kyei-Boahen et al., 2001). Hence, agrochemicals are much needed in countries like India, which has a large population that depends on agriculture; otherwise, this crop loss could result in shutting down farmers' business, which will contribute to the world's food scarcity. Agrochemicals include pesticides and herbicides insecticides, biopesticides, and other agrochemicals compound, which help farmers to control the adverse effects of pests and weeds on crops and increase total yield by inhibiting the growth of weeds around crops and reducing pest attack on the standing crops (Table 2). We can define these agrochemicals in the following different classes based on their roles in agriculture.

4 Crop protectors

1. Pesticides—Pesticides such as DDT (dichlorodiphenyltrichloroethane) and HCH (hexachlorocyclohexane) were used to kill the pests related to the destruction of agriculture as well as cause human diseases. However, with the increased uses of these synthetic chemical-based pesticides and insecticides, food standards are compromised along with increasing health problems in human due to their bioaccumulation in the human body as a result of regular exposure of human to these chemicals present in the soil, water, and air through various routes such as inhalation, dermal contacts, etc. (Abhilash & Singh, 2009). With the development of biochemistry and molecular biology-based techniques, these chemical-based pesticides are now being replaced by biopesticides. Biological pesticides or biopesticides are defined as the use of living organisms or organisms derived from natural materials to inhibit the growth or kill the pest, which is related to the loss of agricultural economics or human health. Biopesticides are classified into three major categories based on their origin and mode of action—microbial pesticides, plant-incorporated protectants, and biochemical pesticides (Singh, Singh, & Prabha, 2017). Besides biocontrol agents and plant-derived products microbial, pesticides are used widely to control mosquito-borne diseases (Dhanker, Kumar, & Hwang, 2013; Dhanker, Kumar, & Raghvendra, 2014). One of the most commonly used microbial pesticides is *Bacillus thuringienesis*; it controls the mosquito *Aedes aegypti* that acts as a transmitting vector of mosquito-borne diseases in humans (Dhanker et al., 2014). Popp, Pető, and Nagy (2013) have reported a list of microbial and biochemical pesticides used in agriculture to increase the yield by controlling the pest-based loss of agricultural products.
2. Herbicides—Herbicides are a class of agrochemicals used in agriculture to reduce or inhibit the growth of other herbs and weeds around crops. As we already know that weeds inhibit the growth of the crops by utilizing the available nutrients and fertilizers from the soil leading to nutritional scarcity for main crops. The basic principle behind using herbicides is based on selectivity, which kills only weeds associated with the crop without

targeting the crops. Thus, it can be used for the selective removal of the weeds growing around the cultivating crops. Herbicides are distinguished based on its applications, either the residual soil herbicides or the foliar herbicides in which the former are directly mixed in the soil before plantation and can show their effect during an extended period, whereas the latter is used in acute cases and show selective removal of weeds when mixed with soil gets degraded in some time and include glyphosate and paraquat (Poonia et al., 2019; Shiri et al., 2020). There are nearly 200 chemical compounds used as herbicides, but out of these, only a few have been checked for their toxicity and for relative concentration to be used in agriculture. Few examples of herbicides are inorganic herbicides and organic arsenicals, 2,4-D or phenoxy acid derivatives, bipyridyl derivatives, ureas and thioureas, phosphonomethyl amino acids or inhibitors of aromatic acid biosynthesis, etc. (Gupta, Snehi, & Singh, 2017). The negative consequences of herbicides have emphasized the development of more advanced techniques and delivery systems for the herbicides so that the maximum efficacy of herbicides can be achieved with the lowest amount of dose during their use (Maruyama et al., 2016).

3. Insecticides—Any substances that can kill or inhibit insects' growth and reproduction are defined as insecticides. These insecticides have an extended application in farming as nearly 67,000 species are associated with the loss in agricultural yields throughout the globe (Campos et al., 2019). Due to the growing population, it is recommended to increase the total crop yield to use growing food pesticides and insecticides. Insecticides are classified based on their toxicity mode mediated by penetration into, contact poison, ingestion poison, and fumigants. Based on chemical synthesis and origin, these insecticides are further classified into various types such as organophosphates, chlorinated hydrocarbons, carbamates, botanical insecticides, etc. In the beginning, synthetic pesticides such as DDT, a chlorinated hydrocarbon, are used to manage pests in agriculture, but with the extended use, it has shown toxicity towards humans and contributed to the disruption of the ecosystem. However, these bioaccumulative insecticides such as DDT, HCH, Toxaphene, Aldrin, and Dieldrin were removed from use in 2002 and replaced by biodegradable and nonbioaccumulative pesticides (Carvalho, 2017). Contact poison works by penetrating in the skin of the arthropods; one of the most common natural contact poisons is pyrethrum that is derived from *Chrysanthemum cinerariafolium* and *Tanacetum coccineum* and rotenone derived from the roots of *Derris* species and related plants.
4. Fungicides—Fungicides are synthetic chemicals used to control the growth of fungus in agricultural fields by the farmers, which cause damage to the overall yield of crops. But with the increased demands and frequent use of fungicides in agricultural fields, many fungal species have become resistant to commonly used fungicides. Some of the resistant strains of fungus can

also escape and cause human health-related problems. Alvarez-Moreno et al. (2017) reported that fungicides have resulted in resistance development in the *A. fumigantus* by the mutation in TR46/Y121F/T289A, causing disease in human and shows resistance against standard antifungal drugs. Hence, to overcome this kind of issues, synthetic fungicides are replaced by bio fungicides. Biofungicides are microorganism or plant-derived metabolites help in reducing the growth and reproduction of fungus in agri-fields. Many microbial species viz. *Trichoderma* spp., *Ulocladium* spp., *Bacillus subtilis*, plant extracts, and their essential oils have shown efficacy while using them as fungicides (Calmes et al., 2017; Hussain & Khan, 2020b; Jahanban, Panahpour, Gholami, Davari, & Lotfifar, 2018; McGehee, Raudales, Elmer, & McAvoy, 2019).

Soil supplements: Organic or inorganic molecules that are added to improve soil structure, quality, water holding capacity, and nutrient value are known as soil supplements or soil amendments. Further, these help to improve the quality of crops along with total yield by modifying the uptake of nutrient and root micro-environment. Soil supplements include fertilizers, hormones, and growth promoters that help in the growth and development of plants after addition in soil by enhancing crops' metabolic and reproductive pathways.

For example, inorganic fertilizers help to improve soil nutrients strength by increasing the concentration of micronutrients and macronutrients in the soil, thereby improving the soil fertility (Ahmed, Rauf, Mukhtar, & Saeed, 2017; Gülser et al., 2019; Willoughby, 2019). These inorganic fertilizers help to yield enhancement and crop quality improvement, but the major concern associated with these fertilizers are environmental contamination and human and animal health-related issues (Pan, Lam, Mosier, Luo, & Chen, 2016; van Zwieten, 2018). Nitrogen and phosphorous accumulate in the surrounding environment and water bodies by leaching and cause the contamination of soil and water.

Several inorganic compounds show detrimental effects on soil health by interacting with humic constituents of soil. Thus, organic compounds were introduced in agriculture to reduce the environmental complications caused by these inorganic fertilizers, which also aids in increasing the humic acid of soil. Organic fertilizers include complex organic compounds derived from plant and plants products, animal dungs, and waste disposal or biogas plants, which when decomposing, releases macro- and micronutrient, thereby improving the fertility of the soil (Zhang, Bei, et al., 2019; Zhang, Wang, et al., 2019). Various organic fertilizers are used as a substrate for the cultivation of soil friendly microorganisms that help to improve the quality of soil (Table 1). For example, cyanobacteria have been cultured in the soil to improve agricultural lands nitrogen content (Fernandez et al., 2016).

TABLE 1 Types of agrochemicals and their applications.

S. no.	Agrochemicals		Application	References
1	Herbicides	Glyphosate Paraquat Bipyridyl derivatives	Use to control the growth of weeds around the crop	Fernandez et al. (2016) Sekhotha et al. (2016) Elkoca et al. (2008)
2	Insecticides	DDT HCH Toxaphene Aldrin Plants-derived extract: *Chrysanthemum cinerariafolium* *Tanacetum coccineum*	Use to kill insects responsible for crop destruction as well as human disease causative agents	Silo-Suh et al. (1994) Hungria, Nogueira, and Araujo (2013)
3	Pesticides	Synthetics: DDT, HCH Biochemical: plant-incorporated protectants (phosphatidylinositol phosphate) Microbial: *Bacillus*	Use to control the growth of Pests	Ahemad and Kibret (2014) Zhan, Feng, Fan, and Chen (2018)
4	Fungicides	Plant extracts microbes	Use to control the growth of mycoses	Backer et al. (2018) Gupta et al. (2017) Ahemad and Khan (2011)
5	Soil supplements	Fertilizers, growth promoters, hormones, etc.	Use to provides the macro and micronutrient to the agricultural crops	Rajkumar, Ma, and Freitas (2018) Khan, Zaidi, and Aamil (2002) Rezzonico et al. (2007) Din et al. (2019)

5 Effects of agrochemicals on microbial ecosystem

Agrochemicals are used throughout the world to aid in managing crop productivity; however, the use of agrochemicals in greater concentration for an extended period results in the establishment of various environmental and human health-related concerns. The leaching of these agrochemicals into the groundwater and nearby water bodies causes a shift in microbe's biodiversity, causing a disturbance in the whole water ecosystem. Undoubtedly, the application of these agrochemicals in agriculture has helped to increase the overall crop yield, but it is not sustainable. It is evident that only 0.01% of these agrochemicals reaching the target pests, and the rest are accumulated in the soil, influencing the nontarget microorganisms present in the soil (Chenseng et al., 2006). These accumulated agrochemicals are environmental contaminants and hence must be mineralized. The excessive use of these agrochemicals has resulted in the inhibition of soil microbes indicating nontargeted killing of the soil microorganisms, thereby ultimately causing the loss of soil structural integrity.

Soil is composed of both living and nonliving components. The nonliving part of soil includes salts, sand, clay, and organic compounds, whereas the living part of soil includes different types of microorganisms such as bacteria, fungi, and algae, which includes nematodes, insects, earthworms, etc. contributing to the fertility of the soil (Darby & Neher, 2016). These microbes help in maintaining soil fertility and also help in intoxicating agrochemicals that are being applied onto the agricultural fields constantly by cycling the nutrients and degrading the organic matters into their monomolecular structure so that they can be mobilized by plants and crops and finally can be incorporated for their growth and development (Tejada, Rodríguez-Morgado, Gómez, & Parrado, 2014). This degradation of toxicants is performed with various microbial processes, including complete mineralization or detoxification of the toxic components. The nontoxic bioproducts are then released in the environment where it is further degraded by plants or/and by other microorganisms, making them immobilized by bonding them to the soil matrix. These microbial life forms are found to attach in roots surrounding, where they help in the mobilization of these compounds to the plants. The accumulation of these agrochemicals in soil affects microbes functioning greatly by preventing their enzymatic activities (Zhang et al., 2015). Soil microbes overcome the toxic effects of pesticides by making them less toxic. Certain microbial enzymes such as urease and alkaline phosphatase help in organic matter decomposition, thereby aiding the overall soil fertility; however, certain agrochemicals negatively influence the functions of these enzymes thereby inhibiting their capacity to improve soil structure and maintain its fertility (Jin et al., 2015).

5.1 Herbicides

Agrochemicals such as herbicides adversely affect the microflora of soil after 7–30 days of its application. These chemical compounds target basic microbial biosynthetic pathways and alter their physiology, which in turn affects the soil enzymatic efficacy. Adverse effects of herbicides on soil microbiota depend upon the herbicides' chemical properties such as its bioavailability, biodegradability, bioactivity, and its persistence in combination with living and nonliving constituents of soil (Yadav et al., 2017). For instance, herbicide glyphosate detrimental effects have been observed on soil microbiota within few days after its applicability. The combination of glyphosate with surfactant causing a rapid decrease in soil bacterium population (Tsui & Chu, 2003). Some herbicides such as Fomesafen are not so sensitive under conventional tillage but become more sensitive when applied with either a combination of herbicide or a heavy metal such as cadmium (Hussain, Siddique, Saleem, Arshad, & Khalid, 2009). The most common herbicide used in India is DDT. The use of DDT has been banned in India, but due to its low price and market availability, it is frequently used by farmers, which cause detrimental effects on nitrogen-fixing bacteria *Rhizobium* sp. This is governed by disturbing the biological association of biological nitrogen fixation bacteria and plant root, thereby hindering the N-fixation and effecting the growth of crops. Other than DDT, 2,4-D, triazene, glyphosate, and paraquat also inhibit the biological function of *Rhizobium* sp. (Zawoznik & Tomaro, 2005). Further, herbicides such as glyphosate, oryzalin, trifluralin, and oxadiazon also affect the soil microbiota by inhibiting the sporulation of arbuscular mycorrhizal fungi (AMF) (Pasaribu et al., 2013; Zaller, Heigl, Ruess, & Grabmaier, 2014).

5.2 Fungicides

Soil structure is composed of various endophytic fungi such as AMF, which help in the growth and development of plants and crops. Several fungicides, including Bavistin, are used to control the growth of unwanted fungi, which can hamper the growth of crops, but these also have a significant effect on the soil microfauna. The growth of AMF is also affected by several fungicides in the farm soil, such as Benzoyl, Emisan, and Carbendazim, which can inhibit the mutualism between legume and AMF and can also cause the killing of AMF by damaging the hyphae of AMF (Cycoń, Piotrowska-Seget, Kaczyńska, & Kozdrój, 2006). The negative impact of several fungicides such as mancozeb, captan, carbendazim, and organomercurial verdean on various microorganisms including bacterial species such as *Rhizobium* sp. and *Bacillus* sp. has already been reported (Chalam, Sasikala, Ramana, Uma, & Rao, 1997; Kyei-Boahen et al., 2001). Further, some of the fungicides can also act as an inhibitor of various enzymes, thereby causing a detrimental impact on the survival of soil microorganisms and soil fertility (Zhang, Bei, et al., 2019; Zhang, Wang, et al., 2019).

5.3 Insecticides

Insecticides are used to control insects' growth, but due to their persistence, they can also inhibit the growth of some farmer-friendly microbial life. DDT is the most commonly used insecticide, but its longer persistence, it affects the total microbial biomass and enzymatic activity of soil microflora. Several insecticides, including phosphamidon, parathion, and methyl phosphorothioate, are affecting the azotobacterial population in the soil (Černohlávková, Jarkovský, & Hofman, 2009). The insecticides such as Chlorfluazuron, Cypermethrin, Phoxim, and Chlorpyrifos and its derivatives can interfere with other soil bacterial populations of soil, including *Bacillus subtilis*, *Fusarium oxysporum*, *Mycobacterium phlei*, *Penicillium expansum*, *Pseudomonas fluorescences*, and *Trichoderma harzianum* (Pandey & Singh, 2004; Virág, Naár, & Kiss, 2007). Several insecticides such as Cypermethrin and Monocrotophosare are highly toxic for the bacterial population in the agricultural land (Madhuri & Rangaswamy, 2002).

6 Role of soil microbes in agrochemical degradation

The management of the soil and its microflora and fauna can be achieved by inducing the growth of such microorganisms that can detoxify the agrochemicals. Microbes are part of the soil and play a major role in detoxifying a large number of agrochemicals. Microorganisms remain in association with plants roots and play a significant role in crop growth and development by nutrient cycling and tolerating the disease-causing pests and fungi (Jin et al., 2015). Biodiversity of soil microbes is affected by several factors, including soil pH, heavy metals, organic matters, as well as agrochemicals, and it has been shown that *Actinobacteria* and *Proteobacteria*are bacterial phyla found most abundantly in the soil having high a concentration of herbicides such as atrazine (Liu et al., 2016). This bacterial population help in the degradation of these toxic compounds into nontoxic simpler molecules, which can be later assimilated in plants and soil. Further, *Arthrobacter* is found in huge population in atrazine contaminated soil with overexpressed trzN, atzB, and atzC genes associated with the degradation of atrazine (Bosso, Scelza, Testa, Cristinzio, & Rao, 2015). Moreover, pentachlorophenol (PCP) is an organochlorine pesticide that is a major contaminant of the ecosystem. Several fungi phyla can degrade this PCP viz. *Phanerochaetechryso sporium*, *Antracophyllum discolor*, *Trametesversicolor*, *Ganodermalucidum*, *Armillariamellea*, and *Gloeophyllum striatum* (Chenseng et al., 2006). Hence, several soil microbes can be utilized for the degradation and intoxication of agrochemicals that can be further explored as preventive soil management measures. The use of biocompost and biofertilizers and organic waste, and several plant-derived products also help to overcome the use of these agrochemicals in agriculture (Liu et al., 2016). Based on the plant growth-promoting bacteria, their roles can be divided into direct impact and indirect impact on plant roots (Fig. 1). Table 2 depicts the details of beneficial microbes, i.e., PGPR, and their role in plant growth enhancement and soil productivity.

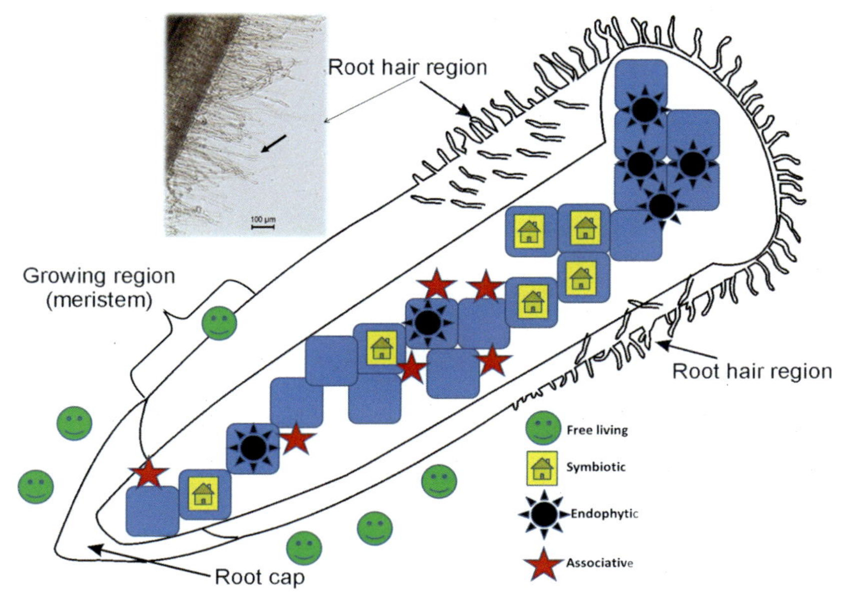

FIG. 1 Association of PGPRs associated with a plant root.

7 Conclusion and future prospects

The continuous use of agrochemical fertilizers is causing serious harm to the environment and human health. The use of beneficial microbes is a sustainable approach for maintaining soil fertility and plant growth. A wide range of microbes may reduce the need of agrochemicals (fertilizers and pesticides) and can improve soil fertility through a variety of mechanisms that include the formation of soil structure, decomposition of organic matter, recycling of essential elements, mineral nutrients, production of multiple regulators of plant growth, and organic pollutant matter degradation. Microbes can stimulate plant root growth, control pathogens of plants and seeds, and promote vegetation changes. At present, the biodegradation potential of microorganisms is an important and promising approach for maintaining environmental sustainability. The use of beneficial microbe strains for pesticide biodegradation would be one of the best strategies adopted not only for pesticide decontamination but also to restrict the use of widely used fertilizers for plant growth. Microbes establish mutual relationships with plants and promote plant growth by providing multiple benefits to the plants. The development of new technologies having a favorable cost and benefit analysis may increase inoculums efficiency and survival rates of microbes. However, limited studies have been done on microbe's role in the biodegradation of agrochemicals and soil aggregations so far. More in-depth studies are needed to understand microbe's beneficial function as

TABLE 2 Plant growth promote regulator and their role in plant growth enhancement and soil productivity.

PGPR	Plant growth-promoting traits	References
Rhizobium, *Bradyrhizobium*	P-solubilization	Hungria et al. (2013)
Bradyrhizobium japonicum *Rhizobium ciceri* *Kluyvera ascorbate* *Mesorhizobium* spp. *Bravibacterium* sp.	Siderophopre	Zhan et al. (2018) Ahemad and Kibret (2014) Sharma, Sharma, and Prasad (2019) Babalola (2010)
Rhizobium leguminosarum	Cytokinin	Backer et al. (2018)
Bradyrhizobium, *Rhizobium*, *Sphingomonas* sp., *Mycobacterium* sp., *Bacillus* sp., *Rhodococcus* sp., *Cellulomonas*	Indole acetic acid	Abd-Alla (1994) Ahemad and Khan (2011)
Pseudomonas aeruginosa *Pseudomonas chlororaphis* *Bacillus firmus* HussainT: Lab. 66 *Bacillus subtilis* HussainT: AMU *Bacillus siamensis*: AMU03	Antifungal activity	Tomar et al. (2014) Rajkumar et al. (2018) Hussain and Khan (2018) Hussain and Khan (2020b) Hussain and Khan (2020a)
Pseudomonas putida *Pseudomonas aeruginosa* *Klebsiella* sp. *Enterobacter asburiae* *Acinetobacter* spp. *Rhizobium* sp. (lentil)	Hydrogen cyanide, ammonia, *exo*-polysaccharides, phosphate solubilization	Burd, Dixon, and Glick (2000) Whipps (2011) Rezzonico et al. (2007) Saraf, Pandya, and Thakkar (2014) Zain, Yasmin, and Hafeez (2019)
Burkholderia *Pseudomonas jessenii*	ACC deaminase, IAA, siderophore, heavy metal solubilization, phosphate solubilization	Chowdhury and Bagchi (2017) Siddiqui, Shaukat, Sheikh, and Khan (2006)

biofertilizers, biocontrol agents, and bioremediation agents who may enhance crop productivity, ecosystem functioning, and sustainability.

Conflict of interest

Authors have no conflict of interest.

References

Abd-Alla, M. H. (1994). Solubilization of rock phosphates by rhizobium and Bradyrhizobium. *Folia Microbiologica*, *39*, 53–56.

Abhilash, P. C., & Singh, N. (2009). Pesticide use and application: An Indian scenario. *Journal of Hazardous Materials*, *165*(1–3), 1–2.

Ahemad, M., & Khan, M. S. (2011). Toxicological assessment of selective pesticides towards plant growth promoting activities of phosphate solubilizing *Pseudomonas aeruginosa*. *Acta Microbiologica et Immunologica Hungarica*, *58*, 169–187.

Ahemad, M., & Kibret, M. (2014). Mechanisms and applications of plant growth promoting rhizobacteria: Current perspective. *Journal of King Saud University—Science*, *26*(1), 1–20.

Ahmed, M., Rauf, M., Mukhtar, Z., & Saeed, N. A. (2017). Excessive use of nitrogenous fertilizers: An unawareness causing serious threats to environment and human health. *Environmental Science and Pollution Research*, *24*(35), Z26983–Z26987.

Alvarez-Moreno, C., Lavergne, R. A., Hagen, F., Morio, F., Meis, J. F., & Le, P. P. (2017). Azole-resistant Aspergillus fumigatus harboring TR_{34}/L98H, TR_{46}/Y121F/T289A and TR_{53} mutations related to flower fields in Colombia. *Scientific Reports*, *7*, 45631. https://doi.org/10.1038/srep45631.

Babalola, O. (2010). Beneficial bacteria of agricultural importance. *Biotechnology Letters*, *32*(11), 1559–1570.

Backer, R., Rokem, J., Ilangumaran, G., Lamont, J., Praslickova, D., Ricci, E., et al. (2018). Plant growth-promoting rhizobacteria: Context, mechanisms of action, and roadmap to commercialization of biostimulants for sustainable agriculture. *Frontiers in Plant Science*, *9*, 1473.

Bosso, L., Scelza, R., Testa, A., Cristinzio, G., & Rao, M. A. (2015). Depletion of pentachlorophenol contamination in an agricultural soil treated with *Byssochlamys nivea*, *Scopulariopsis brumptii* and urban waste compost: A laboratory microcosm study. *Water, Air, and Soil Pollution*, *226*(6), 1–9.

Burd, G. I., Dixon, D. G., & Glick, B. R. (2000). Plant growth promoting bacteria that decrease heavy metal toxicity in plants. *Canadian Journal of Microbiology*, *46*, 237–245.

Calmes, B., N'Guyen, G., Dumur, J., Brisach, C. A., Campion, C., Iacomi, B., et al. (2017). Glucosinolate-derived isothiocyanates impact mitochondrial function in fungal cells and elicit an oxidative stress response necessary for growth recovery. *Frontiers in Plant Science*, *6*, 414.

Campos, E. V., Proença, P. L., Oliveira, J. L., Bakshi, M., Abhilash, P. C., & Fraceto, L. F. (2019). Use of botanical insecticides for sustainable agriculture: Future perspectives. *Ecological Indicators*, *105*, 483–495.

Carvalho, F. P. (2017). Pesticides, environment, and food safety. *Food and Energy Security*, *6*(2), 48–60.

Černohlávková, J., Jarkovský, J., & Hofman, J. (2009). Effects of fungicides mancozeb and dinocap on carbon and nitrogen mineralization in soils. *Ecotoxicology and Environmental Safety*, *72*, 80–85.

Chalam, A. V., Sasikala, C., Ramana, C. V., Uma, N. R., & Rao, P. R. (1997). Effect of pesticides on the diazotrophic growth and nitrogenase activity of purple nonsulfur bacteria. *Bulletin of Environmental Contamination and Toxicology*, *58*, 463–468.

Chenseng, L., Toepel, K., Irish, R., Richard, A. F., Dana, B. B., & Bravo, R. (2006). Organic diets significantly lower children's dietary exposure to organo-phosphorous pesticides. *Environmental Health Perspectives*, *114*(2), 250–263.

Chowdhury, N., & Bagchi, A. (2017). Structural insight into the gene expression profiling of the hcn operon in *Pseudomonas aeruginosa*. *Applied Biochemistry and Biotechnology*, *182*(3), 1144–1157.

Cycoń, M., Piotrowska-Seget, Z., Kaczyńska, A., & Kozdrój, J. (2006). Microbiological characteristics of a sandy loam soil exposed to tebuconazole and λ-cyhalothrin under laboratory conditions. *Ecotoxicology*, *15*, 639–646.

Darby, B. J., & Neher, D. A. (2016). Microfauna within biological soil crusts. In B. Weber, et al. (Eds.), *226. Biological soil crusts: An organizing principle in drylands. Ecological Studies* (pp. 139–157). Cham: Springer. https://doi.org/10.1007/978-3-319-30214-0_8.

Devi, P. I., Thomas, J., & Raju, R. K. (2017). Pesticide consumption in India: A spatiotemporal analysis. *Agricultural Economics Research Review*, *30*(1), 163–172.

Dhanker, R., Kumar, R., & Hwang, J. S. (2013). How effective are *Mesocyclops aspericornis* (Copepoda: Cyclopoida) in controlling mosquito immatures in the environment with an application of phytochemicals? *Hydrobiologia*, *716*, 147–162.

Dhanker, R., Kumar, R., & Raghvendra, K. (2014). Efficiency of copepods to control *Aedes aegypti* larvae in medium applied with insecticides (Bti and temephos). In R. Kumar (Ed.), *Climate change, aquatic community structure and disease: Proceedings of the National Symposium on biodiversity, biotechnology and man: Interdependence and future challenges, 26–28 March, 2010, Lansdowne (Uttarakhand)*. Delhi, India: Ilavart Publication. ISBN: 81-904058-7-X.

Din, B. U., Sarfraz, S., Xia, Y., Kamran, M. A., Javed, M. T., Sultan, T., et al. (2019). Mechanistic elucidation of germination potential and growth of wheat inoculated with exopolysaccharide and ACC-deaminase producing *Bacillus* strains under induced salinity stress. *Ecotoxicology and Environmental Safety*, *183*, 109466.

Elkoca, E., Kantar, F., & Sahin, F. (2008). Influence of nitrogen fixing and phosphorus solubilizing bacteria on the nodulation, plant growth, and yield of chickpea. *Journal of Plant Nutrition*, *31*, 157–171.

Fernandez, A. L., Sheaffer, C. C., Wyse, D. L., Staley, C., Gould, T. J., & Sadowsky, M. J. (2016). Structure of bacterial communities in soil following cover crop and organic fertilizer incorporation. *Applied Microbiology and Biotechnology*, *100*(21), 9331–9341.

Gülser, C., Zharlygasov, Z., Kızılkaya, R., Kalimov, N., Akça, I., & Zharlygasov, Z. (2019). The effect of NPK foliar fertilization on yield and macronutrient content of grain in wheat under Kostanai-Kazakhstan conditions. *Eurasian Journal of Soil Science*, *8*(3), 275–281.

Gupta, G., Snehi, S. K., & Singh, V. (2017). Role of PGPR in biofilm formations and its importance in plant health. In *Biofilms in plant and soil health*. https://doi.org/10.1002/9781119246329.ch2.

Hungria, M., Nogueira, M. A., & Araujo, R. S. (2013). Co-inoculation of soybeans and common beans with rhizobia and azospirilla: Strategies to improve sustainability. *Biology and Fertility of Soils*, *49*, 791–801.

Hussain, T., & Khan, A. A. (2018). *Bacillus firmusHussainT: Lab. 66*: A new biosurfactant producing bacteria for the biocontrol of late blight of potato caused by *Phytophthora infestans* (Mont.) de Bary. In *National Seminar on new paradigms of plant health management: Sustainable food security under climatic scenario from 17thNov. To 19th Nov. 2018 at Bihar agricultural university, Sabour (Bhagalpur), Bihar, India under Indian phytopathological society, Eastern zone region annual meeting, OP* (p. 60).

Hussain, T., & Khan, A. A. (2020a). Determining the antifungal activity and characterization of *Bacillus siamensis AMU03* against *Macrophomina phaseolina* (Tassi) Goid. *Indian Phytopathology*. https://doi.org/10.1007/s42360-020-00239-6.

Hussain, T., & Khan, A. A. (2020b). *Bacillus subtilis HussainT-AMU* and its antifungal activity against Potato black scurf caused by *Rhizoctonia solani*. *Biocatalysis and Agricultural Biotechnology*, *23*, 101433. http://www.careratings.com/upload/NewsFiles/Studies/Agrochemicals.pdf.

Hussain, S., Siddique, T., Saleem, M., Arshad, M., & Khalid, A. (2009). Chapter 5: Impact of pesticides on soil microbial diversity, enzymes, and biochemical reactions. *Advances in Agronomy*, *102*, 159–200.

Jahanban, L., Panahpour, E., Gholami, A., Davari, M. R., & Lotfifar, O. (2018). Combined effect of chickpea cultivation and type of fertilizer on growth, yield and mineral element concentration of corn (*Zea mays* L.). *Applied Ecology and Environmental Research*, *16*(3), 3159–3169.

Jin, Z., Li, Z., Li, Q., Hu, Q., Yang, R., Tang, H., et al. (2015). Canonical correspondence analysis of soil heavy metal pollution, microflora and enzyme activities in the Pb–Zn mine tailing dam collapse area of Sidi village, SW China. *Environment and Earth Science*, *73*, 267–274.

Khan, M. S., Zaidi, A., & Aamil, M. (2002). Biocontrol of fungal pathogens by the use of plant growth promoting rhizobacteria and nitrogen fixing microorganisms. *Indian Journal of Botanical Society*, *81*, 255–263.

Kyei-Boahen, S., Slinkard, A. E., & Walley, F. L. (2001). Rhizobial survival and nodulation of chickpea as influenced by fungicide seed treatment. *Canadian Journal of Microbiology*, *47*, 585–589.

Liu, X., Hui, C., Bi, L., Romantschuk, M., Kontro, M., Strömmer, R., et al. (2016). Bacterial community structure in atrazine treated reforested farmland in Wuying China. *Applied Soil Ecology*, *98*, 39–46.

Madhuri, R. J., & Rangaswamy, V. (2002). Influence of selected insecticides on phosphatase activity in groundnut (*Arachishypogeae* L.) soils. *Journal of Environmental Biology*, *23*(4), 393–397.

Maruyama, C. R., Guilger, M., Pascoli, M., Bileshy-José, N., Abhilash, P. C., Fraceto, L. F., et al. (2016). Nanoparticles based on chitosan as carriers for the combined herbicides imazapic and imazapyr. *Scientific Reports*, *6*, 19768.

McGehee, C. S., Raudales, R. E., Elmer, W. H., & McAvoy, R. J. (2019). Efficacy of biofungicides against root rot and damping-off of microgreens caused by *Pythium* spp. *Crop Protection*, *121*, 96–102.

Pan, B., Lam, S. K., Mosier, A., Luo, Y., & Chen, D. (2016). Ammonia volatilization from synthetic fertilizers and its mitigation strategies: A global synthesis. *Agriculture, Ecosystems & Environment*, *232*, 283–289.

Pandey, S., & Singh, D. K. (2004). Total bacterial and fungal population after chlorpyrifos and quinalphos treatments in groundnut (*Arachishypogaea* L.) soil. *Chemosphere*, *55*, 197–205.

Pasaribu, A., Mohamad, R. B., Hashim, A., Rahman, Z. A., Omar, D., Morshed, M. M., et al. (2013). Effect of herbicide on sporulation and infectivity of vesicular arbuscularmycorrhizal (*Glomusmosseae*) symbiosis with peanut plant. *Journal of Animal and Plant Sciences*, *23*, 1671–1678.

Poonia, S. P., Kumar, V., & Malik, R. K. (2019). Herbicides and their improved spraying techniques: Efficient weed control under conservation agriculture. In *Conservation agriculture for climate resilient farming & doubling farmers' income* (p. 246). ICAR Research Complex for Eastern Region. Patna Training Manual No. 2.

Popp, J., Pető, K., & Nagy, J. (2013). Pesticide productivity and food security. A review. *Agronomy for Sustainable Development*, *33*, 243–255. https://doi.org/10.1007/s13593-012-0105-x.

Rajkumar, M., Ma, Y., & Freitas, H. (2018). Characterization of metal resistant plant-growth promoting *Bacillus weihenstephanensis* isolated from serpentine soil in Portugal. *Journal of Basic Microbiology*, *48*, 500–508.

Ramasamy, M., Geetha, T., & Yuvaraj, M. (2020). Role of biofertilizers in plant growth and soil health. In *Nitrogen fixation*. https://doi.org/10.5772/intechopen.87429.

Rezzonico, F., Zala, M., Keel, C., Duffy, B., Moënne-Loccoz, Y., & Défago, G. (2007). Is the ability of biocontrol fluorescent pseudomonads to produce the antifungal metabolite 2, 4 diacetylphloroglucinol really synonymous with higher plant protection? *New Phytologist*, *73*(4), 861–872.

Saraf, M., Pandya, U., & Thakkar, A. (2014). Role of allelochemicals in plant growth promoting rhizobacteria for biocontrol of phytopathogens. *Microbiological Research*, *169*(1), 18–29.

Sekhotha, M. M., Monyeki, K. D., & Sibuyi, M. E. (2016). Exposure to agrochemicals and cardiovascular disease: A review. *International Journal of Environmental Research and Public Health*, *13*(2), 229.

Sharma, K., Sharma, S., & Prasad, S. (2019). PGPR: Renewable tool for sustainable agriculture. *International Journal of Current Microbiology and Applied Sciences*, *8*(01), 525–530.

Shiri, T., Kumar, A., Priyta, V., Kumar, A., Singh, G., Rashmi, et al. (2020). Stimulus of panchgavya bio-manure (PGBM) on developmental growth as well as harvest of *Pisum sativum*. *Journal of Pharmacognosy and Phytochemistry*, *9*(3), 905–910.

Siddiqui, I. A., Shaukat, S. S., Sheikh, I. H., & Khan, A. (2006). Role of cyanide production by *Pseudomonas fluorescens* CHA0 in the suppression of root-knot nematode, *Meloidogyne javanica* in tomato. *World Journal of Microbiology and Biotechnology*, *22*(6), 641–650.

Silo-Suh, L. A., Lethbridge, B. J., & Raffel, S. J. (1994). Biological activities of two fungistatic antibiotics produced by *Bacillus cereus UW85*. *Applied and Environmental Microbiology*, *60*, 2023–2030.

Singh, D. P., Singh, H. B., & Prabha, R. (Eds.). (2017). *Plant-microbe interactions in agroecological perspectives*. New Delhi: Springer.

Tejada, M., Rodríguez-Morgado, B., Gómez, I., & Parrado, J. (2014). Degradation of chlorpyrifos using different biostimulants/biofertilizers: Effects on soil biochemical properties and microbial community. *Applied Soil Ecology*, *84*, 158–165.

Tomar, S., Singh, B. P., Lal, M., Khan, M. A., Hussain, T., Sharma, S., et al. (2014). Screening of novel microorganism for biosurfactant and biocontrol activity against *Phytophthora infestans*. *Journal of Environmental Biology*, *35*, 893–899.

Tsui, M. T. K., & Chu, L. M. (2003). Aquatic toxicity of glyphosate-based formulations: Comparison between different organisms and the effects of environmental factors. *Chemosphere*, *52*, 1189–1197.

van Zwieten, L. (2018). The long-term role of organic amendments in addressing soil constraints to production. *Nutrient Cycling in Agroecosystems*, *111*, 99–102. https://doi.org/10.1007/s10705-018-9934-6.

Virág, D., Naár, Z., & Kiss, A. (2007). Microbial toxicity of pesticide derivatives produced with UV-photodegradation. *Bulletin of Environmental Contamination and Toxicology*, *79*, 356–359.

Whipps, J. M. (2011). Microbial interactions and biocontrol in the rhizosphere. *Journal of Experimental Botany*, *52*(Suppl. 1), 487–511.

Willoughby, G. L. (2019). *Industrial fertilizers in agriculture*. Oxford Research Encyclopedia of Environmental Science.

Yadav, G. S., Datta, R., Imran Pathan, S., Lal, R., Meena, R. S., Babu, S., et al. (2017). Effects of conservation tillage and nutrient management practices on soil fertility and productivity of rice (*Oryzasativa* L.)—rice system in north eastern region of India. *Sustainability*, *9*(10), 1816.
Yadav, S., & Dutta, S. (2019). A study of pesticide consumption pattern and farmer's perceptions towards pesticides: A case of Tijara Tehsil, Alwar (Rajasthan). *International Journal of Current Microbiology and Applied Sciences*, *8*(4), 96–104.
Zain, M., Yasmin, S., & Hafeez, F. Y. (2019). Isolation and characterization of plant growth promoting antagonistic bacteria from cotton and sugarcane plants for suppression of phytopathogenic *Fusarium* species. *Iranian Journal of Biotechnology*, *17*(2), 61–70.
Zaller, J. G., Heigl, F., Ruess, L., & Grabmaier, A. (2014). Glyphosate herbicide affects belowground interactions between earthworms and symbiotic mycorrhizal fungi in a model ecosystem. *Scientific Reports*, *4*, 5634.
Zawoznik, M. S., & Tomaro, M. L. (2005). Effect of chlorimuron-ethyl on *Bradyrhizobium japonicum* and its symbiosis with soybean. *Pest Management Science*, *61*, 1003–1008.
Zhan, H., Feng, Y., Fan, X., & Chen, S. (2018). Recent advances in glyphosate biodegradation. *Applied Microbiology and Biotechnology*, *102*(12), 5033–5043.
Zhang, J., Bei, S., Li, B., Zhang, J., Christie, P., & Li, X. (2019). Organic fertilizer, but not heavy liming, enhances banana biomass, increases soil organic carbon and modifies soil microbiota. *Applied Soil Ecology*, *136*, 67–79.
Zhang, X., Dong, W., Dai, X., Schaeffer, S., Yang, F., Radosevich, M., et al. (2015). Responses of absolute and specific soil enzyme activities to long term additions of organic and mineral fertilizer. *Science of the Total Environment*, *536*, 59–67.
Zhang, M., Wang, W., Zhang, Y., Teng, Y., & Xu, Z. (2019). Effects of fungicide iprodione and nitrification inhibitor 3, 4-dimethylpyrazole phosphate on soil enzyme and bacterial properties. *Science of the Total Environment*, *599–600*, 254–263.

Chapter 11

Potential role of heavy metal-resistant plant growth-promoting rhizobacteria in the bioremediation of contaminated fields and enhancement of plant growth essential for sustainable agriculture

Krishnendu Pramanik[a,b], Tushar Kanti Maiti[b], and Narayan Chandra Mandal[a]

[a]*Mycology and Plant Pathology Laboratory, Department of Botany, Visva-Bharati, Santiniketan, West Bengal, India,* [b]*Microbiology Laboratory, Department of Botany, The University of Burdwan, Golapbag, Purba Bardhaman, West Bengal, India*

Chapter outline

Recent Advancement in Microbial Biotechnology. https://doi.org/10.1016/B978-0-12-822098-6.00014-8

1 Introduction

Agriculture has a direct influence on the economy of a nation, whether it is developing or developed. However, with the rapid progress of human civilization, many anthropogenic activities viz., extensive mining, widespread industrialization, and unplanned agricultural practices lead to the intensification of soil, air, and water pollution. These factors, in addition to global warming, environmental degradation, etc., result in an extensive agricultural loss in terms of productivity and quality of crops. Besides, it has now been widely accepted that the human population explosion is another important factor in accelerating this problem, leading to a global food crisis. The worldwide human population is alarming to reach 9.8 billion by 2050 and 11.2 billion in 2100 (UN, 2017). Therefore, to ensure food security and quality of food, much attention has been paid to the development of eco-friendly, inexpensive, and sustainable ways for the agricultural benefit and reducing human health risks.

Heavy metals/metalloids are among the most life-threatening materials that impart several health hazards of almost all life forms worldwide. Plant growth reduction due to heavy metal/metalloid stress has been reported by several earlier workers (Sirari, Kashyap, & Mehta, 2016; Tran & Popova, 2013; Wan et al., 2012). Several approaches for heavy metal detoxification have already been proposed and practiced (Donald, 2003; Hongbo et al., 2011; Khan, Zaidi, Wani, & Oves, 2009; Macek, Macková, & Káš, 2000; Suresh & Ravishankar, 2004). In this context, utilization of various heavy metal-resistant plant growth-promoting rhizobacteria (PGPR) has proven very promising toward the bioremediation of metal-contaminated agricultural fields as well as for plant growth promotion (Ahmad et al., 2014; Chen et al., 2016; Danish et al., 2019; Kartik, Jinal, & Amaresan, 2016; Mitra et al., 2018; Mitra et al., 2018; Pramanik, Ghosh, Ghosh, Sarkar, & Maiti, 2016; Pramanik, Mitra, Sarkar, & Maiti, 2018; Pramanik, Mitra, Sarkar, Soren, & Maiti, 2017; Roman-Ponce et al., 2017; Singh, Pathak, & Fulekar, 2015; Wang et al., 2020; Wu et al., 2020). Crop improvement due to such PGPR inoculation could be a ray of hope in the future to feed the world (Glick, 2014). The present review illustrates the concept of heavy metals/metalloids (with emphasis on three heavy metals viz. Cr, Cd, Pb, and one metalloid viz., As), their sources and impact on living beings. Moreover, this review also covers the possible remedies or ways to get rid of these heavy metals/metalloids, heavy metal-resistant PGPR in alleviating metal toxicity, their role in plant growth promotion, and the mechanisms involved to tackle such metals.

2 Definitions of heavy metals and metalloids

Although the term "heavy metal" is not well-defined by International Union of Pure and Applied Chemistry (IUPAC), the criteria used to define heavy metals include density, atomic weight, atomic number, or periodic table position (Duffus, 2002). The most widely accepted definition of heavy metals is heavy metals with a density above 5 g/cm^3 and specific gravity at least five times higher than the water (Nies, 1999). On the other hand, a *metalloid* is any chemical element with properties between metals and nonmetals, often called "*semi-metal*." Unlike other metals and nonmetals, heavy metals are toxic even at very low concentrations. Heavy metal-driven soil pollution is leading to a great environmental problem and ecological risk nowadays. Interestingly some heavy metals (Zn, Co, Ni, Mn, and Fe) are essential for the organism as a micronutrient; however, few others (Cr, Cd, Hg, Pb), including metalloid (As) have no positive biological role rather very harmful for the organism at mild doses.

3 Sources of heavy metals and metalloids

The sources of heavy metals/metalloids can be categorized into two sections viz., lithogenic source and anthropogenic source (Alloway, 2013). Lithogenic sources, i.e., the geological minerals in the soil parent material, are among the most prevailing sources of heavy metals/metalloids (Alloway, 2013). Besides, there are various anthropogenic sources of heavy metals/metalloids that badly affect the agricultural lands. However, the localized contamination from a predominant source such as metal industries has remarkable effects on the vegetation and the health, especially of the local population due to consumption of locally cultivated crops (Alloway, 2013). The problem is more prominent, especially in those countries with inadequate emission controls of industrial by-products and improper maintenance of soil quality standards (Alloway, 2013). The degree of contamination of specific heavy metals/metalloids depends on the product, by-products, and the wastes of the metal industry. The sources of different heavy metals/metalloids are summarized in Fig. 1.

4 Effects of heavy metals on organisms and microorganisms

Selected heavy metals and metalloids at limited doses play pivotal roles in various cellular and metabolic processes of organisms. These are known to work either as micronutrients, enzyme cofactors, or sometimes as osmotic pressure regulators. Most of them have no known biological function; instead, they are toxic even at very low concentrations (Fashola, Ngole-Jeme, & Babalola,

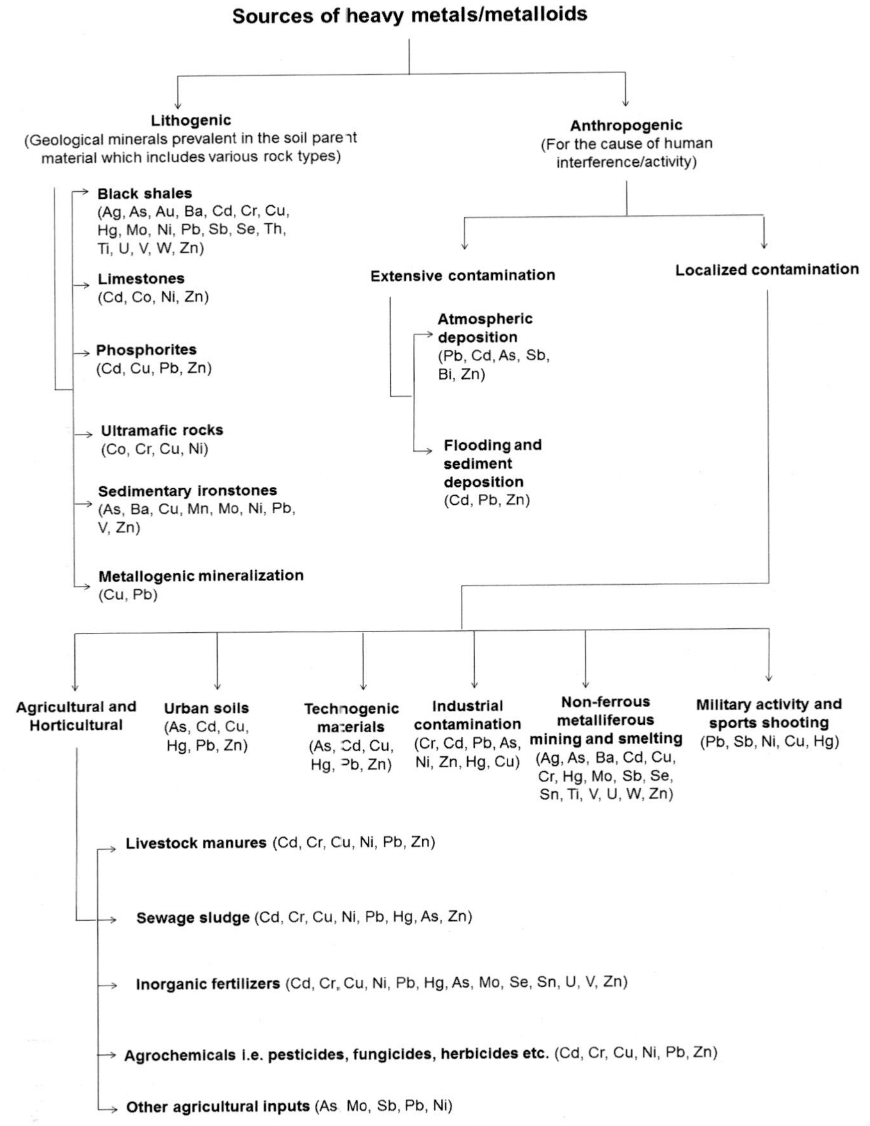

FIG. 1 Different sources of heavy metals and metalloids.

2016). Heavy metals such as Cr, Cd, Pb, and metalloids such as As cause many undesirable health effects on humans, plants, and microorganisms.

4.1 Effects of heavy metals/metalloids on animals

Cr, Cd, Pb, and As are four of the most toxic heavy metals/metalloids that impart several health hazards to human beings. Cr causes diarrhea, chronic bronchitis, bronchopneumonia, headache, emphysema, skin irritation, liver diseases, itching of the respiratory tract, nausea, lung cancer, reproductive toxicity, renal failure, and vomiting (Ayangbenro & Babalola, 2017). Moreover, Cd is responsible for various bone diseases, emphysema, headache, coughing, itai-itai, kidney diseases, lymphocytosis, lung and prostate cancers, microcytic hypochromic anemia, testicular atrophy, hypertension, etc. (Ayangbenro & Babalola, 2017). Besides, Pb-induced health risks include anorexia, damage to neurons, chronic nephropathy, high blood pressure, hyperactivity, learning deficits, insomnia, renal system, reduced fertility, risk factor for Alzheimer's disease, etc. (Ayangbenro & Babalola, 2017). On the other hand, As causes severe brain damage, cardiovascular and respiratory disorders, skin cancer, dermatitis, conjunctivitis, etc. (Ayangbenro & Babalola, 2017).

4.2 Effects of heavy metals/metalloids on plants

Apart from animals, plants are also the major victims of various heavy metals. In particular, Cr causes chlorosis, delayed senescence, oxidative stress, wilting, biochemical lesions, stunted growth, reduced biosynthesis germination, etc., while Cd is responsible for reduced seed germination, decrease in plant nutrient content, growth inhibition, etc. (Ayangbenro & Babalola, 2017; Khan et al., 2009). Moreover, Pb affects photosynthesis and growth, causes chlorosis, inhibits seed germination and enzyme activities, and creates oxidative stress (Ayangbenro & Babalola, 2017; Khan et al., 2009). However, the heavy metal "As" is associated with damages in the cell membrane, inhibition of plant growth, alteration of roots extension and proliferation, physiological disorders, oxidative stress, crop yield, and fruit production (Ayangbenro & Babalola, 2017; Khan et al., 2009).

4.3 Effects of heavy metals/metalloids on microorganisms

Heavy metals also exhibit their toxic behavior to susceptible microorganisms, which directly or indirectly help plants to grow better. Many beneficial microflora are directly affected by various heavy metals such as Cr, Cd, Pb, As, etc. present in their habitat. Cr is responsible for preventing the activation of enzymes for DNA replication, thus occurs elongation of lag phase, growth inhibition, inhibition of oxygen uptake, while Cd is also known to denature protein, damage nucleic acid, inhibit carbon and nitrogen mineralization, inhibit cell division and transcription, etc. (Ayangbenro & Babalola, 2017). Moreover,

Pb and As are responsible for denatures of nucleic acid and proteins, inhibiting vital cellular enzymes activities and transcription processes (Ayangbenro & Babalola, 2017).

5 Causes of heavy metal toxicity

Most of the heavy metal(loid)s are dangerously toxic in their cationic forms (except Hg, which is toxic even at elemental form). The toxicity lies on three major factors: (i) due to strong affinity of metal cations for sulfur, found in proteins (e.g., enzymes); (ii) sulfhydryl groups, —SH, in many enzymes, react with ingested metal ions; (iii) can deactivate the enzyme that stops or alters metabolic processes even leading to cell death.

6 Approaches of heavy metals/metalloids remediation

There are several physical and chemical approaches for heavy metal/metalloid remediation of soil but the success is still being questioned in their effectiveness. These approaches can be categorized as follows:

6.1 Conventional approaches

6.1.1 Physical method

Physical methods involve the ex situ method that involves the treatment of contaminated material "off-site" such as leaching, landfilling, incineration, etc., whereas in situ method contaminated material treated "on-site" such as an application of electrokinetics, vitrification, volatilization, etc.

6.1.2 Chemical method

Use of some metal chelators facilitates both the bioavailability and solubility of heavy metals (Chelate extraction method). But the above-mentioned physical and chemical methods are restricted due to several *disadvantages* like these methods are labor-sensitive and expensive (Donald, 2003), excessive use of nonbiodegradable chelators such as ethylenediamine-tetraacetic acid (EDTA) can harm biotic components of the ecosystem (Hongbo et al., 2011), can alter nature of soil viz., structure, texture, fertility, etc. (Hongbo et al., 2011). Moreover, binding reagents can make heavy metal/metalloid ions more soluble and mobile, which can cause groundwater pollution (Hongbo et al., 2011). Due to the deficiency of selectivity of a few metal chelators, loss of beneficial ions often occurs (Hongbo et al., 2011).

6.2 Phytoremediation

Phytoremediation is an eco-friendly, cost-effective in situ bioremediation method by which plants accumulate, sequester, or degrade contaminants from soil by their innate capabilities. Phytoremediation methods are successful in remediating contaminated industrial environments (Macek et al., 2000; Suresh & Ravishankar, 2004). The plants involved in phytoremediation often called "Phytoremediator" or "Hyperaccumulator". Only 0.2% of angiosperm flora (Baker & Brooks, 1989) reported as hyperaccumulators such as *Typha* spp. *Brassica juncea*, *Arabidopsis halleri*, *Viola calaminaria*, *Astragalus racemosus*, etc. Phytoremediation technique broadly involves phytoextraction/phytostabilization, phytostabilization/phytostabilization, phytostimulation, and phytodegradation, phytoassimilation, phytotransformation, phytoreduction, phytooxidation, etc. (Ma, Prasad, Rajkumar, & Freitas, 2011). The molecular mechanism underlying hyperaccumulation is attributed to some metal-specific transporters, chelators such as phytochelatins, metallothioneins, and organic acids such as citrate, and antioxidants such as glutathione (Krämer, 2010).

6.3 Bioremediation and rhizoremediation

Bioremediation is a process of detoxifying or degrading contaminants present in the soil, wastewater, or industrial sludge by biological means. Microorganisms (bacteria, fungi, etc.) can take part in this process, although plant-assisted bioremediation is often termed as phytoremediation (Ma et al., 2011). On the other hand, microorganisms in the rhizospheric niches when taking part in the remediation of the pollutants specifically present in their niche are called rhizoremediation (Ma et al., 2011). Many PGPR are known to involve in the rhizoremediation process (Table 1).

7 Plant growth-promoting (PGP) traits and their role

PGPR is a group of bacteria that are defined as having some properties such as root colonization, antagonistic action against phytopathogens, and promotion of plant growth (Kloepper, 1994). PGPR exhibit their role in plant growth and development by various mechanisms. Based on the availability of pathogen, the mechanisms can be classified into two as follows:

7.1 Direct mechanism: Absence of pathogen

In this case, the PGPR act as a biofertilizer and phytostimulator. The mechanism includes nitrogen fixation, phosphate solubilization, siderophore production, phytohormone production, 1-aminocyclopropane-1-carboxylate (ACC) deaminase activity, etc.

TABLE 1 Heavy metal/metalloid (Cr, Cd, Pb, and As only) resistant plant growth promoting rhizobacteria (PGPR) (2005 onward).

PGPR strains	PGP traits[a]	Resistance to heavy metal(loid)s with (MIC)	Plant growth promotion studied on	References
Enterobacter bugandensis TJ6	Urease, IAA, siderophore	Cd—400 mg/L, Pb −1700 mg/L	*Lactuca sativa* (Lettuce)	Wang et al. (2020)
Serratia marcescens SNB6	Phosphate, IAA, siderophore	Cd-ND	*Chrysopogon zizanioides*	Wu et al. (2020)
Agrobacterium fabrum SDW6	Phosphate, siderophore, IAA, potassium	Cr-ND	*Zea mays*	Danish et al. (2019)
Klebsiella sp. CPSB4	IAA, PO_4, ammonia, siderophore, HCN	Cr^{6+} (ND)	ND	Gupta, Kumar, Usmani, Rani, and Chandra (2018)
Pseudomonas sp. CPSB21	IAA, PO_4, ammonia, siderophore, HCN	Cr^{6+} (700 mg/L)	Sunflower and tomato	Gupta, Rani, Chandra, and Kumar (2018)
Azotobacter chroococcum	IAA, ACCD, ammonia, siderophore, HCN	Pb^{2+} (2000 μg/mL)	*Zea mays* L.	Rizvi & Khan, 2018
Pseudomonas gessardii BLP141, *Pseudomonas fluorescens* A506, and *Pseudomonas fluorescens* LMG 2189	ND	Pb^{2+} (ND)	Sunflower	Saleem, Asghar, Zahir, and Shahid (2018)

Bacillus vietnamensis AB403 *Kocuria flava* AB402	IAA, siderophore	As^{3+} (20 mM, 35 mM)	Rice	Mallick et al. (2018)
Enterobacter sp. S2 MCC 3090	IAA, ACCD, N_2, PO_4	Cd^{2+} (3500 μg/mL), Pb^{2+} (2500 μg/mL), As^{3+} (1050 μg/mL)	*Oryza sativa* L. (rice)	Mitra, Pramanik, Sarkar, et al. (2018)
Klebsiella michiganensis S8 MCC 3089	IAA, ACCD, N_2, PO_4	Cd^{2+} (3500 μg/mL), Pb^{2+} (3000 μg/mL), As^{3+} (1050 μg/mL)	*Oryza sativa* L. (rice)	Mitra, Pramanik, Ghosh, et al. (2018)
Enterobacter aerogenes K6 MCC 3092	IAA, ACCD, N_2, PO_4, ammonia, siderophore, HCN	Cd^{2+} (4000 μg/mL), Pb^{2+} (3800 μg/mL), As^{3+} (1500 μg/mL)	*Oryza sativa* L. (rice)	Pramanik et al. (2018)
Klebsiella pneumoniae K5 MCC 3091	IAA, ACCD, N_2, PO_4, EPS, ammonia, siderophore	Cd^{2+} (4000 μg/mL), Pb^{2+} (4000 μg/mL), As^{3+} (1500 μg/mL)	*Oryza sativa* L. (rice)	Pramanik et al. (2017)
Bacillus cereus and *Pseudomonas moraviensis*	ND	Cr^{6+} (100 ppm) Cd^{2+} (50 ppm) Pb^{2+} (50 ppm)	Wheat	Hassan, Bano, and Naz (2017)
Bacteroidetes bacterium, *Pseudomonas fluorescens*, and *Variovorax* sp.	ND	Cd^{2+} (ND) Pb^{2+} (ND)	*Brassica napus*	Dąbrowska, Hrynkiewicz, Trejgell, and Baum (2017)
Bacillus sp. MNU16	IAA, PO_4, siderophore	Cr^{6+} (900 mg/L), Cd^{2+} (150 mg/L), As^{3+} (600 mg/L)	ND	Upadhyay et al. (2017)
Bacillus sp., *Alcaligens* sp., *Curtobacterium* sp., and *Microbacterium* sp.	IAA, siderophore	Pb^{2+} (24 mmol/L), As^{5+} (480 mmol/L)	*Brassica nigra*	Roman-Ponce et al. (2017)

Continued

TABLE 1 Heavy metal/metalloid (Cr, Cd, Pb, and As only) resistant plant growth promoting rhizobacteria (PGPR) (2005 onward)—cont'd

PGPR strains	PGP traits	Resistance to heavy metal(loid)s with (MIC)	Plant growth promotion studied on	References
Citrobacter freundii N52, *Acinetobacter lwoffii* T24, *Bacillus subtilis* T23, and *Klebsiella* spp.	IAA, PO_4, HCN	As^{3+} (175–200 mM) As^{5+} (500–550 mM)	*Triticum aestivum*	Qamar, Rehman, and Hasnain (2017)
Pseudomonas sp	IAA	As^{3+} (50 mM), As^{5+} (800 mM), Cd^{2+} (8 mM), Cr^{6+} (2.5 mM)	*Lens culinaris*	Biswas et al. (2017)
Enterobacter sp. EG16	IAA, siderophore	Cd^{2+} (>250 mg/L)	ND	Chen et al. (2016)
Raoultella sp. CrS2	IAA, N_2, PO_4, ammonia	Cr^{6+} (25 mM), Cd^{2+} (1.25 mM), Pb^{2+} (0.75 mM), As^{5+} (40 mM), As^{3+} (2.5 mM)	ND	Pramanik et al. (2016)
Bacillus (CIK-517, CIK-519), *Klebsiella* (CIK-518), *Leifsonia* (CIK-521), and *Enterobacter* (CIK-521R)	IAA, PO_4, EPS	Cd^{2+} (1.78–4.45 mmol/L)	*Zea mays* L.	Ahmad, Akhtar, Asghar, Ghafoor, and Shahid (2016)
Bacillus megaterium ASNF3	N_2	Cr^{6+} (1000 μg/mL)	Wheat	Aslam, Hussain, and Qazi (2016)
Bacillus sp. *Providencia* sp. *Morganella* sp. *Stenotrophomonas* sp.	IAA, PO_4, K, siderophore,	Cd^{2+} (15 mM)	*Sesbania bispinosa*	Kartik et al. (2016)

Exiguobacterium sp. As-9	IAA, PO_4, EPS	As^{3+} (ND), As^{5+} (ND)	*Vigna radiata*	Pandey and Bhatt (2016)
Serratia marcescens	ND	Cd^{2+} (ND)	*Populus euramericana* (Dode) Guinier	Cocozza et al. (2015)
Bacillus safensis KM39, *Pseudomonas putida* GG29	IAA, PO_4, HCN, siderophore, ammonia	Cd^{2+} (ND)	ND	Singh et al. (2015)
Ralstonia sp. TISTR 2219 and *Arthrobacter* sp. TISTR 2220	IAA	Cd^{2+} (ND)	*Ocimum gratissimum* L.	Prapagdee & Khonsue, 2015
Pseudomonas putida ATCC 39213	ACCD	Cd^{2+} (ND)	*Eruca sativa*	Kamran, Syed, Eqani, Munis, and Chaudhary (2015)
Arthrobacter sp., *Pseudomonas aeruginosa, Bacillus licheniformis, Pseudomonas stutzeri, Aerococcus* sp.	IAA, ACCD, PO_4, siderophore,	Cr^{6+} (500–3000 mg/L), Cd^{2+} (100 mg/L), Pb^{2+} (100 mg/L)	*Lolium multiflorum* L. (ryegrass)	Khan et al. (2015)
Rhodococcus erythropolis EC 34, *Achromobacter* sp. 1AP2, *Arthrobacter* sp. EC 10 and *Microbacterium* sp. 3ZP2	IAA, ACCD, PO_4, ammonia, siderophore, HCN	Cd^{2+} (ND)	*Trifolium repens*	Pereira, Barbosa, and Castro (2015)
Acinetobacter sp., *Bacillus* sp., *Comamonas* sp., *Enterobacter* sp., *Geobacillus* sp., and *Paenibacillus* sp.	IAA, ACCD, PO_4, siderophore	As^{3+} (upto 30 mM) As^{5+} (upto 320 mM)	ND	Das, Jean, Kar, Chou, and Chen (2014)
Ralstonia eutropha and *Chryseobacterium humi*	ND	Cd^{2+} (500 mg/L)	*Zea mays* L.	Moreira, Marques, Franco, Rangel, and Castro (2014)

Continued

TABLE 1 Heavy metal/metalloid (Cr, Cd, Pb, and As only) resistant plant growth promoting rhizobacteria (PGPR) (2005 onward)—cont'd

PGPR strains	PGP traits	Resistance to heavy metal(loid)s with (MIC)	Plant growth promotion studied on	References
Bradyrhizobium sp. YL-6	IAA, ACCD, PO_4, siderophore	Cd^{2+} (100 mg/L)	*Lolium multiflorum* Lam. and *Glycine max* (L.) Merr.	Guo and Chi (2014)
Pseudomonas sp. Lk9	Biosurfactant, siderophore, organic acids	Cd^{2+} (ND)	*Solanum nigrum* L.	Chen, Luo, and Li (2014)
Azotobacter vinelandii, *Achromobacter xylosoxidans*, and *Azospirillum lipoferum*	IAA, PO_4	Cr^{6+} (upto 2 mM)	ND	Mohan, Devi, Srinivasan, and Sushamani (2014)
Klebsiella sp., *Stenotrophomonas* sp., *Bacillus* sp., and *Serratia*	IAA, EPS, siderophore	Cd^{2+} (500 mg/L)	Wheat and Maize	Ahmad et al. (2014)
Ochrobactrum sp. CdSP9 *Bacillus* sp. PbSP6 *Bacillus* sp. AsSP9	ACCD, siderophore	Cd^{2+} (100 μg/mL) Pb^{2+} (400 μg/mL) As (180 μg/mL)	*Oryza sativa* L. (rice)	Pandey, Ghosh, Ghosh, De, and Maiti (2013)
Micrococcus sp. MU1 and *Klebsiella* sp. BAM1	IAA, ACCD	Cd^{2+} (ND)	*Helianthus annuus*	Prapagdee, Chanprasert, and Mongkolsuk (2013)

Enterobacter intermedius MH8b	IAA, ACCD, PO_4, HCN	Cd^{2+} (10 mM)	*Sinapis alba* L.	Płociniczak, Sinkkonen, Romantschuk, and Piotrowska-Seget (2013)
Pseudomonas sp. VRK3	IAA, PO_4, siderophore	Cr^{6+} (500 mg/L)	ND	Hemambika, Balasubramanian, Rajesh Kannan, and Arthur James (2013)
Staphylococcus arlettae	IAA, ACCD	As^{3+} (40 mM) As^{5+} (366 mM)	*Brassica juncea* (L.) Czern.	Shrivastava and Kumar (2013)
Bacilllus sp. VRK1	IAA, PO_4, siderophore	Cr^{6+} (500 mg/L)	ND	Hemambika and Kannan (2012)
Bacillus sp. (BA1–BA8)	IAA, PO_4, HCN, ammonia, antifungal	Cr^{6+} (200 μg/mL)	ND	Karuppiah and Rajaram (2011)
Bradyrhizobium sp. 750, *Pseudomonas* sp. Az13, and *Ochrobactrum cytisi* Azn6-2	N_2	Cd^{2+} (upto 1.5 mM) Pb^{2+} (upto 6 mM) As^{3+} (upto 5 mM)	*Lupinus luteus*	Dary, Chamber-Pérez, Palomares, and Pajuelo (2010)
Brevibacterium sp.	ND	Cr^{6+} (ND)	*Helianthus annuus*	Faisal and Hasnain (2010)
Cellulosimicrobium cellulans strain KUCr3	IAA, PO_4	Cr^{6+} (450 mM) As^{3+} (5 mM) Cd^{2+} (3 mM)	Chilli	Chatterjee et al. (2009)
Pseudomonas aeruginosa MKRh3	IAA, ACCD, PO_4, siderophore	Cd^{2+} (7 mM)	*Vigna mungo* (Black gram)	Ganesan (2008)
Mesorhizobium RC3	IAA, siderophore	Cr^{6+} (500 μg/mL)	Chickpea	Wani et al. (2008a)

Continued

TABLE 1 Heavy metal/metalloid (Cr, Cd, Pb, and As only) resistant plant growth promoting rhizobacteria (PGPR) (2005 onward)—cont'd

PGPR strains	PGP traits	Resistance to heavy metal(loid)s with (MIC)	Plant growth promotion studied on	References
Pseudomonas tolaasii ACC23, *Pseudomonas fluorescens* ACC9, *Alcaligenes* sp. ZN4, and *Mycobacterium* sp. ACC14	IAA, ACCD, siderophore	Cd^{2+} (0.5–2.5 mM)	*Brassica napus*	Dell'Amico, Cavalca, and Andreoni (2008)
Bacillus spp. (PSB1, PSB7, PSB10)	IAA, ACCD, PO_4, siderophore, SA	Cr^{6+} (400–550 μg/mL)	ND	Wani et al. (2007a)
Ochrobactrum intermedium CrT-1 and *Bacillus cereus* S-6	ND	Cr^{6+} (50 mg/mL)	*Vigna radiata*	Faisal and Hasnain (2006)
Variovorax paradoxus, Rhodoccus sp. and *Flavobacterium* sp.	IAA, ACCD, siderophore	Cd^{2+} (0.2–3.5 mM)	*Brassica juncea* L. Czern.	Belimov et al. (2005)

[a]IAA, *indole-3-acetic acid, phosphate=phosphate solubilization, siderophore=siderophore production, ACCD=1-aminocyclopropane-1-carboxylic acid deaminase activity, potassium=potassium solubilization,* N_2*=nitrogen fixation, HCN=hydrocyanic acid production, ammonia=ammonia production, urease=urease activity, ND=not determined, MIC = minimum inhibitory concentrations.*

7.1.1 Nitrogen fixation

Although nitrogen (N_2) is the most abundant element (about 78%) in the atmosphere, growing plants cannot take it up directly. Nitrogen, however, is considered an essential nutrient for the growth, development, heredity, as well as metabolism of green plants. Some biological nitrogen-fixing (BNF) microorganisms can resolve this vital problem that can convert atmospheric nitrogen to ammonia by a unique and complex enzyme known as *Nitrogenase* (Kim & Rees, 1994). BNFs fix about two-thirds of atmospheric nitrogen, while the rest of the nitrogen is industrially synthesized by the Haber–Bosch process (Rubio & Ludden, 2008). BNFs have now been used as an alternative to chemical fertilizer. They are of two types: (1) symbiotic BNFs (Family: Rhizobiaceae) and (2) nonsymbiotic BNFs (includes free-living, associative, and endophytic bacteria).

7.1.2 Phosphate solubilization

Next to N_2, phosphorus (P) is the major macronutrient essential for both structural (compulsory for constructing DNA backbone) and metabolic processes such as energy transfer, signal transduction, macromolecular biosynthesis, photosynthesis, respiration, etc. Soil is rich in bound forms of phosphate (generally aluminum and iron phosphates in acid soils and calcium phosphates in alkaline soils), but those bound forms are highly insoluble in nature; therefore, plants fail to utilize it because they can absorb phosphate only in two soluble forms—monobasic ($H_2PO_4^-$) and dibasic (HPO_4^{2-}) ions (Bhattacharyya & Jha, 2012; Glass, 1989). Chemical phosphate fertilizers have been widely practiced for years, but frequent applications in highly unenviable to the environment. Therefore, to conquer the P deficiency in agricultural fields, it is recommended to employ ecologically safe and economically sound phosphate fertilizers. In this context, phosphate solubilizing bacteria (PSB) can be a promising and possible substitute for chemical phosphate solubilizers. They used to accomplish this ability by the production of some organic acids such as gluconic acid, 2-ketogluconic acid, citric acid, lactic acid, succinic acid, etc., which chelate the bound cations from the insoluble phosphates and make the P available to the crop plants and by the secretion of some enzymes such as phytases, phosphatases, etc.

7.1.3 Siderophore production

Iron (Fe) is one of the key nutrients indispensable for almost all forms of life. It plays a significant role in various physiological and biochemical processes such as photosynthesis, chlorophyll synthesis, and respiration, more specifically in the electron transport chain, as a cofactor for many enzymes (Litwin & Calderwood, 1993), nitrate reduction, nitrogen fixation, etc. However, the fact is that the insolubility of iron compounds is the utmost hurdle to plants to take it up directly. The hydroxide and oxyhydroxide polymer of

Fe^{3+} in an aerobic environment are inaccessible to both plants and microorganisms (Rajkumar, Ae, Prasad, & Freitas, 2010), but the more reduced form of iron-Fe^{2+} is accessible to both. Some PGPR can form a low molecular weight (<10 kD) iron-chelating compound called siderophore that aids in the acquisition of iron. Under iron-deficient conditions, to sequester and solubilize iron, the synthesis, and excretion of siderophore in bacteria is exceeded even more than that of their own dry cell weight. Siderophores are water-soluble molecule can be excreted extracellularly or produced intracellularly. There are three main types of siderophores, yet known—hydroxamate, catecholate, and carboxylate.

7.1.4 Phytohormone production

The most important phytohormone classes are auxins (especially indole-3 acetic acid (IAA)), cytokinins, gibberellins, etc., responsible for plant growth, development, and yield. Nearly 80% of rhizobacteria can produce auxins as secondary metabolites (Patten & Glick, 1996). IAA production by bacteria is well-documented various times by many researchers, but reports of gibberellin producing PGPR (*Bacillus licheniformis* and *Bacillus pumilus* reported by Gutiérrez-Mañero et al., 2001) are limited. IAA secreted by rhizobacteria interferes with many plant developmental processes such as root initiation, cell division, cell enlargement, tissue differentiation, apical dominance, pigment formation, stimulation of nitrogen fixation because the endogenous pool of plant, IAA may be altered by the acquisition of IAA that has been secreted by rhizobacteria (Glick, 2012; Spaepen, Vanderleyden, & Remans, 2007). Cytokinins promote cell enlargement, cell divisions, and morphogenesis (Timmusk, Nicander, Granhall, & Tillberg, 1999), and gibberellins are known to amend plant morphology by the extension of the plant, mainly stem tissue.

7.1.5 ACC deaminase activity

Like all other phytohormones, ethylene is also an important and essential endogenously produced plant growth regulator required for normal growth and development. But ethylenes effect depends on the concentration in root tissues as high concentration leads to defoliation, inhibition of root-shoot growth that results in poor crop yield (Li, Saleh-Lakha, & Glick, 2005). This happens in the emergence of different environmental stresses such as temperature stress, drought stress, flooding stress, salinity stress, heavy metal stress, pathogenic stress, etc. In response to these stresses, the plant synthesizes ACC, the immediate precursor of "stress hormone" (Abeles, Morgan, & Saltveit, 1992)—ethylene (Chen, Randlett, Findell, & Schaller, 2002; Farwell et al., 2007), and this overproduction causes poor root growth and damage. Some PGPR are reported to cope with such mess by producing ACC deaminase enzyme, which cleaves ACC to α-ketobutyrate and ammonium (Glick, Penrose, & Li,

1998; Grichko & Glick, 2001; Mayak, Tirosh, & Glick, 2004; Pramanik et al., 2017; Pramanik et al., 2018), thereby significantly lowering the level of stress ethylene.

7.2 Indirect mechanism: Presence of pathogen

In this case, PGPR act as biocontrol agents. The mechanism encompasses antibiotic production, antifungal activity, hydrocyanic acid (HCN) production, induced systematic resistance, etc.

7.2.1 Antibiotic production

Antibiotics are a heterogeneous group of low molecular weight compounds, and their production by the microorganisms is associated with the biocontrol of phytopathogen (Raaijmakers, Vlami, & De Souza, 2002). Under laboratory conditions, various PGPR-produced antibiotics were effective against many phytopathogenic agents (Bowen & Rovira, 1999).

7.2.2 Antifungal activity

Some PGPR are known to produce antifungal antibiotics, which can inhibit phytopathogenic fungi (Nowak-Thompson, Gould, Kraus, & Loper, 1994). The production of chitinases, laminarinase, and β-1,3-glucanases were associated with the degradation of the fungal cell wall (Mauch, Mauch-Mani, & Boller, 1988; Potgieter & Alexander, 1966; Velazhahan, Samiyappan, & Vidhyasekaran, 1999). Moreover, siderophore producing heavy metal-resistant isolates was shown to have biocontrol properties *Fusarium oxysporum* and *Rhizoctonia solani*, causing vascular wilts and peanut stem rot of peanut (Sindhu, Suneja, & Dadarwal, 1997). Sayyed and Patel (2011) also reported the biocontrol potential of siderophore producing *Alcaligenes* sp. and *Pseudomonas aeruginosa* having heavy metal-resistant properties against various fungal species.

7.2.3 Hydrocyanic acid (HCN) production

HCN is a volatile, secondary metabolite produced by many PGPR known to suppress the growth of other microorganisms (Siddiqui, 2005). HCN also acts as an inhibitor of metal enzymes formed from glycine by HCN synthetase enzyme, found to be associated with the rhizobacterial plasma membrane (Blumer & Haas, 2000).

7.2.4 Induced systematic resistance (ISR)

The amplified level of resistance using external agents without modifying the plant genome, is known as induced or acquired systematic resistance (ISR). The expression of ISR can be local or systemic depending upon the exposure to the inducer agents such as chemical activators, microorganisms, or extracts of cells of living organisms (Stadnik, 2000). ISR is associated with several benefits of the plant, including protection against various pathogens, stability, energy economy, etc. (Liu, Kloepper, & Tuzun, 1995; Raj et al., 2003).

8 PGPR: The dual players

The microbial communities grown in rhizosphere, which are beneficial for plant growth, yield, and crop quality have been called as "plant growth-promoting rhizobacteria (PGPR)" (Kloepper, 1978). They can be further classified as biofertilizer (capable of accelerating the accessibility of nutrients to the plant), phytostimulator (capable of facilitating the plant growth by phytohormones), rhizoremediator (those involved in the degradation of pollutants), and biopesticides (capable of producing antimicrobial metabolites that manage plant diseases). In contrast to the non-PGPR group (Davolos & Pietrangeli, 2013; Liu, Guo, Li, & Xiang, 2013; Muneer, Rehman, Shakoori, & Shakoori, 2009; Rehman, Zahoor, Muneer, & Hasnain, 2008; Shakya, Pradhan, Smith, Shrestha, & Tuladhar, 2012) PGPR group play a dual role in heavy metal bioremediation and plant growth promotion. However, PGPR is one-way produced phytostimulating substances (phytohormones) that act as biofertilizer (fixed nitrogen, solubilized, and mineralized phosphate), biocontrolling agent (produced siderophore, HCN), etc. Besides, they play a pivotal role in environmental cleanup by decreasing toxic heavy metals or metalloids as bioremediating agents and stress alleviating agents by producing ACC deaminase that helps plants grow better in heavy metal contaminated soil. To date, several heavy metal-resistant PGPR have been isolated, and their plant growth promotion also studied on several crops (Table 1). The level of heavy metal tolerance is highly variable from strain to strain, and their level of PGP activities are also diversified (Table 1).

9 Mechanisms of metal resistance

Bacteria have evolved a range of survival strategies to cope up with heavy metal extreme soils that include (1) extrusion of metal ions out of the cell, (2) metal bioaccumulation and sequestration, (3) biotransformation of toxic states of certain metal ions (Wani, Khan, & Zaidi, 2008a), and (4) metal adsorption/desorption (Mamaril, Paner, & Alpante, 1997).

PGPR with such properties improved the overall growth and yield of chickpea (*Cicer arietinum*) (Gupta, Rai, Bagdwal, & Goel, 2005) and pea (*Pisum sativum*) (Wani, Khan, & Zaidi, 2008b). Besides, PGPR can increase soil fertility and enhance crop productivity (Zaidi & Khan, 2006; Zaidi, Khan, & Aamil, 2004; Zaidi, Khan, & Amil, 2003) as well as act as potent plant growth regulators (Wani, Khan, & Zaidi, 2007a, 2007b, 2007c). Promotion of plant growth under abiotic or biotic stress is often mediated by ACC-deaminase producing PGPR (Belimov et al., 2005; Glick et al., 1998; Uchiumi et al., 2004). Moreover, phytoremediation is another cleanup method to treat contaminated soils that engages plants to remove, transfer, or stabilize the metals, which is considered as a time-consuming process (Brooks, 1998; Wenzel, Adriano, Salt, & Smith, 1999). However, soil microflora directly or indirectly influences the efficiency of the phytoremediation technique (Wang et al., 1989). For instance, the use of PGPR enhanced the phytoremediation capacity of maize (*Zea mays* L.) plants (Lippmann, Leinhos, & Bergmann, 1995). Cardoso, Gratão, Gomes-Junior, Medici, and Azevedo (2005) also showed enhanced phytoprotective capacity under metal stress with a positive influence from associated bacteria.

As previously stated, many bacteria have been reported to detoxify heavy metals and metalloids by developing various survival strategies (Davolos & Pietrangeli, 2013; Liu et al., 2013; Muneer et al., 2009; Rehman et al., 2008; Shakya et al., 2012). These bacteria are very promising in terms of application in the metal-contaminated areas for bioremediation. But some rhizospheric bacteria that exhibit one or more plant growth-promoting traits such as IAA production, N_2 fixation, phosphate solublization, ACC deaminase activity, siderophore production, antifungal activity, HCN production, etc., have shown more promising especially for better plant growth promotion under metal stress condition (Ahmad et al., 2014; Kartik et al., 2016; Mitra, Pramanik, Ghosh, et al., 2018; Mitra, Pramanik, Sarkar, et al., 2018; Pramanik et al., 2016, 2017; Pramanik et al., 2018; Roman-Ponce et al., 2017; Singh et al., 2015). They performed to remediate the heavy metals effectively and enhanced plant growth promotion under various metal stresses. Moreover, heavy metal-resistant bacteria combat with various types of toxic metals in the following ways:

9.1 Bioaccumulation

Bioaccumulation is an intrinsic property of some metal-resistant microorganisms by which they accumulate heavy metals in their cationic forms inside the cells. Many earlier workers have reported bioaccumulation of various heavy metals while working on different heavy metal-resistant PGPR (Chen et al., 2016; Mitra, Pramanik, Ghosh, et al., 2018; Mitra, Pramanik, Sarkar, et al., 2018; Pramanik et al., 2017; Pramanik et al., 2018; Treesubsuntorn, Dhurakit, Khaksar, & Thiravetyan, 2018).

9.2 Metal-binding proteins and peptides

Like plants, few bacteria also possess metallothionein (MT), phytochelatin (PC), and some metal-binding natural chelators, which play a very crucial role in microbe-metal interaction. Overexpression of PC synthase helps accumulate and tolerate metal ions (Sriprang et al., 2003). Besides, some novel metal-binding peptides containing histidines or cysteines are also reported and important for their selectivity, higher affinity, and specificity for metal ions.

9.3 Valence transformation

Oxidation or reduction of metal ions by bacterial enzymes is found to be associated with lessening the level of intracellular toxicity. Hexavalent chromium reduction to the trivalent state by bacteria is studied widely (Chatterjee, Sau, & Mukherjee, 2009; Pramanik et al., 2016), while arsenic biotransformation is reported by Ghosh et al. (2018) with a change from As (V) to As (III). In contrast, As-oxidizing plant growth-promoting bacteria (PGPB) has been reported by Das, Jean, Chou, Rathod, and Liu (2016).

9.4 Extracellular chemical precipitation by EPS production

Extracellular polymeric substances (EPS) are natural high molecular weight polymers secreted by many soil bacteria have a profound role in entrapping metal ions (Ha, Gélabert, Spormann, & Brown, 2010; Kenney, 2011; Pramanik et al., 2017). In fact, they are also adsorbed metal sulfides and oxides by an extracellular mixture of polysaccharides, mucopolysaccharides, and proteins (Lugtenberg, de Weger, & Bennett, 1991). It is found that peptidoglycan carboxyl groups are the principal cation-binding sites for Gram-positive bacterial cell walls, whereas phosphate groups for Gram-negative microbes. Rhizobacterial EPSs constitute several biological functions such as quorum sensing, root colonization with host, biofilm development, protects from environmental stressors such as heavy metals (Bramhachari, Nagaraju, & Kariali, 2018). Hence, plant-microbe-EPS interaction is a useful and reliable component in sustainable agriculture (Bramhachari et al., 2018).

9.5 Siderophore complexation

Siderophores are iron-chelating low molecular weight organic molecules. Their biological function is to concentrate iron in the environment where their concentration is very low and to transport iron into the cell. Siderophores, however, may interact with other metals that are chemically similar to iron viz., aluminum, gallium, and chromium. By binding to metals, siderophores reduce metal bioavailability and thereby metal toxicity. For example, siderophores reduce copper toxicity in cyanobacteria (Roane, Pepper, & Gentry, 2015).

9.6 Biosurfactant complexation

Biosurfactants are compounds that are produced by many bacteria that, in some cases, are excreted outside the cell. Biosurfactants are known complex with metals such as cadmium, lead, and zinc (Miller, 1995). Biosurfactant complexation apparently increases metals' solubility, but such a complex metal is nontoxic to the cell.

10 Conclusion

Unlike many conventional heavy metal remediation methods, bioremediation is a more eco-friendly and sustainable approach toward the effective reclamation of agricultural lands. Many heavy metal-resistant PGPR strains have already been isolated, characterized, and applied for metal detoxification. However, every heavy metal-resistant PGPR strains have their limitations in terms of their specific metal tolerance, tolerance level, acclimatization to the specific environmental conditions, plant specification for root colonization, susceptibility to soil-borne pathogen attack, level of PGP activities under stress condition, etc. Therefore, global demand for the isolation of more and more heavy metal-resistant PGPR strains is currently a very emerging area of research to apply in the heavy metal contaminated fields for effective reclamation and reduce the health risks of human beings.

Acknowledgments

KP is thankful to University Grants Commission (UGC), New Delhi, India for the award of UGC—Dr. D. S. Kothari Post-Doctoral Fellowship [No.F.4-2/2006 (BSR)/BL/19-20/0072 dated October 21, 2019].

References

Abeles, F. B., Morgan, P. W., & Saltveit, M. E., Jr. (1992). *Ethylene in plant biology*. Academic Press.

Ahmad, I., Akhtar, M. J., Asghar, H. N., Ghafoor, U., & Shahid, M. (2016). Differential effects of plant growth-promoting rhizobacteria on maize growth and cadmium uptake. *Journal of Plant Growth Regulation*, *35*(2), 303–315.

Ahmad, I., Akhtar, M. J., Zahir, Z. A., Naveed, M., Mitter, B., & Sessitsch, A. (2014). Cadmium-tolerant bacteria induce metal stress tolerance in cereals. *Environmental Science and Pollution Research*, *21*(18), 11054–11065.

Alloway, B. J. (2013). Sources of heavy metals and metalloids in soils. In *Heavy metals in soils* (pp. 11–50). Dordrecht: Springer.

Aslam, S., Hussain, A., & Qazi, J. I. (2016). Dual action of chromium-reducing and nitrogen-fixing *Bacillus megaterium*-ASNF3 for improved agro-rehabilitation of chromium-stressed soils. *3 Biotech*, *6*(2), 125.

Ayangbenro, A. S., & Babalola, O. O. (2017). A new strategy for heavy metal polluted environments: A review of microbial biosorbents. *International Journal of Environmental Research and Public Health*, *14*(1), 94.

Baker, A. J., & Brooks, R. (1989). Terrestrial higher plants which hyperaccumulate metallic elements. A review of their distribution, ecology and phytochemistry. *Biorecovery*, *1*(2), 81–126.

Belimov, A. A., Hontzeas, N., Safronova, V. I., Demchinskaya, S. V., Piluzza, G., Bullitta, S., & Glick, B. R. (2005). Cadmium-tolerant plant growth-promoting bacteria associated with the roots of Indian mustard (*Brassica juncea* L. Czern.). *Soil Biology and Biochemistry*, *37*(2), 241–250.

Bhattacharyya, P. N., & Jha, D. K. (2012). Plant growth-promoting rhizobacteria (PGPR): Emergence in agriculture. *World Journal of Microbiology and Biotechnology*, *28*(4), 1327–1350.

Biswas, J. K., Mondal, M., Rinklebe, J., Sarkar, S. K., Chaudhuri, P., Rai, M., ... Rizwan, M. (2017). Multi-metal resistance and plant growth promotion potential of a wastewater bacterium *Pseudomonas aeruginosa* and its synergistic benefits. *Environmental Geochemistry and Health*, *39* (6), 1583–1593.

Blumer, C., & Haas, D. (2000). Mechanism, regulation, and ecological role of bacterial cyanide biosynthesis. *Archives of Microbiology*, *173*(3), 170–177.

Bowen, G. D., & Rovira, A. D. (1999). The rhizosphere and its management to improve plant growth. In *Vol. 66. Advances in agronomy* (pp. 1–102). Academic Press.

Bramhachari, P. V., Nagaraju, G. P., & Kariali, E. (2018). Current perspectives on rhizobacterial-EPS interactions in alleviation of stress responses: Novel strategies for sustainable agricultural productivity. In *Role of rhizospheric microbes in soil* (pp. 33–55). Singapore: Springer.

Brooks, R. R. (1998). Geobotany and hyperaccumulators. In *Plant that hyperaccumulate heavy metals* (pp. 55–94). Wallingford, UK: CAB International.

Cardoso, P. F., Gratão, P. L., Gomes-Junior, R. A., Medici, L. O., & Azevedo, R. A. (2005). Response of *Crotalaria juncea* to nickel exposure. *Brazilian Journal of Plant Physiology*, *17* (2), 267–272.

Chatterjee, S., Sau, G. B., & Mukherjee, S. K. (2009). Plant growth promotion by a hexavalent chromium reducing bacterial strain, *Cellulosimicrobium cellulans* KUCr3. *World Journal of Microbiology and Biotechnology*, *25*(10), 1829–1836.

Chen, L., Luo, S., Li, X., et al. (2014). Interaction of Cd-hyperaccumulator *Solanum nigrum* L. and functional endophyte *Pseudomonas* sp. Lk9 on soil heavy metals uptake. *Soil Biology and Biochemistry*, *68*, 300–308.

Chen, Y., Chao, Y., Li, Y., Lin, Q., Bai, J., Tang, L., ... Qiu, R. (2016). Survival strategies of the plant-associated bacterium *Enterobacter* sp. strain EG16 under cadmium stress. *Applied and Environmental Microbiology*, *82*(6), 1734–1744.

Chen, Y. F., Randlett, M. D., Findell, J. L., & Schaller, G. E. (2002). Localization of the ethylene receptor ETR1 to the endoplasmic reticulum of Arabidopsis. *Journal of Biological Chemistry*, *277*(22), 19861–19866.

Cocozza, C., Trupiano, D., Lustrato, G., Alfano, G., Vitullo, D., Falasca, A., ... Scippa, S. (2015). Challenging synergistic activity of poplar–bacteria association for the Cd phytostabilization. *Environmental Science and Pollution Research*, *22*(24), 19546–19561.

Dąbrowska, G., Hrynkiewicz, K., Trejgell, A., & Baum, C. (2017). The effect of plant growth-promoting rhizobacteria on the phytoextraction of Cd and Zn by Brassica napus L. *International Journal of Phytoremediation*, *19*(7), 597–604.

Danish, S., Kiran, S., Fahad, S., Ahmad, N., Ali, M. A., Tahir, F. A., ... Mubeen, M. (2019). Alleviation of chromium toxicity in maize by Fe fortification and chromium tolerant ACC deaminase producing plant growth promoting rhizobacteria. *Ecotoxicology and Environmental Safety*, *185*, 109706.

Dary, M., Chamber-Pérez, M. A., Palomares, A. J., & Pajuelo, E. (2010). "In situ" phytostabilisation of heavy metal polluted soils using *Lupinus luteus* inoculated with metal resistant plant-growth promoting rhizobacteria. *Journal of Hazardous Materials*, *177*(1–3), 323–330.

Das, S., Jean, J. S., Chou, M. L., Rathod, J., & Liu, C. C. (2016). Arsenite-oxidizing bacteria exhibiting plant growth promoting traits isolated from the rhizosphere of *Oryza sativa* L.: Implications for mitigation of arsenic contamination in paddies. *Journal of Hazardous Materials, 302*, 10–18.

Das, S., Jean, J. S., Kar, S., Chou, M. L., & Chen, C. Y. (2014). Screening of plant growth-promoting traits in arsenic-resistant bacteria isolated from agricultural soil and their potential implication for arsenic bioremediation. *Journal of Hazardous Materials, 272*, 112–120.

Davolos, D., & Pietrangeli, B. (2013). A molecular study on bacterial resistance to arsenic-toxicity in surface and underground waters of Latium (Italy). *Ecotoxicology and Environmental Safety, 96*, 1–9.

Dell'Amico, E., Cavalca, L., & Andreoni, V. (2008). Improvement of *Brassica napus* growth under cadmium stress by cadmium-resistant rhizobacteria. *Soil Biology and Biochemistry, 40*(1), 74–84.

Donald, A. M. (2003). The use of environmental scanning electron microscopy for imaging wet and insulating materials. *Nature Materials, 2*(8), 511–516.

Duffus, J. H. (2002). "Heavy metals" a meaningless term? (IUPAC technical report). *Pure and Applied Chemistry, 74*(5), 793–807.

Faisal, M., & Hasnain, S. (2006). Growth stimulatory effect of *Ochrobactrum intermedium* and *Bacillus cereus* on *Vigna radiata* plants. *Letters in Applied Microbiology, 43*(4), 461–466.

Faisal, M., & Hasnain, S. (2010). Plant growth promotion by *Brevibacterium* under chromium stress. *Research Journal of Botany, 5*(1), 43–48.

Farwell, A. J., Vesely, S., Nero, V., Rodriguez, H., McCormack, K., Shah, S., … Glick, B. R. (2007). Tolerance of transgenic canola plants (*Brassica napus*) amended with plant growth-promoting bacteria to flooding stress at a metal-contaminated field site. *Environmental Pollution, 147*(3), 540–545.

Fashola, M. O., Ngole-Jeme, V. M., & Babalola, O. O. (2016). Heavy metal pollution from gold mines: Environmental effects and bacterial strategies for resistance. *International Journal of Environmental Research and Public Health, 13*(11), 1047.

Ganesan, V. (2008). Rhizoremediation of cadmium soil using a cadmium-resistant plant growth-promoting rhizopseudomonad. *Current Microbiology, 56*(4), 403–407.

Ghosh, P. K., Maiti, T. K., Pramanik, K., Ghosh, S. K., Mitra, S., & De, T. K. (2018). The role of arsenic resistant *Bacillus aryabhattai* MCC3374 in promotion of rice seedlings growth and alleviation of arsenic phytotoxicity. *Chemosphere, 211*, 407–419.

Glass, A. D. (1989). *Plant mineral nutrition. An introduction to current concepts*. Jones and Bartlett Publishers, Inc.

Glick, B. R. (2012). Plant growth-promoting bacteria: Mechanisms and applications. *Scientifica, 2012*, 963401.

Glick, B. R. (2014). Bacteria with ACC deaminase can promote plant growth and help to feed the world. *Microbiological Research, 169*(1), 30–39.

Glick, B. R., Penrose, D. M., & Li, J. (1998). A model for the lowering of plant ethylene concentrations by plant growth-promoting bacteria. *Journal of Theoretical Biology, 190*(1), 63–68.

Grichko, V. P., & Glick, B. R. (2001). Amelioration of flooding stress by ACC deaminase-containing plant growth-promoting bacteria. *Plant Physiology and Biochemistry, 39*(1), 11–17.

Guo, J., & Chi, J. (2014). Effect of Cd-tolerant plant growth-promoting rhizobium on plant growth and Cd uptake by *Lolium multiflorum* Lam. and *Glycine max* (L.) Merr. in Cd-contaminated soil. *Plant and Soil, 375*(1–2), 205–214.

Gupta, P., Kumar, V., Usmani, Z., Rani, R., & Chandra, A. (2018). Phosphate solubilization and chromium (VI) remediation potential of *Klebsiella* sp. strain CPSB4 isolated from the chromium contaminated agricultural soil. *Chemosphere*, *192*, 318–327.

Gupta, A., Rai, V., Bagdwal, N., & Goel, R. (2005). In situ characterization of mercury-resistant growth-promoting fluorescent pseudomonads. *Microbiological Research*, *160*(4), 385–388.

Gupta, P., Rani, R., Chandra, A., & Kumar, V. (2018). Potential applications of *Pseudomonas* sp. (strain CPSB21) to ameliorate Cr^{6+} stress and phytoremediation of tannery effluent contaminated agricultural soils. *Scientific Reports*, *8*(1), 1–10.

Gutiérrez-Mañero, F. J., Ramos-Solano, B., Probanza, A. N., Mehouachi, J., Tadeo, R., & F., & Talon, M. (2001). The plant-growth-promoting rhizobacteria *Bacillus pumilus* and *Bacillus licheniformis* produce high amounts of physiologically active gibberellins. *Physiologia Plantarum*, *111*(2), 206–211.

Ha, J., Gélabert, A., Spormann, A. M., & Brown, G. E., Jr. (2010). Role of extracellular polymeric substances in metal ion complexation on *Shewanella oneidensis*: Batch uptake, thermodynamic modeling, ATR-FTIR, and EXAFS study. *Geochimica et Cosmochimica Acta*, *74*(1), 1–15.

Hassan, T. U., Bano, A., & Naz, I. (2017). Alleviation of heavy metals toxicity by the application of plant growth promoting rhizobacteria and effects on wheat grown in saline sodic field. *International Journal of Phytoremediation*, *19*(6), 522–529.

Hemambika, B., Balasubramanian, V., Rajesh Kannan, V., & Arthur James, R. (2013). Screening of chromium-resistant bacteria for plant growth-promoting activities. *Soil and Sediment Contamination: An International Journal*, *22*(7), 717–736.

Hemambika, B., & Kannan, V. R. (2012). Intrinsic characteristics of Cr^{6+}-resistant bacteria isolated from an electroplating industry polluted soils for plant growth-promoting activities. *Applied Biochemistry and Biotechnology*, *167*(6), 1653–1667.

Hongbo, S., Liye, C., Gang, X., Kun, Y., Lihua, Z., & Junna, S. (2011). Progress in phytoremediating heavy-metal contaminated soils. In *Detoxification of heavy metals* (pp. 73–90). Berlin, Heidelberg: Springer.

Kamran, M. A., Syed, J. H., Eqani, S. A. M. A. S., Munis, M. F. H., & Chaudhary, H. J. (2015). Effect of plant growth-promoting rhizobacteria inoculation on cadmium (Cd) uptake by *Eruca sativa*. *Environmental Science and Pollution Research*, *22*(12), 9275–9283.

Kartik, V. P., Jinal, H. N., & Amaresan, N. (2016). Characterization of cadmium-resistant bacteria for its potential in promoting plant growth and cadmium accumulation in *Sesbania bispinosa* root. *International Journal of Phytoremediation*, *18*(11), 1061–1066.

Karuppiah, P., & Rajaram, S. (2011). Exploring the potential of chromium reduction *Bacillus* sp. and there plant growth promoting activities. *Journal of Microbiology Research*, *1*, 17–23.

Kenney, J. P. (2011). *Metal adsorption to bacterial cells and their products*. University of Notre Dame.

Khan, M. U., Sessitsch, A., Harris, M., Fatima, K., Imran, A., Arslan, M., … Afzal, M. (2015). Cr-resistant rhizo-and endophytic bacteria associated with *Prosopis juliflora* and their potential as phytoremediation enhancing agents in metal-degraded soils. *Frontiers in Plant Science*, *5*, 755.

Khan, M. S., Zaidi, A., Wani, P. A., & Oves, M. (2009). Role of plant growth promoting rhizobacteria in the remediation of metal contaminated soils. *Environmental Chemistry Letters*, *7*(1), 1–19.

Kim, J., & Rees, D. C. (1994). Nitrogenase and biological nitrogen fixation. *Biochemistry*, *33*(2), 389–397.

Kloepper, J. W. (1978). Plant growth-promoting rhizobacteria on radishes. In *Vol. 2. Proc. of the 4th Internet. Conf. on Plant Pathogenic Bacter, Station de Pathologie Vegetale et Phytobacteriologie, INRA, Angers, France8* (pp 879–882).

Kloepper, J. W. (1994). Plant growth-promoting rhizobacteria. In Y. Okon (Ed.), *Azospirillum/plant associations* (pp. 137–166). Boca Raton, FL: CRC Press.

Krämer, U. (2010). Metal hyperaccumulation in plants. *Annual Review of Plant Biology*, *61*, 517–534.

Li, Q., Saleh-Lakha, S., & Glick, B. R. (2005). The effect of native and ACC deaminase-containing *Azospirillum brasilense* Cd1843 on the rooting of carnation cuttings. *Canadian Journal of Microbiology*, *51*(6), 511–514.

Lippmann, B., Leinhos, V., & Bergmann, H. (1995). Influence of auxin producing rhizobacteria on root morphology and nutrient accumulation of crops. I: Changes in root morphology and nutrient accumulation in maize (*Zea mays* L.) caused by inoculation with indole-3 acetic acid (IAA) producing *Pseudomonas* and *Acinetobacter* strains or IAA applied exogenously. *Angewandte Botanik*, *69*(1–2), 31–36.

Litwin, C. M., & Calderwood, S. B. (1993). Role of iron in regulation of virulence genes. *Clinical Microbiology Reviews*, *6*(2), 137–149.

Liu, Q., Guo, H., Li, Y., & Xiang, H. (2013). Acclimation of arsenic-resistant Fe (II)-oxidizing bacteria in aqueous environment. *International Biodeterioration & Biodegradation*, *76*, 86–91.

Liu, L., Kloepper, J. W., & Tuzun, S. (1995). Induction of systemic resistance in cucumber against *Fusarium* wilt by plant growth-promoting rhizobacteria. *Phytopathology*, *85*(6), 695–698.

Lugtenberg, B. J., de Weger, L. A., & Bennett, J. W. (1991). Microbial stimulation of plant growth and protection from disease. *Current Opinion in Biotechnology*, *2*(3), 457–464.

Ma, Y., Prasad, M. N. V., Rajkumar, M., & Freitas, H. (2011). Plant growth promoting rhizobacteria and endophytes accelerate phytoremediation of metalliferous soils. *Biotechnology Advances*, *29*(2), 248–258.

Macek, T., Macková, M., & Káš, J. (2000). Exploitation of plants for the removal of organics in environmental remediation. *Biotechnology Advances*, *18*(1), 23–34.

Mallick, I., Bhattacharyya, C., Mukherji, S., Dey, D., Sarkar, S. C., Mukhopadhyay, U. K., & Ghosh, A. (2018). Effective rhizoinoculation and biofilm formation by arsenic immobilizing halophilic plant growth promoting bacteria (PGPB) isolated from mangrove rhizosphere: A step towards arsenic rhizoremediation. *Science of the Total Environment*, *610*, 1239–1250.

Mamaril, J. C., Paner, E. T., & Alpante, B. M. (1997). Biosorption and desorption studies of chromium (iii) by free and immobilized *Rhizobium* (BJVr 12) cell biomass. *Biodegradation*, *8*(4), 275–285.

Mauch, F., Mauch-Mani, B., & Boller, T. (1988). Antifungal hydrolases in pea tissue: II. Inhibition of fungal growth by combinations of chitinase and β-1, 3-glucanase. *Plant Physiology*, *88*(3), 936–942.

Mayak, S., Tirosh, T., & Glick, B. R. (2004). Plant growth-promoting bacteria confer resistance in tomato plants to salt stress. *Plant Physiology and Biochemistry*, *42*(6), 565–572.

Miller, R. M. (1995). Biosurfactant-facilitated remediation of metal-contaminated soils. *Environmental Health Perspectives*, *103*(suppl 1), 59–62.

Mitra, S., Pramanik, K., Ghosh, P. K., Soren, T., Sarkar, A., Dey, R. S., … Maiti, T. K. (2018). Characterization of Cd-resistant *Klebsiella michiganensis* MCC3089 and its potential for rice seedling growth promotion under Cd stress. *Microbiological Research*, *210*, 12–25.

Mitra, S., Pramanik, K., Sarkar, A., Ghosh, P. K., Soren, T., & Maiti, T. K. (2018). Bioaccumulation of cadmium by *Enterobacter* sp. and enhancement of rice seedling growth under cadmium stress. *Ecotoxicology and Environmental Safety*, *156*, 183–196.

Mohan, V., Devi, K. S., Srinivasan, R., & Sushamani, K. (2014). In-vitro evaluation of chromium tolerant plant growth promoting bacteria from tannery sludge sample, Dindugal, Tamil Nadu, India. *International Journal of Current Microbiology and Applied Sciences*, *3*(10), 336–344.

Moreira, H., Marques, A. P., Franco, A. R., Rangel, A. O., & Castro, P. M. (2014). Phytomanagement of Cd-contaminated soils using maize (*Zea mays* L.) assisted by plant growth-promoting rhizobacteria. *Environmental Science and Pollution Research, 21*(16), 9742–9753.

Muneer, B., Rehman, A., Shakoori, F. R., & Shakoori, A. R. (2009). Evaluation of consortia of microorganisms for efficient removal of hexavalent chromium from industrial wastewater. *Bulletin of Environmental Contamination and Toxicology, 82*(5), 597–600.

Nies, D. H. (1999). Microbial heavy-metal resistance. *Applied Microbiology and Biotechnology, 51*(6), 730–750.

Nowak-Thompson, B., Gould, S. J., Kraus, J., & Loper, J. E. (1994). Production of 2, 4-diacetylphloroglucinol by the biocontrol agent *Pseudomonas fluorescens* Pf-5. *Canadian Journal of Microbiology, 40*(12), 1064–1066.

Pandey, N., & Bhatt, R. (2016). Role of soil associated *Exiguobacterium* in reducing arsenic toxicity and promoting plant growth in *Vigna radiata*. *European Journal of Soil Biology, 75*, 142–150.

Pandey, S., Ghosh, P. K., Ghosh, S., De, T. K., & Maiti, T. K. (2013). Role of heavy metal resistant *Ochrobactrum* sp. and *Bacillus* spp. strains in bioremediation of a rice cultivar and their PGPR like activities. *Journal of Microbiology, 51*(1), 11–17.

Patten, C. L., & Glick, B. R. (1996). Bacterial biosynthesis of indole-3-acetic acid. *Canadian Journal of Microbiology, 42*(3), 207–220.

Pereira, S. I. A., Barbosa, L., & Castro, P. M. L. (2015). Rhizobacteria isolated from a metal-polluted area enhance plant growth in zinc and cadmium-contaminated soil. *International Journal of Environmental Science and Technology, 12*(7), 2127–2142.

Płociniczak, T., Sinkkonen, A., Romantschuk, M., & Piotrowska-Seget, Z. (2013). Characterization of *Enterobacter intermedius* MH8b and its use for the enhancement of heavy metals uptake by *Sinapis alba* L. *Applied Soil Ecology, 63*, 1–7.

Potgieter, H. J., & Alexander, M. (1966). Susceptibility and resistance of several fungi to microbial lysis. *Journal of Bacteriology, 91*(4), 1526–1532.

Pramanik, K., Ghosh, P. K., Ghosh, A., Sarkar, A., & Maiti, T. K. (2016). Characterization of PGP traits of a hexavalent chromium resistant *Raoultella* sp. isolated from the rice field near industrial sewage of Burdwan District, WB, India. *Soil and Sediment Contamination: An International Journal, 25*(3), 313–331.

Pramanik, K., Mitra, S., Sarkar, A., & Maiti, T. K. (2018). Alleviation of phytotoxic effects of cadmium on rice seedlings by cadmium resistant PGPR strain *Enterobacter aerogenes* MCC 3092. *Journal of Hazardous Materials, 351*, 317–329.

Pramanik, K., Mitra, S., Sarkar, A., Soren, T., & Maiti, T. K. (2017). Characterization of cadmium-resistant *Klebsiella pneumoniae* MCC 3091 promoted rice seedling growth by alleviating phytotoxicity of cadmium. *Environmental Science and Pollution Research, 24*(31), 24419–24437.

Prapagdee, B., Chanprasert, M., & Mongkolsuk, S. (2013). Bioaugmentation with cadmium-resistant plant growth-promoting rhizobacteria to assist cadmium phytoextraction by *Helianthus annuus*. *Chemosphere, 92*(6), 659–666.

Prapagdee, B., & Khonsue, N. (2015). Bacterial-assisted cadmium phytoremediation by *Ocimum gratissimum* L. in polluted agricultural soil: A field trial experiment. *International Journal of Environmental Science and Technology, 12*(12), 3843–3852.

Qamar, N., Rehman, Y., & Hasnain, S. (2017). Arsenic-resistant and plant growth-promoting Firmicutes and γ-Proteobacteria species from industrially polluted irrigation water and corresponding cropland. *Journal of Applied Microbiology, 123*, 748–758.

Raaijmakers, J. M., Vlami, M., & De Souza, J. T. (2002). Antibiotic production by bacterial biocontrol agents. *Antonie Van Leeuwenhoek, 81*(1–4), 537.

Raj, S. N., Chaluvaraju, G., Amruthesh, K. N., Shetty, H. S., Reddy, M. S., & Kloepper, J. W. (2003). Induction of growth promotion and resistance against downy mildew on pearl millet (*Pennisetum glaucum*) by rhizobacteria. *Plant Disease*, *87*(4), 380–384.

Rajkumar, M., Ae, N., Prasad, M. N. V., & Freitas, H. (2010). Potential of siderophore-producing bacteria for improving heavy metal phytoextraction. *Trends in Biotechnology*, *28*(3), 142–149.

Rehman, A., Zahoor, A., Muneer, B., & Hasnain, S. (2008). Chromium tolerance and reduction potential of a *Bacillus* sp. ev3 isolated from metal contaminated wastewater. *Bulletin of Environmental Contamination and Toxicology*, *81*(1), 25–29.

Rizvi, A., & Khan, M. S. (2018). Heavy metal induced oxidative damage and root morphology alterations of maize (*Zea mays* L.) plants and stress mitigation by metal tolerant nitrogen fixing *Azotobacter chroococcum*. *Ecotoxicology and Environmental Safety*, *157*, 9–20.

Roane, T. M., Pepper, I. L., & Gentry, T. J. (2015). Microorganisms and metal pollutants. In *Environmental microbiology* (pp. 415–439). Academic Press.

Roman-Ponce, B., Reza-Vázquez, D. M., Gutierrez-Paredes, S., María de Jesús, D. E., Maldonado-Hernandez, J., Bahena-Osorio, Y., … Vásquez-Murrieta, M. S. (2017). Plant growth-promoting traits in rhizobacteria of heavy metal-resistant plants and their effects on *Brassica nigra* seed germination. *Pedosphere*, *27*(3), 511–526.

Rubio, L. M., & Ludden, P. W. (2008). Biosynthesis of the iron-molybdenum cofactor of nitrogenase. *Annual Review of Microbiology*, *62*, 93–111.

Saleem, M., Asghar, H. N., Zahir, Z. A., & Shahid, M. (2018). Impact of lead tolerant plant growth promoting rhizobacteria on growth, physiology, antioxidant activities, yield and lead content in sunflower in lead contaminated soil. *Chemosphere*, *195*, 606–614.

Sayyed, R. Z., & Patel, P. R. (2011). Biocontrol potential of siderophore producing heavy metal resistant *Alcaligenes* sp. and *Pseudomonas aeruginosa* RZS3 vis-a-vis organophosphorus fungicide. *Indian Journal of Microbiology*, *51*(3), 266–272.

Shakya, S., Pradhan, B., Smith, L., Shrestha, J., & Tuladhar, S. (2012). Isolation and characterization of aerobic culturable arsenic-resistant bacteria from surfacewater and groundwater of Rautahat District, Nepal. *Journal of Environmental Management*, *95*, S250–S255.

Shrivastava, U. P., & Kumar, A. (2013). Characterization and optimization of 1-aminocyclopropane-1-carboxylate deaminase (ACCD) activity in different rhizospheric PGPR along with *Microbacterium* sp. strain ECI-12A. *International Journal of Applied Sciences and Biotechnology*, *1*(1), 11–15.

Siddiqui, Z. A. (2005). PGPR: Prospective biocontrol agents of plant pathogens. In *PGPR: Biocontrol and biofertilization* (pp. 111–142). Dordrecht: Springer.

Sindhu, S. S., Suneja, S., & Dadarwal, K. R. (1997). Plant growth promoting rhizobacteria and their role in crop productivity. In *Biotechnological approaches in soil microorganisms for sustainable crop production* (pp. 149–193). Jodhpur: Scientific Publisher.

Singh, R., Pathak, B., & Fulekar, M. H. (2015). Characterization of PGP traits by heavy metals tolerant *Pseudomonas putida* and *Bacillus safensis* strain isolated from rhizospheric zone of weed (*Phyllanthus urinaria*) and its efficiency in Cd and Pb removal. *International Journal of Current Microbiology and Applied Sciences*, *4*(7), 954–975.

Sirari, K., Kashyap, L., & Mehta, C. M. (2016). Stress management practices in plants by microbes. In *Microbial inoculants in sustainable agricultural productivity* (pp. 85–99). New Delhi: Springer.

Spaepen, S., Vanderleyden, J., & Remans, R. (2007). Indole-3-acetic acid in microbial and microorganism-plant signaling. *FEMS Microbiology Reviews*, *31*(4), 425–448.

Sriprang, R., Hayashi, M., Ono, H., Takagi, M., Hirata, K., & Murooka, Y. (2003). Enhanced accumulation of Cd^{2+} by a *Mesorhizobium* sp. transformed with a gene from *Arabidopsis thaliana* coding for phytochelatin synthase. *Applied and Environmental Microbiology*, *69*(3), 1791–1796.

Stadnik, M. J. (2000). Induc¸a˜o de resiste ̂ncia a Oı'dios. *Summa Phytopathologica*, *26*, 175–177.

Suresh, B., & Ravishankar, G. A. (2004). Phytoremediation—A novel and promising approach for environmental clean-up. *Critical Reviews in Biotechnology*, *24*(2–3), 97–124.

Timmusk, S., Nicander, B., Granhall, U., & Tillberg, E. (1999). Cytokinin production by *Paenibacillus polymyxa*. *Soil Biology and Biochemistry*, *31*(13), 1847–1852.

Tran, T. A., & Popova, L. P. (2013). Functions and toxicity of cadmium in plants: Recent advances and future prospects. *Turkish Journal of Botany*, *37*(1), 1–13.

Treesubsuntorn, C., Dhurakit, P., Khaksar, G., & Thiravetyan, P. (2018). Effect of microorganisms on reducing cadmium uptake and toxicity in rice (*Oryza sativa* L.). *Environmental Science and Pollution Research*, *25*(26), 25690–25701.

Uchiumi, T., Ohwada, T., Itakura, M., Mitsui, H., Nukui, N., Dawadi, P., ... Saeki, K. (2004). Expression islands clustered on the symbiosis island of the *Mesorhizobium loti* genome. *Journal of Bacteriology*, *186*(8), 2439–2448.

UN. (2017). *United Nations World population prospects: The 2017 revision*. https://www.un.org/development/desa/en/news/population/world-population-prospects-2017.html (Accessed 10 April 2018).

Upadhyay, N., Vishwakarma, K., Singh, J., Mishra, M., Kumar, V., Rani, R., ... Sharma, S. (2017). Tolerance and reduction of chromium (VI) by *Bacillus* sp. MNU16 isolated from contaminated coal mining soil. *Frontiers in Plant Science*, *8*, 778.

Velazhahan, R., Samiyappan, R., & Vidhyasekaran, P. (1999). Relationship between antagonistic activities of *Pseudomonas fluorescens* isolates against *Rhizoctonia solani* and their production of lytic enzymes/Beziehungen zwischen antagonistischer Aktivität von *Pseudomonas fluorescens*-Isolaten gegen *Rhizoctonia solani* und ihrer Produktion lytischer enzyme. *Zeitschrift für Pflanzenkrankheiten und Pflanzenschutz/Journal of Plant Diseases and Protection*, *106*, 244–250.

Wan, Y., Luo, S., Chen, J., Xiao, X., Chen, L., Zeng, G., ... He, Y. (2012). Effect of endophyte-infection on growth parameters and Cd-induced phytotoxicity of Cd-hyperaccumulator *Solanum nigrum* L. *Chemosphere*, *89*(6), 743–750.

Wang, P. C., Mori, T., Komori, K., Sasatsu, M., Toda, K., & Ohtake, H. (1989). Isolation and characterization of an *Enterobacter cloacae* strain that reduces hexavalent chromium under anaerobic conditions. *Applied and Environmental Microbiology*, *55*(7), 1665–1669.

Wang, T., Wang, S., Tang, X., Fan, X., Yang, S., Yao, L., ... Han, H. (2020). Isolation of urease-producing bacteria and their effects on reducing Cd and Pb accumulation in lettuce (*Lactuca sativa* L.). *Environmental Science and Pollution Research*, *27*(8), 8707–8718.

Wani, P. A., Khan, M. S., & Zaidi, A. (2007a). Chromium reduction, plant growth–promoting potentials, and metal solubilizatrion by Bacillus sp. isolated from alluvial soil. *Current Microbiology*, *54*(3), 237–243.

Wani, P., Khan, M., & Zaidi, A. (2007b). Co-inoculation of nitrogen-fixing and phosphate-solubilizing bacteria to promote growth, yield and nutrient uptake in chickpea. *Acta Agronomica Hungarica*, *55*(3), 315–323.

Wani, P. A., Khan, M. S., & Zaidi, A. (2007c). Synergistic effects of the inoculation with nitrogen-fixing and phosphate-solubilizing rhizobacteria on the performance of field-grown chickpea. *Journal of Plant Nutrition and Soil Science*, *170*(2), 283–287.

Wani, P. A., Khan, M. S., & Zaidi, A. (2008a). Chromium-reducing and plant growth-promoting *Mesorhizobium* improves chickpea growth in chromium-amended soil. *Biotechnology Letters*, *30*(1), 159–163.

Wani, P. A., Khan, M. S., & Zaidi, A. (2008b). Effects of heavy metal toxicity on growth, symbiosis, seed yield and metal uptake in pea grown in metal amended soil. *Bulletin of Environmental Contamination and Toxicology*, *81*(2), 152–158.

Wenzel, W. W., Adriano, D. C., Salt, D., & Smith, R. (1999). Phytoremediation: A plant—Microbe-based remediation system. In D. C. Adriano, et al. (Eds.), *Vol. 37. Bioremediation of contaminated soils* (pp. 457–508). The American Society of Agronomy, Inc. Crop Science Society of America, Inc. Soil Science Society of America, Inc.

Wu, B., He, T., Wang, Z., Qiao, S., Wang, Y., Xu, F., & Xu, H. (2020). Insight into the mechanisms of plant growth promoting strain SNB6 on enhancing the phytoextraction in cadmium contaminated soil. *Journal of Hazardous Materials*, *385*, 121587.

Zaidi, A., & Khan, M. S. (2006). Co-inoculation effects of phosphate solubilizing microorganisms and *Glomus fasciculatum* on green gram-*Bradyrhizobium* symbiosis. *Turkish Journal of Agriculture and Forestry*, *30*(3), 223–230.

Zaidi, A., Khan, M. S., & Aamil, M. (2004). Bioassociative effect of rhizospheric microorganisms on growth, yield, and nutrient uptake of greengram. *Journal of Plant Nutrition*, *27*(4), 601–612.

Zaidi, A., Khan, M. S., & Amil, M. D. (2003). Interactive effect of rhizotrophic microorganisms on yield and nutrient uptake of chickpea (*Cicer arietinum* L.). *European Journal of Agronomy*, *19* (1), 15–21.

Chapter 12

Nanotechnology: Recent trends in microbial nanotechnology

Hina Zain, Nazia Kanwal, Hareem Mohsin, Anum Ishaq, Unsa Bashir, and Syed Abdul Qadir Shah

Department of Allied Health Sciences, The Superior College Lahore, Lahore, Pakistan

Chapter outline

1 Introduction

Particles having one dimension ranging up to 100 nm are said to be nanoparticles (NPs). NPs are of great interest in the present world due to their attractive and noble properties i.e., size, shape, surface area, and optical properties.

Recent Advancement in Microbial Biotechnology. https://doi.org/10.1016/B978-0-12-822098-6.00007-0

Nanomaterials can be classified into different groups such as carbon-based nanomaterials (graphene, single-walled carbon nanotube, and multiwalled carbon nanotube), metals (gold, palladium, and cadmium), and metal-oxide nanomaterials (including titanium dioxide, cerium oxide, and zinc oxide) (Mouhib et al., 2019). NPs synthesized by the green approach have high surface area, greater catalytic activity, enhanced interaction between the metal salt and enzyme (Li, Xu, Chen, & Chen, 2011a). During the last decade inorganic nanomaterials that include silver, gold, and selenium are synthesized by bacteria with remarkable characteristics for the formulation of third-generation biosensors and voltammetric sensoristic devices for potential diagnostic applications such as biolabeling and cell imaging, and helpful in the formation of thin films and annealing without coating the surface (Grasso, Zane, & Dragone, 2020). The applications of NPs are beneficial corresponding to their bulk counterparts (Daniel & Astruc, 2004; Li et al., 2011a). Generally, there are various methods of nanoparticle synthesis such as biological, chemical, top-down, bottom-up techniques and composite methods for the formation of NPs of various types (Liu, Qiao, Hu, & Lu, 2011). Though the physical and synthetic methods to synthesize NPs are of great interest, frequent use of toxic chemicals limits their use in biomedical practices and health care units. Thus, there is a need to develop nontoxic, reliable, and sustainable methods for the formation of NPs.

Environmental sustainability is the major challenge of the nanomaterials market globally (Grasso et al., 2020; Mishra, Dixit, & Soni, 2015). Another option for the synthesis of NPs for environmental sustainability is to utilize microorganisms (Li et al., 2011a). Sustainability of the environment, economic production, and biocompatible procedures are required for the formation of NPs. So, NPs biosynthesize by means of microbes including yeasts, bacteria, and fungi present harmless, cheap, and environment-friendly substitution method (Ahmed & Ikram, 2016; Fariq, Khan, & Yasmin, 2017). It is an eco-friendly and energy-free process as compared to the chemical method (Li et al., 2011a). The association of microbiology, biotechnology, and nanotechnology has emerged as a new discipline of microbial nanobiotechnology in which microbes are playing a key role for the synthesis of NPs. Microbes are proved as nanofactories (Narayanan & Sakthivel, 2010). Microbial synthesis of NPs seems to be very innovative and fascinating for nanomanufacturing as compared to the traditional physical and chemical approaches. Bionanotechnology has revolutionized the way of forming NPs and it has encountered considerable attention owing to the increasing demand for environment-friendly approaches for the synthesis of nanomaterial (Li et al., 2011a). Microbes are preferable owing to their benefits of rapid growth rate, easy and simple culturing, and their capability to grow at ambient conditions (pH, pressure, and temperature). Synthesis of NPs by biogenic enzymatic procedure provides distinct shape and size in short duration. Utilization of an enzyme-mediated process removes the exploitation of expensive chemicals and lead to "green" means of nanoparticle synthesis also termed as "green nanotechnology." Thus, microbes use enzymatic properties to entrap metal ions from surrounding and transform it into elemental form (Li et al., 2011a).

Some drawbacks that we encounter from bacteria-based nanoparticle synthesis is less control on the distribution, shape, and size extra step of laborious bacterial culturing (Jeevanandam, Chan, & Danquah, 2016). Fungi also have some extracellular and intracellular enzymes that can synthesize NPs with definite sizes and geometry. Fungi are preferable for the production of NPs because of large mass as compared to bacteria (Jeevanandam et al., 2016). The increasing demand of nanomaterials market during the period of 2016 till 2022 is up to 20% or more (Inshakova & Inshakov, 2017).

Overreaching the best method for synthesis of a nanoparticle is significant for future utilization of nanostructured-based techniques and their utilization such as diagnostic and therapeutic application, electrochemical and optical sensoristics appliances, antimicrobial activity and, bioimaging in vivo and in vitro (Dragone, Grasso, Muccini, & Toffanin, 2017; Grasso et al., 2020; Kiessling, Mertens, Grimm, & Lammers, 2014).

2 Biosynthesis of nanoparticles

Living bodies and inorganic materials have coexisted since the beginning of life on the earth. This regular interaction has ensured the sustainability of life on earth with rich mineral deposition. The interest of scientists in studying the interaction between biological entities and inorganic molecules have dominated the research. Recent studies have found that inorganic NPs can be produced using microorganisms. Metallic nanoparticles (MtNPs) contributed a major portion to the development of the macroeconomic industry and results in increasing demand. During the last couple of decades, mostly chemical and physical methods are used for synthesis. These physical and chemical methods are not only expensive but also involve unsafe and hazardous chemicals. Toxic chemicals produce side effects especially in biomedical applications of MtNPs. To avoid this, scientists explored biological resources for the synthesis of MtNPs, commonly known as green synthesis. Green synthesis is economical, nontoxic, simple, and eco-friendly. Therefore, both plants and microorganisms gained great consideration as biofactories for the synthesis of MtNPs (Patra et al., 2015). Researchers are giving more importance to microbial synthesis due to the enhanced growth of microorganisms under controlled conditions and its easy cultivation. The existence of life is possible when there is a well-established relationship between inorganic and biomolecules. There is a broad range of microorganisms that react differently with the metal ions. Microorganism's machinery itself acts as stabilizing agents essentially needed for the synthesis of NPs. Due to the wide applications of gold and silver NPs, various microbes have been used for the synthesis of the best quality of Au and Ag NPs. Other metals such as Zn, Cu, Se, Ti, and Te NPs synthesis were also recently reported in microbes. MtNPs of different sizes and shapes are synthesized in different strains of bacteria, fungi, algae, and yeast. Microbial synthesis can be intracellular and extracellular. In the intracellular method, ion transport system is responsible for the synthesis of MtNPs. In an extracellular method, bioreduction of metal ions is carried out by the reductase enzymes secreted by microbial cells. NPs synthesized by microbes are widely used in bioapplications (Ovais et al., 2018).

2.1 Biosynthesis of MTNPs in bacteria and cyanobacteria

Bacteria are readily available nanofactories for the synthesis of MtNPs. Both extracellular and intercellular methods have been applied. Extracellular biosynthesis includes diversified techniques using either bacterial biomass or cell-free extracts or supernatant of culture. This method is preferred over the intracellular method because of its simple processing.

2.2 Mycosythesis of MTNPs

Fungi are proven to be more resourceful than bacteria with more active metabolites, high production, and greater accumulation. Fungi are excellent candidates owing to the greater variety of enzymes. In fungi, like bacteria, both extracellular and intracellular approach is observed (Molnár et al., 2018). Extracellular synthesis is done by fungal extracts, whereas mycelia convert metal salts into less toxic metal for the production of MtNPs in the intracellular method (Alghuthaymi, Almoammar, Rai, Said-Galiev, & Abd-Elsalam, 2015; Castro-Longoria, Vilchis-Nestor, & Avalos-Borja, 2011; Singh, Kim, Zhang, & Yang, 2016). Extracellular synthesis has also been reported using many strains such as *Volvariella volvacea*. CdS NPs biosynthesized using *Fusarium oxysporum* (Ahmad et al., 2002a, 2002b), but not all the strains of *Fusarium* are productive (Durán, Marcato, Alves, Souza, & Esposito, 2005a, 2005b). It was seen that *Fusarium moniliforme* was unable to synthesize silver nanoparticles (AgNPs). *V. volvacea* is used for the synthesis of bimetallic NPs (Vigneshwaran et al., 2007).

2.3 Algae as nanofactories

Algae are now popularly used for the biosynthesis of NPs. *Sargassum* sp. were used to synthesize monodispersed gold and zinc NPs (Sanaeimehr, Javadi, & Namvar, 2018). Bimetallic nanoparticle biosynthesis has also been reported using *Spirulina platensis*; 7 nm of CuNPs has been reported using *Cystoseira trinodis* (Koopi & Buazar, 2018).

2.4 Biosynthesis of nanoparticles using yeast

Yeast is the most studied species among the eukaryotes for biosynthesis. Different species show different mechanisms for the synthesis of MtNPs. Yeast has developed a unique mechanism to overcome the toxic effect of heavy metals accumulated in insignificant amounts (Breierová et al., 2002). These metals are enzymatically oxidized or reduced as present in the cytoplasm or cell wall. Some yeast strains chelate toxic metals with extracellular peptides. Yeast is also known as semiconductor crystals for its ability to synthesize semiconductor NPs (Sharma, Sharma, & Chaudhary, 2020). Cadmium salts and γ-glutamyl peptide of *Schizosaccharomyces pombe* biosynthesized cadmium NPs (Agnihotri, Joshi, Kumar, Zinjarde, & Kulkarni, 2009). Lead NPs have also been recently reported in *Torulopsis* (Kowshik et al., 2002). *Yarrowia lipolytica* can withstand heavy metals and have the ability to degrade hydrocarbons (Bankar, Kumar, & Zinjarde, 2009).

2.5 Biosynthesis of gold nanoparticles

Gold nanoparticles (AuNPs) are in use for centuries, when nobody knew about the term "nanoparticles," the colloidal solution of gold had been used in glass stains and curing of diseases. Michael Faraday was the first one who found, 150 years ago, that the properties of the colloidal solution of gold are different from the gold element (Hayat, 1989). In the present era, AuNPs are widely used as a catalyst in many chemical reactions, for example, reduction of azo dyes (Najafinejad, Mohammadi, Mehdi Afsahi, & Sheibani, 2019) and degradation of aromatic pollutants (Qu et al., 2019). Microbial synthesis of it is reported both extracellularly and intercellularly. *Rhodopseudomonas capsulata* synthesized AuNPs of size ranging 10–20 nm at pH 7 (He, Guo, Zhang, Zhang, & Ning, 2007). Morphology of any NPs is controlled by altering the reaction conditions (Husseiny, El-Aziz, Badr, & Mahmoud, 2007). Gold NPs were also synthesized using *Lyngbya majuscule* (Bakir, Younis, Mohamed, & El Semary, 2018); 9–20 nm size of gold particles were found in the *Chlorella vulgaris* cells (Ovais et al., 2018) and 8–12 nm of AuNPs were extracellularly synthesized using *Sargassum wightii greville* (Gu, Chen, Chen, Zhou, & Parsaee, 2018; Singaravelu, Arockiamary, Kumar, & Govindaraju, 2007). *Thermomonospora* sp. synthesized 8 nm AuNPs in extremely alkaline conditions (Ahmad, Senapati, Khan, Kumar, & Sastry, 2003). AuNPs of size ranging 5–15 nm were synthesized intracellularly in alkaline tolerant *Rhdococcus sp (A.* Ahmad et al., 2003*)*. AuNPs are also synthesized extracellularly by the reaction of AuCl ions with the biomass. AuNPsNPs were synthesized in *Pichia jadinii* and *Candida* sp. both intracellularly and extracellularly (Gericke & Pinches, 2006).

2.6 Biosynthesis of silver nanoparticles

AgNPs are extensively used against Gram-positive and Gram-negative bacteria and many other highly resistant strains like *Staphylococcus aureus*. Due to the vast applications of AgNPs, scientists made efforts to develop an eco-friendly method. These researchers have tried many microbes for their biosynthesis. Bacterial strain *Pseudomonas stutzeri* produced silver nanoparticle of size

200 nm (Joerger, Klaus, & Granqvist, 2000; Klaus, Joerger, Olsson, & Granqvist, 1999), whereas dried cells of *Cornebacterium* sp. produced silver nanoparticle of size 10–15 nm (Shahverdi, Minaeian, Shahverdi, Jamalifar, & Nohi, 2007). In recent years, eight different strains of cyanobacteria were investigated for the production of AgNPs (Lengke, Fleet, & Southam, 2007). A highly monodispersed silver nanoparticle was produced using *Phormidium fragile* (Satapathy & Shukla, 2017). Intracellular reduction of silver ions was reported using the biomass of *verticillium* sp. The Ag^+ ion is trapped electrostatically on the surface of mycelia. Ag + ions are then reduced by the reductase enzyme secreted by the cell wall to produce AgNPs (Mukherjee et al., 2001). Monodispersed AgNPsNPs were produced extracellularly by *Aspergillus fumigatus*. It has also been observed that *A. fumigatus* is used for large-scale production because it takes less time (Bhainsa & D'Souza, 2006). Quicker synthesis has also been found in *Penicillium fellutanum* (Kathiresan, Manivannan, Nabeel, & Dhivya, 2009).

Gold and silver NPs have a very rich history. They are vastly used in biomedicines and electronic technologies. For these reasons, over the last many years, research has been done to obtain good-quality NPs of these metals. But recently much work has been done in exploring the applications of NPs of other metals such as selenium, zirconium, tellurium, titanium, palladium, etc., and it is also found that NPs of other metals could be efficiently synthesized in microbes. Some of them are discussed in the following section.

2.7 Oxide nanoparticles

Until now it is reported that there are two types of oxide NPs synthesized using microbes, magnetic oxide, and nonmagnetic oxide.

2.8 Magnetic oxide nanoparticles

NPs that show response to the applied magnetic fields are called magnetic NPs. Size ranges from 1 to 100 nm, but the best size range for various applications is 10–20 nm (Gobbo, Sjaastad, Radomski, Volkov, & Prina-Mello, 2015). These are recently developed unique materials possessing superparamagnetic properties with high coercive force. Due to this nature, these are widely used in biomedicine fields (Kandasamy & Maity, 2015). These magnetic particles are synthesized intracellularly in magnetotactic bacteria and are referred to as bacterial magnetic particles (BacMPs). These are not only oxides (Fe_3O_4, Fe_2O_3) but also sulfides (Fe_3S_4) and are easily dispersed in an aqueous phase (Kandasamy & Maity, 2015). Magnetic NPs are synthesized as crystalline particles surrounded by an organic membrane forming a vesicle known as a magnetosome. Magnetosomes are arranged in linear chains and orient along the geomagnetic lines.

Magnetotactic bacteria morphologically and metabolically belong to a diverse group of Gram-negative prokaryotes. They move with the flagella according to the earth's magnetic field. Magnetotactic bacteria were first reported in 1975 (Blakemore, 1975). They possess a variety of morphology such as cocci, spirilla, vibrios, ovoid, and rod-shaped. They inhabit different types of aquatic environments. Magnetotactic coccus strains are frequently found on the surface of aquatic and are microaerophilic. Magnetotactic vibrio strains are isolated from the salt marshes (Li, Xu, Chen, & Chen, 2011b; Thornhill, Burgess, & Matsunaga, 1995). Spirillacea strain is found in fresh water. Cultured strains grow at 30°C, while uncultured ones grow below 30°C.

2.9 Nonmagnetic oxide nanoparticles

All the oxide NPs that do not show magnetic nature known as nonmagnetic oxide NPs such as Sb_2O_3, SiO_2, ZnO, ZrO_2, TiO_2, etc.

CuO NPs proved to be having efficient antimicrobial activity and is a promising tool in biotechnology. Biomass extracts of the strains of *Streptomyces zaomyceticus* and *Streptomyces pseudogriseolus* were used for its biosynthesis (Hassan et al., 2019). The algal extract is a very good bioreducer and stabilizing agent for the production of oxide NPs. *Sargassum ilicifolium* extract produced aluminum oxide NPs (Koopi & Buazar, 2018). The supernatant and biomass of zinc tolerant *Lactobacillus plantarum* synthesized ZnO NPs (Mohd Yusof, Abdul Rahman, Mohamad, Zaidan, & Samsudin, 2020). Selenium is also a very well-known metal used as semiconductors and in solar cells. Selenium oxide NPs are recovered from *Klebsiella pneumonia*. Selenium oxyanions can investigate the toxicity in the body and also act as anticarcinogenic (Cruz, Wang, & Liu, 2019). Biosynthesis of Sb_2O_3 NPs of size 2–10 nm has been reported using *Saccharomyces cerevisiae* biomass (Mandal, Bolander, Mukhopadhyay, Sarkar, & Mukherjee, 2006). Trigonal NPs of BaTiO3 of size ranging 4–5 nm and quasi-spherical zirconia oxide NPs of size 3–11 nm were biosynthesized using *F. oxysporum* (Bansal, Poddar, Ahmad, & Sastry, 2006; Bansal, Rautaray, Ahmad, & Sastry, 2004). SiO_2 and TiO_2 NPs are also biosynthesized using the same strain (Bansal et al., 2005). Cadmium oxide NPs are synthesized using *Penicillium oxalicum* (Asghar et al., 2020).

2.10 Platinum nanoparticle

Platinum NPs have great reductive properties, which is extensively used in many chemical industries for the reduction of chemical components. However, due to less availability of platinum in earth's crust and high cost and less stability, it is not much used in the synthesis of NPs. Most of the platinum NPs are synthesized either using chemical methods or leaf extracts. Microbial synthesis of platinum NPs is reported in *F. oxysporum* (Gupta & Chundawat, 2019).

2.11 Sulfide nanoparticles

Sulfide NPs have technical applications as biomarkers and cell-labeling agents (Yang, Santra, & Holloway, 2005). Cadmium sulfide (CdS) nanoparticle is the most typical type. It was found that CdS precipitated on the cell surface of *Clostridium thermoacetium* in the presence of cysteine hydrochloride. Cysteine is the major source of sulfide (Romero, Gatti, & Bruno, 1999); 20–200 nm CdNPs synthesized using *K. pneumonia*. Intracellular production of cadmium NPs was observed in *Escherichia coli* (Bai & Zhang, 2009). Monodispersed ZnS NPs of size range 2–5 nm was successfully obtained by *Rhodobacter sphaeroides*. Fungi produce stable sulfide NPs of zinc, lead, molybdenum, and cadmium extracellularly (Ahmad et al., 2002a, 2002b).

2.12 Alloy nanoparticles

Alloy NPs are either bimetallic or trimetallic. They are widely used in the electronic and biomedical industry because of their very strong characteristics with novel abilities. They possess unique structural, catalytic, and optical properties that make them different from the original elements. The synthesis of alloy NPs by chemical method involves a complex process that needs many reducing agents. These chemicals are hazardous. For these reasons eco-friendly protocol is needed to be developed but not many successful results are obtained. Very few alloy NPs are reported in literature synthesized from microorganisms. A highly stable 8–14 nm Au-Ag alloy was formed by the reductase enzyme secreted by the solution of *F. oxysporum* (Senapati, Ahmad, Khan, Sastry, & Kumar, 2005). Ag-Ni alloy particles are biosynthesized using leaf extracts of *Ocimum sanctum* (Akinsiku et al., 2018). Some of the trimetallic NPs reported in the literature are (SnZnCu) NPs, (PtNiCu) NPs, (AuPtPd) NPs, and (AuPtAg) NPs (Nasrollahzadeh, Sajjadi, Iravani, & Varma, 2020).

2.13 Other miscellaneous nanoparticles

Fullerols C60(OH)24 nanoparticles (FNPs) are biosynthesized using *Aspergillus flavus*. Bismuth selenide (Bi_2Se_3) NPs were synthesized in different strains of *pseudomonas and stenotrophomonas*. Microbes reduced the selenite to organoselenides. It is used as a semiconductor material in generators (Kuroda et al., 2019).

2.14 Mechanistic approach of nanoparticles

The insight detail of mechanism of biosynthesis of NPs in microbes is still not very well understood though much work has been done in this field. Earlier it was assumed that metals have a very toxic effect on microorganisms. To counter these toxic effects microorganisms developed genetic responses for their survival. Microorganisms start producing specific proteins to convert these metals into less toxic substances (Nies, 1999). Microbes contain different biological components intracellularly and extracellularly, which facilitate the formation of nanoparticle metal ions. It is found that metal ion is first trapped on the surface or within the cell and then reduced. There are two approaches regarding biosynthesis; the first approach is that metal modifies the composition of cell solution that facilitates the synthesis and the second is that the production of organic polymers may result in nucleation (Benzerara et al., 2011).

The general intercellular mechanism was clearly explained using *Verticillium* sp. (Mukherjee et al., 2001). It involves the trapping of ions, their bioreduction, and capping. The trapping of ions is due to the electrostatic interaction of the cell wall. Cell wall secretes the enzymes that cause reduction of the trapped ions into NPs. These synthesized NPs then get diffused. It is also found that different strains show little deviation from the general mechanism. For example, it was observed in *Lactobacillus* sp. (Nair & Pradeep, 2002) that there is a cluster of metal ion that interacts with the cell wall and results in the formation of nanoclusters, which are then later diffused off through the bacterial wall.

The extracellular synthesis was first studied in AgNPs (Kalishwaralal, Venkataraman, Ramkumarpandian, Nellaiah, & Sangiliyandi, 2008). It was discovered that nitrate reductase is an enzyme that is responsible for the biosynthesis of NPs. The role of nitrate reductase was then confirmed with the assay conducted by reacting nitrate with 2,3-diaminophthalene, which produced 2,3-diaminonapthotriazole (Mikhailov & Mikhailova, 2019). The same mechanism is also found with gold and other metals using different microbes. However, it is also observed that *F. moniliforme* is not able to produce NPs even though it releases the nitrate reductase enzyme (Durán et al., 2005a, 2005b). It was further investigated and proved that the combination of cofactor NADPH (nicotinamide adenine dinucleotide phosphate), electron carrier, and stabilizing proteins are needed for the biosynthesis of NPs. Other biological metabolites such as organic acids and polysaccharides also affect the biosynthesis of NPs.

The mechanism of biosynthesis of NPs is very different in magnetotactic bacteria. Its mechanism is also not clearly understood yet. Different studies have been done in this field. One possible suggested mechanism is that it is a multistep process. In the first step, there is an invagination of the cell membrane forming a vesicle. Then ferrous ions are translocated into the vesicle with the help of chaperons. The iron ion is under the control of the redox system in the vesicle. The bacterial proteins present in the vesicles then cause nucleation of nanocrystals (Arakaki, Nakazawa, Nemoto, Mori, & Matsunaga, 2008). The other suggested mechanism is that first there would be active production of ferrous ions, which are then localized on the negatively charged membrane resulting in the formation of magnetosome (Perez Gonzalez et al., 2010).

Different mediums and variations in physical conditions such as temperature and pH, strains of the same microbes produced a variety of NPs (Gurunathan et al., 2009). Sizes varied from 2 to 5 nm to 100 nm, and the morphology can be triangular, hexagonal, spherical, and other shapes. It is observed that spherical particles are smaller than hexagonal or triangular ones. However, it was observed that magnetotactic bacteria produce magnetic NPs of uniform sizes and morphology. Mms6 is the protein of magnetotactic bacteria that is responsible for the formation of uniform iron oxide magnetic NPs (A. Arakaki, Masuda, Amemiya, Tanaka, & Matsunaga, 2010).

3 Applications of microbially synthesized nanoparticles

The utilization of microorganisms for the production of eco-friendly NPs has magnificently compared with chemical synthesis, which is expensive, and toxic method in synthesizing size-dependent nanomaterials. Microbes including bacteria, fungi, and yeast can biosynthesize MtNPs in an aqueous solution, making the separation of NPs very simple, eco-friendly, and cheap. Moreover, these microbes are considered as a never-ending supply that can be stored, conserved, and reused.

4 Microbially synthesized nanoantibiotics

NPs synthesized by different microbes are emerging candidates to develop antibiotics. These eco-friendly NPs attach to the bacterial cell wall, fragment the cell wall, and cause leakage of cellular content leading to cell death. Microbially synthesized nanoparticles (MsNPs) bear strong antibacterial activities. Their efficiencies are subject to small size, specific morphology, and large surface areas. The large surface areas of MsNPs help to enhance communications of these particles among microbial surfaces. These small-sized MsNPs do not only stick to the membranes but can also enter into the cell, diffuse into the DNA, obstruct its replication and inhibit the synthesis of respiratory enzymes of pathogens. This type of mechanism is exhibited by bacterially synthesized AgNPs against different strains of pathogenic bacteria (Fariq et al., 2017).

The potential of eco-friendly synthesized MtNPs against multidrug-resistant bacterial species is an attractive area of research for the development of new antibiotics. Gold and silver NPs synthesized by eco-friendly or green technology are being considered for experimental trials in disease diagnostics and treatments. Different studies of microbially synthesized gold and silver NPs against pathogenic and resistant bacterial species are listed in Table 1. It is found that AgNPs have equivalent potential counter to both Gram-positive and Gram-negative bacterial strains (Ramalingam et al., 2014; Sudha, Rajamanickam, & Rengaramanujam, 2013), while microbially synthesized gold NPs showed miscellaneous trends. In a few studies, greater activity was found against Gram-negative bacteria including *Pseudomonas aeruginosa*, which is a human pathogen and causes respiratory infections (Syed et al., 2016). In contrast, few studies have depicted very toxic activity against Gram-positive bacteria. The positively charged AuNPs attached to the negatively charged bacterial cell wall

TABLE 1 Antibacterial activities of microbially synthesized nanoparticles.

No.	Microbes used to synthesize NPs	Type of NPs	Pathogens tested	Reference
1	*P. fluorescens*	Gold	*P. aeruginosa* *E. coli* *S. aureus* *B. subtilis* *K. pneumoniae*	Syed, Prasad, and Satish (2016)
	Garcenia combogia	Gold	*Bacillus subtilis, E. coli, L. monocytogenes, Proteus vulgaris, Vibrio parahaemolyticus*	Nithya and Jayachitra (2016)
	A. flavus *E. nidulans*	Silver	*E. coli* *P. aeruginosa* *S. aureus*	Barapatre, Aadil, and Jha (2016)
	Enterococcus sp.	Cadmium	*S. nematodiphila, E. coli, K. planticola,* and *Vibrio* sp.	Rajeshkumar, Ponnanikajamideen, Malarkodi, Malini, and Annadurai (2014)
	Streptomyces sp. Al-Dhabi-87	Silver	*E. coli, Acinetobacter baumannii, S. aureus*	Al-Dhabi, Mohammed Ghilan, and Arasu (2018)
	Candida sp. VITDKGB	Silver	*Staphylococcus aureus* and *Klebsiella pneumoniae*	Kumar, Karthik, Kumar, and Roa (2011)
	Planomicrobium sp.	Titanium oxide	*B. subtilis, K. planticola*	Nadeem et al. (2018)
	Fusarium oxysporum	Iron	*Bacillus subtilis, E. coli, Staphylococcus*	Abdeen and Praseetha (2013)
	Aspergillus clavatus	Silver	*E. coli, Pseudomonas flurorescens*	Verma, Kharwar, and Gange (2010)

membranes; these electrostatic interactions cause pits in the cell wall and induce permeability and lead to cell death.

Bacteria are among the oldest living thing on this earth due to rapid evolutionary amendments taken to adopt changing environmental conditions. With the same potential, they have learned to confer resistance against commercially available antibiotics. Thus, in the quest of new and novel antibiotics different types of MtNPs are being tested, for example, oxides of cerium and sulfides of cadmium synthesized from microbes have shown promising results in the inhibition of drug-resistant bacterial growth (Patil & Kim, 2018). The antibacterial activity of biogenic cerium oxide NPs can be increased by using different organic or inorganic amalgamates (Schröfel et al., 2014).

The antimicrobial action of MtNPs designed by eco-friendly approach increases by hundreds of times if used in combination with existing antibiotics. A synergistic effect of AgNPs synthesized from black bread mold *Rhizopus stolonifer* with different antibiotics include carbenicillin, nitrofurantoin, and ciprofloxacin reported by Banu and colleagues. The highest inhibitory activity up to 50% was observed by the combination of AgNPs and nitrofurantoin antibiotics (Banu, Rathod, & Ranganath, 2011). Another study conducted by Singh and colleagues examined the enhanced activity of AgNPs synthesized from various microbes in combination with a variety of antibiotics (Singh et al., 2015). Most of the time, this type of combinatorial study was conducted by using biogenic silver particles. However, a few studies have reported that a combination of TiO_2 NPs synthesized from *Planomicrobium* sp. along with different antibiotics significantly increased the antibacterial activity against *Bacillus subtilis* and *Klebsiella planticola* (Nadeem et al., 2018). Like this, iron NPs produced by a fungus *F. oxysporum* revealed antibacterial properties against *B. subtilis*, *E. coli*, and *Staphylococcus* sp. The small size of iron NPs hinders oxygen transportation and thus attacks the respiratory system of pathogenic bacteria (Abdeen & Praseetha, 2013).

The rapid evolutionary changes in the bacterial genome that have rendered antibiotic resistance strongly appeal to test microbially synthesized, safe, less toxic, and eco-friendly NPs for the development of new rationales. The use of MtNPs in biomedical science gives a new approach to explore new antibiotics in replacement of existing ineffective antibiotics available in the market.

5 Microbially synthesized nanoantifungals

The fungi are the most important human pathogens and continue to be the focus of extensive research. The high infection rate of fungal pathogens is due to inappropriate diagnostic ways, treatments, and dosage of drugs (Morrell, Fraser, & Kollef, 2005). The overuse and side effects of available antifungal drugs have evolved resistant strains of fungi and contributed to the development of serious health problems, respectively. The shortlist of available antifungal drugs and resistant strains urged the need to develop new and novel antifungals with unusual mechanisms (Turecka, Chylewska, Kawiak, & Waleron, 2018).

Fungi can synthesize NPs of definite size, shape, and geometries. To perform this task, fungi have many intra/extracellular enzymes. They have larger biomass as compared to bacteria and produce a high yield/number of NPs. This is the reason fungi are named as "nano-factories." A variety of fungi species *Verticillium*, *Collitotrichum*, *Fusarium*, *Trichothecium*, *Aspergillus*, *Alternata*, and *Trichoderma* are used to biologically manufacture NPs of different sizes and shapes (Fariq et al., 2017). Fungal synthesized NPs exhibited antifungal properties against different pathogenic fungi species including *Candida albicans* (human pathogen), *Aspergillus*, *Fusarium* (plant pathogen), and *Malassezia* sp. (dandruff-causing). A few examples of antifungal activities of Ms-NPs are listed in Table 2. The biogenic AgNPs attach to the polymers of the fungal cell walls, particularly chitin and beta-glucan, penetrate deep into the cell, attach to phosphorous-containing DNA and RNA, thus resisting replication machinery of fungal cells (Reidy, Haase, Luch, Dawson, & Lynch, 2013). Moreover, MsNPs of cerium oxide kills fungi by switching increased production of free radicals and reactive oxygen species, which alter the structure and physiology of fungi cell, causing cell death (Fariq et al., 2017). The efficiency

TABLE 2 Antifungal activities of microbially synthesized nanoparticles.

Microbes used to synthesize NPs	Types of nanoparticles	Pathogens tested	References
Enterococcus sp.	Cadmium	*Aspergillus niger*, *Aspergillus flavus*	Rajeshkumar et al. (2014)
Nocardiopsis sp.	Silver	*Aspergillus niger*, *A. brasiliensis*, *A. fumigates*, *Candida albicans*	Manivasagan, Venkatesan, Senthilkumar, Sivakumar, and Kim (2013)
Brevundimonas sp.	Cadmium	*Candida albicans*, *C. tropicalis*, *C. kruzei*	Rajamanickam et al. (2013)
Alternaria alternata	Silver	*Phoma glomerata*, *Phoma herbarum*, *Fusarium semitectum*, *Trichoderma* sp., *Candida albicans*	Gajbhiye, Kesharwani, Ingle, Gade, and Rai (2009)
Trichoderma atroviride	Selenium	*Pyricularia grisea* *Colletotrichum capsici* and *Alternaria solani*	Joshi, De Britto, Jogaiah, and Ito (2019)

of existing ineffective antifungal drugs increases by using drugs in amalgamation with AgNPs. The mycogenically synthesized AgNPs have increased the antifungal potential of fluconazole drug in combination therapy against resistant *Candida albicans* (Khandel & Shahi, 2018; Turecka et al., 2018). The efficiency of AgNPs is because of strong interaction with thiol groups of fungal enzymes. All these examples suggest that AgNPs are very toxic to fungal species and can be used for the designing of antifungal drugs.

6 Microbially synthesized nanopesticides

Large quantities of chemical pesticides are used extensively to boost agriculture yields. Only 0.1% of the total amount of pesticide applied in the field reached the targets, which hereby require an increased concentration of chemical pesticide enough to control pests. The frequent use of chemical pesticides has dangerous effects on human and environmental health (Camara et al., 2019). Precision farming is necessary to create novel formulations of effective pesticides, but with the least environmental risks. This could be attained by the use of nanotechnology in the development of eco-friendly pesticides.

A variety of reported plant pathogens can be controlled by using Ms-NPs-based bioformulations. These bioformulations include a combination of fungicides/pesticides along with NPs (Singh & Arora, 2016). Significant growth inhibition of a dangerous plant pathogen (*Colletotrichum gloesporioides*) was found by using microbially synthesized AgNPs (Aguilar-Méndez, San Martín-Martínez, Ortega-Arroyo, Cobián-Portillo, & Sánchez-Espíndola, 2011). Moreover, it was observed that in the presence of AgNPs *Colletotrichum gloesporioides* showed much-delayed growth. This fungal pathogen is a serious threat to a wide range of fruit crops. A very significant antifungal activity of AgNPs was observed on another plant pathogen i.e., *Sphaerotheca pannosa*. This pathogen causes a very serious disease called powdery mildew. A spray of double-capsulated AgNPs has killed 95% of *S. pannosa* from the affected area in 2 days. Moreover, this treatment inhibited the reoccurrence of fungal attacks (Kim et al., 2008). Hence, it is concluded that MsNPs are an emerging alternative to toxic chemically synthesized pesticides/fungicides.

7 Factors affecting antimicrobial activities of microbially synthesized nanoparticles

This antimicrobial potential is ascribed to the distinguishing superficial surface chemistry, nanosize, polyvalent, and photothermic nature (Boisselier & Astruc, 2009). The MtNPs react with sulfur- and phosphorous-containing compounds, attach to thiol groups of respiratory enzymes, accelerate the production of unstable atoms of hydrogen and oxygen generally called reactive oxygen species that brought a cell to death (Cui et al., 2012). The antimicrobial potential is highly dependent upon the size of the NPs, smaller the size increased antimicrobial activity, and vice versa. Small-sized AuNPs have confined trans-membrane proton pump in the fungal cell more effectively as compared to large-sized AuNPs (Ahmad et al., 2013).

The antimicrobial activity of eco-friendly NPs also depends upon the cell wall composition of the target microbe. AuNPs exhibited profound activity against Gram-negative bacteria as compared to Gram-positive bacteria because the cell wall of Gram-positive bacteria has a dense layer of peptidoglycans (Kaviya, Santhanalakshmi, Viswanathan, Muthumary, & Srinivasan, 2011).

Moreover, surface reforms and alterations, the concentration of NPs used, and purification methods also affect the antimicrobial capacities of MtNPs. Their efficacy can be increased by coating with existing antibiotics (Payne et al., 2016; Zhang, Shareena Dasari, Deng, & Yu, 2015).

The bactericidal property of cerium oxide NPs is credited to electrostatic interactions. The robust electrostatic potential activates the interaction of NPs with the thiol groups of membrane-bound proteins, deteriorating the three-dimensional structure and stability of proteins. Loss of biological activity of proteins leads to membrane impermeability and results in microbial death (Nadeem et al., 2020).

8 Microbially synthesized anticancer NPs

Cancer, a group of ailments is one of the leading causes of death all over the globe. Initial diagnosis and localized drug delivery to the target proliferating organ has not been achieved yet. Moreover, existing treatment methods have many side effects (Teixeira, Ten Dijke, & Zhu, 2020). These limitations result in an increasing number of disease incidences and it is important to find out effective alternative ways to make out the previous issues. Nanomedicine has efficaciously deployed for the detection of proliferation, targeted drug delivery, and cancer treatments (Sutradhar & Amin, 2014). Biogenic NPs bearing intrinsic capabilities have auspicious potential to develop molecular interactions. Due to their very small size, these NPs can enter into the target cell without harming normal cells (Fariq et al., 2017). Microbiologically synthesized NPs of platinum (Buzea, Pacheco, & Robbie, 2007), silver, gold, and selenium have cytotoxic activities against epidermoid carcinoma cell lines, mammary epithelial cell lines, human breast cancer cell lines, hepatic carcinoma cell lines, and immortal human liver cell lines. AgNPs induced cellular apoptosis (Ortega et al., 2015) while AuNPs triggered mitochondrial apoptosis, penetrated the nuclei, and caused DNA impairment. AuNPs inhibits cell proliferation by arresting cytokinesis (El-Batal, Mona, & Al-Tamie, 2015; Hamed & Abdelftah, 2019). Moreover, selenium NPs involved the recruitment of chromatin-bound copper followed by accumulation of reactive oxygen species leading to cell death. Details of their synthesis and cytotoxic activity against different cell lines are listed in Table 3.

TABLE 3 Anticancer activities of microbially synthesized nanoparticles.

No.	Microbes used to synthesize NPs	Type of NPs	Tested cell lines	Reference
1	*S. boulardii*	Platinum	Epidermoid carcinoma cell lines, Mammary epithelial cell lines	Buzea et al. (2007)
2	*C. laurentii*	Silver	Breast cancer cell lines	Ortega et al. (2015)
3	*S. cyaneus*	Gold	Hepatic carcinoma cell lines, Breast cancer cell lines, Colon carcinoma cell lines	El-Batal et al. (2015) and Hamed and Abdelftah (2019)
4	*S. bikiniensis*	Selenium	Human liver cell lines Breast cancer cell lines	Fariq et al. (2017)
5	*Pleurotus ostreatus*	Silver	Human breast cancer cell lines	Yehia and Al-Sheikh (2014)

9 Microbially synthesized antimalarial nanoparticles

Mycogenic-synthesized gold and silver NPs have been reported to hold the potential to control malarial infections. Multiple studies have shown that these biogenic MtNPs successfully killed *Aedes aegepti*, *Anopheles stephensi*, and vectors of dengue (Banu & Balasubramanian, 2014). These NPs can be used for the control of malaria and dengue in the future.

10 Microbially synthesized nanobiosensors

The diagnosis of the disease depends on the selection of an accurate method. Healthcare professionals and scientists can work better for the progress of therapeutics, respectively. The diagnosis of disease and disease-causing agents helps to understand the condition of the patient and the progression of a particular disease in the human body. Several methods of diagnosis are available in the market but due to labor-intensive, time-consuming, and volume of samples result in a lack of attraction for routine diagnostics (Merkoci & Chamorro, 2016). The involvement of nanotechnology has revolutionized a vast number of fields of basic and applied sciences (Chandra, Mahato, & Maurya, 2018). Quick analysis, robust features, onset detection, and a small amount of sample requirement of these nanodiagnostics generally make them better in contrast to laboratory-based techniques (Chandra et al., 2018; Mahato, Kumar, Kumar Maurya, & Chandra, 2018). The main components of a biosensor include a detector, transducer, biorecognition element, and amplifier (Mahato, Prasad, Maurya, & Chandra, 2016). It is demonstrated that diagnostic devices are composed of silicon nanowires and carbon nanotubes and are capable to detect femtomolar (FM) concentrations of target analytes (Guy et al., 2012). Nanoscale biosensors can offer early recognition and diagnosis of disease and allow real-time monitoring of patient's conditions. But the use of microbially synthesized MtNPs in diagnostic devices is a very early stage. AuNPs synthesized from a fungus *Candida albicans* were used to identify liver cancer cells. These AuNPs bind with liver cancer cell-specific antibodies present particularly on the cancer cell lines only and, in this way, help to distinguish between normal and cancer cells (Chauhan et al., 2011). The potential of biosynthesized NPs for the diagnosis of different ailments is an attractive field of research and their use will be beneficial for robust and accurate diagnosis at early stages.

11 Microbially synthesized nanoparticles in drug delivery

The drug delivery system is responsible for the efficient transportation of NPs through blood capillaries and the lymphatic system. It increases the blood circulation time and greater accumulation at the target tissue. Rapid development in the NPs for drug delivery is due to increased biocompatibility and less inflammatory and immune responses. Physiochemical effects of nanomedicines are due to their formulation or composition, which affects the efficacy and toxicity. However, the in vivo distribution of nanomedicine is dependent on its absorption capacity with the biomolecules. Recently, macrobiotic nanomedicine was found much more efficient in the delivery system. Recent research studies evidenced that microbially synthesized NPs are good candidates if used as vehicles for drug delivery at target sites, especially in the treatment of different cancers. For instance, zinc oxide NPs synthesized by *Rhodococcus pyridinivorans* laden with anthraquinone showed profound cytotoxic activity against colon cancer cells. The cytotoxic activity increased by increasing the concentration of anthraquinone (Kundu, Hazra, Chatterjee, Chaudhari, & Mishra, 2014). The microbially synthesized polar NPs interact electrostatically with the membranes and enable the entry of the drug into a cell at ease. In other studies, AuNPs synthesized from two different strains of fungi, *Helminthosporum solani* and

Trichoderma reesei, conjugated with anticancer drug doxorubicin, exhibited significant cytotoxic potential against human kidney cell lines and hepatic cancer cell lines, respectively (Kumar, Peter, & Nadeau, 2008; Syed, Raja, Kundu, Gambhir, & Ahmad, 2013). The microbially synthesized AuNPs facilitates the absorption of drugs into a cell. Similarly, the efficiency and toxicity of taxol drug against cancer cell lines were increased by using the drug in conjugation with gadolinium oxide NPs biosynthesized by *Humicola* sp. of fungi (Khan, Gambhir, & Ahmad, 2014). These examples emphasize the credibility of biosynthesized NPs in the arena of site-specific drug delivery.

NPs synthesized from bacteria, yeast, and fungi cultures are safe, environment friendly, and have many applications in healthcare, medicine, agriculture, environment, drug delivery, cancer treatments, and diagnosis. Future challenges include commercial-scale production and purification of MtNPs of desirable size, shape, and morphology.

12 Future perspective

According to the reported data, microbial synthesis of NPs has shown remarkable advancement in the last few years. These NPs have been tested for a number of applications. However, many efforts needed to optimize number of parameters, which include protocol efficiency, size, and mono-dispersity of NPs. Microbial synthesis usually demands a lot of time that can last from several hours to days. It can be made more attractive especially for industrial applications if scientists are able to reduce the synthesis period to compete with physical and chemical methods. Stability of these NPs is another issue as number of studies showed the decomposition of NPs after a certain period of time. Optimization in key parameters such as growth media, pH, temperature, and growth phase of microbes can address these issues. Most importantly, understanding the synthesis mechanism at the cellular and molecular levels can revolutionize microbial nanotechnology. A comprehensive understanding of the synthesis process along with the methods of isolation and purification can ensure an efficient synthesis of NPs within a short reaction time.

References

Abdeen, S., & Praseetha, P. (2013). Diagnostics and treatment of metastatic cancers with magnetic nanoparticles. *Journal of Nanomedicine & Biotherapeutic Discovery*, *3*(2), 115.

Agnihotri, M., Joshi, S., Kumar, A. R., Zinjarde, S., & Kulkarni, S. (2009). Biosynthesis of gold nanoparticles by the tropical marine yeast *Yarrowia lipolytica* NCIM 3589. *Materials Letters*, *63*(15), 1231–1234. https://doi.org/10.1016/j.matlet.2009.02.042.

Aguilar-Méndez, M. A., San Martín-Martínez, E., Ortega-Arroyo, L., Cobián-Portillo, G., & Sánchez-Espíndola, E. (2011). Synthesis and characterization of silver nanoparticles: Effect on phytopathogen *Colletotrichum gloesporioides*. *Journal of Nanoparticle Research*, *13*(6), 2525–2532.

Ahmad, A., Mukherjee, P., Mandal, D., Senapati, S., Khan, M. I., Kumar, R., & Sastry, M. (2002a). Enzyme mediated extracellular synthesis of CdS nanoparticles by the fungus, *Fusarium oxysporum. Journal of the American Chemical Society*, *124*(41), 12108–12109. https://doi.org/10.1021/ja027296o.

Ahmad, A., Mukherjee, P., Mandal, D., Senapati, S., Khan, M. I., Kumar, R., & Sastry, M. (2002b). Enzyme mediated extracellular synthesis of CdS nanoparticles by the fungus, *Fusarium oxysporum. Journal of the American Chemical Society*, *124*(41), 12108–12109.

Ahmad, A., Senapati, S., Khan, M. I., Kumar, R., Ramani, R., Srinivas, V., & Sastry, M. (2003). Intracellular synthesis of gold nanoparticles by a novel alkalotolerant actinomycete, *Rhodococcus* species. *Nanotechnology*, *14*(7), 824–828. https://doi.org/10.1088/0957-4484/14/7/323.

Ahmad, A., Senapati, S., Khan, M. I., Kumar, R., & Sastry, M. (2003). Extracellular biosynthesis of monodisperse gold nanoparticles by a novel Extremophilic Actinomycete, *Thermomonospora* sp. *Langmuir*, *19*(8), 3550–3553. https://doi.org/10.1021/la026772l.

Ahmad, T., Wani, I. A., Lone, I. H., Ganguly, A., Manzoor, N., Ahmad, A., ... Al-Shihri, A. S. (2013). Antifungal activity of gold nanoparticles prepared by solvothermal method. *Materials Research Bulletin*, *48*(1), 12–20.

Ahmed, S., & Ikram, S. (2016). Biosynthesis of gold nanoparticles: A green approach. *Journal of Photochemistry and Photobiology B: Biology*, *161*, 141–153.

Akinsiku, A. A., Dare, E. O., Ajanaku, K. O., Ajani, O. O., Olugbuyiro, J. A. O., Siyanbola, T. O., ... Emetere, M. E. (2018). Modeling and synthesis of Ag and Ag/Ni allied bimetallic nanoparticles by green method: Optical and biological properties. *International Journal of Biomaterials*, *1*, 9658080.

Al-Dhabi, N. A., Mohammed Ghilan, A. K., & Arasu, M. V. (2018). Characterization of silver nanomaterials derived from marine *Streptomyces* sp. Al-Dhabi-87 and its in vitro application against multidrug resistant and extended-spectrum beta-lactamase clinical pathogens. *Nanomaterials*, *8*(5), 279–291.

Alghuthaymi, M. A., Almoammar, H., Rai, M., Said-Galiev, E., & Abd-Elsalam, K. A. (2015). Myconanoparticles: Synthesis and their role in phytopathogens management. *Biotechnology and Biotechnological Equipment*, *29*(2), 221–236.

Arakaki, A., Masuda, F., Amemiya, Y., Tanaka, T., & Matsunaga, T. (2010). Control of the morphology and size of magnetite particles with peptides mimicking the Mms6 protein from magnetotactic bacteria. *Journal of Colloid and Interface Science*, *343*(1), 65–70.

Arakaki, A., Nakazawa, H., Nemoto, M., Mori, T., & Matsunaga, T. (2008). Formation of magnetite by bacteria and its application. *Journal of the Royal Society, Interface/the Royal Society*, *5*, 977–999. https://doi.org/10.1098/rsif.2008.0170.

Asghar, M., Habib, S., Zaman, W., Hussain, S., Ali, H., & Saqib, S. (2020). Synthesis and characterization of microbial mediated cadmium oxide nanoparticles. *Microscopy Research and Technique*, *83*(12), 1574–1584. https://doi.org/10.1002/jemt.23553.

Bai, H.-J., & Zhang, Z.-M. (2009). Microbial synthesis of semiconductor lead sulfide nanoparticles using immobilized *Rhodobacter sphaeroides*. *Materials Letters*, *63*, 764–766. https://doi.org/10.1016/j.matlet.2008.12.050.

Bakir, E. M., Younis, N. S., Mohamed, M. E., & El Semary, N. A. (2018). Cyanobacteria as nanogold factories: Chemical and anti-myocardial infarction properties of gold nanoparticles synthesized by *Lyngbya majuscula*. *Marine Drugs*, *16*(6), 217–237.

Bankar, A. V., Kumar, A. R., & Zinjarde, S. S. (2009). Removal of chromium (VI) ions from aqueous solution by adsorption onto two marine isolates of *Yarrowia lipolytica*. *Journal of Hazardous Materials*, *170*(1), 487–494. https://doi.org/10.1016/j.jhazmat.2009.04.070.

Bansal, V., Poddar, P., Ahmad, A., & Sastry, M. (2006). Room-temperature biosynthesis of ferroelectric barium titanate nanoparticles. *Journal of the American Chemical Society*, *128*(36), 11958–11963.

Bansal, V., Rautaray, D., Ahmad, A., & Sastry, M. (2004). Biosynthesis of zirconia nanoparticles using the fungus *Fusarium oxysporum*. *Journal of Materials Chemistry*, *14*, 3303–3305. https://doi.org/10.1039/b407904c.

Bansal, V., Rautaray, D., Bharde, A., Ahire, K., Sanyal, A., Ahmad, A., & Sastry, M. (2005). Fungus-mediated biosynthesis of silica and titania particles. *Journal of Materials Chemistry*, *15*, 2583–2589. https://doi.org/10.1039/b503008k.

Banu, A. N., & Balasubramanian, C. (2014). Myco-synthesis of silver nanoparticles using *Beauveria bassiana* against dengue vector, *Aedes aegypti* (Diptera: Culicidae). *Parasitology Research*, *113*(8), 2869–2877.

Banu, A., Rathod, V., & Ranganath, E. (2011). Silver nanoparticle production by *Rhizopus stolonifer* and its antibacterial activity against extended spectrum β-lactamase producing (ESBL) strains of Enterobacteriaceae. *Materials Research Bulletin*, *46*(9), 1417–1423. https://doi.org/10.1016/j.materresbull.2011.05.008.

Barapatre, A., Aadil, K. R., & Jha, H. (2016). Synergistic antibacterial and antibiofilm activity of silver nanoparticles biosynthesized by lignin-degrading fungus. *Bioresources and Bioprocessing*, *3*(1), 8.

Benzerara, K., Miot, J., Morin, G., Ona-Nguema, G., Skouri-Panet, F., & Férard, C. (2011). Significance, mechanisms and environmental implications of microbial biomineralization. *Comptes Rendus Geoscience*, *343*(2), 160–167. https://doi.org/10.1016/j.crte.2010.09.002.

Bhainsa, K. C., & D'Souza, S. F. (2006). Extracellular biosynthesis of silver nanoparticles using the fungus *Aspergillus fumigatus*. *Colloids and Surfaces. B, Biointerfaces*, *47*(2), 160–164.

Blakemore, R. (1975). Magnetotactic bacteria. *Science*, *190*(4212), 377–379.

Boisselier, E., & Astruc, D. (2009). Gold nanoparticles in nanomedicine: Preparations, imaging, diagnostics, therapies and toxicity. *Chemical Society Reviews*, *38*(6), 1759–1782.

Breierová, E., Vajczikova, I., Sasinková, V., Stratilová, E., Fisera, I., Gregor, T., & Sajbidor, J. (2002). Biosorption of cadmium ions by different yeast species. *Zeitschrift fur Naturforschung C: Journal of Biosciences*, *57*(7–8), 634–639. https://doi.org/10.1515/znc-2002-7-815.

Buzea, C., Pacheco, I. I., & Robbie, K. (2007). Nanomaterials and nanoparticles: Sources and toxicity. *Biointerphases*, *2*(4), MR17–MR71.

Camara, M. C., Campos, E. V. R., Monteiro, R. A., do Espirito Santo Pereira, A., de Freitas Proença, P. L., & Fraceto, L. F. (2019). Development of stimuli-responsive nano-based pesticides: Emerging opportunities for agriculture. *Journal of Nanbiotechnology*, *17*(1), 019–0533.

Castro-Longoria, E., Vilchis-Nestor, A. R., & Avalos-Borja, M. (2011). Biosynthesis of silver, gold and bimetallic nanoparticles using the filamentous fungus *Neurospora crassa*. *Colloids and Surfaces. B, Biointerfaces*, *83*(1), 42–48.

Chandra, P., Mahato, K., & Maurya, P. K. (2018). Fundamentals and commercial aspects of nanobiosensors in point-of-care clinical diagnostics. *3 Biotech*, *8*, 1–14.

Chauhan, A., Zubair, S., Tufail, S., Sherwani, A., Sajid, M., Raman, S. C., … Owais, M. (2011). Fungus-mediated biological synthesis of gold nanoparticles: Potential in detection of liver cancer. *International Journal of Nanomedicine*, *6*, 2305.

Cruz, L. Y., Wang, D., & Liu, J. (2019). Biosynthesis of selenium nanoparticles, characterization and X-ray induced radiotherapy for the treatment of lung cancer with interstitial lung disease. *Journal of Photochemistry and Photobiology. B*, *191*, 123–127. https://doi.org/10.1016/j.jphotobiol.2018.12.008.

Cui, Y., Zhao, Y., Tian, Y., Zhang, W., Lü, X., & Jiang, X. (2012). The molecular mechanism of action of bactericidal gold nanoparticles on Escherichia coli. *Biomaterials*, *33*(7), 2327–2333.

Daniel, M.-C., & Astruc, D. (2004). Gold nanoparticles: Assembly, supramolecular chemistry, quantum-size-related properties, and applications toward biology, catalysis, and nanotechnology. *Chemical Reviews*, *104*(1), 293–346.

Dragone, R., Grasso, G., Muccini, M., & Toffanin, S. (2017). Portable bio/chemosensoristic devices: Innovative systems for environmental health and food safety diagnostics. *Frontiers in Public Health*, *5*, 80.

Durán, N., Marcato, P. D., Alves, O. L., Souza, G. I., & Esposito, E. (2005a). Mechanistic aspects of biosynthesis of silver nanoparticles by several *Fusarium oxysporum* strains. *Journal of Nanbiotechnology*, *3*(8), 1477–3155.

Durán, N., Marcato, P. D., Alves, O. L., Souza, G. I. H. D., & Esposito, E. (2005b). Mechanistic aspects of biosynthesis of silver nanoparticles by several *Fusarium oxysporum* strains. *Journal of Nanbiotechnology*, *3*, 8. https://doi.org/10.1186/1477-3155-3-8.

El-Batal, A., Mona, S., & Al-Tamie, M. (2015). Biosynthesis of gold nanoparticles using marine *Streptomyces cyaneus* and their antimicrobial, antioxidant and antitumor (in vitro) activities. *Journal of Chemical and Pharmaceutical Research*, *7*(7), 1020–1036.

Fariq, A., Khan, T., & Yasmin, A. (2017). Microbial synthesis of nanoparticles and their potential applications in biomedicine. *Journal of Applied Biomedicine*, *15*(4), 241–248.

Gajbhiye, M., Kesharwani, J., Ingle, A., Gade, A., & Rai, M. (2009). Fungus-mediated synthesis of silver nanoparticles and their activity against pathogenic fungi in combination with fluconazole. *Nanomedicine: Nanotechnology, Biology and Medicine*, *5*(4), 382–386. https://doi.org/10.1016/j.nano.2009.06.005.

Gericke, M., & Pinches, A. (2006). Microbial production of gold nanoparticles. *Gold Bulletin*, *39*(1), 22–28. https://doi.org/10.1007/bf03215529.

Gobbo, O. L., Sjaastad, K., Radomski, M. W., Volkov, Y., & Prina-Mello, A. (2015). Magnetic nanoparticles in cancer theranostics. *Theranostics*, *5*(11), 1249–1263.

Grasso, G., Zane, D., & Dragone, R. (2020). Microbial nanotechnology: Challenges and prospects for green biocatalytic synthesis of nanoscale materials for sensoristic and biomedical applications. *Nanomaterials*, *10*(1), 11.

Gu, H., Chen, X., Chen, F., Zhou, X., & Parsaee, Z. (2018). Ultrasound-assisted biosynthesis of CuO-NPs using brown alga *Cystoseira trinodis*: Characterization, photocatalytic AOP, DPPH scavenging and antibacterial investigations. *Ultrasonics Sonochemistry*, *41*, 109–119.

Gupta, K., & Chundawat, T. S. (2019). Bio-inspired synthesis of platinum nanoparticles from fungus *Fusarium oxysporum*: Its characteristics, potential antimicrobial, antioxidant and photocatalytic activities. *Materials Research Express*, *6*. https://doi.org/10.1088/2053-1591/ab4219.

Gurunathan, S., Kalishwaralal, K., Vaidyanathan, R., Venkataraman, D., Pandian, S. R., Muniyandi, J., … Eom, S. H. (2009). Biosynthesis, purification and characterization of silver nanoparticles using *Escherichia coli*. *Colloids and Surfaces. B, Biointerfaces*, *74*(1), 328–335.

Guy, O. J., Burwell, G., Tehrani, Z., Castaing, A., Walker, K. A., & Doak, S. (2012). Graphene nanobiosensors for detection of cancer risk. In *Paper presented at the Materials Science Forum*.

Hamed, M. M., & Abdelftah, L. S. (2019). Biosynthesis of gold nanoparticles using marine *Streptomyces griseus* isolate (M8) and evaluating its antimicrobial and anticancer activity. *Egyptian Journal of Aquatic Biology and Fisheries*, *23*(1), 173–184.

Hassan, S. E., Fouda, A., Radwan, A. A., Salem, S. S., Barghoth, M. G., Awad, M. A., … El-Gamal, M. S. (2019). Endophytic actinomycetes *Streptomyces* spp mediated biosynthesis of copper oxide nanoparticles as a promising tool for biotechnological applications. *Journal of Biological Inorganic Chemistry*, *24*(3), 377–393. https://doi.org/10.1007/s00775-019-01654-5.

Hayat, M. A. (1989). *Colloidal gold: Principles, methods, and applications*. San Diego: Academic Press.

He, S., Guo, Z., Zhang, Y., Zhang, S., & Ning, G. (2007). Biosynthesis of gold nanoparticles using the bacteria *Rhodopseudomonas capsulata*. *Materials Letters*, *61*, 3984–3987. https://doi.org/10.1016/j.matlet.2007.01.018.

Husseiny, M. I., El-Aziz, M. A., Badr, Y., & Mahmoud, M. A. (2007). Biosynthesis of gold nanoparticles using *Pseudomonas aeruginosa*. *Spectrochimica Acta. Part A, Molecular and Biomolecular Spectroscopy*, *67*(3–4), 1003–1006.

Inshakova, E., & Inshakov, O. (2017). World market for nanomaterials: Structure and trends. In *Paper presented at the MATEC web of conferences*.

Jeevanandam, J., Chan, Y. S., & Danquah, M. K. (2016). Biosynthesis of metal and metal oxide nanoparticles. *ChemBioEng Reviews*, *3*(2), 55–67.

Joerger, R., Klaus, T., & Granqvist, C. G. (2000). Biologically produced silver-carbon composite materials for optically functional thin-film coatings. *Advanced Materials*, *12*(6), 407–409.

Joshi, S. M., De Britto, S., Jogaiah, S., & Ito, S.-I. (2019). Mycogenic selenium nanoparticles as potential new generation broad spectrum antifungal molecules. *Biomolecules*, *9*(9), 419.

Kalishwaralal, K., Venkataraman, D., Ramkumarpandian, S., Nellaiah, H., & Sangiliyandi, G. (2008). Extracellular biosynthesis of silver nanoparticles by the culture supernatant of *Bacillus licheniformis*. *Materials Letters*, *62*, 4411–4413. https://doi.org/10.1016/j.matlet.2008.06.051.

Kandasamy, G., & Maity, D. (2015). Recent advances in superparamagnetic iron oxide nanoparticles (SPIONs) for in vitro and in vivo cancer nanotheranostics. *International Journal of Pharmaceutics*, *496*(2), 191–218.

Kathiresan, K., Manivannan, S., Nabeel, M. A., & Dhivya, B. (2009). Studies on silver nanoparticles synthesized by a marine fungus, *Penicillium fellutanum* isolated from coastal mangrove sediment. *Colloids and Surfaces. B, Biointerfaces*, *71*(1), 133–137.

Kaviya, S., Santhanalakshmi, J., Viswanathan, B., Muthumary, J., & Srinivasan, K. (2011). Biosynthesis of silver nanoparticles using *Citrus sinensis* peel extract and its antibacterial activity. *Spectrochimica Acta Part A: Molecular and Biomolecular Spectroscopy*, *79*(3), 594–598.

Khan, S. A., Gambhir, S., & Ahmad, A. (2014). Extracellular biosynthesis of gadolinium oxide (Gd2O3) nanoparticles, their biodistribution and bioconjugation with the chemically modified anticancer drug taxol. *Beilstein Journal of Nanotechnology*, *5*(1), 249–257.

Khandel, P., & Shahi, S. K. (2018). Mycogenic nanoparticles and their bio-prospective applications: Current status and future challenges. *Journal of Nanostructure in Chemistry*, *8*(4), 369–391.

Kiessling, F., Mertens, M. E., Grimm, J., & Lammers, T. (2014). Nanoparticles for imaging: Top or flop? *Radiology*, *273*(1), 10–28.

Kim, K.-J., Sung, W. S., Moon, S.-K., Choi, J.-S., Kim, J. G., & Lee, D. G. (2008). Antifungal effect of silver nanoparticles on dermatophytes. *Journal of Microbiology and Biotechnology*, *18*(8), 1482–1484.

Klaus, T., Joerger, R., Olsson, E., & Granqvist, C. G. (1999). Silver-based crystalline nanoparticles, microbially fabricated. *Proceedings of the National Academy of Sciences of the United States of America*, *96*(24), 13611–13614.

Koopi, H., & Buazar, F. (2018). A novel one-pot biosynthesis of pure alpha aluminum oxide nanoparticles using the macroalgae *Sargassum ilicifolium*: A green marine approach. *Ceramics International*, *44*(8), 8940–8945. https://doi.org/10.1016/j.ceramint.2018.02.091.

Kowshik, M., Deshmukh, N., Vogel, W., Urban, J., Kulkarni, S. K., & Paknikar, K. M. (2002). Microbial synthesis of semiconductor CdS nanoparticles, their characterization, and their use in the fabrication of an ideal diode. *Biotechnology and Bioengineering*, *78*(5), 583–588.

Kumar, D., Karthik, L., Kumar, G., & Roa, K. (2011). Biosynthesis of silver anoparticles from marine yeast and their antimicrobial activity against multidrug resistant pathogens. *Pharmacology Online*, *3*, 1100–1111.

Kumar, S. A., Peter, Y.-A., & Nadeau, J. L. (2008). Facile biosynthesis, separation and conjugation of gold nanoparticles to doxorubicin. *Nanotechnology*, *19*(49), 495101.

Kundu, D., Hazra, C., Chatterjee, A., Chaudhari, A., & Mishra, S. (2014). Extracellular biosynthesis of zinc oxide nanoparticles using *Rhodococcus pyridinivorans* NT2: Multifunctional textile finishing, biosafety evaluation and in vitro drug delivery in colon carcinoma. *Journal of Photochemistry and Photobiology B: Biology*, *140*, 194–204.

Kuroda, M., Suda, S., Sato, M., Ayano, H., Ohishi, Y., Nishikawa, H., … Ike, M. (2019). Biosynthesis of bismuth selenide nanoparticles using chalcogen-metabolizing bacteria. *Applied Microbiology and Biotechnology*, *103*(21 − 22), 8853–8861. https://doi.org/10.1007/s00253-019-10160-2.

Lengke, M. F., Fleet, M. E., & Southam, G. (2007). Biosynthesis of silver nanoparticles by filamentous cyanobacteria from a silver(I) nitrate complex. *Langmuir*, *23*(5), 2694–2699.

Li, X., Xu, H., Chen, Z.-S., & Chen, G. (2011a). Biosynthesis of nanoparticles by microorganisms and their applications. *Journal of Nanomaterials*, *2011*, 1–16.

Li, X., Xu, H., Chen, Z.-S., & Chen, G. (2011b). Biosynthesis of nanoparticles by microorganisms and their applications. *Journal of Nanomaterials*, *2011*, 270974. https://doi.org/10.1155/2011/270974.

Liu, J., Qiao, S. Z., Hu, Q. H., & Lu, G. Q. (2011). Magnetic nanocomposites with mesoporous structures: Synthesis and applications. *Small*, *7*(4), 425–443.

Mahato, K., Kumar, A., Kumar Maurya, P., & Chandra, P. (2018). Shifting paradigm of cancer diagnoses in clinically relevant samples based on miniaturized electrochemical nanobiosensors and microfluidic devices. *Biosensors & Bioelectronics*, *100*, 411–428.

Mahato, K., Prasad, A., Maurya, P., & Chandra, P. (2016). Nanobiosensors: Next generation point-of-care biomedical devices for personalized diagnosis. *Journal of Analytical & Bioanalytical Techniques*, *7*, 421–431.

Mandal, D., Bolander, M. E., Mukhopadhyay, D., Sarkar, G., & Mukherjee, P. (2006). The use of microorganisms for the formation of metal nanoparticles and their application. *Applied Microbiology and Biotechnology*, *69*(5), 485–492. https://doi.org/10.1007/s00253-005-0179-3.

Manivasagan, P., Venkatesan, J., Senthilkumar, K., Sivakumar, K., & Kim, S.-K. (2013). Biosynthesis, antimicrobial and cytotoxic effect of silver nanoparticles using a novel *Nocardiopsis* sp. MBRC-1. *BioMed Research International*, *2013*, 1–9.

Merkoci, A., & Chamorro-Garcia, A. (2016). Nanobiosensors in diagnostics. *Nanobiomedicine*, *3*, 26.

Mikhailov, O. V., & Mikhailova, E. O. (2019). Elemental silver nanoparticles: Biosynthesis and bio applications. *Materials (Basel, Switzerland)*, *12*(19), 3177. https://doi.org/10.3390/ma12193177.

Mishra, S., Dixit, S., & Soni, S. (2015). Methods of nanoparticle biosynthesis for medical and commercial applications. In *Bio-nanoparticles: Biosynthesis and sustainable biotechnological implications* (pp. 141–154). Wiley.

Mohd Yusof, H., Abdul Rahman, N., Mohamad, R., Zaidan, U. H., & Samsudin, A. A. (2020). Biosynthesis of zinc oxide nanoparticles by cell-biomass and supernatant of *Lactobacillus plantarum* TA4 and its antibacterial and biocompatibility properties. *Scientific Reports*, *10*(1), 19996. https://doi.org/10.1038/s41598-020-76402-w.

Molnár, Z., Bódai, V., Szakacs, G., Erdélyi, B., Fogarassy, Z., Sáfrán, G., … Lagzi, I. (2018). Green synthesis of gold nanoparticles by thermophilic filamentous fungi. *Scientific Reports*, *8*(1), 018–22112.

Morrell, M., Fraser, V. J., & Kollef, M. H. (2005). Delaying the empiric treatment of *Candida* bloodstream infection until positive blood culture results are obtained: A potential risk factor for hospital mortality. *Antimicrobial Agents and Chemotherapy*, *49*(9), 3640–3645.

Mouhib, M., Antonucci, A., Reggente, M., Amirjani, A., Gillen, A. J., & Boghossian, A. A. (2019). Enhancing bioelectricity generation in microbial fuel cells and biophotovoltaics using nanomaterials. *Nano Research*, *12*, 1–16.

Mukherjee, P., Ahmad, A., Mandal, D., Senapati, S., Sainkar, S. R., Khan, M. I., ... Kumar, R. (2001). Bioreduction of AuCl(4)(−) ions by the fungus, *Verticillium* sp. and surface trapping of the gold nanoparticles formed D.M. and S.S. thank the Council of Scientific and Industrial Research (CSIR), Government of India, for financial assistance. *Angewandte Chemie (International Ed. in English)*, *40*(19), 3585–3588.

Nadeem, M., Khan, R., Afridi, K., Nadhman, A., Ullah, S., Faisal, S., ... Abbasi, B. H. (2020). Green synthesis of cerium oxide nanoparticles (CeO_2 NPs) and their antimicrobial applications: A review. *International Journal of Nanomedicine*, *15*, 5951.

Nadeem, M., Tungmunnithum, D., Hano, C., Abbasi, B. H., Hashmi, S. S., Ahmad, W., & Zahir, A. (2018). The current trends in the green syntheses of titanium oxide nanoparticles and their applications. *Green Chemistry Letters and Reviews*, *11*(4), 492–502.

Nair, B., & Pradeep, T. (2002). Coalescence of nanoclusters and formation of submicron crystallites assisted by *Lactobacillus* strains. *Crystal Growth & Design*, *2*(4), 293–298. https://doi.org/10.1021/cg0255164.

Najafinejad, M. S., Mohammadi, P., Mehdi Afsahi, M., & Sheibani, H. (2019). Biosynthesis of Au nanoparticles supported on Fe_3O_4@polyaniline as a heterogeneous and reusable magnetic nanocatalyst for reduction of the azo dyes at ambient temperature. *Materials Science & Engineering. C, Materials for Biological Applications*, *98*, 19–29. https://doi.org/10.1016/j.msec.2018.12.098.

Narayanan, K. B., & Sakthivel, N. (2010). Biological synthesis of metal nanoparticles by microbes. *Advances in Colloid and Interface Science*, *156*(1–2), 1–13.

Nasrollahzadeh, M., Sajjadi, M., Iravani, S., & Varma, R. (2020). Trimetallic nanoparticles: Greener synthesis and their applications. *Nanomaterials*, *10*, 1784. https://doi.org/10.3390/nano10091784.

Nies, D. H. (1999). Microbial heavy-metal resistance. *Applied Microbiology and Biotechnology*, *51*(6), 730–750.

Ortega, F. G., Fernández-Baldo, M. A., Fernández, J. G., Serrano, M. J., Sanz, M. I., Díaz-Mochón, J. J., ... Raba, J. (2015). Study of antitumor activity in breast cell lines using silver nanoparticles produced by yeast. *International Journal of Nanomedicine*, *10*, 2021.

Ovais, M., Khalil, A. T., Ayaz, M., Ahmad, I., Nethi, S. K., & Mukherjee, S. (2018). Biosynthesis of metal nanoparticles via microbial enzymes: A mechanistic approach. *International Journal of Molecular Sciences*, *19*(12), 4100. https://doi.org/10.3390/ijms19124100.

Patil, M. P., & Kim, G.-D. (2018). Marine microorganisms for synthesis of metallic nanoparticles and their biomedical applications. *Colloids and Surfaces B: Biointerfaces*, *172*, 487–495.

Patra, S., Mukherjee, S., Barui, A. K., Ganguly, A., Sreedhar, B., & Patra, C. R. (2015). Green synthesis, characterization of gold and silver nanoparticles and their potential application for cancer therapeutics. *Materials Science & Engineering. C, Materials for Biological Applications*, *53*, 298–309.

Payne, J. N., Waghwani, H. K., Connor, M. G., Hamilton, W., Tockstein, S., Moolani, H., ... Dakshinamurthy, R. (2016). Novel synthesis of kanamycin conjugated gold nanoparticles with potent antibacterial activity. *Frontiers in Microbiology*, *7*, 607.

Perez Gonzalez, T., Jimenez-Lopez, C., Neal, A., Rull, F., Rodriguez-Navarro, A., Fernandez-Vivas, A., & Iañez-Pareja, E. (2010). Magnetite biomineralization induced by *Shewanella oneidensis*. *Geochimica et Cosmochimica Acta*, *74*, 967–979. https://doi.org/10.1016/j.gca.2009.10.035.

Qu, Y., Li, X., Lian, S., Dai, C., Jv, Z., Zhao, B., & Zhou, H. (2019). Biosynthesis of gold nanoparticles using fungus *Trichoderma* sp. WL-Go and their catalysis in degradation of aromatic pollutants. *IET Nanobiotechnology*, *13*(1), 12–17. https://doi.org/10.1049/iet-nbt.2018.5177.

Rajamanickam, K., Sudha, S., Francis, M., Sowmya, T., Rengaramanujam, J., Sivalingam, P., & Prabakar, K. (2013). Microalgae associated *Brevundimonas* sp. MSK 4 as the nano particle synthesizing unit to produce antimicrobial silver nanoparticles. *Spectrochimica Acta Part A: Molecular and Biomolecular Spectroscopy*, *113*, 10–14.

Rajeshkumar, S., Ponnanikajamideen, M., Malarkodi, C., Malini, M., & Annadurai, G. (2014). Microbe-mediated synthesis of antimicrobial semiconductor nanoparticles by marine bacteria. *Journal of Nanostructure in Chemistry*, *4*(2), 96.

Ramalingam, V., Rajaram, R., Premkumar, C., Santhanam, P., Dhinesh, P., Vinothkumar, S., & Kaleshkumar, K. (2014). Biosynthesis of silver nanoparticles from deep sea bacterium *Pseudomonas aeruginosa* JQ989348 for antimicrobial, antibiofilm, and cytotoxic activity. *Journal of Basic Microbiology*, *54*(9), 928–936.

Reidy, B., Haase, A., Luch, A., Dawson, K. A., & Lynch, I. (2013). Mechanisms of silver nanoparticle release, transformation and toxicity: A critical review of current knowledge and recommendations for future studies and applications. *Materials*, *6*(6), 2295–2350.

Romero, M. C., Gatti, E. M., & Bruno, D. E. (1999). Effects of heavy metals on microbial activity of water and sediment communities. *World Journal of Microbiology and Biotechnology*, *15*(2), 179–184. https://doi.org/10.1023/a:1008834725272.

Sanaeimehr, Z., Javadi, I., & Namvar, F. (2018). Antiangiogenic and antiapoptotic effects of green-synthesized zinc oxide nanoparticles using *Sargassum muticum* algae extraction. *Cancer Nanotechnology*, *9*(1), 018–0037.

Satapathy, S., & Shukla, S. P. (2017). Application of a marine cyanobacterium *Phormidium fragile* for green synthesis of silver nanoparticles. *Indian Journal of Biotechnology*, *16*, 110–113.

Schröfel, A., Kratošová, G., Šafařík, I., Šafaříková, M., Raška, I., & Shor, L. M. (2014). Applications of biosynthesized metallic nanoparticles—A review. *Acta Biomaterialia*, *10*(10), 4023–4042.

Senapati, S., Ahmad, A., Khan, M. I., Sastry, M., & Kumar, R. (2005). Extracellular biosynthesis of bimetallic Au-Ag alloy nanoparticles. *Small*, *1*(5), 517–520.

Shahverdi, A. R., Minaeian, S., Shahverdi, H. R., Jamalifar, H., & Nohi, A.-A. (2007). Rapid synthesis of silver nanoparticles using culture supernatants of Enterobacteria: A novel biological approach. *Process Biochemistry*, *42*(5), 919–923. https://doi.org/10.1016/j.procbio.2007.02.005.

Sharma, D., Sharma, R., & Chaudhary, A. (26 June 2020). In S. G. Sharma, N. R. Sharma, & M. Sharma (Eds.), *Microbial cell factories in nanotechnology*. Singapore: Springer.

Singaravelu, G., Arockiamary, J. S., Kumar, V. G., & Govindaraju, K. (2007). A novel extracellular synthesis of monodisperse gold nanoparticles using marine alga, *Sargassum wightii* Greville. *Colloids and Surfaces. B, Biointerfaces*, *57*(1), 97–101.

Singh, R., & Arora, N. K. (2016). Bacterial formulations and delivery systems against pests in sustainable agro-food production. *Food Science*, *1*, 1–11.

Singh, P., Kim, Y. J., Singh, H., Wang, C., Hwang, K. H., Farh, M. E.-A., & Yang, D. C. (2015). Biosynthesis, characterization, and antimicrobial applications of silver nanoparticles. *International Journal of Nanomedicine*, *10*, 2567.

Singh, P., Kim, Y. J., Zhang, D., & Yang, D. C. (2016). Biological synthesis of nanoparticles from plants and microorganisms. *Trends in Biotechnology*, *34*(7), 588–599.

Sudha, S., Rajamanickam, K., & Rengaramanujam, J. (2013). Microalgae mediated synthesis of silver nanoparticles and their antibacterial activity against pathogenic bacteria. *Indian Journal of Experimental Biology*, *51*, 393–399.

Syed, B., Prasad, N. M., & Satish, S. (2016). Endogenic mediated synthesis of gold nanoparticles bearing bactericidal activity. *Journal of Microscopy and Ultrastructure*, *4*(3), 162–166.

Sutradhar, K. B., & Amin, M. (2014). Nanotechnology in cancer drug delivery and selective targeting. *International Scholarly Research Notices*, *2014*, 1–12.

Syed, A., Raja, R., Kundu, G., Gambhir, S., & Ahmad, A. (2013). Extracellular biosynthesis of monodispersed gold nanoparticles, their characterization, cytotoxicity assay, biodistribution and conjugation with the anticancer drug doxorubicin. *Journal of Nanomedicine & Nanotechnology*, *4*(1), 156–161.

Teixeira, A. F., Ten Dijke, P., & Zhu, H.-J. (2020). On-target anti-TGF-β therapies are not succeeding in clinical cancer treatments: What are remaining challenges? *Frontiers in Cell and Development Biology*, *8*, 605.

Thornhill, R., Burgess, J., & Matsunaga, T. (1995). PCR for direct detection of indigenous uncultured magnetic cocci in sediment and phylogenetic analysis of amplified 16S ribosomal DNA. *Applied and Environmental Microbiology*, *61*, 495–500. https://doi.org/10.1128/aem.61.2.495-500.1995.

Turecka, K., Chylewska, A., Kawiak, A., & Waleron, K. F. (2018). Antifungal activity and mechanism of action of the Co(III) coordination complexes with diamine chelate ligands against reference and clinical strains of *Candida* spp. *Frontiers in Microbiology*, *9*(1594). https://doi.org/10.3389/fmicb.2018.01594.

Verma, V. C., Kharwar, R. N., & Gange, A. C. (2010). Biosynthesis of antimicrobial silver nanoparticles by the endophytic fungus *Aspergillus clavatus*. *Nanomedicine*, *5*(1), 33–40.

Vigneshwaran, N., Ashtaputre, N. M., Varadarajan, P. V., Nachane, R. P., Paralikar, K. M., & Balasubramanya, R. H. (2007). Biological synthesis of silver nanoparticles using the fungus *Aspergillus flavus*. *Materials Letters*, *61*(6), 1413–1418. https://doi.org/10.1016/j.matlet.2006.07.042.

Yang, H., Santra, S., & Holloway, P. (2005). Syntheses and applications of Mn-doped II-VI semiconductor nanocrystals. *Journal of Nanoscience and Nanotechnology*, *5*, 1364–1375. https://doi.org/10.1166/jnn.2005.308.

Yehia, R. S., & Al-Sheikh, H. (2014). Biosynthesis and characterization of silver nanoparticles produced by *Pleurotus ostreatus* and their anticandidal and anticancer activities. *World Journal of Microbiology and Biotechnology*, *30*(11), 2797–2803.

Zhang, Y., Shareena Dasari, T. P., Deng, H., & Yu, H. (2015). Antimicrobial activity of gold nanoparticles and ionic gold. *Journal of Environmental Science and Health. Part C, Environmental Carcinogenesis & Ecotoxicology Reviews*, *33*(3), 286–327.

Chapter 13

Nif genes: Tools for sustainable agriculture

Debmalya Dasgupta[a], Amrita Kumari Panda[b], Rojita Mishra[c], Arabinda Mahanty[d], Surajit De Mandal[e], and Satpal Singh Bisht[f]
[a]*Department of Biotechnology, National Institute of Technology, Yupia, Arunachal Pradesh, India,* [b]*Department of Biotechnology, Sant Gahira Guru University, Ambikapur, Chhattisgarh, India,* [c]*Department of Botany, Polasara Science College, Polasara, Ganjam, Odisha, India,* [d]*Crop Protection Division, National Rice Research Institute, Cuttack, Odisha, India,* [e]*College of Plant Protection, South China Agricultural University, Laboratory of Bio-Pesticide Innovation and Application of Guangdong Province, Guangzhou, PR China,* [f]*Department of Zoology, Kumaun University, Nainital, Uttarakhand, India*

Chapter outline

1 Introduction

The agricultural sector is one of the most vital wheels that drive the world economy. More than 70% of the rural livelihood depends upon agriculture. It not only provides food and livelihood security but also responsible for the overall development of the country, especially its GDP. It is estimated that the agriculture sector contributes 3% of the world's GDP and 43% of the exported goods (Food and Agricultural Organization, 2000). South Asian countries like India, Nepal, and Myanmar have the majority of agricultural production as well as income compared to the Middle East countries like Oman and Yemen. To meet the needs of the world's fastest growing population, increasing agricultural production is required, and this can be achieved by providing plants with adequate nutrients. The available nitrogen is vital in agriculture since it is a constituent of nucleic acids, protein, and other essential biomolecules in all the organisms. Nitrogen is the third important factor for the growth and development of the

Recent Advancement in Microbial Biotechnology. https://doi.org/10.1016/B978-0-12-822098-6.00012-4

plant (Thilakarathna, McElroy, Chapagain, Papadopoulos, & Raizada, 2016). Plants, however, are unable to use atmospheric N_2, and they largely depend on the reduced or oxidized form of nitrogen. In general, the nitrogen is present in limited content in most of the cropping conditions and is supplemented in the form of fertilizer (Beringer & Hirsch, 1984). Modern technological advancements in the 20th century have made this possible via the Haber-Bosch process that synthesizes artificial nitrogen fertilizer. However, these artificially produced chemical nitrogen fertilizer comes with the shortcomings viz. low stability in soils, more energy consumption, and a rise in greenhouse gas, i.e., nitrous oxide due to the leeching effect (Bloch, Ryu, Ozaydin, & Broglie, 2020). Hence, this method is unsuitable for large-scale applications. Therefore, an environmentally friendly alternative is the need of the hour. The search for natural nitrogen-fixing ability among the plants and microbes has shifted the attention of agronomists to microorganisms with nitrogen-fixing ability, including the synthesis of nitrogenase and other proteins essential for nitrogen fixation. During the late 70s, when recombinant DNA technology became known, efforts were made either to modify genes of cereal crops with an increase in the ability of nitrogen-fixing genes or utilization of diazotrophs associated with roots that can supply nitrogen via nitrogen fixation. The first approach is quite difficult owing to the plant genome complexities as well as complicated nitrogen-fixing machinery. The other approach exploits the function of *nif* genes in rhizospheric microorganisms that can fix atmospheric nitrogen to other nitrogen forms that can be used by crop plants. Nitrogenase, the key enzyme encoded by the *nif* genes, is associated with this process and found only in prokaryotes. However, with a successful plant-microbe association, eukaryotes, such as plants, can interact with the nitrogen-fixing microorganisms and fulfills the nutritional requirements. The present chapter provides an insight into the understanding of the origin, evolution, and functional mechanism of the *nif* genes in different organisms and identifies areas where most scientific explorations are required as per the current scenario of biotechnological and molecular interventions.

2 Biological nitrogen fixation and agricultural sustainability

Agriculture and agricultural practices render huge pressure on the environment and natural resources. Sustainable agricultural practices prevent the loss of natural resources, improve soil quality and productivity, and help in increasing yield. From the economist's point of view, sustainability is measured as a ratio of output to input considering the stock depletion. In agriculture, stock means soil, water, nonrenewable energy resources, and environmental quality. Modern agricultural practices are away from the goals of sustainability since it is based on maximum output without input efficiency and maintenance of the environment (Odum, 1989). Nitrogen fixing microorganisms and biological nitrogen fixation are the primary components of sustainable agricultural systems (Fig. 1).

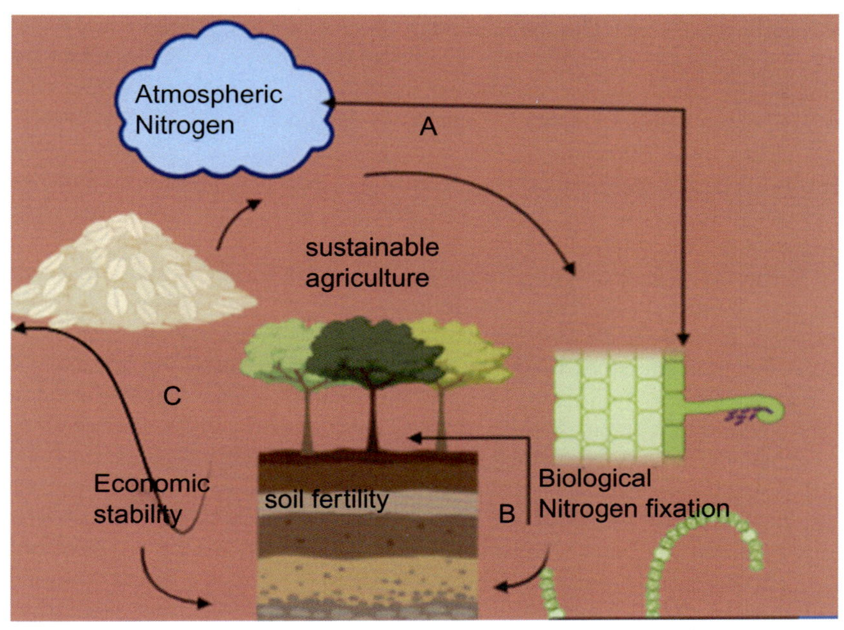

FIG. 1 Diagrammatic representation of sustainable agriculture and its relationships with biological nitrogen fixation and economic stability. (A) Sustainability in agriculture; (B) biological nitrogen fixation; (C) economic stability. *(Created with BioRender.com.)*

The availability of nitrogen intrinsically modulates the plant physiology and biochemical processes and affects crop production. Nitrogen is an integral component of proteins (amino acids), which are indispensable for the plant's metabolic processes. For example, the availability of nitrogen for plant stimulates leaf growth and photosynthesis. A well-regulated metabolism is necessary for growth, development, and good yield of the crop. Thus, the availability of nitrogen is necessary for plant growth and productivity. However, the application of the excess amount of nitrogen fertilizer for an increase in productivity leads to deposition of nitrogen in the soil, and its loss from the agricultural field causes eutrophication in water bodies and increases the nitric oxide in the atmosphere. Increasing the diversity of leguminous plants in agricultural systems reduces nitrogen pollution. Constant use of legume increases total soil organic carbon and nitrogen and decreases environmental pollution, providing strong evidence of the relationship between nitrogen fixation and sustainable agriculture (Blesh, 2019). The present century has a great challenge concerning sustainability and sustainable practices. The nitrogen-fixing microbes living in symbiosis with leguminous plants efficiently perform the nitrogen fixation process. The use of legumes, along with crops in the agricultural system, can provide healthy food and a sustainable ecosystem. Two major groups of bacteria associated with the symbiotic nitrogen fixation within the plant nodules are *Rhizobia* and *Frankia*. The *Rhizobia* are mainly associated with the leguminous plants under the superfamily of angiosperms (Fabaceae), whereas and *Frankia* is linked with diverse plant families (Franche, Lindström, & Elmerich, 2009; Huss-Danell, 1997; Sprent & Platzmann, 2001; Vessey, Pawlowski, & Bergman, 2005). The third important group of nitrogen-fixing bacteria is the cyanobacteria that can form a symbiotic association with different plants, fungi, and algae (Meeks & Elhai, 2002).

However, most of the cereal crops lack symbiotic nitrogen fixation and thus are dependent on the external availability of nitrogen. Two models have been proposed to enhance the nitrogen fixation in cereal crops; the first one is the heterologous expression of *nif* genes in the plants itself, and the other one is to recreate the cereal plant-microbe symbiosis similar to the legumes and rhizobia symbiosis in cereal crops. Because of the complexity of the plant genome, it is challenging to insert and express a complex cluster of *nif* genes in a plant cell. Thus, more emphasis is being given to the second approach. Unlike the leguminous plants, the symbiotic relationship between cereal and rhizospheric bacteria is less specialized, and efforts are being made for heterologous expression of *nif* genes in these bacteria to enhance the nitrogen fixation process (Bloch et al., 2020). Thus, an in-depth understanding of the *nif* genes and its functions and tools to improve the genetic potential for higher and sustainable agricultural production is needed.

3 The family of *nif* genes and their functions

Nitrogen fixation (*nif*) genes are present in diazotrophs viz. nitrogen-fixing bacteria, cyanobacteria, and symbiotic bacteria to fix atmospheric nitrogen into plant accessible, functional forms. They are also found in the plasmids of some bacteria, as well as in association with some other genes such as those required by the bacteria to communicate with the host plant. For the first time, in 1980, the *nif* genes were cloned from *Rhizobium* by Gary Ruvkun and Sharon R. Long in Frederick M. Ausubel's laboratory (Spaink, Kondorosi, & Hooykaas, 2012). The *nif* gene is clustered together in an operon called *nif* gene operon. *Nif* gene cluster is around 20–24 kbp with seven operons that code 20 different proteins (Glick, 2012). Recent studies reported that the *nif* gene cluster ranges from 11 kbp in *Paenibacillus* to 64 kbp in *Azorhizobium caulinodans* (Ryu et al., 2020). The operon is regulated and controlled by several *nif* genes, each having different assigned roles. It is a set of seven operons comprising 17 *nif* genes viz. *nif* A, D, L, K, F, H S, U, Y, W, Z having both positive and negative regulators, unlike in the case of *Anabaena* where the *nif* genes are reportedly scattered (Mazur, Rice, & Haselkorn, 1980). The nitrogenase (*nif*) genes include the nitrogenase structural genes, genes involved in the activation of the Fe protein, biosynthesis of molybdenum nitrogenase, iron-molybdenum cofactor biosynthesis, electron donation to nitrogenase, and the regulatory genes required for the expression of *nif* genes (Ahemad, 2014). The *nif* regulon contains factors, which have the potential to turn on, and off the production of proteins needed for nitrogen fixation according to the conditions that it may find suitable. In addition to the nitrogenase enzyme, the *nif* genes also encode several regulatory proteins involved in nitrogen fixation. The *nif* genes are triggered when fixed nitrogen and oxygen are at minimal concentration. The *nif* genes primarily code for a nitrogenase complex, which converts atmospheric nitrogen to other forms that can be used by plants depending on their needs. The individual role of each *nif* gene, either experimentally established or proposed, is given in Table 1. It is, however, pertinent to mention here that the majority of the information regarding *nif* genes and its products along with function have been determined from *Klebsiella pneumoniae* and *Azotobacter vinelandii*. The nitrogenase structural genes (*nif*HDK) and cofactor biosynthesis genes remain conserved among taxa (Dos Santos, Fang, Mason, Setubal, & Dixon, 2012). The enzyme complex nitrogenase is mainly an iron-sulfur molybdoenzyme (Boyd, Costas, Hamilton, Mus, & Peters, 2015), while other alternative nitrogenases may exist with iron-sulfur-vanadium or iron-sulfur cofactors (Eady, 1996). The molybdenum dependent nitrogenase made up of two proteins: MoFe protein (*Nif*DK): contains an active site and Fe protein (*Nif*H): transfers an electron to the P-cluster of the MoFe protein (Hoffman, Lukoyanov, Yang, Dean, & Seefeldt, 2014).

4 The evolution of *nif* gene in the different model organism

Biological nitrogen fixation is mostly catalyzed by a molybdenum-dependent enzyme nitrogenase consists of two-component proteins: MoFe protein (active site meant for substrate binding and reduction) and Fe protein (donates an electron to MoFe component) (Li, Liu, Zhang, & Chen, 2019). Nitrogenase appeared in anaerobes and afterward expanded into facultative anaerobes and aerobes. This transition leads to extensive augmentation in the number of *nif* genes from 7 to 20 genes (Boyd et al., 2015). The simplest *nif* gene organization is identified in methanogenic archaea *Methanococcus maripaludis* composed of 6–8 *nif* genes (*nif*H, *nif*I1, *nif*I2, *nif*D, *nif*K, *nif*E, *nif*N, and *nif*X) in a single operon (Kessler, Blank, & Leigh, 1998) and *nif*B gene, which is effectively required for nitrogenase, is located outside of the *nif* gene cluster in this archaea. The *nif* gene organization in obligate anaerobe *Clostridium acetobutylicum* and facultative anaerobe *Paenibacillus graminis* shares somewhat common genes except the regulation gene orf1 emerged in the gene cluster in diazotrophic *Paenibacillus* (Fig. 2). The *nif*HDKENXorf1 cluster was also identified in other aerobic microorganisms but was separated into two clusters: *nif*HDK and ENXorf1 in Proteobacteria *Azotobacter vinelandii* (Fig. 2) with *Nif*A as a positive transcription regulator (Setubal et al., 2009).

Nif genes were first analyzed in *Klebsiella pneumoniae* in which 20 *nif* genes (*nif*JHDKTYENXUSVWZMFLABQ) organized in several transcriptional

TABLE 1 Individual *nif* genes and their functions.

Sl. no.	Gene	Known/proposed role	Reference
1.	*nifH*	Dinitrogenase reductase. Obligate electron donor to dinitrogenase during nitrogenase turnover. Also required for FeMo-co biosynthesis and apodinitrogenase maturation	Ludden (1993)
2.	*nifD*	α-Subunit of dinitrogenase. Forms an $a_2\beta_2$ tetramer with β-subunit FeMo-co, the site of substrate reduction, is present buried within the α subunit of dinitrogenase	Rubio and Ludden (2005)
3.	*nifK*	β-Subunit of dinitrogenase. P-cluster are present at the β-subunit-interface	Rubio and Ludden (2005)
4.	*nifT*	Unknown	Rubio and Ludden (2005)
5.	*nifY*	In *K. pneumonia,* aids in the insertion of Fe-Mo-co into apodinitrogenase	Rubio and Ludden (2005)
6.	*nifE*	Forms a_2B_2 tetramer with *NifN*. Required for FeMo-co synthesis Proposed to function as a scaffold on which FeMo-co is synthesized	Rubio and Ludden (2005)
7.	*nifN*	Required for FeMo-cosynthesis	Rubio and Ludden (2005)
8.	*nifX*	Involved in FeMo-co synthesis. Specific role is not known	Rubio and Ludden (2005)
9.	*nifU*	Involved in mobilization of Fe for Fe-S cluster synthesis and repair	Rubio and Ludden (2005)
10.	*nifS*	Involved in mobilization of S for Fe-S cluster synthesis and repair	Ludden (1993)
11.	*nifV*	Homocitrate synthase, involve in FeMo-cosynthesis	Ludden (1993)

12.	*nifW*	Involved in stability of dinitrogenase. Proposed to protect dinitrogenase from O_2 inactivation	Rubio and Ludden (2005)
13.	*nifZ*	Unknown	Rubio and Ludden (2005)
14.	*nifM*	Required for maturation of *NifH*	Rubio and Ludden (2005)
15.	*nifF*	Flavodoxin. Physiologic electron donor of *NifH*	Ludden (1993)
16.	*nifL*	Negative regulatory element	Ludden (1993)
17.	*nifA*	Positive regulatory element	Ludden (1993)
18.	*nifB*	Required for FeMo-co synthesis. Metabolic product, *NifB*-co is the specific Fe and S donor to FeMo-co	Ludden (1993)
19.	*fdxN*	Ferredoxin. In *R. capsulatus*, serves as electron donor to nitrogenase	Rubio and Ludden (2005)
20.	*nifQ*	Involved in FeMo-co synthesis. Proposed to function in early $MoO_4^{2\cdot}$ processing	Ludden (1993)
21.	*nifJ*	Pyruvate:flavodoxin (ferredoxin) oxidoreductase. Involved in electron transport to nitrogenase	Ludden (1993)
22.	*nifKD*	Dinitrogenase (MoFe protein); $\alpha_2\beta_2$ tetramer, contains 2 FeMo-co,2'P' clusters/tetramer. Substrate reduction	Ludden (1993)
23.	*nifNE*	FeMo-co biosynthesis $\alpha_2\beta_2$ tetramer with significant similarity to *NifKD*. Oxygen-labile FeS protein	Ludden (1993)

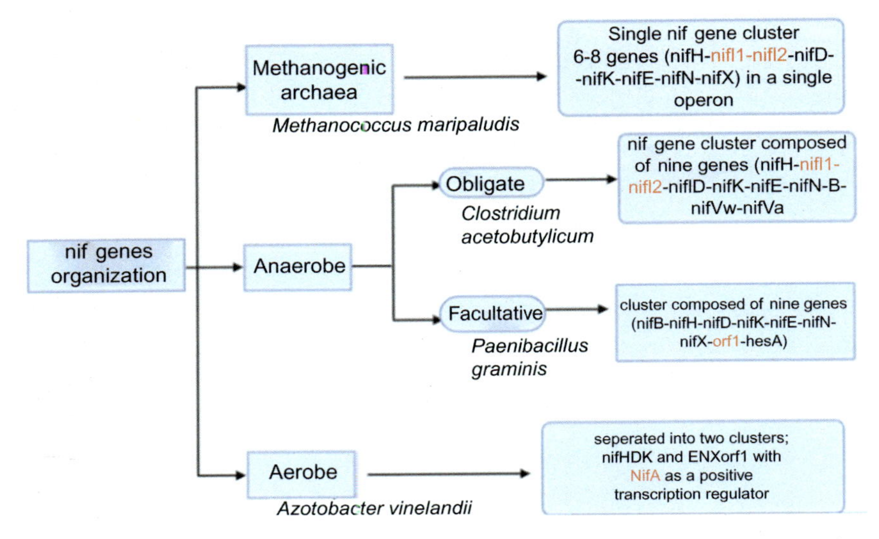

FIG. 2 Evolution of *nif* genes from anaerobic to aerobic organisms. *(Created with BioRender.com.)*

units were clustered together (Rubio & Ludden, 2008). The physiological conditions of the organism affect the *nif* gene cluster, e.g., *Klebsiella pneumoniae* fixes N_2 under anaerobic conditions have all the *nif* genes in one linkage group, whereas *Azotobacter vinelandii*, an obligate aerobe, clustered the *nif* genes into two different linkage groups. The *nif*A, *nif*B, and *nif*Q genes are contained in one, and the *nif*H, *nif*D, *nif*K, *nif*E, *nif*N, *nif*U, *nif*S, *nif*V, and *nif*F genes are within the other linkage group (Jacobson et aL, 1989). The *K. pneumoniae nif*-specific genes encode the following components: *nif*H (Fe protein subunit); *nif*D (MoFe protein a subunit); *nif*K (MoFe protein subunit); *nif*F and *nif*J (electron transport components); *nif*E, *nif*N, *nif*V, *nif*B, and *nif*Q (FeMo cofactor biosynthetic enzymes); *nif*M (Fe protein maturation component); *nif*L (negative regulatory element); *nif*A (positive regulatory element); and niJT, *nif*Y, nmX, *nif*U, *nif*S, *nif*W, and *nif*Z (functions not known) (Jacobson et al., 1989). There is evidence that *nif* clusters transfer laterally between species (Kechris, Lin, Bickel, & Glazer, 2006; Yan et al., 2008); still, the transfer of the *nif* cluster possesses many confront. It has also been reported that the *nif* gene can be acquired by the nonnitrogen-fixing heterologous host through horizontal gene transfer and confers the ability to fix nitrogen (Dixon & Postgate, 1972; Fox et al., 2016; Setten et al., 2013).

The first successful transfer of *nif* genes was reported in gram-negative bacterium *K. pneumoniae* strain M5a1 by both transduction (Streicher, Gurney, & Valentine, 1971) as well as conjugation approaches (Dixon & Postgate, 1972). Both the approaches, however, confirmed that the *nif* genes were located close to the histidine (his) biosynthetic operon in one particular region of the chromosome. A few years later, Dunican and Tierney reported genetic transfer of nitrogen-fixing genes from *Rhizobium trifolii* to *K. aerogenes*, a nonnitrogen fixing strain (Dunican & Tierney, 1974). Further, in vivo studies led to the finding of many recombinant plasmids that has the capability of carrying the *his-nif* region of *K. pneumoniae* to *Agrobacterium tumefaciens* and *Rhizobium meliloti* (Dixon, Cannon, & Kondorosi, 1976). These plasmids not only had a broad host range, but it was also found to be stable in *K. pneumoniae*, and as such, it could be used for cloning purposes. Linkage map analysis of *nif* genes revealed that it is composed of two major clusters viz. a *his*-proximal cluster comprising *nifBA(L)F* and a *his*-distal group of *nifEKDH* (Kennedy, 1977).

Due to the conserved nature of nitrogen fixation (*nif*) genes, fragments taken from *Klebsiella pneumoniae* was cloned and used as probes for the identification and cloning of *Anabaena nif* genes. Results showed that the sequence homology was similar, but the gene order may change depending upon the microorganism (Mazur et al., 1980). Some *nif* gene cluster has also been cloned (Quiviger et al., 1982; Robson, Woodley, & Jones, 1986). Mutations in the *nifX* gene of *K. pneumoniae* reported to negatively regulate the *nif* regulon (Gosink, Franklin, & Roberts, 1990). Further studies established that the *glnF* (*ntrA*) gene of *K. pneumoniae* plays a vital role in the positive regulation of many nitrogen assimilation genes, including the nitrogen fixation (*nif*) gene cluster (De Bruijn & Ausubel, 1983). Signal transduction studies in nitrogen sensory protein GlnK via protein-protein interaction have provided molecular insight on how the *nif* genes are expressed in *K. pneumoniae* (Glöer, Thummer, Ullrich, & Schmitz, 2008). The mechanism of function of the *nif* gene regulon in *K. pneumoniae* is shown in Fig. 3.

Besides *K. pneumoniae*, *nif* gene clusters have been extensively studied in *Enterobacter agglomerans*, *Rhodobacter capsulatus*, *Bradyrhtzobtum japonicum* USDA 110, *Anabaena variabilis*, *Leptolyngbya boryana* strain dg5, and *Cyanothece* sp. ATCC51142 (Bradburne, Mathis, & Israel, 1994; Hübner, Masepohl, Klipp, & Bickle, 1993; Klingmüller, 1991; Thiel, 2019). *nif* gene rearrangements have also been reported in some microorganisms such as diazotroph (Golden, Robinson, & Haselkorn, 1985), bacteria (Haselkorn, 1992), *Bacillus subtilis* (Stragier, Kunkel, Kroos, & Losick, 1989), and in heterocystous cyanobacteria viz. *Anabaena* (Apte & Prabhavathi, 1994), *Rhodobacter capsulatus nifA* mutants (Paschen, Drepper, Masepohl, & Klipp, 2001), and *Leptospirillum ferrooxidans* (Parro & Moreno-Paz, 2003). In the late 1980s, most of the research focused on the regulation of *nif* genes, such as the *nif* gene regulation pattern in *Azospirillum brasilense* (Pedrosa & Yates, 1984). Similar studies were done in *Anabaena variabilis* (Helber, Johnson, Yarbrough, & Hirschberg, 1988). In some Klebacteria residing in arctic seawater, high gene diversity of *nifH* has been observed (Díez, Bergman, Pedrós-Alió, Antó, & Snoeijs, 2012).

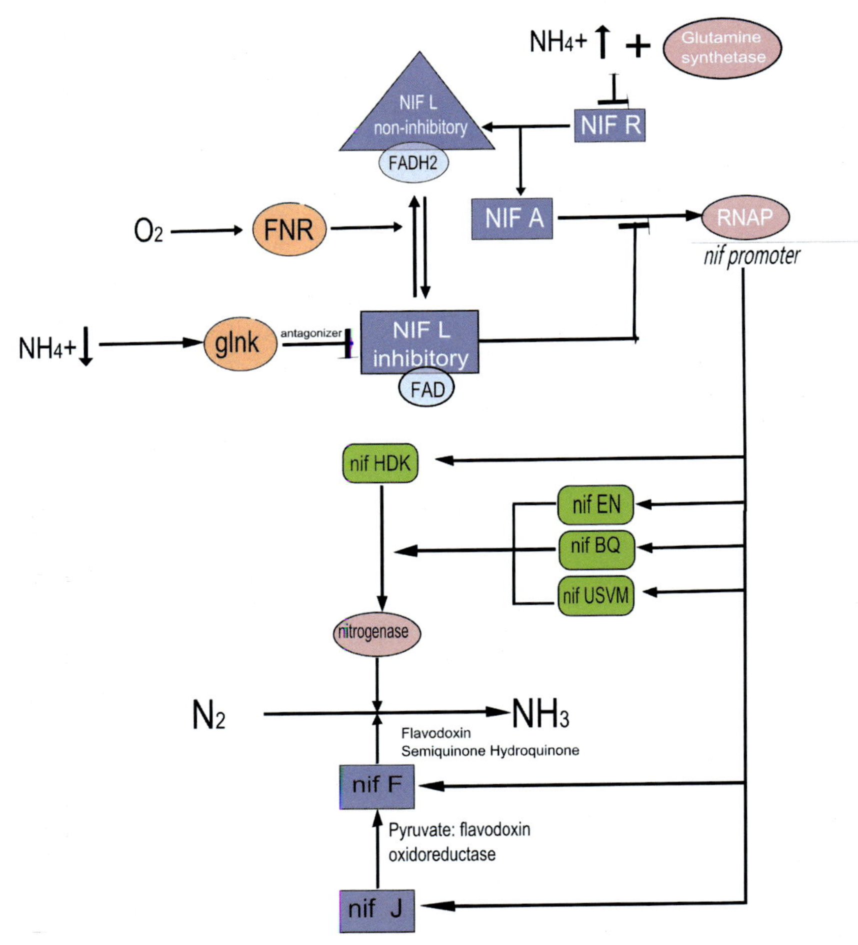

FIG. 3 Diagrammatic representation of *nif regulon* in *K. pneumonia.*

Cyanobacteria is the only oxygen-producing microorganisms capable of fixing atmospheric nitrogen (N_2) and representing a wide range of morphologies. Moreover, they comprise a large polycistronic gene cluster that is easier to manipulate and express in the plant cell (Thiel, 2019). Typically, cyanobacterial nitrogenases are organized in distinct operons such as *nif*B-fdxN-*nif*SU, *nif*HDK, *nif*ENXW, and *nif*VZT. However, the nitrogenase complex is sensitive to oxygen, and diazotrophic microorganisms have evolved with multiple strategies to protect the nitrogenase enzymes. They created specialized nitrogen-fixing cells known as heterocyst that contains an anaerobic site well protected from external oxygen by a thick membrane. An 11 kb excision element was reported in the *nif*HDK operon of *Anabaena* spp. This excision is performed during the differentiation of vegetative cells to the heterocysts (Brusca, Chastain, & Golden, 1990; Esteves-Ferreira et al., 2017; Golden & Wiest, 1988). It has also been reported that the intrinsic capacity of cyanobacterial N_2 fixation depends on the nitrogenase enzyme system, with the molybdenum nitrogenase (Mo-nitrogenase). This complex contains about 10% of the total cellular protein of many diazotrophs and catalyzes the synthesis of nearly half of all of the fixed N_2 on Earth (Betancourt, Loveless, Brown, & Bishop, 2008; Esteves-Ferreira et al., 2017; Falkowski, 1997).

Studies of *nif* genes in *Rhodospirillum rubrum* have revealed that the nitrogenase activity of the gene is related to the activation of the Fe-protein that, in turn, can cause a quick, reversible, "switch off" effect. This effect can be induced by various agents such as the addition of ammonium ion, asparagine, or glutamine to *nif* cultures (Falk & Johansson, 1983). *Nif* genes have been identified in large plasmids of *Rhizobium* strains viz. *Acacia melanoxylon*, *Acacia cyanophylla*, *Prosopis chilensis*, and *Sophora microphylla* present in legume tree root nodules (Toro, Herrera, & Olivares, 1984). In addition, the *nif*A and ntrC genes were characterized by the diazotrophic bacterium ORS 571 that was isolated from the annual stem nodules of the citrus plant (Pawlowski, Ratet, Schell, & De Bruijn, 1987). The similarity in *nif* genes in *Rhizobium meliloti* and *Klebsiella pneumonia* was used to design a gene probe for identifying conserved portions of the genome of all *Frankia* strains (Simonet, Normand, & Bardin, 1988).

Interaction between different promoter regions of *nif* genes with upstream activator sequence (UAS) as well as the upstream regulatory sequence of the *fdhF* gene from *Escherichia coli* (URS) established a direct correlation between *nifA* and RNA polymerase at the *ntrA*-dependent promoter (Birkmann, Hennecke, & Bock, 1989). Studies also confirmed that *groEL* protein could both perform as a regulatory as well as a structural role in *nif* gene regulon (Govezensky, Greener, Segal, & Zamir, 1991). Certain *nif* sequences have been obtained, such as that of *Azotobacter vinelandii* and *Enterobacter agglomerans* plasmid pEA9 (Jacobson et al., 1989; Selenska & Klingmüller, 1991). The *nif* regulon can also be used to study the differential expression of several other genes, such as lacZ gene fusions (Milcamps & Banderleyden, 1991). In some prokaryotes viz. *Synechococcus* sp. RF-1, presence of nitrate, represses the N_2-fixing ability following light-dark circles (Huang & Chou, 1991).

A few years later, the *nif* genes of *Rhodobacter capsuiatus* responsible for expression, especially the promoter elements viz. *rpoN*, *nifA1*, and *nifA2* were mapped by exonuclease-lll mediated deletions and by primer extension analysis (Preker, Hübner, Schmehl, Klipp, & Bickle, 1992). Similarly, *nif* genes viz. *nifH* and *nifD* were also found out in *Bacillus polymyxa* and *Bacillus macerans* (Oliveira, Seldin, & Bastos, 1993). Studies on *nif* genes of *Rhodobacter capsulatus* have resulted in the finding of 10 genes that are essential for its nitrogenase activity (Schüddekopf, Hennecke, Liese, Kutsche, & Klipp, 1993). Subsequently, *nif* genes have also been documented in *Azospirillum brasilense* (De Zamaroczy, Delorme, & Elmerich, 1989; Hartmann, Fu, & Burris, 1986; Zhang, Burris, Ludden, & Roberts, 1997). The *nif* genes were also found to be structurally conserved between nitrogen-fixing species. Studies have suggested the possible mechanism by which *nif* genes could be made to express in the plastids of plant cells by restricting the synthesis of nitrogenase to the dark period, thus protecting the enzyme from oxidative damage (Dixon, Cheng, Shen, Day, & Dowson-Day, 1997).

Dominic, Zani, Chen, Mellon, and Zehr (2000), successfully cloned and sequenced a 16 kb fragment of a nonheterocystous filamentous cyanobacterium viz. *Trichodesmium sp. IMS101*. They found similarity in the gene clusters of *Anabaena variabilis* in *nifB* and *nifW* with differences of two ORFs, i.e., ORF3 and ORF1, that was absent in *Trichodesmium* sp., unlike that of *A. variabilis* (Dominic et al., 2000). In the same year, the regulatory role of GlnK was determined. It was found that it regulates *NifL*-mediated inhibition of *NifA* activity in response to the nitrogen status because of the presence of two residues in the T-Loop of the GlnK (Arcondéguy, Lawson, & Merrick, 2000). The H-NS-like protein HvrA was reported to modulate the expression of nitrogen fixation genes in the phototrophic nonsulfur purple bacterium *Rhodobacter capsulatus* by binding to the *nifH* promoter region (Raabe, Drepper, Riedel, Masepohl, & Klipp, 2002). A year later, it was found that the endophytic bacterium *Herbaspirillum seropedicae* expresses *nif* genes in gramineous plants viz. maize, sorghum, wheat, and rice (Roncato-Maccari et al., 2003). Studies were done to identify *nifH* and *nifHDKYE′* genes in 20 heterotrophic bacterial strains isolated from rice fields in the Yangtze River Plain, China (Xie et al., 2006). Some genes related to a cAMP receptor protein, AnCrpA in *Anabaena* sp. strain PCC 7120, were found to be expressed, highlighting the role of AnCrpA protein in regulating the collective expressions of gene clusters related to nitrogen fixation in the presence of nitrate (Suzuki, Yoshimura, Ehira, Ikeuchi, & Ohmori, 2007). Some cloned genes viz. *frxC-ORF469* have similarities in sequence with *nifH*, suggesting some kind of evolutionary genetic transfer (Ogura et al., 1992). High expression levels in the *nifH* gene in dry beans have also shed light on selecting plants having a high nitrogen-fixing ability (Akter, Pageni, Lupwayi, & Balasubramanian, 2014).

5 Regulation of *nif* genes

The mechanism of *nif* gene regulation has been quite elusive in unicellular cyanobacteria are concerned. In *K. pneumoniae*, some of the gene products in can serve *nif* specific regulation by activating certain transcription factors (Merrick et al., 1982). However, in some diazotrophs, *nif* gene is either regulated by O_2-responsive regulatory system (Grabbe & Schmitz, 2003; Martinez-Argudo, Little, Shearer, Johnson, & Dixon, 2004) or by the anaerobic sensory system (Elsen, Dischert, Colbeau, & Bauer, 2000; Elsen, Swem, Swem, & Bauer, 2004; Grabbe, Klopprogge, & Schmitz, 2001; Joshi & Tabita, 1996). In some reports, the availability of nitrogen and the energy status of diazotrophic cells also regulates *nif* genes (Rabouille, Staal, Stal, & Soetaert, 2006) or by assessing the levels of some transcriptional activators involved in nitrogen control (Steunou et al., 2008). Studies of the *nif* gene in some microorganisms viz. *Azospirillum brasilense* has revealed that environmental factors play a role in determining the nitrogenase activity (Mugnai et al., 1994). Kinetic studies of *nif* genes in nondiazotroph *Azotobacter vinelandii* have revealed that the entire *nif* regulon is very tightly controlled and regulated (Poza-Carrión, Jiménez-Vicente, Navarro-Rodríguez, Echavarri-Erasun, & Rubio, 2014). Studies have found that P_{II}-like proteins such as GlnB, GlnK, and GlnY have specific roles that help in mediating nitrogen and oxygen control of *nif* gene transcription and nitrogenase activity and, in a way, helps in indirect regulation of *nif* genes (Martin & Reinhold-Hurek, 2002). Some *nif* genes such as *nifl1* and *nifl2* (Enkh-Amgalan, Kawasaki, & Seki, 2006), as well as *nifD-nifK* (Mishra et al., 2015), have been used to study phylogenetic and evolutionary relationships at bacterium as well as gene level. Similar line studies have been carried out taking DNA probes corresponding to *nif* genes of cyanobacterium *Anabaena* PCC 7120 to study evolutionary relationships between symbiotic organisms like *Azolla* (Van Coppenolle, McCouch, Watanabe, Huang, & Van Hove, 1995).

Some in vitro synthesized proteins have also been reported to be activating *nif* gene transcriptionally by binding to its promoter region for its expression (Kim, Hidaka, Masaki, Beppu, & Uozumi, 1991). The *nif* genes are reportedly highly conserved in certain organisms. However, between closely related organisms, some region is variable but well-structured (Jackman & Mulligan, 1995). During the last few years, several novel approaches have been undertaken, such as adding leader sequences to nuclear-encoded *Nif* proteins, facilitating their delivery to mitochondria of yeast *Saccharomyces cerevisiae* (Burén, Young, et al., 2017; Burén, Jiang, López-Torrejón, Echavarri-Erasun, & Rubio, 2017; Lopez-Torrejón et al., 2016) as well as *Nicotiana benthamiana* (Allen et al., 2017; Yang et al., 2018). *nifS* gene-based primer has been used to amplify *nif* genes present in cyanobacteria (Anand, Singh, Kumari, & Kumar, 2016). During the last 50 years, much of the effort has been made to transfer the nitrogenase gene from *Klebsiella oxytoca* to nondiazotrophic hosts (Thiel, 2019). The main drawback of this method is the complex gene regulation of the *nif* gene cluster of *K. oxytoca*. Recently many attempts have been made to express *nif* genes from various Cyanobacteria and bacteria in other heterologous hosts (Table 2). Few reports stated the expression of *nif* gene clusters yielded low levels of nitrogenase activity in engineered hosts (Tsujimoto et al., 2018). Thus expression of *nif* genes in heterologous hosts requires a better understanding of *nif* gene regulation and optimization of conditions for nitrogenase activity in the host cell.

TABLE 2 Successful expression of nitrogenase genes in heterologous hosts.

Source of *nif* gene	Expressed in	No. of genes transferred for nitrogenase activity	Name of vector/ promoter used for the expression	Nitrogenase activity	Reference
Leptolyngbya boryana	*Synechocystis* sp. PCC 6803	25 *nif* and *nif*-related genes along with the regulatory gene cnfR	*nifB* and *nifP* promoters	0.26% of *L. boryana* under anaerobic conditions	Tsujimoto et al. (2018)
Cyanothece sp. ATCC 51142	*Synechocystis* sp. PCC 6803	35 nitrogen fixation (*nif*) genes	Native promoters in the *nif* cluster	2% of wild-type *Cyanothece*	Liu, Liberton, Yu, Pakrasi, and Bhattacharyya-Pakrasi (2010) and Liu et al. (2018)
Pseudomonas stutzeri	*Pseudomonas protegens* Pf-5	Entire nitrogenase island	Native promoter from *Pseudomonas stutzeri* DSM4166	Significant nitrogenase activity	Jing et al. (2020)
Pseudomonas stutzeri DSM4166	*Pseudomonas protegens* CHA0	49-kb *Nif* gene island	pBeloBAC11 vector via linear-linear homologous recombination	Improved biological nitrogen-fixation function	Yu et al. (2019)
Azotobacter vinelandii	*S. cerevisiae*	*nif*H, *nif*M, *nif*U, and *nif*S genes	GAL1 or GAL10 promoters	Accumulation of active *Nif*U and Fe protein in Yeast mitochondria despite their extreme O_2 sensitivity	Lopez-Torrejón et al. (2016)
Azotobacter vinelandii	*S. cerevisiae*	*nif* genes (*nif*HDKUSMBEN)	Galactose inducible promoters	*Nif*DK tetramer formation in Yeast mitochondria	Burén, Young, et al. (2017)
Azotobacter vinelandii	*Nicotiana benhamiana*	*nifH* and *nifM* protein coding regions	pMON253685 vector/ribosomal RNA promoter *Prrn*	Expression of active Fe subunit of nitrogenase in plant plastids	Ivleva, Groat, Staub, and Stephens (2016)

The expression and regulation of *nif* genes have also been studied in *Azotobacter vinelandii* (Poza-Carrión, Echavarri-Erasun, & Rubio, 2015). Other approaches, such as transferring the *nif* gene cluster from one pseudomonas strain viz. *P. stutzeri* DSM4166 to another viz. *P. protegens* Pf-5 having both the properties of biological fertilizer as well as biocontrol agent, have also been employed (Jing et al., 2020). Factors such as ferrous ion (Fe^{2+}) ion reportedly upregulates the expression of *nif* genes in *Rhodobacter sphaeroides* (Liu, Liberton, et al., 2018; Liu, Zheng, et al., 2018). Similarly, factors such as osmotic stress can activate *nif* genes in *Rhizobium tropici* CIAT 899, resulting in Nod factor production (Del Cerro et al., 2019). Similar line studies were carried out for *Rhizobium sp.* IRBG74 that could be genetically engineered by transferring a *nif* cluster from either *Rhodobacter sphaeroides* or *Klebsiella oxytoca* resulting in attaining nitrogen-fixing ability under free-living conditions (Ryu et al., 2020). Analysis at the transcriptional level has shown that the *orf1* gene is a crucial factor for the nitrogen fixation in the gram-positive bacteria *Paenibacillus* (Li et al., 2019). Synthetic approaches have also been carried out recently, which targets the mitochondrial-processing peptidases (MMPs) of yeast, *Arabidopsis thaliana, Nicotiana tabacum, Oryza sativa*, and in *Escherichia coli*. These MMPs are responsible for cleaving the *nifD* subunit in MoFe protein, which renders instability in gene transfer to higher organisms. Such modification in MMPs could help in the stability of *nif* proteins required for plant expression (Xiang et al., 2020).

6 Conclusion

Rapidly growing population, coupled with diverse climatic changes, has generated an exacerbating condition across the globe. This has necessitated the demand for a larger quantity of crop production and hence requires more amount of nitrogen for it. However, excess nitrogen can cause serious catastrophic effects on the atmosphere, with an increase in nitrogen pollution as well as greenhouse gas emission. Although many significant developments have been done as per as nitrogen-fixing bacterium and its symbiotic plant is concerned, more novel strategies are required. More microorganisms need to be explored that could have the potential to fix nitrogen more effectively. The prospect of direct *nif* gene transfer remains open, but challenges such as the nitrogenase sensitivity to O_2 and the complexities of nitrogenase biosynthesis are need to be addressed (Curatti & Rubio, 2014). Genetic engineering of cereals with high nitrogenase enzyme expression is a remarkable option but not quite appropriate for the rapid outcome. This has shifted the attention of scientists, researchers, and agronomists to genetically modified microorganisms that could create an immediate impact. Since the inception of *nif* genes, it is clear that a mere inserting gene of interest won't sort out the issue. Certain environmental factors such as soil type and oxygen level could play a decisive role in the successful expression of the gene of interest. Hence, strategies to optimize such potential constraints must be given equal importance for the successful engineering of cereal crops with enhanced nitrogenase expression.

Moreover, other gene analogues expressing nitrogenase must be extensively investigated that could be more effective and provide an in-depth study of the *nif* gene in different model organisms. This may lead to the discovery of some novel bacterium inoculant that can minimize all the current issues that the researchers and scientists are facing. More in-depth understanding, especially that of structural based information of nitrogenase proteins, is required. Governments should also adopt effective strategies to ensure the best management of sustainable utilization of nitrogen viz. cost-effective high-yielding nitrogen fertilizers.

Acknowledgements

The authors are thankful to Mr. Nirmalya Das Gupta for proofreading the manuscript thoroughly. The support extended by Director, NIT Arunachal Pradesh, Yupia is duly acknowledged.

References

Ahemad, M. (2014). Mechanisms and applications of plant growth promoting rhizobacteria : Current perspective. *Journal of King Saud University – Science*, *26*(1), 1–20. https://doi.org/10.1016/j.jksus.2013.05.001.

Akter, Z., Pageni, B. B., Lupwayi, N. Z., & Balasubramanian, P. M. (2014). Biological nitrogen fixation and *nif* H gene expression in dry beans (*Phaseolus vulgaris* L.). *Canadian Journal of Plant Science*, *94*(2), 203–212.

Allen, R. S., Tilbrook, K., Warden, A. C., Campbell, P. C., Rolland, V., Singh, S. P., & Wood, C. C. (2017). Expression of 16 nitrogenase proteins within the plant mitochondrial matrix. *Frontiers in Plant Science*, *8*, 287.

Anand, R., Singh, B., Kumari, D., & Kumar, D. (2016). Identification of nitrogen fixing cyanobacteria by PCR amplification of *Nif* genes. *Indian Journal of Applied Research*, *6*, 109–111.

Apte, S. K., & Prabhavathi, N. (1994). Rearrangements of nitrogen fixation (*nif*) genes in the heterocystous cyanobacteria. *Journal of Biosciences*, *19*(5), 579–602.

Arcondéguy, T., Lawson, D., & Merrick, M. (2000). Two residues in the T-loop of GlnK determine *Nif*L-dependent nitrogen control of *nif* gene expression. *Journal of Biological Chemistry*, *275* (49), 38452–38456.

Beringer, J. E., & Hirsch, P. R. (1984). Genetic engineering and nitrogen fixation. *Biotechnology and Genetic Engineering Reviews*, *1*(1), 65–88.

Betancourt, D. A., Loveless, T. M., Brown, J. W., & Bishop, P. E. (2008). Characterization of diazotrophs containing Mo-independent nitrogenases, isolated from diverse natural environments. *Applied and Environmental Microbiology*, *74*(11), 3471–3480.

Birkmann, A., Hennecke, H., & Bock, A. (1989). Construction of chimaeric promoter regions by exchange of the upstream regulatory sequences from fdhF and *nif* genes. *Molecular Microbiology*, *3*(6), 697–703.

Blesh, J. (2019). Feedbacks between nitrogen fixation and soil organic matter increase ecosystem functions in diversified agroecosystems. *Ecological Applications: A Publication of the Ecological Society of America*, *29*(8). https://doi.org/10.1002/eap.1986, e01986.

Bloch, S. E., Ryu, M. H., Ozaydin, B., & Broglie, R. (2020). Harnessing atmospheric nitrogen for cereal crop production. *Current Opinion in Biotechnology*, *62*, 181–188.

Boyd, E. S., Costas, A. M. G., Hamilton, T. L., Mus, F., & Peters, J. W. (2015). Evolution of molybdenum nitrogenase during the transition from anaerobic to aerobic metabolism. *Journal of Bacteriology*, *197*, 1690–1699.

Bradburne, J. A., Mathis, J. N., & Israel, D. W. (1994). *nif* Gene expression in a *Nif*+, Fix− *Bradyrhizobium japanicum* variant. *FEMS Microbiology Letters*, *123*(1–2), 91–98.

Brusca, J. S., Chastain, C. J., & Golden, J. W. (1990). Expression of the *Anabaena* sp. strain PCC 7120 xisA gene from a heterologous promoter results in excision of the *nif*D element. *Journal of Bacteriology*, *172*(7), 3925–3931.

Burén, S., Young, E. M., Sweeny, E. A., Lopez-Torrejón, G., Veldhuizen, M., Voigt, C. A., & Rubio, L. M. (2017). Formation of nitrogenase *Nif*DK tetramers in the mitochondria of *Saccharomyces cerevisiae*. *ACS Synthetic Biology*, *6*(6), 1043–1055.

Burén, S., Jiang, X., López-Torrejón, G., Echavarri-Erasun, C., & Rubio, L. M. (2017). Purification and in vitro activity of mitochondria targeted nitrogenase cofactor maturase *Nif*B. *Frontiers in Plant Science*, *8*, 1567.

Curatti, L., & Rubio, L. M. (2014). Challenges to develop nitrogen-fixing cereals by direct *nif*-gene transfer. *Plant Science*, *225*, 130–137.

De Bruijn, F. J., & Ausubel, F. M. (1983). The cloning and characterization of the glnF (ntrA) gene of *Klebsiella pneumoniae*: Role of glnF (ntrA) in the regulation of nitrogen fixation (*nif*) and other nitrogen assimilation genes. *Molecular and General Genetics MGG*, *192*(3), 342–353.

De Zamaroczy, M., Delorme, F., & Elmerich, C. (1989). Regulation of transcription and promoter mapping of the structural genes for nitrogenase (*nif*HDK) of *Azospirillum brasilense* Sp7. *Molecular and General Genetics MGG*, *220*(1), 33–42.

Del Cerro, P., Megias, M., López-Baena, F. J., Gil-Serrano, A., Perez-Montano, F., & Ollero, F. J. (2019). Osmotic stress activates *nif* and fix genes and induces the *Rhizobium tropici* CIAT 899 Nod factor production via NodD2 by up-regulation of the nodA2 operon and the nodA3 gene. *PLoS One*, *14*(3), e0213298.

Díez, B., Bergman, B., Pedrós-Alió, C., Antó, M., & Snoeijs, P. (2012). High cyanobacterial *nif*H gene diversity in Arctic seawater and sea ice brine. *Environmental Microbiology Reports*, *4*(3), 360–366.

Dixon, R. A., & Postgate, J. R. (1972). Genetic transfer of nitrogen fixation from *Klebsiella pneumoniae* to *Escherichia coli*. *Nature*, *237*(5350), 102–103.

Dixon, R. A. Y., Cannon, F., & Kondorosi, A. (1976). Construction of a P plasmid carrying nitrogen fixation genes from *Klebsiella pneumoniae*. *Nature*, *260*(5548), 268–271.

Dixon, R., Cheng, Q., Shen, G. F., Day, A., & Dowson-Day, M. (1997). *Nif* gene transfer and expression in chloroplasts: Prospects and problems. *Plant and Soil*, *194*(1–2), 193–203.

Dominic, B., Zani, S., Chen, Y. B., Mellon, M. T., & Zehr, J. P. (2000). Organization of the *nif* genes of the nonheterocystous cyanobacterium *Trichodesmium* sp. IMS101. *Journal of Phycology*, *36* (4), 693–701.

Dos Santos, P. C., Fang, Z., Mason, S. W., Setubal, J. C., & Dixon, R. (2012). Distribution of nitrogen fxation and nitrogenase-like sequences amongst microbial genomes. *BMC Genomics*, *13*, 162.

Dunican, L. K., & Tierney, A. B. (1974). Genetic transfer of nitrogen fixation from *Rhizobium trifolii* to *Klebsiella aerogenes*. *Biochemical and Biophysical Research Communications*, *57*(1), 62–72.

Eady, R. R. (1996). Structure−function relationships of alternative nitrogenases. *Chemical Reviews*, *96*(7), 3013–3030.

Elsen, S., Dischert, W., Colbeau, A., & Bauer, C. E. (2000). Expression of uptake hydrogenase and molybdenum nitrogenase in *Rhodobacter capsulatus* is coregulated by the RegB-RegA two-component regulatory system. *Journal of Bacteriology*, *182*(10), 2831–2837.

Elsen, S., Swem, L. R., Swem, D. L., & Bauer, C. E. (2004). RegB/RegA, a highly conserved redox-responding global two-component regulatory system. *Microbiology and Molecular Biology Reviews*, *68*(2), 263–279.

Enkh-Amgalan, J., Kawasaki, H., & Seki, T. (2006). Molecular evolution of the *nif* gene cluster carrying *nif*I1 and *nif*I2 genes in the Gram-positive phototrophic bacterium *Heliobacterium chlorum*. *International Journal of Systematic and Evolutionary Microbiology*, *56*(1), 65–74.

Esteves-Ferreira, A. A., Cavalcanti, J. H. F., Vaz, M. G. M. V., Alvarenga, L. V., Nunes-Nesi, A., & Araújo, W. L. (2017). Cyanobacterial nitrogenases: Phylogenetic diversity, regulation and functional predictions. *Genetics and Molecular Biology*, *40*(1), 261–275.

Falk, G., & Johansson, B. C. (1983). Complementation of a nitrogenase Fe-protein mutant of *Rhodospirillum rubrum* with the *nif*-plasmid pRD1 containing *nif* genes of *Klebsiella pneumoniae*. *FEMS Microbiology Letters*, *19*(2–3), 145–149.

Falkowski, P. G. (1997). Evolution of the nitrogen cycle and its influence on the biological sequestration of CO_2 in the ocean. *Nature*, *387*(6630), 272–275.

Food and Agricultural Organization. (2000). *The state of the world fisheries and aquaculture*. Food and Agricultural Organization.

Fox, A. R., Soto, G., Valverde, C., Russo, D., Lagares, A., Jr., Zorreguieta, Á., ... Ayub, N. D. (2016). Major cereal crops benefit from biological nitrogen fixation when inoculated with the nitrogen-fixing bacterium *Pseudomonas protegens* Pf-5 X940. *Environmental Microbiology*, *18*(10), 3522–3534.

Franche, C., Lindström, K., & Elmerich, C. (2009). Nitrogen-fixing bacteria associated with leguminous and non-leguminous plants. *Plant and Soil*, *321*(1–2), 35–59.

Glick, B. R. (2012). Plant growth-promoting bacteria: Mechanisms and applications. *Scientifica*, *2012*, 963401. https://doi.org/10.6064/2012/963401.

Glöer, J., Thummer, R., Ullrich, H., & Schmitz, R. A. (2008). Towards understanding the nitrogen signal transduction for *nif* gene expression in *Klebsiella pneumoniae*. *The FEBS Journal*, *275* (24), 6281–6294.

Golden, J. W., & Wiest, D. R. (1988). Genome rearrangement and nitrogen fixation in Anabaena blocked by inactivation of xisA gene. *Science*, *242*(4884), 1421–1423.

Golden, J. W., Robinson, S. J., & Haselkorn, R. (1985). Rearrangement of nitrogen fixation genes during heterocyst differentiation in the cyanobacterium Anabaena. *Nature*, *314*(6010), 419–423.

Gosink, M. M., Franklin, N. M., & Roberts, G. P. (1990). The product of the *Klebsiella pneumoniae nif*X gene is a negative regulator of the nitrogen fixation (*nif*) regulon. *Journal of Bacteriology*, *172*(3), 1441–1447.

Govezensky, D., Greener, T., Segal, G., & Zamir, A. (1991). Involvement of GroEL in *nif* gene regulation and nitrogenase assembly. *Journal of Bacteriology*, *173*(20), 6339–6346.

Grabbe, R., & Schmitz, R. A. (2003). Oxygen control of *nif* gene expression in *Klebsiella pneumoniae* depends on *Nif*L reduction at the cytoplasmic membrane by electrons derived from the reduced quinone pool. *European Journal of Biochemistry*, *270*(7), 1555–1566.

Grabbe, R., Klopprogge, K., & Schmitz, R. A. (2001). Fnr is required for *Nif*L-dependent oxygen control of *nif* gene expression in *Klebsiella pneumoniae*. *Journal of Bacteriology*, *183*(4), 1385–1393.

Hartmann, A., Fu, H., & Burris, R. (1986). Regulation of nitrogenase activity by ammonium chloride in *Azospirillum* spp. *Journal of Bacteriology*, *165*(3), 864–870.

Haselkorn, R. (1992). Developmentally regulated gene rearrangements in prokaryotes. *Annual Review of Genetics*, *26*(1), 113–130.

Helber, J. T., Johnson, T. R., Yarbrough, L. R., & Hirschberg, R. (1988). Effect of nitrogenous compounds on nitrogenase gene expression in anaerobic cultures of *Anabaena variabilis*. *Journal of Bacteriology*, *170*(2), 558–563.

Hoffman, B. M., Lukoyanov, D., Yang, Z. Y., Dean, D. R., & Seefeldt, L. C. (2014). Mechanism of nitrogen fixation by nitrogenase: The next stage. *Chemical Reviews*, *114*(8), 4041–4062.

Huang, T. C., & Chou, W. M. (1991). Setting of the circadian N2-fixing rhythm of the prokaryotic *Synechococcus* sp. RF-1 while its *nif* gene is repressed. *Plant Physiology*, *96*(1), 324–326.

Hübner, P., Masepohl, B., Klipp, W., & Bickle, T. A. (1993). *nif* gene expression studies in *Rhodobacter capsulatus*: nfrC-independent repression by high ammonium concentrations. *Molecular Microbiology*, *10*(1), 123–132.

Huss-Danell, K. (1997). Tansley review no. 93. *Actinorhizal symbioses* and their N 2 fixation. *New Phytologist*, *136*, 375–405.

Ivleva, N. B., Groat, J., Staub, J. M., & Stephens, M. (2016). Expression of active subunit of nitrogenase via integration into plant organelle genome. *PLoS ONE*, *11*(8), e0160951.

Jackman, D. M., & Mulligan, M. E. (1995). Characterization of a nitrogen-fixation (*nif*) gene cluster from *Anabaena azollae* 1a shows that closely related cyanobacteria have highly variable but structured intergenic regions. *Microbiology*, *141*(9), 2235–2244.

Jacobson, M. R., Brigle, K. E., Bennett, L. T., Setterquist, R. A., Wilson, M. S., Cash, V. L., … Dean, D. R. (1989). Physical and genetic map of the major *nif* gene cluster from *Azotobacter vinelandii*. *Journal of Bacteriology*, *171*(2), 1017–1027.

Jing, X., Cui, Q., Li, X., Yin, J., Ravichandran, V., Pan, D., … Zhang, Y. (2020). Engineering *Pseudomonas protegens* Pf-5 to improve its antifungal activity and nitrogen fixation. *Microbial Biotechnology*, *13*(1), 118–133.

Joshi, H. M., & Tabita, F. R. (1996). A global two component signal transduction system that integrates the control of photosynthesis, carbon dioxide assimilation, and nitrogen fixation. *Proceedings of the National Academy of Sciences*, *93*(25), 14515–14520.

Kechris, K. J., Lin, J. C., Bickel, P. J., & Glazer, A. N. (2006). Quantitative exploration of the occurrence of lateral gene transfer by using nitrogen fxation genes as a case study. *Proceedings of the National Academy of Sciences of the United States of America*, *103*, 9584–9589.

Kennedy, C. (1977). Linkage map of the nitrogen fixation (*nif*) genes in *Klebsiella pneumoniae*. *Molecular and General Genetics MGG*, *157*(2), 199–204.

Kessler, P. S., Blank, C., & Leigh, J. A. (1998). The *nif* gene operon of the methanogenic archaeon *Methanococcus maripaludis*. *Journal of Bacteriology*, *180*, 1504–1511.

Kim, Y. M., Hidaka, M., Masaki, H., Beppu, T., & Uozumi, T. (1991). *Nif*A protein synthesized in vitro binds to the upstream activator sequence in the promoter of the *Klebsiella oxytoca nif*B gene. *Agricultural and Biological Chemistry*, *55*(12), 3121–3123.

Klingmüller, W. (1991). Nitrogen-fixing Enterobacter: A cornerstone in *nif*-gene group development. *Naturwissenschaften*, *78*(1), 16–20.

Li, Q., Liu, X., Zhang, H., & Chen, S. (2019). Evolution and functional analysis of orf1 within *nif* gene cluster from *Paenibacillus graminis* RSA19. *International Journal of Molecular Sciences*, *20*(5), 1145.

Liu, D., Liberton, M., Yu, J., Pakrasi, H. B., & Bhattacharyya-Pakrasi, M. (2018). Engineering nitrogen fixation activity in an oxygenic phototroph. *MBio*, *9*. https://doi.org/10.1128/mBio.01029-18, e01029-18.

Liu, S., Zheng, Z., Tie, J., Kang, J., Zhang, G., & Zhang, J. (2018). Impacts of Fe2+ on 5-aminolevulinic acid (ALA) biosynthesis of *Rhodobacter sphaeroides* in wastewater treatment by regulating *nif* gene expression. *Journal of Environmental Sciences*, *70*, 11–19.

Lopez-Torrejón, G., Jiménez-Vicente, E., Buesa, J. M., Hernandez, J. A., Verma, H. K., & Rubio, L. M. (2016). Expression of a functional oxygen-labile nitrogenase component in the mitochondrial matrix of aerobically grown yeast. *Nature Communications*, *7*(1), 1–6.

Ludden, P. W. (1993). Nif gene products and their roles in nitrogen fixation. In *New horizons in nitrogen fixation* (pp. 101–104). Dordrecht: Springer.

Martin, D. E., & Reinhold-Hurek, B. (2002). Distinct roles of PII-like signal transmitter proteins and amtB in regulation of *nif* gene expression, nitrogenase activity, and posttranslational modification of *Nif*H in *Azoarcus* sp. strain BH72. *Journal of Bacteriology*, *184*(8), 2251–2259.

Martinez-Argudo, I., Little, R., Shearer, N., Johnson, P., & Dixon, R. (2004). The *Nif*L-*Nif*A system: A multidomain transcriptional regulatory complex that integrates environmental signals. *Journal of Bacteriology*, *186*(3), 601–610.

Mazur, B. J., Rice, D., & Haselkorn, R. (1980). Identification of blue-green algal nitrogen fixation genes by using heterologous DNA hybridization probes. *Proceedings of the National Academy of Sciences*, *77*(1), 186–190.

Meeks, J. C., & Elhai, J. (2002). Regulation of cellular differentiation in filamentous cyanobacteria in free-living and plant-associated symbiotic growth states. *Microbiology and Molecular Biology Reviews*, *66*(1), 94–121.

Merrick, M., Hill, S., Hennecke, H., Hahn, M., Dixon, R., & Kennedy, C. (1982). Repressor properties of the *nif*L gene product in *Klebsiella pneumoniae*. *Molecular and General Genetics MGG*, *185*(1), 75–81.

Milcamps, A., & Banderleyden, J. (1991). In vitro construction of LacZ gene fusions with the *nif* HDK-operon of *Azospirillum brasilense*. *FEMS Microbiology Letters*, *77*(1), 79–84.

Mishra, A. K., Singh, P. K., Singh, P., Singh, A., Singh, S. S., Srivastava, A., … Sarma, H. K. (2015). Phylogeny and evolutionary genetics of Frankia strains based on 16S rRNA and *nif*D–K gene sequences. *Journal of Basic Microbiology*, *55*(8), 1013–1020.

Mugnai, M., Bazzicalupo, M., Fani, R , Gallori, E., Paffetti, D., & Pastorelli, R. (1994). Factors affecting nitrogen fixation and *nif* gene transcription in *Azospirillum brasilense*. *FEMS Microbiology Letters*, *120*(1–2), 133–136.

Odum, E. P. (1989). Input management of production systems. *Science*, *243*, 177–182.

Ogura, Y., Takemura, M., Oda, K., Yamato, K., Ohta, E., Fukuzawa, H., & Ohyama, K. (1992). Cloning and nucleotide sequence of a frxC-ORF469 gene cluster of Synechocystis PCC6803: Conservation with liverwort chloroplast frxC-ORF465 and *nif* operon. *Bioscience, Biotechnology, and Biochemistry*, *56*(5), 788–793.

Oliveira, S. S., Seldin, L., & Bastos, M. C. F. (1993). Identification of structural nitrogen-fixation (*nif*) genes in Bacillus polymyxa and *Bacillus macerans*. *World Journal of Microbiology and Biotechnology*, *9*(3), 387–389.

Parro, V., & Moreno-Paz, M. (2003). Gene function analysis in environmental isolates: The *nif* regulon of the strict iron oxidizing bacterium *Leptospirillum ferrooxidans*. *Proceedings of the National Academy of Sciences*, *100*(13), 7883–7888.

Paschen, A., Drepper, T., Masepohl, B. & Klipp, W. (2001). *Rhodobacter capsulatus nif*A mutants mediating *nif* gene expression in the presence of ammonium. *FEMS Microbiology Letters*, *200* (2), 207–213.

Pawlowski, K., Ratet, P., Schell, J., & De Bruijn, F. J. (1987). Cloning and characterization of *nif*A and ntrC genes of the stem nodulating bacterium ORS571, the nitrogen fixing symbiont of Sesbaniarostrata: Regulation of nitrogen fixation (*nif*) genes in the free living versus symbiotic state. *Molecular and General Genetics MGG*, *206*(2), 207–219.

Pedrosa, F. D. O., & Yates, M. G. (1984). Regulation of nitrogen fixation (*nif*) genes of *Azospirillum brasilense* by *nif*A and ntr (gln) type gene products. *FEMS Microbiology Letters*, *23*(1), 95–101.

Poza-Carrión, C., Jiménez-Vicente, E., Navarro-Rodríguez, M., Echavarri-Erasun, C., & Rubio, L. M. (2014). Kinetics of *nif* gene expression in a nitrogen-fixing bacterium. *Journal of Bacteriology*, *196*(3), 595–603.

Poza-Carrión, C., Echavarri-Erasun, C., & Rubio, L. M. (2015). Regulation of *nif* gene expression in *Azotobacter vinelandii*. *Biological Nitrogen Fixation*, *1*, 101–107.

Preker, P., Hübner, P., Schmehl, M., Klipp, W., & Bickle, T. A. (1992). Mapping and characterization of the promoter elements of the regulatory *nif* genes rpoN, *nif*A1 and *nif*A2 in *Rhodobacter capsulatus*. *Molecular Microbiology*, *6*(8), 1035–1047.

Quiviger, B., Franche, C., Lutfalla, G., Rice, D., Haselkorn, R., & Elmerich, C. (1982). Cloning of a nitrogen fixation (*nif*) gene cluster of *Azospirillum brasilense*. *Biochimie*, *64*(7), 495–502.

Raabe, K., Drepper, T., Riedel, K. U., Masepohl, B., & Klipp, W. (2002). The H-NS-like protein HvrA modulates expression of nitrogen fixation genes in the phototrophic purple bacterium *Rhodobacter capsulatus* by binding to selected *nif* promoters. *FEMS Microbiology Letters*, *216*(2), 151–158.

Rabouille, S., Staal, M., Stal, L. J., & Soetaert, K. (2006). Modeling the dynamic regulation of nitrogen fixation in the cyanobacterium *Trichodesmium* sp. *Applied and Environmental Microbiology*, *72*(5), 3217–3227.

Robson, R., Woodley, P., & Jones, R. (1986). Second gene (*nif*H*) coding for a nitrogenase iron protein in *Azotobacter chroococcum* is adjacent to a gene coding for a ferredoxin-like protein. *The EMBO Journal*, *5*(6), 1159–1163.

Roncato-Maccari, L. D., Ramos, H. J., Pedrosa, F. O., Alquini, Y., Chubatsu, L. S., Yates, M. G., … Souza, E. M. (2003). Endophytic *Herbaspirillum seropedicae* expresses *nif* genes in gramineous plants. *FEMS Microbiology Ecology*, *45*(1), 39–47.

Rubio, L. M., & Ludden, P. W. (2005). Maturation of nitrogenase: A biochemical puzzle. *Journal of Bacteriology*, *187*(2), 405–414.

Rubio, L. M., & Ludden, P. W. (2008). Biosynthesis of the iron-molybdenum cofactor of nitrogenase. *Annual Review of Microbiology*, *62*, 93–111.

Ryu, M. H., Zhang, J., Toth, T., Khokhani, D., Geddes, B. A., Mus, F., … Voigt, C. A. (2020). Control of nitrogen fixation in bacteria that associate with cereals. *Nature Microbiology*, *5* (2), 314–330.

Schüddekopf, K., Hennecke, S., Liese, U., Kutsche, M., & Klipp, W. (1993). Characterization of anf genes specific for the alternative nitrogenase and identification of *nif* genes required for both nitrogenases in *Rhodobacter capsulatus*. *Molecular Microbiology*, *8*(4), 673–684.

Selenska, S., & Klingmüller, W. (1991). Direct detection of *nif*-gene sequences of *Enterobacter agglomerans* in soil. *FEMS Microbiology Letters*, *80*(2–3), 243–245.

Setten, L., Soto, G., Mozzicafreddo, M., Fox, A. R., Lisi, C., Cuccioloni, M., … Ayub, N. D. (2013). Engineering *Pseudomonas protegens* Pf-5 for nitrogen fixation and its application to improve plant growth under nitrogen-deficient conditions. *PLoS One*, *8*(5), e63666.

Setubal, J. C., dos Santos, P., Goldman, B. S., Ertesvag, H., Espin, G., Rubio, L. M., … et al. (2009). Genome sequence of *Azotobacter vinelandii*, an obligate aerobe specialized to support diverse anaerobic metabolic processes. *Journal of Bacteriology*, *191*, 4534–4545.

Simonet, P., Normand, P., & Bardin, R. (1988). Heterologous hybridization of Frankia DNA to *Rhizobium meliloti* and *Klebsiella pneumoniae nif* genes. *FEMS Microbiology Letters*, *55*(2), 141–146.

Spaink, H. P., Kondorosi, A., & Hooykaas, P. J. (Eds.). (2012). *The rhizobiaceae: Molecular biology of model plant-associated bacteria* Springer Science & Business Media.

Sprent, J. I., & Platzmann, J. (2001). *Nodulation in legumes* (p. 146). Kew: Royal Botanic Gardens.

Steunou, A. S., Jensen, S. I., Brecht, E., Becraft, E. D., Bateson, M. M., Kilian, O., ... Kühl, M. (2008). Regulation of *nif* gene expression and the energetics of N_2 fixation over the diel cycle in a hot spring microbial mat. *The ISME Journal*, *2*(4), 364–378.

Stragier, P., Kunkel, B., Kroos, L., & Losick, R. (1989). Chromosomal rearrangement generating a composite gene for a developmental transcription factor. *Science*, *243*(4890), 507–512.

Streicher, S., Gurney, E., & Valentine, R. C. (1971). Transduction of the nitrogen-fixation genes in *Klebsiella pneumoniae*. *Proceedings of the National Academy of Sciences*, *68*(6), 1174–1177.

Suzuki, T., Yoshimura, H., Ehira, S., Ikeuchi, M., & Ohmori, M. (2007). AnCrpA, a cAMP receptor protein, regulates *nif*-related gene expression in the cyanobacterium *Anabaena* sp. strain PCC 7120 grown with nitrate. *FEBS Letters*, *581*(1), 21–28.

Thiel, T. (2019). Organization and regulation of cyanobacterial *nif* gene clusters: Implications for nitrogenase expression in plant cells. *FEMS Microbiology Letters*, *366*(7), fnz077.

Thilakarathna, M. S., McElroy, M. S., Chapagain, T., Papadopoulos, Y. A., & Raizada, M. N. (2016). Belowground nitrogen transfer from legumes to non-legumes under managed herbaceous cropping systems. A review. *Agronomy for Sustainable Development*, *36*(4), 58.

Toro, N., Herrera, M. A., & Olivares, J. (1984). Location of *nif* genes on large plasmids in *Rhizobium* strains isolated from legume tree root nodules. *FEMS Microbiology Letters*, *24*(1), 113–115.

Tsujimoto, R., Kotani, H., Yokomizo, K., Yamakawa, H., Nonaka, A., & Fujita, Y. (2018). Functional expression of an oxygen-labile nitrogenase in an oxygenic photosynthetic organism. *Scientific Reports*, *8*(1), 1–10.

Van Coppenolle, B., McCouch, S. R., Watanabe, I., Huang, N., & Van Hove, C. (1995). Genetic diversity and phylogeny analysis of *Anabaena azollae* based on RFLPs detected in Azolla-*Anabaena azollae* DNA complexes using *nif* gene probes. *Theoretical and Applied Genetics*, *91*(4), 589–597.

Vessey, J. K., Pawlowski, K., & Bergman, B. (2005). Root-based N 2-fixing symbioses: *Legumes*, actinorhizal plants, *Parasponia* sp. and cycads. In *Root physiology: From gene to function* (pp. 51–78). Dordrecht: Springer.

Xiang, N., Guo, C., Liu, J., Xu, H., Dixon, R., Yang, J., & Wang, Y. P. (2020). Using synthetic biology to overcome barriers to stable expression of nitrogenase in eukaryotic organelles. *Proceedings of the National Academy of Sciences*, *117*(28), 16537–16545.

Xie, G. H., Cui, Z., Yu, J., Yan, J., Hai, W., & Steinberger, Y. (2006). Identification of *nif* genes in N2-fixing bacterial strains isolated from rice fields along the Yangtze River Plain. *Journal of Basic Microbiology*, *46*(1), 56–63.

Yan, Y., et al. (2008). Nitrogen fxation island and rhizosphere competence traits in the genome of root-associated *Pseudomonas stutzeri* A1501. *Proceedings of the National Academy of Sciences of the United States of America*, *105*(7564–7569), 63.

Yang, J., Xie, X., Xiang, N., Tian, Z. X., Dixon, R., & Wang, Y. P. (2018). Polyprotein strategy for stoichiometric assembly of nitrogen fixation components for synthetic biology. *Proceedings of the National Academy of Sciences*, *115*(36), E8509–E8517.

Yu, F., Jing, X., Li, X., Wang, H., Chen, H., Zhong, L., ... Xia, L. (2019). Recombineering *Pseudomonas protegens* CHA0: An innovative approach that improves nitrogen fixation with impressive bactericidal potency. *Microbiological Research*, *218*, 58–65.

Zhang, Y., Burris, R. H., Ludden, P. W., & Roberts, G. P. (1997). Regulation of nitrogen fixation in *Azospirillum brasilense*. *FEMS Microbiology Letters*, *152*(2), 195–204.

Chapter 14

Recent technological advancements in studying biodegradation of polycyclic aromatic hydrocarbons through theoretical approaches

Kunal Dutta[a], Monalisha Karmakar[a], Priyanka Raul[a], Debarati Jana[a], Amiya Kumar Panda[b], and Chandradipa Ghosh[a]
[a]*Department of Human Physiology, Vidyasagar University, Midnapore, West Bengal, India,*
[b]*Department of Chemistry, Vidyasagar University, Midnapore, West Bengal, India*

Chapter outline

1 Introduction

In the present era, petroleum hydrocarbon contamination is considered as a major and widespread environmental crisis due to progressive industrialization. This eventually leads to excessive release of polycyclic aromatic hydrocarbons (PAHs) into the environment, rendering it to be a matter of major concern all over the globe. Anthropogenic activities like disposal of oil sludge and improper management in the industries result in leaks and accidental spills during the exploration, production, refining, transport, and storage of petroleum products. This in turn gets scattered into the atmosphere, terrestrial soil, marine waters, and sediments (Balachandran, Duraipandiyan, Balakrishnan, & Ignacimuthu, 2012). In soils, PAHs diffuse into nanopores of soil particles and bind into

Recent Advancement in Microbial Biotechnology. https://doi.org/10.1016/B978-0-12-822098-6.00006-9

organic matter by a natural process, called ageing. Over time they get accumulated in the surrounding soil sediments and groundwater causing extensive damage to animal tissues due to their carcinogenic, mutagenic, and potentially immune toxicant properties (Bach, Kim, Choi, & Oh, 2005; Mishra, Jyot, Kuhadand, & Lal, 2001). The PAHs are one class of toxic environmental pollutants and perhaps the first recognized environmental carcinogens that get accumulated in the environment either accidentally or due to human activities (Juhasz & Naidu, 2000). To date 16 PAHs are categorized as pollutants of high concern by US Environmental Protection Agency. The World Health Organization also documented 10 PAHs [Fluoranthene, FA ($C_{16}H_{10}$); Pyrene, PY ($C_{16}H_{10}$); Benz[*a*]anthracene, BaA ($C_{18}H_{12}$); Benzo[*b*]fluoranthene, BbFA ($C_{20}H_{12}$); Benzo[*j*]fluoranthene, BjFA ($C_{20}H_{12}$); Benzo[*k*]fluoranthene, BkFA ($C_{20}H_{12}$); Benzo[*a*]pyrene, BaP ($C_{20}H_{12}$); Dibenz(*a*,*h*)anthracene, DBahA ($C_{22}H_{14}$); Benzo[*ghi*]perylene, BghiP ($C_{22}H_{12}$); and Indeno[1,2,3-*cd*]pyrene, IP ($C_{22}H_{12}$)] to be potential carcinogenic pollutants (Bansal & Kim, 2015). PAHs are typically fused bicyclic or polycyclic aromatic ring hydrocarbon compounds those have strong molecular bonds, low volatility and water solubility, high affinity for soil substances, and particulate matters. Because of these structural complexities, PAHs remain highly recalcitrant under normal conditions (Vincent, 2006). Soil is a privileged habitat for many microbial populations and is one of the most biodiverse ecosystems on earth (Tamames, Abellán, Pignatelli, Camacho, & Moya, 2010). The versatility of soil microorganisms enables them to play a critical role in determining soil characteristics, plant growth, balancing the energy flow, and cycle of matter in soil ecosystems (Peng, Zi, & Wang, 2015). Remediation of petroleum-contaminated soils is crucial for maintaining both environmental health and agricultural production. Besides, in agro-industry maintaining soil quality, rehabilitation of PAHs contamination is essential to ensure maximum safety of the consumers (Thakur, 2020). Moreover, in the last few years, bioaccumulation of PAHs in the edible vegetables have been overlooked (Chen et al., 2018). Green technologies such as bioremediation, biodegradation, bioaugmentation, or bioengineered bioremediation offer great advantages in this action compared to conventional techniques (Dutta et al., 2018). Bioremediation is a biological approach that relies on the metabolic potential of microorganisms to remove contaminants (Hara, Kurihara, Nomura, Nakajima, & Uchiyama, 2013). In fact, microbe-soil-contaminants interaction in stressful environments is a highly dynamic and complex process, and multiple metabolic pathways can occur simultaneously (Theron & Cloete, 2000). Soil microorganisms, especially microorganisms with catabolic potential of the contaminants, are directly associated with the biological treatment processes and are essential element for developing bioremediation technology. The success of PAH degradation and bioremediation depends on one's ability to establish and maintain conditions that favour enhanced PAHs biodegradation rates in the contaminated environment that include the employment of correct population of microorganism with the appropriate metabolic abilities (Holliger et al., 1997). But most of the earlier reports on the biodegradation of PAHs were about culturable soil bacteria (Bao et al., 2020; Gou et al., 2020; Posada-Baquero et al., 2020). However, with the improvement of sequencing techniques, the cost per megabases of DNA has decreased dramatically in recent years (Rohland & Reich, 2012). This achievement has intensified the efforts to improve the bioinformatics tools and techniques for studying biodegradation of environmental pollutants that can circumvent the tedious nature and constrains of the experimental procedures in acquiring the complete picture of microbial diversity in samples collected from diverse environmental sites. Without this broad base understanding of microbial diversity the research targets on biodegradation and subssequent cleaning up of polluted sites cannot be fulfilled.

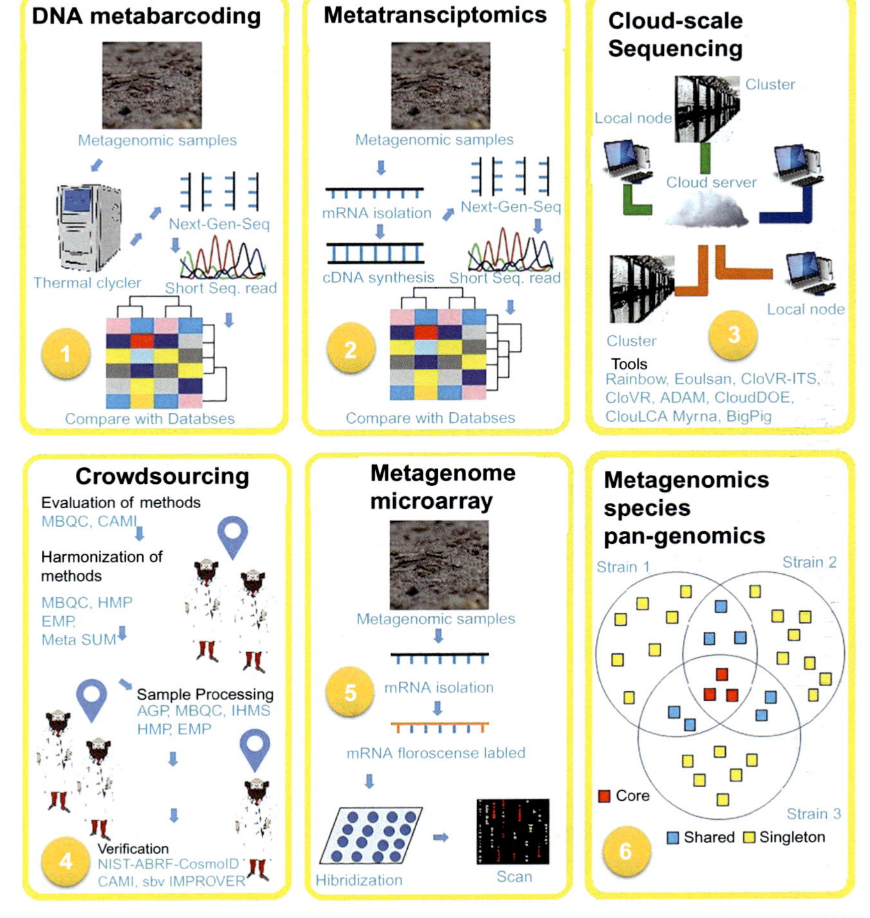

FIG. 1 Modern tools in studying biodegradation of polycyclic aromatic hydrocarbons. (1) DNA metabarcoding, (2) metatranscriptomics, (3) cloud-scale sequencing, (4) crowdsourcing of metagenomics, (5) metagenome microarray, and (6) metagenomics species pan-genomics.

Herein, an effort has been made to summarize available modern techniques for the study of microbial diversity (Fig. 1) to have a complete picture of the aptitude and possible mechanisms for biodegradation of PAHs.

2 Biodegradation and microbial diversity

Environmental pollutants and the indigenous microorganisms generally developed a new relationship between biochemistry and metagenomics (Rohland & Reich, 2012). Metagenomics provides information on all sorts of microorganisms in that environment, while biochemical investigations help in understanding the different metabolic pathways involved in the process of biodegradation (Fang et al., 2014; Kachienga, Jitendra, & Momba, 2018). In order to remediate the PAHs-contaminated environment, understanding is to be developed regarding the growing relationship based on the characterization of the spill microbiome. In spite of numerous efforts to characterize the microenvironment of spill sites (Chemerys et al., 2014), several studies indicate that the pollution level do not affect the taxonomic diversity of bacterial communities (Coupe, Smith, Newman, & Puehmeier, 2003). However, reports are also available where the pollution level has an impact on the taxonomic classification of microbial diversity (Bourceret et al., 2016). Therefore, the effect of pollution level and information about related species could be addressed by increasing the amount of data analysis or by upgrading techniques used in the studies.

2.1 DNA metabarcoding

A high-throughput approach on DNA-based biodiversity assessments of multiple species (a diverse range of higher level taxon) is DNA metabarcoding. Specifically, it is used for multispecies identification using total DNA or degraded metagenomic sample or from bulk samples of entire organisms (Taberlet, Coissac, Pompanon, Brochmann, & Willerslev, 2012). Besides, it plays significant role in identifying unknown species and also the parts of an organism against its reference database.

A short variable gene region, flanked by highly conserved sequences for developing universal PCR primers is known as marker or barcode (Table 1) in relation to identifying different organismal groups for wide taxonomic assignment (Taberlet, Bonin, Coissac, & Zinger, 2018). Moreover, high interspecific and low intraspecific variability are the key features of DNA barcoding (Hebert, Cywinska, Ball, & Dewaard, 2003). Most commonly, 658-bp region of cytochrome *c* oxidase I (COI or COX1) gene is used as a barcode (Taberlet et al., 2012). Thus the methodology is simply based on the detection of DNA in the environment (Herder et al., 2014).

This DNA metabarcoding includes four steps, viz., DNA extraction, PCR amplification, sequencing, and data analysis (Taberlet et al., 2018). Total environmental DNA (eDNA) is usually extracted from different environmental samples using a DNA isolation kit according to the standard extraction protocol

TABLE 1 Different genetic markers for DNA metabarcoding in various microorganisms.

Bacteria	Fungi	Protists
COI *(Cytochrome c oxidase subunit 1)* ***rpoB*** *(Beta subunit of RNA polymerase)* ***16S*** *(gene coding for16S ribosomal RNA)* ***cpn60*** *(60kDa chaperonin protein)* ***tuf*** *(Elongation factor Tu)* ***RIF*** *(Rifampin resistance gene)* ***gnd*** *(6-Phosphogluconate dehydrogenase)*	***ITS*** *(Internal Transcribed Spacer)* ***RPB1 (LSU)*** *[Large subunit of DNA directed RNA polymerase II]* ***RPB2 (LSU)*** *[Large subunit of RNA polymerase II]* ***18S (SSU)*** *[Small subunit of 18S ribosomal RNA]*	***ITS*** *(Internal Transcribed Spacer)* ***COI*** *(Cytochrome c oxidase subunit 1)* ***rbcL*** *(Code for large subunit of Ribulose 1,5 bisphosphate carboxylase/oxygenase)* ***18S*** *(gene coding for 18S ribosomal RNA)* ***28S*** *(gene coding for 28S ribosomal RNA)*

(Roslund et al., 2018). Two-step PCR-based amplification of eDNA is designed on smaller fragment sizes (<200 base pairs) (Piggott, 2016). The barcode standard is set from extracted eDNA using Sanger sequencing for species identification focused on the variability of the amplified region (Taberlet et al., 2012). However, only one species can be identified at a time by the Sanger sequencing method. Conversely, the next-generation sequencing (NGS) technology offers identification of multiple species simultaneously from a bulk of samples containing a mixture of various species (Pavan-Kumar, Gireesh-Babu, & Lakra, 2015; Taberlet et al., 2012). The taxonomic identification in NGS is performed using reference sequence databases, viz., SILVA, RDB-12, etc.

In metabarcoding, species are well-defined as a cluster of similar sequences, and these clusters are known as the operational taxonomic unit (OTU) (Pavan-Kumar et al., 2015). Moreover, unknown species are identified by comparing the sequence of the sample barcode gene with these reference barcode sequence databases (Pavan-Kumar et al., 2015). However, most reference databases cannot cover all species, so new data entries take place continuously (Weigand et al., 2019). There are different databases depending on the organism group and the genetic markers like BOLD, QBOL, etc. (Pavan-Kumar et al., 2015).

Despite being a powerful method, several limitations of this method have also been identified. One such drawback is the dependence on PCR where errors are seen during amplification due to both substitutions and insertions/deletions. Another major limitation is the lack of high-quality reference databases for metabarcoding as it takes more time and resources to create them. A simple and easy way to avoid PCR is to sequence the eDNA directly with NGS technology using the HiSeq 2000 System, Illumina platform, so that millions of sequences can be identified in a single experiment (Taberlet et al., 2012).

DNA metabarcoding has enormous potential to enhance the characterization of species diversity and also assesses the spatial and temporal changes of biodiversity within an ecosystem (Chariton, Court, Hartley, Colloff, & Hardy, 2010; Pavan-Kumar et al., 2015). Soils, sediments, or water are collected from the different industrial areas where PAHs are extensively produced from the incomplete combustion of organic matter; thus its level in the environment is gradually increasing that is harmful to humans as well as all living beings (Ravindra, Sokhi, & Van Grieken, 2008). In addition, utilization of the biodegrading potential of some microorganisms that are capable of using PAHs could be promising alternative (Bonaglia et al., 2020). Microbial community structures during biodegradation of PAHs can be monitored throughout the process by DNA metabarcoding. The relative abundance of different bacteria (specifically, proteobacteria and actinobacteria) or methanogenic archaea in PAH-enriched environmental sample (e.g., oil-polluted sediment, soil, water, etc.) can be identified and compared to the noncontaminated samples (Roslund et al., 2018; Xie et al., 2018). Thus this technology shows an immense prospect for compositional analysis of species and detection of invasive species present in different environmental samples (Taberlet et al., 2012).

2.2 Metatranscriptomics

Advanced molecular techniques help to understand the complex stages of the bioremediation pathway easily, e.g., biodegradation of PAHs. Metatranscriptomics (MT) is a culture-independent approach that reveals total microbial communities and helps to associate metabolic pathways (Singh, Prabha, Gupta, & Verma, 2018). Moreover, MT of PAHs-contaminated soil could offer information regarding the regulation of gene expressions in changing environmental conditions (Moran, 2009). However, the protocol for MT analysis of a microbial community is differed according to the context of the sample. It involves multistep processes that includes RNA isolation and mRNA enrichment, cDNA synthesis, library preparations, sequencing, and finally data analysis. However, it may not be possible to capture the whole metatranscriptome due to the complexity of the communities. In addition, short lifespan of the microbial communities also creates some technical limitations. A particular advantage is that they provide a less-biased perspective of in situ community diversity and function. These approaches are used to describe the direct as well as indirect effects of PAHs on gene expression without any bias towards community responses.

Metatranscritome is a nexus for identifying a new biodegradation pathway related to the catalytic enzymes. Dataset specifically for transcripts is associated with the phenanthrene amendment. Those gene transcripts are classified as dioxygenases. In MG-RAST or the metagenomics RAST (rapid annotation using subsystems technology), RefSeq (Reference Sequence database at NCBI) annotation is utilized for identification and enumeration of dioxygenase transcripts that are irrelevant to PAH metabolism. Most of the transcripts in both phenanthrenes amended and unamended soil change with protein metabolism, RNA metabolism, virulence, and stress responses. Genes of dioxygenases associated with stress response, toxic compound, antibiotic resistance, and detoxification, increased in the presence of phenanthrene. General metabolism is less affected by phenanthrene amendment (de Menezes, Clipson, & Doyle, 2012). The transcripts are classified as coding for stress response/detoxification proteins that may have varying roles in the cell except for heat shock proteins.

2.3 Cloud-scaled sequencing

The microbiome of a PAHs-contaminated environment is essential for understanding biodegradation of PAHs. However, the microbiome profile of the contaminated sites is often overlooked (Vandermaesen et al., 2016). Therefore, a data-intensive metagenomics study should be required.

A data-intensive metagenomics study would generate a massively large amount of sequence data (Keegan, Glass, & Meyer, 2016). In recent days, research groups are coming out with their ideas to handle, analyze, and interpret such big data in a single pipeline. However, while considering massive big sequence datasets, obtaining a meaningful output becomes a monotonous job. Moreover, calculating such massive big data using a local single processing unit sometimes also becomes slow and troublesome. Cloud-scale sequencing is an outstanding platform that offers an analysis of tons of gigabases of sequence (DNA/RNA) per day using cloud-based multiprocessor virtual machines (Zhao et al., 2013). Herein, a discussion is made on available cloud-based sequencing tools/techniques for a high-resolution understanding of the spill site. For example, MT of spill sites provide useful information about the functional genes present in that environment (Ivanova, Wegner, Kim, Liesack, & Dedysh, 2016).

MT data could be obtained either by metagenome microarray (MR) or by RNA sequencing (Eleuterio & Batista, 2010; Zampolli, Zeaiter, Di Canito, & Di Gennaro, 2019). Myrna is a publicly available (bowtie-bio.sf.net/myrna) cloud-computing pipeline for calculating gene expressions of a large volume of RNA-sequence datasets at a differential level. Myrna offers different analyses in a very same computational pipeline viz., integration of aligned short reads with interval calculations, normalization, aggregation, and statistical modeling (Langmead, Hansen, & Leek, 2010). The statistical analysis includes coverage of exons, genes/coding DNA region (CDR), and differential expression using parametric/nonparametric statistical analysis. Soon after analysis, Myrna returns per-gene P-, Q-values for differential expression, etc. (Langmead et al., 2010).

In the past few years with the advancement of technology, the cost of the sequencing has been reduced dramatically (Rohland & Reich, 2012). The lower cost of per gigabase of sequences has stimulated the small/middle financially aided research groups to go for NGS. As mentioned earlier, handling such a vast volume of sequence datasets becomes tiresome for a sequence statistician. However, operating such datasets in a grid-enabled cloud-scale computing platform could be a very easy task.

Rainbow is a cloud-scaled software package that assists the automation of large-scale whole-genome sequence analysis (Zhao et al., 2013). Rainbow could also assist the whole metagenome analysis of the PAHs-contaminated site. Therefore, Rainbow provides a great advantage in the deep understanding of the PAHs-contaminated site before their bioremediation. Quite similarly, CloVR-ITS, an extension of CloVR (White, Maddox, White, Angiuoli, & Fricke, 2013), offers an automated cloud-scale pipeline for metagenomic DNA sequence analyses of fungal microbiota. It includes a variety of common analyses viz., *16S* rRNA amplicon analysis, quality and chimera check of the sequence, clustering the sequences information into OTUs, taxonomic classification, and statistical analysis (White et al., 2013). Cloud Virtual Resource is a hybrid module of a portable computer package that offers minimal local computer resources and optional cloud-scaled resources on-demand (Angiuoli et al., 2011). It is also an automated cloud-based metasequence analysis package and is operated from a local machine and supports the customization of remote virtual machines to improve the quality of the analyses.

On the other hand, Eoulsan is dedicated to high-throughput sequencing data analysis (Angiuoli et al., 2011). Eoulsan users can customize the cloud-scale cluster according to their needs (Jourdren, Bernard, Dillies, & Le Crom, 2012). Eoulsan also offers automatic analysis of the vast volume of sequence data (Jourdren et al., 2012). Similar to Eoulsan, BigPig is also a dedicated sequence analysis flow language. BigPig is a programming that requires less time compared to conventional MPI-based algorithms (Nordberg, Bhatia, Wang, & Wang, 2013). BigPig offers automated analysis up to 500 Gb intensive bioinformatics analysis and can be ported without modification. Overall, BigPig offers a novel data-intensive cloud-based bioinformatics software that is released as an open-source license (Table 2).

ADAM is another high-level computer language that is designed exclusively for genome data formats and pattern processing for cloud-scale computing (Massie et al., 2013). ADAM can offer an analysis of genomic data up to 250 Gb. ADAM is operable with relatively less amount of computing resources compared to Binary Alignment Map (Massie et al., 2013). CloudDOE is a graphical user interface (GUI), a genome mapping stand-alone bioinformatics software that enabled cloud-scaled genome assembly (Chung et al., 2014). Metagenomics species pan-genomics of the polluted site could be accelerated by using CloudDOE. CloudDOE is written in JAVA and offers three-stage operation to users viz., deploy, operate, and extend wizards. CloudBurst,

TABLE 2 Tools available for cloud-scale sequence analysis.

Tools	Analysis type	Availibity
Rainbow	Automated whole-genome	http://s3.amazonaws.com/jnj_rainbow/index.html
CloVR-ITS	DNA sequnce analyses of fungal microbiota	http://clovr.org
CloVR	DNA analysis package	http://clovr.org
Eoulsan	A pipeline and a framework for NGS analysis (RNA-Seq)	http://transcriptome.ens.fr/eoulsan/
BigPig	Upto 500 Gb Omics sciences	https://sites.google.com/a/lbl.gov/biopig/
ADAM	Large scale genome format processing	http://bdgenomics.org/
Myrna	Gene expression of RNA-seq dataset	https://bowtie-bio.sf.net/myrna
CloudDOE	GUI-enabled ulta-large sequence data	http://clouddoe.iis.sinica.edu.tw/
CloudLCA	Taxonomization of metagenomic sequence	http://www.ebiomed.org/CloudLCA-software.zip
eXpress-D	RNA seq and Chip-seq	https://github.com/adarob/express-d
CS-BWAMEM	pair-end genome read	https://github.com/ytchen0323/cloud-scale-bwamem

CloudBrush, and CloudRS are included in the GUI of the CloudDOE. The most attractive feature of CloudDOE is that the MapReduce plug-in application could be installed by the user (as an administrator) (Chung et al., 2014).

Whole metagenome sequencing of a PAHs-contaminated site automatically generates billions of reads (Oniciuc et al., 2018). A deep view of the data-intensive bioinformatics analyses is usually time-consuming and demands high-skilled bioinformaticians (Hampton et al., 2017). CS-Burrows-Wheeler Aligner-Map-Reduced Programming Model (BWA-MEM) is a cloud-scale genome sequence aligner modified version of the BWA-MEM where pair-end genome read (30 ×) can be aligned within just 80 min in a 25-node cluster with 300 cores (Chen, Cong, Lei, & Wei, 2015). Taxonomization of the large volume of metagenomic sequence datasets is considered as a key problem. CloudLCA is a parallel LCA algorithm that can make taxonomization tasks high-throughput, accurate, and very fast (Zhao et al., 2012). Indeed, CloudLCA uses cloud-scale computation resources and with that, it can reach nearly 215 million reads per minute. CloudLCA also offers universal solutions for finding the lowest common ancestor (Zhao et al., 2012). NGS techniques (Illumina HiSeq, MiSeq, etc.) have chopped DNA/RNA into relatively short fragments (Van Nieuwerburgh et al., 2012). In the sequence analysis process, accurate alignment of billions of short sequence fragments is a problematic job. Many read mappers are available to address this problem (Kelly, Wickstead, & Gull, 2011; Li, Ruan, & Durbin, 2008). For example, eXpress-D is a cloud-based read mapper that offers massive-big high-throughput sequence analysis (Roberts, Feng, & Pachter, 2013). eXpress-D is using the expectation-maximization algorithm to exact maximum likelihood assignment of ambiguous fragments. The accuracy of the eXpress-D is very high compared to others (Table 2) (Roberts et al., 2013).

2.4 Crowdsourcing of metagenomics

In biomedical research, crowdsourcing, a natural evolution of web technologies, has emerged as a novel technique that quickly collects, processes, interlinks, and interprets data from large-scale experiments. Crowdsourcing coined in 2006 (Howe, 2006) offers open recruitment of participants form different geo-locations connected by a web-based platform (Afshinnekoo, Ahsanuddin, & Mason, 2016). Crowdsourcing of metagenomics includes four steps (Poussin et al., 2018), i.e., evaluation of methods, randomization of methods, sample processing, and verification (Fig. 1.). Recently, crowdsourcing has arisen as an important tool in many dimensions of research including healthcare and agricultural perspectives (Swan, 2012). Crowdsourced medical research has created ground-breaking opportunities to bridge the gap between science and society. Moreover, public is now participating in sample collection and funding efforts. Additionally, the crowdsourced data collections have become integrated research efforts for students and researchers (Afshinnekoo et al., 2016).

Environmental genomics is well-known as metagenomics. In other words, genomic studies of environmental samples are metagenomics. Metagenomics explores the culturable and nonculturable organisms and their functional properties. In metagenomics, culture-based techniques are replaced by direct genomic sequence-based techniques. The background information of the sampling environment is as important as the taxonomic profiling of the bacterial isolates. Metagenomics can bridge the gap between background information of the environment and the taxonomic properties of the microbial organisms. Besides, the interlinking of the contextual information and sequence data is important for accurate interpretation and comparisons of the metagenomics data. Moreover, rich information about microbial strains is available in the literature and microbial culture collection center (Global Bioresource Center). Crowdsourcing can significantly contribute curation of such a vast volume of databases. Besides, handling information about the bacterial strains that are not culturable or not classified taxonomically is very challenging. For example, limited information on viable bacteria that are not culturable in laboratory settings is available in the literature (Hirschman et al., 2016). Moreover, limited information is available about the environment from which the metagenomic sampling was performed. Crowdsourcing could again be very useful in cases where there is some information available in the publication. The information about the ecosystem from where soil samples were isolated plays an important role in the absence of other metadata. GOLD, a metagenomic crowdsourcing tool that has been developed based on habitat classification could be useful for this purpose.

2.5 Metagenome microarray

In the earth's biological diversity, the microorganisms have significant contribution. Though few of the microorganisms have been characterized, which are culturable in nature. Soil metagenomics is a culture-independent approach to unfold the microbial diversity of soil microbial communities. However, analysis of the metagenomic samples by the conventional polymerase chain reaction hinders limitation of complete identification of the microbial community due to availibility of the few primers (Reysenbach, Giver, Wickham, & Pace, 1992; Suzuki & Giovannoni, 1996; Wang & Wang, 1996; Wintzingerode, Göbel, & Stackebrandt, 1997). Purified metagenomic DNA samples are used to construct small-insert libraries. Identifiaction of new genes, small operons participating in the metabolic pathways could be easily performed by constructing small-insert DNA libraries. However, large-insert DNA libraries would be useful to identifiy complex metabolic pathways. The complex metabolic pathways could provide useful information for agricultural perspective. To avoid conventional polymerase chain reaction, cloning of metagenomic DNA samples will be more effective. Besides, taxonomic information about the organism could be obtained where genomic fragments are more than 100 kb long. Such metagenomic libraries have been used to identify novel genes from uncultivated species (Béja et al., 2000; Béja, Spudich, Spudich, Leclerc, & DeLong, 2001; Eilers, Pernthaler, Glöckner, & Amann, 2000; Hebert et al., 2003) and to isolate novel enzymes that are used in pharmaceutical industries (Brady, Chao, & Clardy, 2004; Gillespie et al., 2002; MacNeil et al., 2001).

Metagenomic library could be utilized for identification of new enzymes with agricultural and industrial importance. For example, high-throughput end sequencing is presently utilized by a research group to characterize metagenomic DNA samples (Breitbart et al., 2002). Although it is not practical to assemble a complete bacterial genome from a metagenome, there is still a need for new functional genomic approaches that systematically yield information about many of the elements in a metagenomic library. These new approaches are allowed to identify novel nonculturable bacteria with agricultural importance. In addition, DNA samples derived from nonculturable organims can be identified by metagenomic MR. Moreover, "metagenomic profiling" involves classification of clones based on hybridization of short DNA insert from environmental sources. DNA MR provides a platform for determining and measuring the contents of gene in entire genomes and its expressions (Joyce, Chan, Salama, & Falkow, 2002). Thus, DNA MR ensures a promising technique for quantitative detection as well as characterization of genes from environments such as soil and water (Bavykin et al., 2001; Cho & Tiedje, 2001). However, there is a limitation in *16S* rRNA markers because there are no specific target available for the nonculturable species in the environment. Metagenomic profiling, a cultivation-independent assessment, offers an effective approach for rapidly characterizing many clones and identifying the clones corresponding to unidentified species of the largely untapped genetic reservoir of soil (PAHs-contaminated) microbial communities.

2.6 Metagenomics species pan-genomics

Pan-genomics of a species describes all sets of genes in a species by sequencing all strains of a species (Alcaraz, 2014). The analysis of complete genome datasets has categorized pan-genomics into three subdatasets namely, core, shared, and singleton (Fig. 1) (de Jesus Sousa et al., 2020). The core regions describe mainly housekeeping genes, genes for metabolism, and other genes for survival and defense (Naz et al., 2020). Conversely, shared and singleton described genes have resulted from genome plasticity (Jaiswal et al., 2020). Genome elasticity is an evolutionary phenomenon resulted from horizontal gene transfer, gene deletion/insertion, or epigenetic impacts. In the case of some species, a large-gene repertoire is described as open pan-genomics. Conversely, the limited gene repertoire of species is described as closed pan-genomics (Jaiswal et al., 2020).

To date, pan-genomics is widely studied discipline to understand the culturable pathogenic microorganisms for possible drug/vaccine development (Jaiswal et al., 2020). As described earlier, a comprehensive biotic audit of the microbiome of a PAHs-contaminated site is important before cleaning the site. Interestingly, the core region of the pan-genomics of a metagenomics species cloud provides valuable informatics regarding the metabolic genes (Jaiswal et al., 2020). Metabolic genes play a central role in bioremediation or biodegradation of the PAHs (Kraiselburd et al., 2019). Therefore, we should narrow down the focus and concentrate in metabolomics of a pan-metagenomics species. Gas chromatography-mass spectrometry, liquid chromatography-mass spectrometry, electron spray ionization-mass spectrometry, and heated electron spray ionization-mass spectrometry are gold standard tools for studying metabolomics (Pourfadakari et al., 2019). The recent advancement of bioinformatics tools revolutionized all branches of genomics. Tools available for studying pan-genomics could be simultaneously useful for studying a metagenomics species pan-genomics (Caputo, Clogston, Calzolai, Rösslein, & Prina-Mello, 2019). Brief technical differences among available tools have been well described. Here, a brief description about the advantages of the pan-genomic tools has been presented. Roary is a bioinformatic tool that can analyze and identify the core and shared genes from a massive large pan-genomic dataset (Page et al., 2015). The low-profile configuration of a computer such as a single node with 13 GB of RAM makes Roary most popular. Roary is written in Perl, and it is freely available under an open-source license. Therefore, the users can customize the codes according to their needs. PanOTC is written for pan-genomics analysis of closely related to prokaryotic species (Fouts, Brinkac, Beck, Inman, & Sutton, 2012). Information on neighborhood-conserved genes is utilized by PanOTC to differentiate paralogous and orthologous genes. PanOTC apply four graph-based methods that are paved ~70% of the clusters and ~86% of the proteins (Fouts et al., 2012). Whole-genome and metagenome sequencing generates massive large amount of sequence data. As mentioned earlier, handling such a huge amount of data is a mind-numbing job. LS-BSR is a data analysis pipeline that can rapidly compare hundreds to thousands of genome sequence data. The relatedness of the bacterial genomes is finally visualized by LS-BSR (Sahl, Caporaso, Rasko, & Keim, 2014).

3 Conclusion

Environmental pollution has become one of the critical global issues since the industrial revolution that have been neglected over many years. Groundwater and agricultural soil contamination by PAHs, along with its bioaccumulation in vegetables, are the serious concerns severly affecting human health and have been overlooked for many years. Different studies involving the biodegradation of PAHs are only about the improvement of culturable soil bacteria, fungi, either individually or in the form of consortium, and also their immobilization, etc. Some recent investigations warrant the importance of studying microbial diversity for a better understanding of PAHs biodegradation. Here, efforts have been made to give a cursary glance on recent technological advancements in studying microbial diversity through approaches utilizing bioinformatics tools. Recent developments in bioinformatics tools and techniques, i.e., with the development of techniques like, DNA metabarcoding, MT, cloud-scaled sequencing, crowdsourcing of metagenomics, metagenome MR, metagenomic species pan-genomics cloud, and abilities to explore new horizon of biodegradation science through broader understanding of the microbial diversity from diverse environmental sources could straightforwardly be reached. This review work is expected to develop conceptual foundation that could have significant input in research designing for active as well as growing researchers during planning their projects/research studies.

Acknowledgment

Council of Scientific and Industrial Research (CSIR), Govt. of India, New Delhi, India, is sincerely acknowledged by K.D. for Senior Research Fellowship (SRF), sanction letter no. 09/599 (0082) 2K19 EMR-Z. AKP sincerely acknowledges the Department of Biotechnology, Ministry of Science and Technology, Govt. of India, for a research project (No: BT/PR3802/BRB/10/981/2011).

References

Afshinnekoo, E., Ahsanuddin, S., & Mason, C. E. (2016). Globalizing and crowdsourcing biomedical research. *British Medical Bulletin, 120*, 27–33.

Alcaraz, L. D. (2014). *Pan-genomics: Unmasking the gene diversity hidden in the bacteria species. 2* (p. e113v2). Peer J Pre Prints.

Angiuoli, S. V., Matalka, M., Gussman, A., Galens, K., Vangala, M., Riley, D. R., et al. (2011). CloVR: A virtual machine for automated and portable sequence analysis from the desktop using cloud computing. *BMC Bioinformatics, 12*, 356.

Bach, Q. D., Kim, S. J., Choi, S. C., & Oh, Y. S. (2005). Enhancing the intrinsic bioremediation of PAH contaminated anoxic estuarine sediments with biostimulating agents. *Journal of Microbiology, 43*(4), 319–324.

Balachandran, C., Duraipandiyan, D., Balakrishnan, K., & Ignacimuthu, S. (2012). Petroleum and polycyclic aromatic hydrocarbons (PAHs) degradationand naphthalene metabolism in *Streptomyces* sp. (ERI-CPDA-1) isolated from oil contaminated soil. *Bioresource Technology, 112*, 83–89.

Bansal, V., & Kim, K. H. (2015). Review of PAH contamination in food products and their health hazards. *Environment International, 84*, 26–38.

Bao, Y., Guo, Z., Chen, R., Wu, M., Li, Z., Linm, X., et al. (2020). Functional community composition has less environmental variability than taxonomic composition in straw-degrading bacteria Yuanyuan. *Biology and Fertility of Soils, 56*, 869–874.

Bavykin, S. G., Akowski, J. P., Zakhariev, V. M., Barsky, V. E., Perov, A. N., & Mirzabekov, A. D. (2001). Portable system for microbial sample preparation and oligonucleotide microarray analysis. *Applied and Environmental Microbiology, 67*(2), 922–928.

Béja, O., Aravind, L., Koonin, E. V., Suzuki, M. T., Hadd, A., Nguyen, L. P., et al. (2000). Bacterial rhodopsin: Evidence for a new type of phototrophy in the sea. *Science, 289*(5486), 1902–1906.

Béja, O., Spudich, E. N., Spudich, J. L., Leclerc, M., & DeLong, E. F. (2001). Proteorhodopsin phototrophy in the ocean. *Nature, 411*(6839), 786–789.

Bonaglia, S., Broman, E., Brindefalk, B., Hedlund, E., Hjorth, T., Rolff, C., et al. (2020). Activated carbon stimulates microbial diversity and PAH biodegradation under anaerobic conditions in oil-polluted sediments. *Chemosphere, 248*, 126023.

Bourceret, A., Cébron, A., Tisserant, E., Poupin, P., Bauda, P., Beguiristain, T., et al. (2016). The bacterial and fungal diversity of an aged PAH-and heavy metal-contaminated soil is affected by plant cover and edaphic parameters. *Microbial Ecology, 71*(3), 711–724.

Brady, S. F., Chao, C. J., & Clardy, J. (2004). Long-chain N-acyltyrosine synthases from environmental DNA. *Applied and Environmental Microbiology, 70*(11), 6865–6870.

Breitbart, M., Salamon, P., Andresen, B., Mahaffy, J. M., Segall, A. M., Mead, D., et al. (2002). Genomic analysis of uncultured marine viral communities. *Proceedings. National Academy of Sciences. United States of America, 99*(22), 14250–14255.

Caputo, F., Clogston, J., Calzolai, L., Rösslein, M., & Prina-Mello, A. (2019). Measuring particle size distribution of nanoparticle enabled medicinal products, the joint view of EUNCL and NCI-NCL. A step by step approach combining orthogonal measurements with increasing complexity. *Journal of Controlled Release*, *299*, 31–43.

Chariton, A. A., Court, L. N., Hartley, D. M., Colloff, M. J., & Hardy, C. M. (2010). Ecological assessment of estuarine sediments by pyrosequencing eukaryotic ribosomal DNA. *Frontiers in Ecology and the Environment*, *8*(5), 233–238.

Chemerys, A., Pelletier, E., Cruaud, C., Martin, F., Violet, F., & Jouanneau, Y. (2014). Characterization of novel polycyclic aromatic hydrocarbon dioxygenases from the bacterial metagenomic DNA of a contaminated soil. *Applied and Environmental Microbiology*, *80*(21), 6591–6600.

Chen, Y. T., Cong, J., Lei, J., & Wei, P. (2015). A novel high-throughput acceleration engine for read alignment. In *Field-programmable custom computing machines (FCCM). IEEE 23rd annual international symposium* (pp. 199–202).

Chen, Y., Zhang, F., Zhang, J., Zhou, M., Li, F., & Liu, X. (2018). Accumulation characteristics and potential risk of PAHs in vegetable system grow in home garden under straw burning condition in Jilin, Northeast China. *Ecotoxicology and Environmental Safety*, *162*, 647–654.

Cho, J.-C., & Tiedje, J. M. (2001). Bacterial species determination from DNA-DNA hybridization by using genome fragments and DNA microarrays. *Applied and Environmental Microbiology*, *67*(8), 3677–3682.

Chung, W.-C., Chen, C.-C., Ho, J.-M., Lin, C.-Y., Hsu, W.-L., Wang, Y.-C., et al. (2014). CloudDOE: A user-friendly tool for deploying hadoop clouds and analyzing high-throughput sequencing data with MapReduce. *PLoS ONE*, *9*(6), e98146.

Coupe, S. J., Smith, H. G., Newman, A. P., & Puehmeier, T. (2003). Biodegradation and microbial diversity within permeable pavements. *European Journal of Protistology*, *39*(4), 495–498.

de Jesus Sousa, T., Jaiswal, A. K., Hurtado, R. E., Tosta, S. F. O., Soares, S. C., Gomide, A. C. P., et al. (2020). Pan-genomics of veterinary pathogens and its applications author links open overlay panel (Chapter 5). In D. Barh, S. Soares, S. Tiwari, & V. Azevedo (Eds.), *Pan-genomics: Applications, challenges, and future prospects* (pp. 101–119). Academic Press.

de Menezes, A., Clipson, N., & Doyle, E. (2012). Comparative metatranscriptomics reveals widespread community responses during phenanthrene degradation in soil. *Environmental Microbiology*, *14*(9), 2577–2588.

Dutta, K., Shityakov, S., Khalifa, I., Mal, A., Moulik, S. P., Panda, A. K., et al. (2018). Effects of secondary carbon supplement on biofilm-mediated biodegradation of naphthalene by mutated naphthalene 1, 2-dioxygenase encoded by *Pseudomonas putida* strain KD9. *Journal of Hazardous Materials*, *357*, 187–197.

Eilers, H., Pernthaler, J., Glöckner, F. O., & Amann, R. (2000). Culturability and in situ abundance of pelagic bacteria from the North Sea. *Applied Environmental Microbiology*, *66*(7), 3044–3051.

Eleuterio, L., & Batista, J. R. (2010). Biodegradation studies and sequencing of microcystin-LR degrading bacteria isolated from a drinking water biofilter and a fresh water lake. *Toxicon*, *55*(8), 1434–1442.

Fang, H., Cai, L., Yang, Y., Ju, F., Li, X., Yu, Y., et al. (2014). Metagenomic analysis reveals potential biodegradation pathways of persistent pesticides in freshwater and marine sediments. *The Science of the Total Environment*, *470*, 983–992.

Fouts, D. E., Brinkac, L., Beck, E., Inman, J., & Sutton, G. (2012). PanOCT: Automated clustering of orthologs using conserved gene neighborhood for pan-genomic analysis of bacterial strains and closely related species. *Nucleic Acids Research*, *40*, e172.

Gillespie, D. E., Brady, S. F., Bettermann, A. D., Cianciotto, N. P., Liles, M. R., Rondon, M. R., et al. (2002). Isolation of antibiotics turbomycin A and B from a metagenomic library of soil microbial DNA. *Applied and Environmental Microbiology*, *68*(9), 4301–4306.

Gou, Y., Zhao, Q., Yang, S., Qiao, P., Cheng, Y., Song, Y., et al. (2020). Enhanced degradation of polycyclic aromatic hydrocarbons in aged subsurface soil using integrated persulfate oxidation and anoxic biodegradation. *Chemical Engineering Journal*, *394*, 125040.

Hampton, S. E., Jones, M. B., Wasser, L. A., Schildhauer, M. P., Supp, S. R., Brun, J., et al. (2017). Skills and knowledge for data-intensive environmental research. *Bioscience*, *67*, 546–557.

Hara, E., Kurihara, M., Nomura, N., Nakajima, T., & Uchiyama, H. (2013). Bioremediation field trial of oil-contaminated soil with food-waste compost. *Journal of JSCE*, *1*, 125–132.

Hebert, P. D., Cywinska, A., Ball, S. L., & Dewaard, J. R. (2003). Biological identifications through DNA barcodes. *Proceedings of the Royal Society of London. Series B: Biological Sciences*, *270* (1512), 313–321.

Herder, J., Valentini, A., Bellemain, E., Dejean, T., Van Delft, J., Thomsen, P., et al. (2014). *Environmental DNA. A review of the possible applications for the detection of (invasive) species.* Nijmegen: Stichting RAVON. 2013–2104.

Hirschman, L., Fort, K., Boué, S., Kyrpides, N., Doğan, R. I., & Cohen, B. K. (2016). Crowdsourcing and curation: Perspectives from biology and natural language processing. *Database*, *2016*, baw115.

Holliger, C., Gaspard, S., Glod, G., Heijman, C., Schumacher, W., Schwarzenbach, R. P., et al. (1997). Contaminated environments in the subsurface and bioremediation: Organic contaminants. *FEMS Microbiology Reviews*, *20*(3–4), 517–523.

Howe, J. (2006). The rise of crowdsourcing. *Wired Magazine*, *14.06*, 1–4. http://www.wired.com/wired/archive/14.06/crowds_pr.html.

Ivanova, A. A., Wegner, C.-E., Kim, Y., Liesack, W., & Dedysh, S. N. (2016). Identification of microbial populations driving biopolymer degradation in acidic peatlands by metatranscriptomic analysis. *Molecular Ecology*, *25*, 4818–4835.

Jaiswal, A. K., Tiwari, S., Jamal, S. B., de Castro Oliveira, L., Alves, L. G., Azevedo, V., et al. (2020). The pan-genome of *Treponema pallidum* reveals differences in genome plasticity between subspecies related to venereal and non-venereal syphilis. *BMC Genomics*, *21*, 33.

Jourdren, L., Bernard, M., Dillies, M. A., & Le Crom, S. (2012). Eoulsan: A cloud computing-based framework facilitating high throughput sequencing analyses. *Bioinformatics*, *28*, 1542–1543.

Joyce, E. A., Chan, K., Salama, N. R., & Falkow, S. (2002). Redefining bacterial populations: A post-genomic reformation. *Nature Reviews Genetics*, *3*(6), 462–473.

Juhasz, A. L., & Naidu, R. (2000). Bioremediation of high molecular weight polycyclic aromatic hydrocarbons: A review of the microbial degradation of benzo[a]pyrene. *International Biodeterioration and Biodegradation*, *45*(1–2), 57–88.

Kachienga, L., Jitendra, K., & Momba, M. (2018). Metagenomic profiling for assessing microbial diversity and microbial adaptation to degradation of hydrocarbons in two South African petroleum-contaminated water aquifers. *Scientific Reports*, *8*, 7564.

Keegan, K. P., Glass, E. M., & Meyer, F. (2016). MG-RAST, a metagenomics service for analysis of microbial community structure and function. In F. Martin, & S. Uroz (Eds.), *Microbial environmental genomics (MEG)* (pp. 207–233). Switzerland: Springer Nature.

Kelly, S., Wickstead, B., & Gull, K. (2011). Archaeal phylogenomics provides evidence in support of a methanogenic origin of the Archaea and a thaumarchaeal origin for the eukaryotes. *Philosophical Transactions of the Royal Society of London. Series B, Biological Sciences*, *278* (1708), 1009–1018.

Kraiselburd, I., Brüls, T., Heilmann, G., Kaschani, F., Kaiser, M., & Meckenstock, R. U. (2019). Metabolic reconstruction of the genome of candidate Desulfatiglans TRIP_1 and identification of key candidate enzymes for anaerobic phenanthrene degradation. *Environmental Microbiology*, *21*(4), 1267–1286.

Langmead, B., Hansen, K. D., & Leek, J. T. (2010). Cloud-scale RNA-sequencing differential expression analysis with Myrna. *Genome Biology*, *11*, R83.

Li, H., Ruan, J., & Durbin, R. (2008). Mapping short DNA sequencing reads and calling variants using mapping quality scores. *Genome Research*, *18*, 1851–1858.

MacNeil, I., Tiong, C., Minor, C., August, P., Grossman, T., Loiacono, K., et al. (2001). Expression and isolation of antimicrobial small molecules from soil DNA libraries. *Journal of Molecular Microbiology and Biotechnology*, *3*(2), 301–308.

Massie, M., Nothaft, F., Hartl, C., Kozanitis, C., Schumacher, A., Joseph, A. D., et al. (2013). *ADAM: Genomics formats and processing patterns for cloud scale computing. Report No.: UCB/EECS-2013-207*. Berkeley: EECS Department, University of California, Berkeley.

Mishra, S., Jyot, J., Kuhadand, R. C., & Lal, B. (2001). Evaluation of inoculum addition to stimulate in situ bioremediation of oily-sludge-contaminated soil. *Applied and Environmental Microbiology*, *67*(4), 1675–1681.

Moran, M. A. (2009). Metatranscriptomics: Eavesdropping on complex microbial communities. *Microbe*, *4*(7), 329–335.

Naz, A., Obaid, A., Fatima Shahid, F., Dar, H. A., Naz, K., Ullah, N., et al. (2020). Chapter 16—Reverse vaccinology and drug target identification through pan-genomics (Chapter 16). In D. Barh, S. Soares, S. Tiwari, & V. Azevedo (Eds.), *Pan-genomics: Applications, challenges, and future prospects* (pp. 101–119). Academic Press.

Nordberg, H., Bhatia, K., Wang, K., & Wang, Z. (2013). BioPig: A hadoop-based analytic toolkit for large-scale sequence data. *Bioinformatics*, *29*, 3014–3019.

Oniciuc, E. A., Likotrafiti, E., Alvarez-Molina, A., Prieto, M., Santos, J. A., & Alvarez-Ordóñez, A. (2018). The present and future of whole genome sequencing (WGS) and whole metagenome sequencing (WMS) for surveillance of antimicrobial resistant microorganisms and antimicrobial resistance genes across the food chain. *Genes*, *9*, 268.

Page, A. J., Cummins, C. A., Hunt, M., Wong, V. K., Reuter, S., Holden, M. T. G., et al. (2015). Roary: Rapid large-scale prokaryote pan genome analysis. *Bioinformatics*, *31*(22), 3691–3693.

Pavan-Kumar, A., Gireesh-Babu, P., & Lakra, W. (2015). DNA metabarcoding: A new approach for rapid biodiversity assessment. *Journal of Cell Science & Molecular Biology*, *2*(1), 111.

Peng, M., Zi, X., & Wang, Q. (2015). Bacterial community diversity of oil-contaminated soils assessed by high throughput sequencing of 16S rRNA genes. *International Journal of Environmental Research and Public Health*, *12*(10), 12002–12015.

Piggott, M. P. (2016). Evaluating the effects of laboratory protocols on eDNA detection probability for an endangered freshwater fish. *Ecology and Evolution*, *6*(9), 2739–2750.

Posada-Baquero, R., Jiménez-Volkerink, S. N., García, J. L., Vila, J., Cantos, M., & Ortega-Calvo, J. J. (2020). Rhizosphere-enhanced biosurfactant action on slowly desorbing PAHs in contaminated soil. *The Science of the Total Environment*, *720*, 137608.

Pourfadakari, S., Ahmadi, M., Jaafarzadeh, N., Takdastan, A., Neisi, A. A., Ghafari, S., et al. (2019). Remediation of PAHs contaminated soil using a sequence of soil washing with biosurfactant produced by *Pseudomonas aeruginosa* strain PF2 and electrokinetic oxidation of desorbed solution, effect of electrode modification with Fe_3O_4 nanoparticles. *Journal of Hazardous Materials*, *379*, 120839.

Poussin, C., Sierro, N., Boué, S., Battey, J., Scotti, E., Belcastro, V., et al. (2018). Interrogating the microbiome: Experimental and computational considerations in support of study reproducibility. *Drug Discovery Today*, *23*, 1644–1657.

Ravindra, K., Sokhi, R., & Van Grieken, R. (2008). Atmospheric polycyclic aromatic hydrocarbons: Source attribution, emission factors and regulation. *Atmospheric Environment*, *42*(13), 2895–2921.

Reysenbach, A.-L., Giver, L. J., Wickham, G. S., & Pace, N. R. (1992). Differential amplification of rRNA genes by polymerase chain reaction. *Applied and Environmental Microbiology*, *58*(10), 3417–3418.

Roberts, A., Feng, H., & Pachter, L. (2013). Fragment assignment in the cloud with eXpress-D. *BMC Bioinformatics*, *14*(1), 358.

Rohland, N., & Reich, D. (2012). Cost-effective, high-throughput DNA sequencing libraries for multiplexed target capture. *Genome Research*, *22*(5), 939–946.

Roslund, M. I., Grönroos, M., Rantalainen, A.-L., Jumpponen, A., Romantschuk, M., Parajuli, A., et al. (2018). Half-lives of PAHs and temporal microbiota changes in commonly used urban landscaping materials. *PeerJ*, *6*, e4508.

Sahl, J. W., Caporaso, J. G., Rasko, D. A., & Keim, P. (2014). The large-scale blast score ratio (LS-BSR) pipeline: A method to rapidly compare genetic content between bacterial genomes. *Peer J*, *2*, e332.

Singh, D. P., Prabha, R., Gupta, V. K., & Verma, M. K. (2018). Metatranscriptome analysis deciphers multifunctional genes and enzymes linked with the degradation of aromatic compounds and pesticides in the wheat rhizosphere. *Frontiers in Microbiology*, *9*, 1–15.

Suzuki, M. T., & Giovannoni, S. J. (1996). Bias caused by template annealing in the amplification of mixtures of 16S rRNA genes by PCR. *Applied and Environmental Microbiology*, *62*(2), 625–630.

Swan, M. (2012). Health 2050: The realization of personalized medicine through crowdsourcing, the quantified self, and the participatory biocitizen. *Journal of Personalized Medicine*, *2*, 93–118.

Taberlet, P., Bonin, A., Coissac, E., & Zinger, L. (2018). *Environmental DNA for biodiversity research and monitoring*. Oxford, UK: Oxford University Press.

Taberlet, P., Coissac, E., Pompanon, F., Brochmann, C., & Willerslev, E. (2012). Towards next-generation biodiversity assessment using DNA metabarcoding. *Molecular Ecology*, *21* (8), 2045–2050.

Tamames, J., Abellán, J. J., Pignatelli, M., Camacho, A., & Moya, A. (2010). Environmental distribution of prokaryotic taxa. *BMC Microbiology*, *10*(1), 85.

Thakur, M. (2020). Fungi as a biological tool for sustainable agriculture. In *Agriculturally important fungi for sustainable agriculture* (pp. 255–273). Switzerland: Springer Nature.

Theron, J., & Cloete, T. E. (2000). Molecular techniques for determining microbial diversity and community structure in natural environments. *Critical Reviews in Microbiology*, *26*(1), 37–57.

Van Nieuwerburgh, F., Thompson, R. C., Ledesma, J., Deforce, D., Gaasterland, T., Ordoukhanian, P., et al. (2012). Illumina mate-paired DNA sequencing-library preparation using Cre-Lox recombination. *Nucleic Acids Research*, *40*, e24.

Vandermaesen, J., Horemans, B., Bers, K., Vandermeeren, P., Herrmann, S., Sekhar, A., et al. (2016). Application of biodegradation in mitigating and remediating pesticide contamination of freshwater resources: State of the art and challenges for optimization. *Applied Microbiology and Biotechnology*, *100*, 7361–7376.

Vincent, M. (2006). Microbial bioremediation of polycyclic aromatic hydrocarbons (PAHs) in oily sludge wastes. *The Middle East Journal*, 1–13.

Wang, G. C., & Wang, Y. (1996). The frequency of chimeric molecules as a consequence of PCR co-amplification of 16S rRNA genes from different bacterial species. *Microbiology*, *142*(5), 1107–1114.

Weigand, H., Beermann, A. J., Čiampor, F., Costa, F. O., Csabai, Z., Duarte, S., et al. (2019). *DNA barcode reference libraries for the monitoring of aquatic biota in Europe: Gap-analysis and recommendations for future work* (p. 576553). BioRxiv.

White, J., Maddox, C., White, O., Angiuoli, S., & Fricke, F. (2013). CloVR-ITS: Automated internal transcribed spacer amplicon sequence analysis pipeline for the characterization of fungal microbiota. *Microbiome*, *1*, 6.

Wintzingerode, F. V., Göbel, U. B., & Stackebrandt, E. (1997). Determination of microbial diversity in environmental samples: Pitfalls of PCR-based rRNA analysis. *FEMS Microbiology Reviews*, *21*(3), 213–229.

Xie, Y., Zhang, X., Yang, J., Kim, S., Hong, S., Giesy, J. P., et al. (2018). eDNA-based bioassessment of coastal sediments impacted by an oil spill. *Environmental Pollution*, *238*, 739–748.

Zampolli, J., Zeaiter, Z., Di Canito, A., & Di Gennaro, P. (2019). Genome analysis and-omics approaches provide new insights into the biodegradation potential of *Rhodococcus*. *Applied and Environmental Microbiology*, *103*, 1069–1080.

Zhao, G., Bu, D., Liu, C., Li, J., Yang, J., Liu, Z., et al. (2012). CloudLCA: Finding the lowest common ancestor in metagenome analysis using cloud computing. *Protein & Cell*, *3*(2), 148–152.

Zhao, S., Prenger, K., Smith, L., Messina, T., Fan, H., Jaeger, E., et al. (2013). Rainbow: A tool for large-scale whole-genome sequencing data analysis using cloud computing. *BMC Genomics*, *14* (1), 425.

Index

Note: Page numbers followed by *f* indicate figures and *t* indicate tables.

C

K

L

M

R

S

Printed in the United States
by Baker & Taylor Publisher Services